VOLUME NINETY EIGHT

SEMICONDUCTORS AND SEMIMETALS

Nanowires for Energy Applications

SERIES EDITORS

VOLUME NINETY EIGHT

SEMICONDUCTORS AND SEMIMETALS

Nanowires for Energy Applications

Edited by

SUDHA MOKKAPATI
School of Physics and Astronomy,
Cardiff University, Cardiff,
United Kingdom

CHENNUPATI JAGADISH
Research School of Physics and Engineering,
Australian National University,
Canberra, ACT, Australia

ELSEVIER

ACADEMIC PRESS
An imprint of Elsevier

Academic Press is an imprint of Elsevier
50 Hampshire Street, 5th Floor, Cambridge, MA 02139, United States
525 B Street, Suite 1650, San Diego, CA 92101, United States
The Boulevard, Langford Lane, Kidlington, Oxford OX5 1GB, United Kingdom
125 London Wall, London, EC2Y 5AS, United Kingdom

First edition 2018

ISBN: 978-0-12-815139-6
ISSN: 0080-8784

Publisher: Zoe Kruze
Acquisition Editor: Jason Mitchell
Editorial Project Manager: Shellie Bryant
Production Project Manager: Abdulla Sait
Cover Designer: Alan Studholme

Typeset by SPi Global, India

CONTENTS

CONTRIBUTORS

Nina Andrejevic
Department of Materials Science and Engineering, Massachusetts Institute of Technology, Cambridge, MA, United States. (ch8)

Simone Assali
École Polytechnique de Montréal, Succursale Centre-Ville, Montréal, QC, Canada. (ch3)

Christoph Bohr
Institute of Inorganic Chemistry, University of Cologne, Cologne, Germany. (ch1)

Chess Boughey
University of Cambridge, Cambridge, United Kingdom. (ch7)

Yonatan Calahorra
University of Cambridge, Cambridge, United Kingdom. (ch7)

Hung-Ling Chen
Centre de Nanosciences et de Nanotechnologies, UMR 9001 CNRS, Université Paris Sud, Université Paris-Saclay, Avenue de la Vauve, 91120 Palaiseau, France. (ch4)

Jayce J. Cheng
Department of Materials Science and Engineering, Massachusetts Institute of Technology, Cambridge, MA, United States. (ch8)

Stéphane Collin
Centre de Nanosciences et de Nanotechnologies, UMR 9001 CNRS, Université Paris Sud, Université Paris-Saclay, Avenue de la Vauve, 91120 Palaiseau, France. (ch4)

Thomas Fischer
Institute of Inorganic Chemistry, University of Cologne, Cologne, Germany. (ch1)

Robert Frohnhoven
Institute of Inorganic Chemistry, University of Cologne, Cologne, Germany. (ch1)

Gerard Gadea
Catalonia Institute for Energy Research (IREC), Barcelona, Barcelona, Spain. (ch5)

Silvija Gradečak
Department of Materials Science and Engineering, Massachusetts Institute of Technology, Cambridge, MA, United States. (ch8)

Olivia D. Hentz
Department of Materials Science and Engineering, Massachusetts Institute of Technology, Cambridge, MA, United States. (ch8)

Sohini Kar-Narayan
University of Cambridge, Cambridge, United Kingdom. (ch7)

Tim Ludwig
Institute of Inorganic Chemistry, University of Cologne, Cologne, Germany. (ch1)

Lorenzo Mancini
Centre de Nanosciences et de Nanotechnologies, UMR 9001 CNRS, Université Paris Sud, Université Paris-Saclay, Avenue de la Vauve, 91120 Palaiseau, France. (ch4)

Sanjay Mathur
Institute of Inorganic Chemistry, University of Cologne, Cologne, Germany. (ch1)

Alex Morata
Catalonia Institute for Energy Research (IREC), Barcelona, Barcelona, Spain. (ch5)

Oussama Moutanabbir
École Polytechnique de Montréal, Succursale Centre-Ville, Montréal, QC, Canada. (ch3)

Samik Mukherjee
École Polytechnique de Montréal, Succursale Centre-Ville, Montréal, QC, Canada. (ch3)

Canlin Ou
University of Cambridge, Cambridge, United Kingdom. (ch7)

Giovanni Pennelli
Università di Pisa, Pisa, Italy. (ch6)

Valerio Piazza
Centre de Nanosciences et de Nanotechnologies, UMR 9001 CNRS, Université Paris Sud, Université Paris-Saclay, Avenue de la Vauve, 91120 Palaiseau, France. (ch4)

Albert Queraltó
Institute of Inorganic Chemistry, University of Cologne, Cologne, Germany. (ch1)

Paul H. Rekemeyer
Department of Materials Science and Engineering, Massachusetts Institute of Technology, Cambridge, MA, United States. (ch8)

Stefano Roddaro
NEST, Scuola Normale Superiore and Istituto Nanoscienze—CNR; Università di Pisa, Pisa, Italy. (ch6)

Francesco Rossella
NEST, Scuola Normale Superiore and Istituto Nanoscienze—CNR, Pisa, Italy. (ch6)

Amit Solanki
Harvard John A. Paulson School of Engineering and Applied Sciences, Harvard University, Cambridge, MA, United States. (ch2)

Albert Tarancon
Catalonia Institute for Energy Research (IREC); Catalan Institution for Research and Advanced Studies (ICREA), Barcelona, Barcelona, Spain. (ch5)

Maria Tchernycheva
Centre de Nanosciences et de Nanotechnologies, UMR 9001 CNRS, Université Paris Sud, Université Paris-Saclay, Avenue de la Vauve, 91120 Palaiseau, France. (ch4)

Handon Um
Harvard John A. Paulson School of Engineering and Applied Sciences, Harvard University, Cambridge, MA, United States. (ch2)

PREFACE

Nanowires are large aspect ratio structures with two dimensions ranging from few nanometers to few hundreds of nanometers and one relatively large dimension (few microns to few 10s of microns). Nanowires in a variety of materials (e.g., metals, group IV elements, III–V compounds, II–VI compounds, metal oxides) offer access to a range of properties not found in the same material in bulk form. The mechanical and thermoelectric properties of materials in the nanowire configuration are very different to the properties observed in the bulk material. Some materials like hexagonal Si, hexagonal GaP, or wurtzite InP that cannot be grown/synthesized in bulk can be relatively easily grown in the nanowire configuration, giving access to a wider range of bandgaps for device applications. The shape anisotropy of nanowires allows for new device configurations/designs that cannot be realized from bulk semiconductors. The large surface areas of nanowires are especially attractive for applications that rely on surface-assisted processes like catalysis and charge exchange. All of the above properties make nanowires promising building blocks for many existing and emerging applications. Currently, there is a lot of research interest in nanowires for energy harvesting, conversion, and storage applications. Nanowires are expected to push the frontiers of the field of "nanoenergy."

This book serves as an introduction to students and researchers interested in energy harvesting applications of nanowires. We have identified innovative emerging research directions in the context of nanowire energy harvesting devices, and brought together leading researchers in this field to contribute to this book project. Readers will benefit from chapters covering all aspects of this topic including nanowire growth/fabrication, nanoscale characterization, and thermoelectric, piezoelectric, and photovoltaic energy harvesting applications of nanowires.

The first few chapters of this book highlight the recent developments in growth/fabrication/synthesis of nanowires that find applications in energy harvesting devices. Electrospinning of carbon-based, metal and metal oxide nanowires, top-down fabrication of elemental group IV (Si) nanowires, and finally epitaxial growth of group IV compound semiconductor nanowires are discussed in detail in Chapters 1–3 of this book. Synthesis/fabrication of nanowires with user-defined dimensions and positions, compositions, and heterointerfaces is discussed in relevant chapters.

The dimensions and structural anisotropy of nanowires demand new characterization tools for assessing their quality or suitability for device applications. The conventional tools used in the context of bulk devices are no longer suitable for nanowires or nanowire devices. Chapter 4 of this book introduces the principle of and discusses in detail techniques like electron beam-induced current suitable for assessing individual nanowires for device applications, with special focus on solar cells and LEDs/lasers.

Thermoelectric and piezoelectric applications of nanowires have received little attention compared to photovoltaic applications of nanowires. Chapters 5 and 6 of this book are dedicated to introducing the thermoelectric properties of nanowires, introducing different configurations in which nanowires can be used as thermoelectric generators and methods for assessing the performance of nanowire-based thermoelectric generators. Chapter 7 is dedicated to piezoelectric applications of nanowires. Chapters 5–7 emphasize the fundamentals or basic physics of thermoelectric and piezoelectric power convertors to serve the nonexpert or beginner reader. In addition to discussing in detail, the synthesis of ZnO nanowire arrays with control on their structural properties, Chapter 8 discusses their application in organic and II–VI compound quantum dot-based solar cells.

We hope that this book proves to be a valuable resource to young researchers keen on leveraging the unconventional physical properties of nanostructured material for energy harvesting applications. Finally, and most importantly, we are grateful to all contributing authors. This book has been made possible only due to their expertise, enthusiasm, participation, support, and hard work.

Sudha Mokkapati
Cardiff University
Chennupati Jagadish
Australian National University
Editors

CHAPTER ONE

Inorganic Nanofibers by Electrospinning Techniques and Their Application in Energy Conversion and Storage Systems

Tim Ludwig, Christoph Bohr, Albert Queraltó, Robert Frohnhoven, Thomas Fischer, Sanjay Mathur[1]
Institute of Inorganic Chemistry, University of Cologne, Cologne, Germany
[1]Corresponding author: e-mail address: sanjay.mathur@uni-koeln.de

Contents

Semiconductors and Semimetals, Volume 98
ISSN 0080-8784
https://doi.org/10.1016/bs.semsem.2018.04.003

ABBREVIATIONS

2-MOE	2-methoxyethanol
ALD	atomic layer deposition
APS	ammonium persulfate
ASR	area-specific resistance
BA	benzoic acid
BET	Brunauer–Emmett–Teller method
CNF	carbon nanofibers
CNT	carbon nanotubes
CR	Congo red
CVD	chemical vapor deposition
DCNF	double-capillary carbon nanofibers
DE	dairy effluent
DMF	dimethylformamide
DMSO	dimethyl sulfoxide
DSSC	dye-sensitized solar cells
EDLC	electrochemical double layer capacitors
EDTA	ethylenediaminetetraacetic acid
EtOH	ethanol
FF	fill factor
FTO	F-doped SnO_2
HP-CNF	highly porous carbon nanofibers
HRTEM	high-resolution transmission electron microscopy
ITO	indium tin oxide
J	photocurrent density
J_{sc}	short-circuit current density
LCGO	liquid crystal graphene oxide
MB	methylene blue
MO	methyl orange
NPCNF	nitrogen-enriched porous carbon nanofibers
ODA	4,4′-oxydianiline
P(VDF-TrFE)	poly(vinylidene fluoride)-*co*-trifluoroethylene
PAH	polycyclic aromatic hydrocarbon
PAN	polyacrylonitrile
PANI	polyaniline
PCL	polycaprolactone
PCNF	porous carbon nanofibers
PDMS	polydimethylsiloxane
PEC	photoelectrochemical water splitting
PEO	polyethylene glycol
Phen	phenanthroline
PI	polyimide
PLLA	poly(L-lactic acid)
Pluronic F127	poly(ethylene glycol)$_{106}$–poly(propylene glycol)$_{70}$–poly(ethylene glycol)$_{106}$
Pluronic P123	poly(ethylene glycol)$_{20}$–poly(propylene glycol)$_{70}$–poly(ethylene glycol)$_{20}$
PmAP	poly(m-aminophenol)
PMDA	pyromellitic dianhydride

PMMA	polymethyl methacrylate
PS	polystyrene
PU	polyurethane
PVA	poly(vinyl alcohol)
PVAc	poly(vinyl acetate)
PVDF-HFP	poly(vinylidene fluoride-*co*-hexafluoropropylene)
PVP	polyvinylpyrrolidone
PZT	$Pb[Zr_xTi_{1-x}]O_3$, lead zirconate titanate
R6G	rhodamine 6G
RhB	rhodamine B
SAED	selected area electron diffraction
SEI	solid electrolyte interface
SEM	scanning electron microscopy
SOFC	solid oxide fuel cells
TCE	transparent conductive electrode
TEOS	tetraethyl orthosilicate
THF	tetrahydrofuran
TMAO	trimethylamine oxide
TPA	terephthalic acid
TTIP	titanium-isopropylate
VGCNF	vapor grown carbon nanofibers
ZIF-8	zeolitic imidazolate framework-8
η	overall conversion efficiency

1. INTRODUCTION

Quantum wires and fibers represent quasi one-dimensional (1D) nanostructures that show unique electrical, optical, and chemical behaviors due to dimensional confinement of charge carriers and presence of surface states making them useful as device elements in number of applications such as batteries, piezoelectric energy harvesters, field-effect transistors, sensors, and photovoltaics. The proven potential of inorganic nanowires in various fields also indicates the need of methods enabling simple and scalable synthesis of nanowires and their crossbar intertwined structures that would allow a continuous synthesis and an easy integration of nanowires and networks thereof in device structures. For the production of nano- and submicron fibers, the electrospinning technique, developed in the early 20th century, has demonstrated significant scientific and technological innovation to become relevant for producing nanofibers and meshes of different compositions. Besides a precise control over fiber dimensions (radial and axial), orientation (random, parallel, crossbar), and configuration (solid, porous, core–shell), high-strength electrospun webs with encapsulated or surface-anchored active components offer entirely new possibilities for creating multimaterial junctions. Nanofiber networks represent a 2D configuration formed by entangled arrangements of 1D elements that are promising architectures due to the mechanical stability and flexibility of the meshes essential for device integration and the possibility of activating the functional performance through an external stimulus.

The principle of electrospinning emerged from the pioneering work of Cooley, who has developed an apparatus for electrostatically dispersing fluids (Cooley, 1902). Following the first patent on electrospinning technique by Cooley (1902), the electrostatic shaping of materials from liquid precursors has emerged to a versatile fabrication technique for anisotropic materials with significant industrial impact. Although limited mostly to polymer fibers in the initial phase, this fabrication technique has rapidly expanded since the mid-1990s to inorganic and hybrid materials (Greiner and Wendorff, 2007; Lu et al., 2009; Wu et al., 2012). The universality of the electroforming process to produce fibers from complex fluids through electrical discharge has resulted in a plethora of different applications where high-surface area anisotropic materials improve the functional properties in textiles (Mirjalili and Zohoori, 2016), sensors (Zhang et al., 2017), actuators (Persano et al., 2015), energy harvesting (Kumar et al., 2014), energy storage (Wang et al., 2015), medical applications (Hasan et al., 2014), catalysis (Xue et al., 2017), and membranes (Liao et al., 2018).

Electrospinning is based on the principle of electrohydrodynamics, which involves fiber production from viscous precursor solutions or sols under application of high electrical fields that cause the deformation of liquid droplets into one-dimensional morphology (Fig. 1). Typically, a precursor/polymer solution or melt is placed in a high-voltage field, which results in the formation of liquid jets ejected from the solution/melt by electrostatic force. The electrospinning setup is typically based on three units: a spinneret, a collector, and a high-voltage DC power supply. The high-voltage power supply generates a high electrical field (several kV/cm) between the spinneret—which is fed by the spinning solution—and the grounded collector. The repulsive electrostatic forces of equal charge carriers injected into the spinning solution lead to a conical deformation of the droplet at the spinneret which is known as Taylor cone. The related phenomenon of electrospraying (droplet formation) is prevented by incorporating suitable

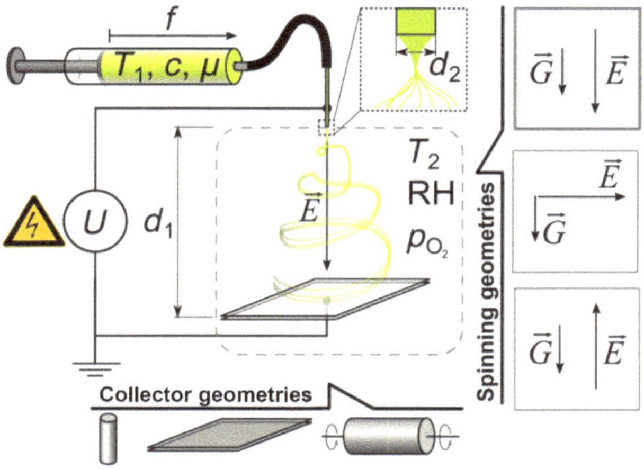

Fig. 1 Generalized features of an electrospinning setup: A precursor solution with temperature T_1, concentration *c*, and viscosity μ is pumped at a constant flow *f* through a metallic needle with the diameter d_2, at which a high-voltage *U* is applied. In the electrical field $\vec{E}$ fibers are formed at a reactor temperature T_2 with a relative humidity RH and oxygen partial pressure p_{O_2} and collected with different collector geometries (i.e., pin, plate, rotating drum), which are placed at a distance d_1. Different spinning geometries can be realized by aligning the electrical field either parallel, perpendicular, or antiparallel to the gravitational field $\vec{G}$.

polymer additives to the precursor solution that also act as structure-guiding agents. Upon evaporation of solvent and/or cooling, solid fibers are formed and collected on a grounded collector. Despite its simplistic nature to process nanomaterials at ambient temperature and pressure with sufficient reliability and reproducibility, the fabrication of homogeneous fibers by electrospinning technique simultaneously depends on various parameters as summarized in Fig. 1: precursor/melt temperature (T_1); concentration (c) and viscosity (μ); temperature (T_2) relative humidity (RH); and oxygen partial pressure (p_{O_2}) of spinning environment. In addition, the geometry of the spinning chamber has significant influence on the obtained fiber structure and alignment: spinning distance (d_1), nozzle diameter (d_2), collector area, and shape. Not only the spinning potential (U) but also its alignment with respect to the gravitational field effects the fiber formation process; one can distinguish between parallel, antiparallel, and perpendicular arrangement of $\vec{E}$ and $\vec{G}$, respectively.

Fig. 1 exemplifies a laboratory spinning setup, while different geometries and configurations have been developed, especially for high-volume production of fibers in industrial settings (Brown et al., 2016). For more detailed description of the fundamental electrospinning process (Huang et al., 2003) and special fabrication techniques (Yarin, 2011) the authors refer to recent literature.

Device applications of one-dimensional inorganic materials have shown their intrinsic advantages in increasing the efficiency of conversion, storage, and release of energy. For instance, nanofibrous morphology of electrode materials in lithium-ion batteries is beneficial in improving the mechanical stability of electrode materials against volume expansion caused by insertion of ions. Moreover, the use of electrospun nanofiber meshes was shown to improve battery power and energy density by making diffusion and displacement reactions faster and more efficient (von Hagen et al., 2012). Moreover, metallic nanowires and their percolating networks have emerged as promising alternative to transparent conductive electrodes and transparent heaters due to their low sheet resistance, but high optical transparency in visible part of electromagnetic spectrum, and stretchability that make them interesting candidates for wearable electronics (Jang et al., 2017). Similarly, the optoelectronic properties of photoelectrocatalysts benefit from the nanostructuring approach that allows to harvest photons from the visible region of the solar spectrum. Since light harvesting and charge-collection efficiencies in heterojunction devices do not show linear correlation, there is always a tradeoff that demands an optimum active layer thickness or nanostructured exciton diffusion pathways. In this context, core–shell nanowires with different electronically coupled junctions are promising multimaterial architectures suitable for creating synergies in optoelectronic (e.g., light absorption and carrier multiplication) and chemical (e.g., multielectron redox reactions) processes.

This chapter covers salient aspects of electrospinning techniques for the processing, assembly and integration of high-axial ratio nanostructures with controlled diameter and aspect ratio. The following account highlights recent advances and current research focus with respect to the application of inorganic electrospun materials in energy harvesting and storage applications from a fundamental and applied perspective. Starting with carbon-based materials, metallic and metal oxide nanofibers, as well as possible fiber architectures (single-, double-, or triple-walled) will be discussed in the following sections. The chapter concludes with current developments in the field of electrospinning that demonstrate the immense potential in the development of new materials architectures and technologies.

2. ELECTROSPUN CARBON NANOFIBERS AND THEIR APPLICATION IN ENERGY STORAGE SYSTEMS

Increasing coverage of today's energy demand by alternative and renewable energy sources like wind and solar power requires more efficient and time-independent storage devices for grid stability. Additionally, the fast-paced development in the field of electrical mobility with integrated energy recovery systems requires improved energy storage systems. Lithium-ion battery and supercapacitor technologies are capable to fulfill most of the aforementioned conditions and must be further improved in their storage capacity as well as power density.

2.1 Working Mechanism of Lithium-Ion Batteries and Supercapacitors

Lithium-ion batteries are nowadays already an integral part of everyday life, powering portable electronic devices such as mobile phones and notebooks, as well as an increasing number of gadgets like headphones, smart-watches, or fitness trackers. These rechargeable power storage devices, consisting of an anode, cathode, separator, and electrolyte, are able to store energy by intercalation of lithium ions into a material depending matrix. Their main advantage, especially compared to conventional Pb–acid or Ni–Cd batteries, is the higher energy-to-weight ratio, lack of memory effect, and longer life span (Dong et al., 2011).

In comparison to lithium-ion batteries, the energy density of supercapacitors is usually 3–30 times lower. On the other hand, supercapacitors are capable of delivering a 100–1000 times higher power in the same material and volume, as well as the ability to charge and discharge in a short time period (Miller et al., 2008). Depending on the charge-storage mechanism, supercapacitors can be classified as EDLC or pseudocapacitors (Conway, 1999).

In EDLCs, the electric charge gets stored at the surface of the electrode through a reversible ion adsorption process, leading to fast energy uptake and delivery. Furthermore, the ions do not need to intercalate into the material, which prevents swelling and provides a better performance and cycle life; however, the specific energy density remains moderate (Lu et al., 2017).

In comparison to the EDLCs, pseudocapacitors provide a higher energy density and specific capacitance. They offer increased storage capability through reversible redox reactions between the electrode material and the electrolyte. As a consequence, redox reactions can swell and shrink the electrode material, like in lithium-ion batteries, lowering the stability and cycle life. Generally, the capacitance of pseudocapacitors can reach up to several hundreds of Farad per gram of active material (Lu et al., 2017). Typically, the electrode is made of conducting polymers or metal oxides, but can also consist of functionalized carbon, combining electrostatic and pseudocapacitive charge-storage mechanisms (González et al., 2016). The hybrid capacitor's design combines the beneficial properties of EDLC and pseudocapacitor systems.

2.2 One-Dimensional Nanoscaled Carbon Materials

Despite their different storage mechanisms, lithium-ion batteries and supercapacitors are based on materials with high electrical conductivity, electrochemical stability, and a large specific surface area. Currently, commercial batteries and supercapacitors use carbon as electrode material. Diverse types of carbon nanomaterials have been investigated to enhance the capacities of energy storage devices.

Besides zero-dimensional (0D) materials such as carbon nanospheres and two-dimensional (2D) materials like graphene or reduced graphene oxide, one-dimensional (1D) materials have attracted increasing attention for their use in energy storage devices due to their large surface area, anisotropic and efficient charge transport properties, as well as chemical and mechanical stability (Wang et al., 2015).

The important distinguished nanoscaled 1D carbon materials include CNTs, built up by individual graphene layers wrapped as hollow cylinders (single- or multiwalled) and CNFs with graphene layers piled up as platelet, herringbone, ribbon-like, or stacked-cup structure (Fig. 2) (Martin-Gullon et al., 2006). These carbon nanofibers can be manufactured in vertically aligned geometry by chemical vapor deposition (CVD) process, whereas randomly oriented amorphous CNFs without graphene layers can be prepared by electrospinning of polymer precursors followed by a pyrolysis step. Among different spinning techniques like wet, gel, melt, and dry spinning, electrospinning is the most effective and versatile method to produce CNFs with controllable diameters in the nano-to-micrometer range (Patil et al., 2017).

Due to the reduced bulk dimensions, both CNFs and CNTs show much higher charge and storage capacities than conventional graphite. Although electrospun CNFs possess lower crystallinity, electrical conductivity, and

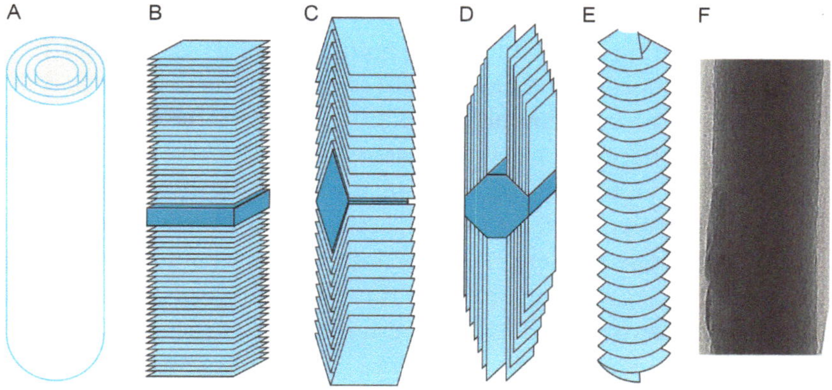

Fig. 2 Structure of (A) multiwalled CNTs, (B) graphene platelets, (C) graphene herringbones, (D) graphene ribbons, (E) stacked cup CNFs, and (F) electrospun amorphous CNF (without graphene layers). *Adapted from Zhang, B., Kang, F., Tarascon, J.M., Kim, J.K., 2016. Prog. Mater. Sci. 76, 319–380. Copyright 2016 Elsevier.*

smaller surface areas compared to CNTs, their ease of fabrication under ambient conditions, good scalability, and reduced manufacturing costs make them favored 1D materials for large-scale integration and application (Zhang et al., 2016a). Possibility of achieving additional morphological features with defined pore sizes and in situ formed composites is further advantages of electrospun CNFs. Furthermore, and in contrast to CNTs and VGCNF, CNFs can be produced without employing an external catalyst or substrate, resulting in a three-dimensional network containing aligned or interwoven 1D fibers with high flexibility and excellent mechanical strength. These properties allow electrospun carbon webs to be directly integrated into batteries or supercapacitors as bendable, freestanding, binder-, and additive-free electrodes without further processing steps or additional dead weight (Liu et al., 2015; Zhou et al., 2014).

2.3 Improvements in the Architectural Design of CNFs

2.3.1 Single-, Multiwalled and Porous Carbon Nanofibers

Due to a high carbon yield and excellent thermal stability PAN-based (polyacrylonitrile) CNFs are one of the most investigated electrospun anode materials for lithium-ion batteries. Carbonized at 1000°C, this electrode delivers a capacity of 450 mAh/g, showing better results than traditional graphite anodes (372 mAh/g) (Kim et al., 2006a). In supercapacitors, the neat PAN-derived CNFs exhibited a specific capacitance of 152 F/g at a scan rate of 2 mV/s (Lai and Lo, 2015). To further increase the electrochemical performance of carbon-based fibers, improvements in the architectural design are necessary. Changes in diameter, pore size, and structure, as well as morphology during the synthesis and posttreatment of electrospun CNFs play an important role in enhancing the capacity of the final network.

Supercapacitor and lithium-ion applications employ the same strategy to create a large surface area electrode material with tailored porosity for efficient charge accumulation and transport management. EDLCs accumulate electrostatic charges on the interface between electrolyte and electrode. Since only the first nanometers of the interface contribute to the storage density, decreasing particle size and thus increasing surface area, while maintaining electron-conducting pathways, is beneficial (Simon and Gogotsi, 2008). The specific capacitance in supercapacitors is attributed to both, pseudocapacitance and electrical double-layer capacitance, where the former is mainly determined by functional groups of the carbon surface. Additional benefits of CNFs in lithium-ion batteries are short lithium-ion insertion/extraction distances and buffering of the volume expansion during intercalation and extraction of lithium ions in the active material. There are many different approaches to produce porous carbon nanofibers (PCNFs), such as the addition of a template polymer precursor or the admixture of metal salts, which are etched away after thermal treatment leaving behind pores (Lai and Lo, 2015; Li et al., 2014a; Ma et al., 2014a; Tan et al., 2017; Wang et al., 2016a). Li and coworkers synthesized highly porous CNFs (HP-CNFs) for lithium-ion batteries with a large number of micro- and mesopores by introducing air into the Ar-flow during carbonization (Li et al., 2015). After stabilization in air at 280°C the fibers were heated to 1000°C and a specific amount of dried air was introduced. Larger surface areas were obtained by increasing amounts of air introduced toward partial oxidation (Fig. 3). Increasing the dry air-flow, the surface area of pristine fibers (120 m^2/g) was enlarged to 405 m^2/g (12.5 mL) and 583 m^2/g (50 mL), respectively, while a collapsed fiber network was obtained when 80 mL of dry air were introduced. The HP-CNFs treated with 50.0 mL air delivered the highest reversible capacity of 1780 mAh/g after 40 cycles at a current density of 50 mA/g. Apparently, the improved porous structure provided good access of the electrolyte to the electrode surface, facilitating a fast charge transfer.

Most of the methods used to fabricate porous nanofibers either need an additional physical (e.g., air, carbon dioxide, water) or chemical (e.g., sodium hydroxide, potassium hydroxide, phosphoric acid) activation process after the stabilization and carbonization step. Self-supported microporous CNFs were synthesized out of PI/PVP-polymer blend electrospinning and thermal treatment without an activation process (Le et al., 2015). PVP was

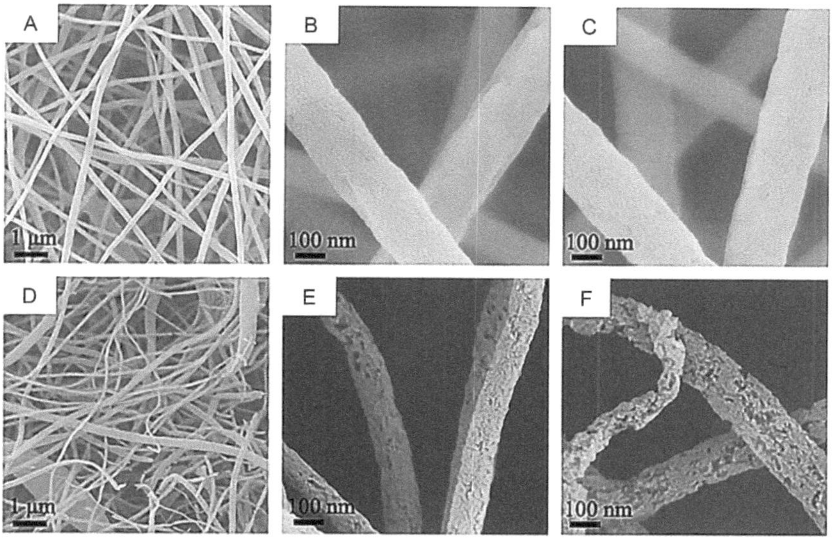

Fig. 3 (A–C) SEM images of dense CNFs, and (D–F) HP-CNFs. *Reprinted with permission from Li, W., Li, M., Wang, M., Zeng, L., Yu, Y., 2015. Nano Energy 13, 693–701. Copyright 2015 Elsevier.*

used as a pore-forming template by controlling the cross-linking between PVP and PI precursor via imidization process. The surface area and mesoporosity raised by increasing PVP content from 20 to 100 mass% (in relation to total masses of polymer precursors) to a value of 804 m^2/g and total pore volume of 0.344 cm^3/g. Due to the decreased weight loss of PVP in the fibers during thermogravimetric studies, a higher mass loading of PVP was tried that resulted in a decreased BET surface and a smaller capacitance performance (Fig. 4). A maximal specific capacitance of 215 F/g was achieved at a current density of 0.2 A/g at the optimized PVP loading (PCNF-100).

In comparison, Go and coworkers reported fibers from PAN-precursor by infrared laser-induced carbonization, employing a diode laser with a maximum laser power of 800 W. This method achieved heating rates of approximately 50 K/s and generated fibers with high specific surface areas and small pore sizes through explosive expulsion of gaseous byproducts (Go et al., 2016). The specific surface area was increased from 11.6 m^2/g (thermally carbonized nanofibers) up to 446 m^2/g (laser-carbonized nanofibers) and was also higher than the KOH-activated nanofibers (139.7 m^2/g). This technique combined the conventional carbonization and activation treatment in one process to reduce energy and time requirements; however, results on the electrochemical performance were not reported.

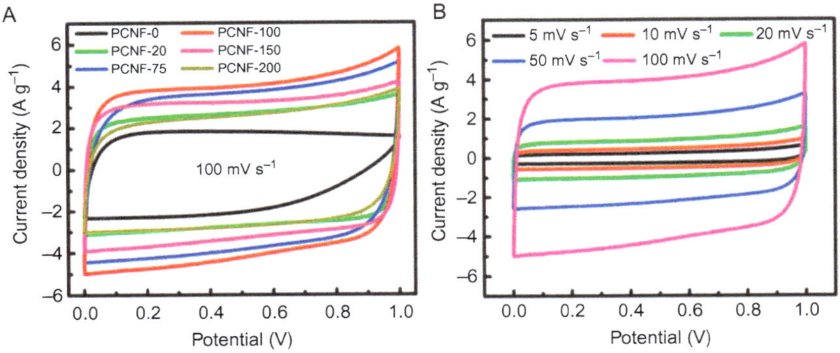

Fig. 4 (A) Cyclic voltammetry plots for PCNFs with different PVP loadings at a scan rate of 100 mV/s, (B) cyclic voltammetry plots of PCNF with 100 mass% loading of PVP (in relation to total masses of other contained polymer precursors) at different scan rates. *Reprinted from Le, T., Yang, Y., Huang, Z., Kang, F., 2015. J. Power Sources 278, 683–692. Copyright 2015 Elsevier.*

Hollow carbon nanofibers with large surface areas can be obtained by using a dual nozzle system. Wang and coworkers designed CNFs with micropores in the inner capillary and mesopores in the outer capillary. This led to a high accessibility to electrolyte, short diffusion lengths for ions, and high flexibility (Wang et al., 2016b). Due to the use of TEOS/ethanol solution in the outer capillary and lignin/DMF solution in the inner one, there is a partial miscibility at the interface of the double capillary. This resulted in the formation of micropores (1.4 nm) in the core and mesopores (2.7–3.7 nm) in the shell. The specific surface area of these double-capillary carbon nanofibers (DCNFs) increased to 870 m^2/g compared to neat (164 m^2/g) and porous (568 m^2/g) CNFs. The DCNFs showed an increased capacitance of 133 F/g at a current density of 1 A/g, compared to neat or porous CNFs with only 16 and 78 F/g, respectively.

2.3.2 Heteroatom-Doped CNFs

Chemical doping of carbon fibers with heteroatoms is another efficient way to tailor their electronic and chemical properties. Incorporation of nitrogen is known to enhance the electrical conductivity and induce *n*-type semiconducting behaviors by introducing donor states near the Fermi level. The fast Faradic redox reaction between nitrogen-containing functional groups and electrolyte ions makes nitrogen-doping an attractive approach for its use in pseudocapacitors or to increase the pseudocapacitance of hybrid capacitors. Either in situ doping or the treatment of the electrospun fibers with nitrogen-containing chemicals leads to nitrogen inclusion (Park et al., 2017).

With its inherent high nitrogen content, PAN delivers a nitrogen amount of 15.24 at% at low calcination temperatures (550°C) in electrospun fibers, which decreases to 3.59 at% when carbonized at 950°C (Zhang et al., 2014). Nitrogen-enriched porous CNFs were successfully prepared via electrospinning of a water-soluble phenolic resin/PVA blend solution and an additional thermal treatment in ammonia at 800°C after the carbonization step. The NH_3 treatment created a rough surface, high conductivity, and a heteroatom-doped surface. In comparison to untreated PCNFs (201.3 F/g), nitrogen-doped fibers, treated for 3 h, showed a nitrogen content of 5.26% and delivered a capacity of 251.2 F/g at 0.1 A/g and a cycle stability of 99% after 2000 cycles (Tian et al., 2015). Besides the increased electrical conductivity, which ensures a rapid electronic diffusion and transmission, especially the nitrogen groups improve the hydrophilic and polar nature of the CNFs, leading to a better interaction between the electrode and aqueous electrolyte solution. Hereby a combination of EDLC and pseudocapacitance contribute to the performance. The redox reactions at the surface of the fibers are thereby mainly ascribed to the pyridinic and pyrrolic nitrogen. Furthermore, nitrogen-doped fibers show extrinsic defects that enhance lithium intercalation. Additionally, the nitrogen functionalities at the surface provide many active sites for lithium-ion adsorption, both highly favorable for lithium-ion batteries.

Polyacrylic acid-based electrospinning with the addition of melamine and the posttreatment with ammonia provided a high-level nitrogen-doped freestanding lithium battery anode. A four times higher reversible capacity of 1150 mAh/g after 50 cycles at 50 mA/g was obtained compared to untreated fibers (Nan et al., 2014). The nitrogen content was increased to 7.9%, and especially the amount of pyridinic nitrogen was greatly enhanced. Pyridinic nitrogen induces a large number of defects, leading to more lithium-ion insertion sites and enhanced lithium storage capacity of the carbon materials. In addition to the increased electrical conductivity and large surface area, this provides a higher capacity and good rate capability of the cell (Fig. 5).

2.3.3 Flexible Electrode Design

In recent years, flexible electronic devices have received tremendous attention due to their potential application in rollup displays and flexible smart mobile devices (Li et al., 2014b). Especially, solid-state supercapacitors

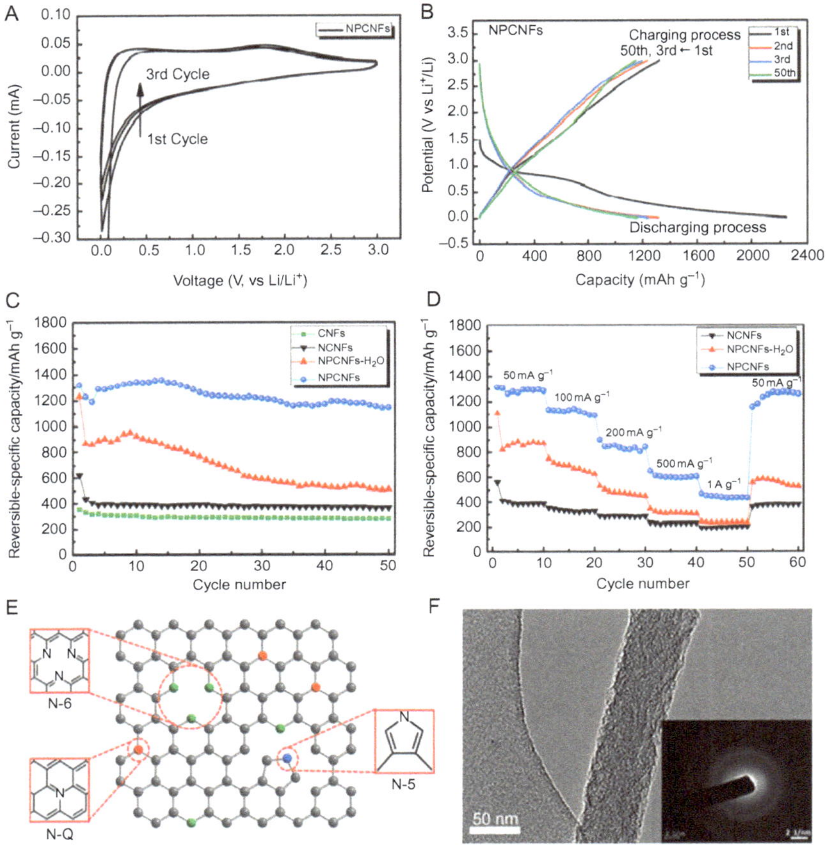

Fig. 5 (A) Cyclic voltammograms and (B) charge–discharge curves of NPCNFs; (C) cycling performance and (D) rate performance of CNFs treated under different processing conditions, (E) schematic illustration of different nitrogen species in N-doped carbon materials. (N-Q = quaternary N atoms, N-6 = pyridinic N atoms, N-5 = pyrrolic N atoms), (F) TEM image and SAED pattern (*inset*) of NPCNFs. *Adapted from Nan, D., Huang, Z.-H., Lv, R., Yang, L., Wang, J.-G., Shen, W., Lin, Y., Yu, X., Ye, L., Sun, H., et al., 2014. J. Mater. Chem. A 2, 19678–19684. Copyright 2014 Royal Society of Chemistry.*

combine high flexibility, stability, even under different mechanical deformation loads, light weight, high specific energy, and power density, as well as reliable safety, which is demonstrated by the increasing number of publications (Cheng et al., 2015a, 2016; Li et al., 2014b). A freestanding network with high specific surface area and good electrical conductivity is essential to fabricate such device for EDLCs. Lawrence et al. developed a binder-free solid-state EDLC, which consisted of a porous CNF mat filled with a blend of silica sol–gel precursor and ionic liquid electrolyte, to form an all-solid matrix after the drying process (Lawrence et al., 2016). Ionic liquids have a large electrochemical potential window with up to 4V and an excellent electrochemical and moisture stability. Thus, they can eliminate the leakage and integrity problems of liquid electrolytes in EDLCs. The supercapacitor network, activated via KOH-etching, exhibited a BET surface area of 2282 m^2/g, and a specific capacitance of 142 F/g at 0.5 A/g was obtained in galvanostatic charge–discharge tests. The resulting energy density of 61 Wh/kg was comparable to that of Pb-acid batteries.

Another solid-state supercapacitor fabrication was inspired by the natural bamboo structure with a well-balanced macro-, meso-, and microporosity, showing excellent mechanical flexibility, foldability, and electrochemical performance (Fig. 6) (Sun et al., 2015). The as-designed device exhibited stable electrochemical performance, even after bending and twisting. The

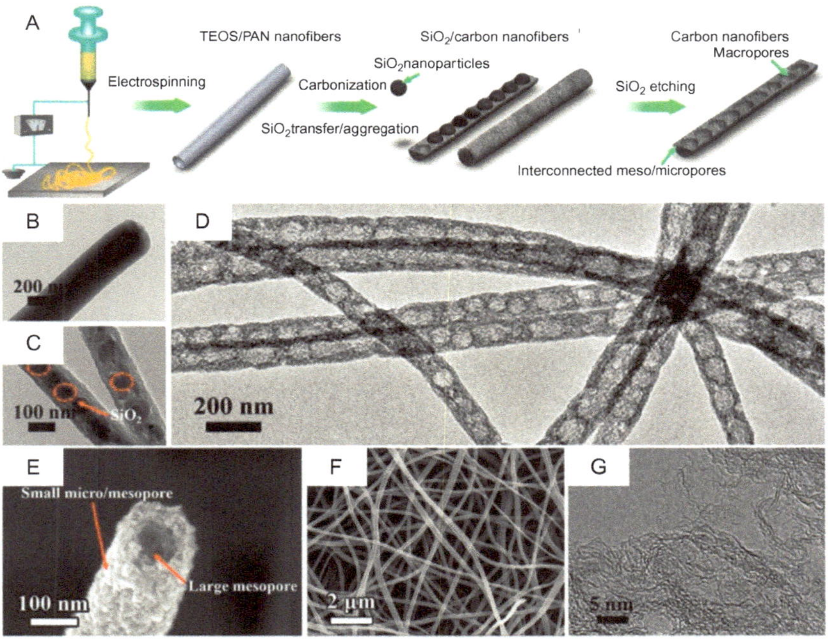

Fig. 6 (A) Schematic illustration of the bamboo-like CNFs fabrication process, (B) TEM image of the initial TEOS/PAN composite nanofibers, (C) SiO_2/carbon composite nanofibers and (D) bamboo-like CNFs, (E, F) SEM images, and (G) HRTEM image of the as-prepared CNFs. *Reprinted from Sun, Y., Sills, R.B., Hu, X., Seh, Z.W., Xiao, X., Xu, H., Luo, W., Jin, H., Xin, Y., Li, T., et al., 2015. Nano Lett. 15, 3899–3906. Copyright 2015 American Society of Chemistry.*

bamboo-like structure obtained by calcination and etching of an electrospun TEOS/PAN-precursor solution led to a periodic distribution of interior holes along the length and a graded pore structure at the cross section. This enhanced the stability of the network under different mechanical deformation conditions, provided a high-surface area up to 1912 m^2/g and a total pore volume of 2.27 cm^3/g. Additionally, a specific capacitance as high as 236 F/g at a current density of 5 A/g and an extraordinarily high stability with nearly 100% of the initial capacitance after 5000 cycles tested at a current density of 10 A/g was reached.

Due to the rapid development of high-tech products in the domain of wearable electronics, development of flexible lithium-ion batteries is highly desirable. The freestanding, foldable, and additive-free structure of electrospun fiber networks are predestined for their application in bendable electronic devices. A simple method that combines the electrospinning of a zinc acetate/PAN/DMF solution with a typical carbonization procedure was employed to fabricate freestanding, pure, and foldable CNFs. Zinc acetate is responsible for the effects of plasticization, catalysis, and activation, which enhanced the mechanical flexibility and porosity of the resulting CNFs (Chen et al., 2017). The as-obtained network was foldable multiple times and showed excellent elastic-recovery properties. Additionally, depending on the stirring temperature of the spinning solution, it is possible to gain more flexible or stronger networks. As self-supported anode, the flexible network exhibited a discharge capacity of 630 mAh/g with a current density of 200 mA/g over 100 cycles with good stability.

Tables 1 and 2 summarize a variety of different strategies demonstrated in recent years to prepare electrospun CNFs. The tables give information about synthesis routes and corresponding electrochemical performances for their application in supercapacitors and lithium-ion batteries.

2.4 Challenges and Perspectives

Despite the improvements of electrospun CNFs, some challenges still remain. For instance, an increased surface area leads to an enlarged contact between electrode and electrolyte, which could be detrimental for lithium-ion batteries due to a thick SEI formation and increasing the irreversible capacity loss during the first cycles (Zhang et al., 2016a). Moreover, a highly porous material can also decrease the mechanical stability of the CNFs, making them more brittle and harder to handle, especially if they are used as free-standing electrodes. Additionally, it must be considered that a high-surface

Table 1 Carbon-Based Nanofiber Anodes for Supercapacitors

Materials	Precursor, Additive, Solvent	Thermal Treatment and Activation	Electrochemical Performance	References
Flexible, cross-linked, nitrogen-doped CNFs	PAN, PVP, TPA, DMF	Preoxidation (300°C, 2h, air), carbonization (800°C, 1h, Ar)	175 F/g at 50 A/g in 1 M H_2SO_4	Cheng et al. (2015a)
Microporous PI-based CNFs	PMDA, ODA, PVP, DMF	Imidization (300°C, 30 min, air), carbonization (900°C, 1h, N_2)	215 F/g at 0.2 A/g in 6 M KOH	Le et al. (2015)
Hierarchical PCNFs	PAN, THF, $CaCO_3$, DMF	Stabilization (250°C, 2h, air), carbonization (800°C, 2h, N_2), immersed in 1.5 M HCl solution to remove CaO (1h)	251 F/g at 0.5 A/g in 6 M KOH	Zhang et al. (2016b)
Nanostructural PCNFs	PAN, PMMA, DMF	Stabilization (250°C, 4h, air), carbonization (800°C, 1h, N_2)	210.0 F/g at 2 mV/s in 2 M KOH	Lai and Lo (2015)
PCNFs	PAN, TEOS, DMF	Stabilization (280°C, 1h, air), activation using H_2O (800°C, 1h), CO_2 (900°C, 3h), or H_2 (800°C, 2h)	161 F/g 1 mA/cm^2 in ionic liquid electrolyte	Kim et al. (2016)
Highly flexible freestanding PCNFs	PAN, $Co(NO_3)_2$, PVP, DMF	Stabilization (220°C, 4h, air), carbonization (800°C, 2h, Ar), immersed into 10 M HCl solution for 48h to etch the cobalt nanoparticles	104.5 F/g at 0.2 A/g in 0.5 M H_2SO_4	Liu et al. (2015)
Porous flexible CNFs	PVA, phenolic resin, H_2O	Stabilization (150°C), carbonization (800°C, N_2), KOH coating, activation (700°C, N_2)	362 F/g 0.2 A/g in 6 M KOH	Ma et al. (2014a)

Continued

Table 1 Carbon-Based Nanofiber Anodes for Supercapacitors—continued

Materials	Precursor, Additive, Solvent	Thermal Treatment and Activation	Electrochemical Performance	References
Nitrogen-doped mesoporous CNFs	PAN, $Mg(OH)_2$, DMF	Preoxidization (280°C, air), carbonization (900°C, N_2), acid treatment to remove MgO template	327.3 F/g at 1.0 A/g in 6 MKOH	Tan et al. (2017)
Mesoporous graphitic CNFs	PVP, Pluronic P123, TEOS, $Ni(NO_3)_2 \times 6H_2O$, EtOH	Carbonization (350°C, 3 h/1000°C, 4 h, N_2), treatment with 2 M HNO_3 solution (2 days) and 10% HF solution (6 h) to remove NiO and silica template	303 F/g at 0.7 A/g	Dong et al. (2015)
Freestanding nonwoven nanofibers	PAN, TMAO, DMF, H_2O	Stabilization (240°C, 6 h, air), carbonization and CO_2 activation (1000°C, 1 h, He)	140 F/g at 10 mV/s in ionic liquid electrolyte	Perananthan et al. (2016)
Mesoporous CNFs with oriented mesopores	PVP, phenolic resin, Pluronic F127, $Mg(NO_3)_2 \times 6H_2O$, EtOH	Curing (100°C, 1 h), carbonization (800°C, 1 h, N_2), treatment with 20 wt.% HCl solution to remove MgO (24 h)	270 F/g at 0.2 A/g in 6 M KOH	Ma et al. (2016)
Nitrogen-enriched hierarchically PCNFs	PAN, ZIF-8, DMF	Stabilization (250°C, 2 h, air) carbonization (800°C, 2 h, Ar) washing with HCl to remove Zn	302 F/g at 0.5 A/g in 6 M KOH	Yao et al. (2017)
Nitrogen-doped porous multi-nano-channel carbon nanorods	PAN, PS, DMF	Stabilization (250°C, 2 h, air), carbonization (800°C, 1 h, Ar)	335 F/g at 0.25 A/g in 1 M H_2SO_4	Ramakrishnan and Shanmugam (2016)

Freestanding nitrogen-doped PCNFs	PAN, PmAP, DMSO	Stabilization (280°C, air), carbonization (800°C, N_2)	347.5 F/g at 0.5 mA/cm^2 in 6 M KOH	Choudhury et al. (2017)
DCNFs	Outer fluid: TEOS, PVP, acetic acid, EtOH Inner fluid: PEO, lignin, DMF	Carbonization (900°C, 2 h, N_2) to obtain C@C/SiO_2 composite nanofibers, treatment with HF to remove SiO_2	245 F/g at 0.5 A/g in ionic liquid electrolyte	Wang et al. (2016b)
Nitrogen-doped arch and hollow nanocarbons	Outer fluid: PAN, DMF Inner fluid: PVP, DMF	Dipping in hot water (75°C, overnight) to remove PVP, stabilization (250°C, 2 h, air) carbonization (800°C, 1 h, Ar)	432 F/g at 0.25 A/g in 1 M H_2SO_4	Ramakrishnan et al. (2015)
All-polymer-based core–sheath nanofibers	Nylon 6.6, formic acid, PANI, APS, HCl	Nylon/PANI core–shell structure by core treatment in an aniline/HCl solution (40°C, 3 h), in situ polymerization by adding an APS/1 M HCl solution	~400 F/g, 1000 cycles at 1 A/g in 1 M H_2SO_4	Lee et al. (2016a)
Nitrogen/phosphorus co-doped nonporous CNFs	PAN, H_3PO_4, DMF	Stabilization (280°C, 2 h, air), carbonization (800°C, 2 h, N_2)	224.9 F/g at 0.5 A/g in 1 M H_2SO_4	Yan et al. (2014)
Nitrogen-enriched meso-macroporous carbon fiber network	PAN, PVP, TEOS, DMF	Stabilization (250°C, 1 h, N_2), carbonization (800°C, 2 h, N_2), soaked in 1 M NaOH (1 h)	242 F/g at 0.2 A/g in 1 M H_2SO_4	Fan et al. (2016)
Nitrogen-doped electrospun CNFs	PVA, phenol, resorcinol, NaOH, H_2O	Curing (150°C, 2h, air), carbonization (800°C, 1 h, Ar), doping (800°C, 3 h, NH_3)	251.2 F/g at 0.1 A/g in 6 M KOH	Tian et al. (2015)

Continued

Table 1 Carbon-Based Nanofiber Anodes for Supercapacitors—continued

Materials	Precursor, Additive, Solvent	Thermal Treatment and Activation	Electrochemical Performance	References
Cellulose-derived, interbonded CNFs	Cellulose acetate, NaOH	Converting of cellulose acetate into cellulose by NaOH solution bath, stabilization (240°C, 1 h, air), carbonization (1000°C, 2 h, N_2), activation (850°C, 1.5 h, CO_2)	241.4 F/g at 1 A/g in 6 M KOH	Cai et al. (2016)
Nanofiber-based solid-state supercapacitors	PAN, Nafion, DMF	Stabilization (280°C, 5 h, air), carbonization (1000°C, 1 h, N_2), soaked in KOH solution, and heated (800°C, 30 min, N_2)	142 F/g at 0.5 A/g in silica-based ionic liquid gel electrolyte	Lawrence et al. (2016)
Hierarchical porous structure for solid-state electrochemical capacitors	PAN, TPA, DMF	Preoxidation (260°C, 2 h, air), carbonization (800°C, 1 h, Ar)	257.6 F/g at 0.5 A/g in 1 M H_2SO_4	Cheng et al. (2016)
Bamboo-inspired nanofibers	PAN, TEOS, DMF	Calcination (1200°C, 12 h, 5 vol% H_2 in Ar), etching in 10 wt% HF solution	236 F/g at 5 A/g in liquid electrolyte	Sun et al. (2015)

Table 2 Carbon-Based Nanofiber Anodes for Lithium Ion Batteries

Materials	Precursor, Additive, Solvent	Thermal Treatment and Activation	Electrochemical Performance	References
Highly porous CNFs	PAN, DMF	Stabilization (280°C, 3 h, air), carbonization (1000°C, 8 h, Ar/air)	1780 mAh/g, 40 cycles at 50 mA/g ($LiPF_6$-based electrolyte)	Li et al. (2015)
Foldable electrospun CNFs	PAN, $Zn(ac)_2$, DMF	Stabilization (280°C, 2 h, air), carbonization (800°C, 6 h, Ar)	630 mAh/g, 100 cycles at 0.2 A/g ($LiPF_6$ liquid electrolyte)	Chen et al. (2017)
Graphitized PCNFs	PAN, DMF	Preoxidation (250°C, 2 h, air), carbonization (1000°C, 2 h, Ar), calcination (400°C, 18 h, air)	473.5 mAh/g, 100 cycles at 0.5 A/g (1 M $LiPF_6$)	Zhu et al. (2016)
Nitrogen-enriched electrospun PCNFs	PAN, DMF, melamine	Stabilization (250°C, 1 h/260°C, 1 h/270°C/2 h, air), carbonization (850°C, 1 h, N_2) activation (30 min, 20 vol% NH_3 in N_2)	1323 mAh/g, 50 cycles at 50 mA/g (1 M $LiPF_6$)	Nan et al. (2014)
Nitrogen-doped CNFs with open channels	PAN, PVB, urea, DMF	Stabilization (280°C, 2 h, air), carbonization (500°C, 2 h/800°C, 2 h, N_2)	734 mAh/g at 0.2°C (1 M $LiPF_6$)	Park et al. (2017)
Nitrogen and boron co-doped PCNFs	PVP, NH_3, BH_3, methanol	Drying (50°C, 6 h), calcination (850°C, 3 h, dynamic N_2)	607 mAh/g at 2000 mA/g (1 M $LiPF_6$)	Zhang et al. (2016c)

Continued

Table 2 Carbon-Based Nanofiber Anodes for Lithium Ion Batteries—continued

Materials	Precursor, Additive, Solvent	Thermal Treatment and Activation	Electrochemical Performance	References
Hierarchical graphene-containing CNFs	PAN, graphene oxide, DMF	Stabilization (250°C, 2h, air), carbonization (90°C, 3h/650°C, 1h, N_2)	275mAh/g, 200cycles at 750mA/g (1M $LiPF_6$)	Dufficy et al. (2016)
Welded mesoporous CNFs	PAN, lignin, Pluronic P123	Stabilization (250°C, 3h, air), carbonization (1200°C, 1h, N_2)	177mAh/g, 400cycles at 1A/g (1.2M $LiPF_6$)	Shi et al. (2017)
PCNFs (Na-ion battery)	PAN, Pluronic F127, DMF	Stabilization (280°C, 3h, air), carbonization (1000°C, 8h, Ar)	140mAh/g, 1000cycles at 500mA/g (1M $NaClO_4$)	Li et al. (2014a)
Nitrogen-doped CNFs (Na-ion battery)	PAN, DMF	Stabilization (250°C, 2h, air), carbonization (800°C, 2h, N_2)	150mAh/g, 200cycles at 1000mA/g (1M $NaClO_4$)	Zhu et al. (2015)
Graphene highly scattered in PCNFs (Na-ion battery)	PVP, graphene oxide, H_2O	Calcination (500°C, 1h, 10vol% H_2 in Ar)	409mAh/g, 100cycles at 100mA/g (1M $NaClO_4$)	Liu et al. (2017)

area not only increases the gravimetric double-layer capacitance but can also reduce the bulk density and the volumetric capacitance (Fan et al., 2016). One of the major challenges remains the upscaling of the electrospun process to meet industrial mass production. The most promising way to achieve higher yields is by using a multiple nozzle system (Theron et al., 2005).

Electrospun CNFs show high potential for applications in energy storage devices. Nowadays, all important parts of a battery can be produced via electrospinning. The high electrical conductivity, large surface area, and large numbers of reaction sites make them favorable for high-performance lithium-ion batteries, as well as for supercapacitors. In comparison to CNTs, electrospun CNFs are easy to produce, scalable, have a low fabrication cost, and make substrates or catalysts obsolete. Additionally, electrospinning allows the direct fabrication of bendable, self-supported electrodes, giving way to the application in flexible electronic devices. Fibers with a diameter down to tens of nanometers result in a high area-to-mass ratio leading to capacities exceeding more than twofold those of traditional graphite. This makes electrospinning a competitive production process for materials used in energy storage devices.

3. METALLIC NANOFIBERS

Metallic nanofibers combine bulk metal characteristics like conductivity, magnetic, or catalytic properties, and the benefits from nanoscale materials like increased surface area, optical properties, short reaction pathways, and economical use of resources in synthesis. Owing to their one-dimensional form, the metal nanowire meshes show very low junction resistance and stretchability in conjunction with high optical transmittance. This special blend of functional attributes made them attractive for applications such as transparent heaters and conductive electrodes, magnetic devices, energy storage, and catalysis.

3.1 Synthesis Routes Toward Metallic Nanofibers

There are two different ways to obtain nanofibrous metallic structures, namely, direct methods (Bognitzki et al., 2006; Hsu et al., 2012; Kim et al., 2015; Wu et al., 2010) and indirect template-assisted methods (An et al., 2016; Hsu et al., 2014; Soltanian et al., 2013; Wu et al., 2013).

Direct methods typically use a metal salt or precursor, which is dissolved in the electrospinning solution and incorporated in the as-spun fiber mats via

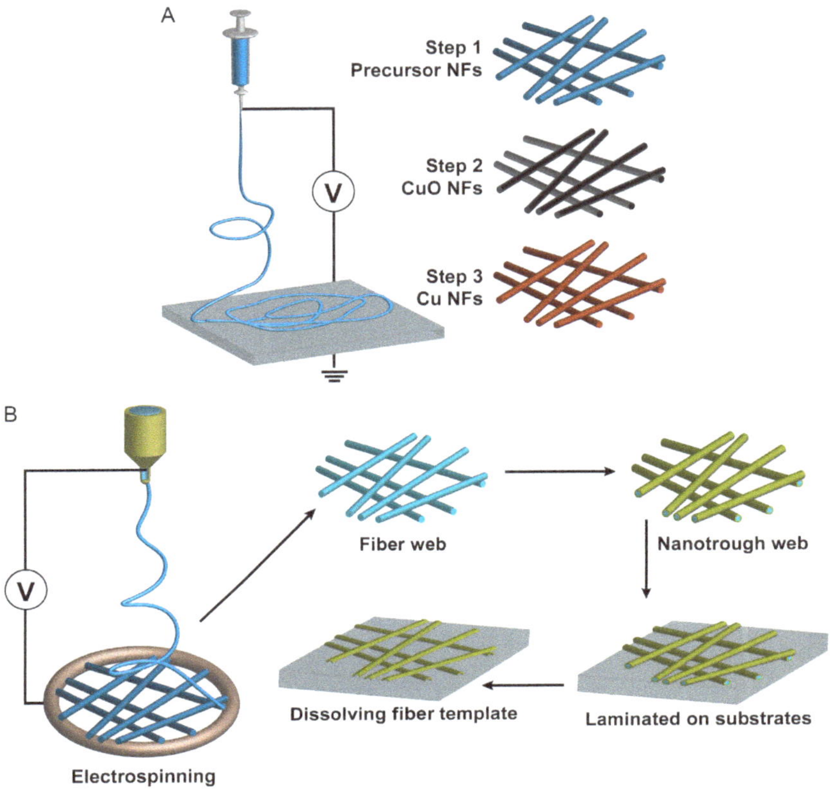

Fig. 7 (A) Direct vs (B) indirect method. *Adapted from Wu, H., Hu, L., Rowell, M.W., Kong, D., Cha, J.J., McDonough, J.R., Zhu, J., Yang, Y., McGehee, M.D., Cui, Y., 2010. Nano Letters 10, 4242–4248. Copyright 2010 American Chemical Society. Adapted from Wu, H., Kong, D., Ruan, Z., Hsu, P.-C., Wang, S.,·Yu, Z., Carney, T.J., Hu, L., Fan, S., Cui, Y., 2013. Nat. Nanotechnology 8, 421–425. Copyright 2013 Springer Nature.*

the electrospinning process. During the following calcination step in air, the polymer matrix is removed completely to obtain either metal oxide or metal fibers, if the decomposition temperature of the oxide is already reached. Metal oxide nanofibers can also be transformed into metallic phases in a subsequent calcination step in a reductive atmosphere (Bognitzki et al., 2006; Hsu et al., 2012; Kim et al., 2015; Wu et al., 2010) (Fig. 7A). Whereas in template-assisted methods, fibers are formed via electrospinning of a pure polymer solution without any metal precursors. These fibers are then used as template and metals are deposited on the surface via physical methods like thermal evaporation (Wu et al., 2013), chemical methods like redox processes on the fiber surface (Hsu et al., 2014), or electroplating (An et al., 2016). In the final processing step, the polymer fiber template is dissolved resulting in metallic nanofibrous structures (Fig. 7B).

3.2 Metallic Nanofibers as Transparent Conductive Electrodes

Nowadays, many optoelectronic devices like solar cells, switchable glass, or displays of any kind use transparent conductive electrodes. With high optical transmittance in combination with suitable electronic properties, especially a low sheet resistance R_S, indium-doped tin oxide (ITO) has been used as the material of choice for these applications in the last decades. On the other hand, with rising demand of ITO in flexible optoelectronic devices and low natural abundance of indium, finding adequate substitutional materials is critically important. In addition, the brittleness of ITO as a ceramic material poses a strong limitation on its use in flexible devices since bending causes cracks and, consequently, results in loss of conductivity in ITO films (Chen et al., 2001; Sierros et al., 2009). These factors have led to a search for materials that can replace ITO without compromising conductivity and transparency and offering at the same time lower resource and fabrication cost as well as the opportunity to use in flexible devices. The most promising TCE materials are listed in Table 3.

Metallic nanofibers preserve their conductivity during bending and can be synthesized and scaled up by straightforward chemical procedures. Applied techniques include sol–gel methods like solvothermal reactions combined with spray deposition, spin coating in case of nanowires, and electrospinning in case of nanofibers. While single-crystalline nanowires

Table 3 Sheet Resistances R_s of Different TCE Materials at Transparency $T = 90\%$ at 550 nm (Hsu et al., 2013)

Material	Sheet Resistance R_s
ITO thin film	5–20 Ω/sq
Metal nanowires	10–30 Ω/sq
Graphene	30–1000 Ω/sq
Carbon nanotubes	60–400 Ω/sq
TiN nanofibers	100–200 Ω/sq
Polymers	100–450 Ω/sq
Oxide nanowires	2000–10,000 Ω/sq

form a percolation network in which the resistivity is highly dependent on the aspect ratio of the wires and the number of wire-to-wire junctions (Hu et al., 2011), the polycrystalline electrospun fiber webs offer nearly endless pathways for electrons without the need to cross a junction. Nanofibrous networks produced via the indirect method are usually obtained with fused junctions (Hsu et al., 2014; Wu et al., 2013), which facilitate electron transport through remarkably low resistivity at these nodal points and thereby offering shorter transport pathways (Fig. 8). Among all the metals in the periodic table of elements, silver and copper exhibit smallest values for bulk electrical resistivity of 1.587 and 1.678 μΩ cm (Lange, 2005), respectively (Table 4). Since the electrical conductivity of copper is comparable to that of silver, it is preferred for industrial scale production due to its higher natural abundance and therefore reduced material cost. (See Table 5.)

The strong tendency of copper to oxidize, associated with a dramatic reduction of conductivity, is a major challenge to fabricate and integrate copper nanowires in devices, and various approaches have been developed to prevent the oxidation of copper nanowire networks. Hsu et al. and Chen et al. presented transparent oxide shells which did not significantly decrease the transparency of metallic cores (Chen et al., 2014; Hsu et al., 2012). Another approach involved coating with graphene or similar structures onto a percolation network of metallic nanowires or nanofibers. Kholmanov et al. used a laminated coating of reduced graphene oxide (Kholmanov et al., 2013), and Deng and coworkers demonstrated a large-scale roll-to-roll

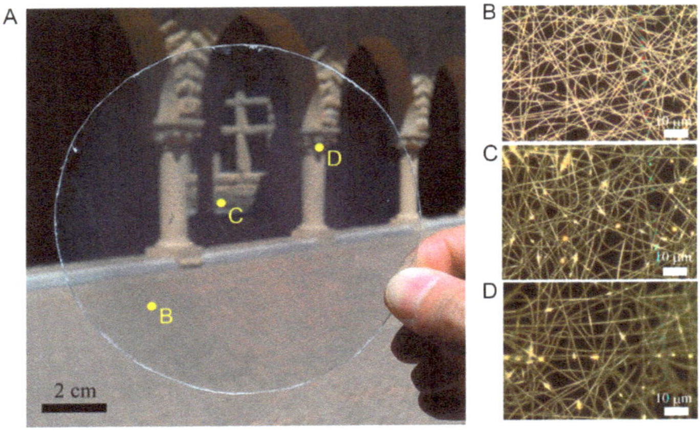

Fig. 8 Uniform coating of template assisted deposited copper nanofiber TCE network. *Reprinted with permission from Hsu, P.-C., Kong, D., Wang, S., Welch, A.J., Wu, H., Cui, Y., 2014. J. Am. Chem. Soc. 136, 10593–10596. Copyright 2014 American Chemical Society.*

Table 4 Bulk Electrical Resistivity of Different Metals at 20°C (Lange, 2005)

Metal	Electrical Resistivity (μΩ cm)
Silver	1.587
Copper	1.678
Gold	2.214
Aluminum	2.6548
Nickel	6.93
Iron	9.61
Chromium	12.5

Table 5 Overview of Electrospun Metal Nanofibers for TCE Application

Metal	Method	Diameter (nm)	Transmission T (550 nm) (%)	Sheet Resistance R_S (Ω/sq)	References
Cu	Direct	270	NA	270	Bognitzki et al. (2006)
Cu	Direct	50–200	90	50	Wu et al. (2010)
Ag	Indirect	406	90	10	Wu et al. (2013)
Au	Indirect	406	90	8	
Cu	Indirect	406	90	2	
Pt, Al, Cr, Ni, alloys	Indirect	406	NA	NA	
Ag	Indirect	200	90	8.5	Hsu et al. (2014)
Cu	Indirect	200	91	11.2	
Cu	Indirect	1000–1800	97	0.42	An et al. (2016)

encapsulation with a CVD-grown graphene film (Deng et al., 2015). Beside the protection against oxidative corrosion, an improvement of conductivity was observed due to coverage and bridging of empty spaces in the networks with another conductive material. An alternative option is the deposition of a more noble metal shell, like silver, gold, or platinum, onto elongated copper nanostructures (Stewart et al., 2015).

Another substantial challenge is the device integration of metal TCEs into, e.g., solar cells (Bailie et al., 2015), electrochromic devices (Deng et al., 2015; Kholmanov et al., 2013), or touchscreens (Lee et al., 2016b).

While ceramic films like ITO are chemically stable, even at increased temperatures up to several 100°C, this is not the case for metal nanowires and nanofibers. One major problem is the drop of conductivity caused by the oxidative transformation of metals into their oxides. Another one is the structural degradation due to reduced melting points of nanoscaled metal structures (Kim et al., 2015; Rathmell and Wiley, 2011). In some cases, the necessary thermal posttreatment of direct method electrospun nanofibers could damage the other components of the device. To avoid these issues, lamination processes are usually applied to transfer preformed metallic TCE layers (Bailie et al., 2015).

3.3 Metal Nanowires for Magnetic Applications

Nanofibers of ferromagnetic transition metals iron (Graeser et al., 2007; Hansen et al., 2012; Shao et al., 2011; Wu et al., 2007), cobalt (Graeser et al., 2007; Hansen et al., 2012; Shao et al., 2011; Wu et al., 2007), and nickel (Barakat et al., 2009; Hansen et al., 2012; Wu et al., 2007), as well as their intermetallic phases (Jing et al., 2016; Lee and Choa, 2012) can be produced via direct methods. The saturation magnetization of these fibers is found to be half the value of the bulk metals, mainly due to the largely oxidized and therefore magnetically inactive surface area (Barakat et al., 2009; Wu et al., 2007). Besides, the hysteresis curve of aligned magnetic metal fibers varies depending on whether the external field is applied parallel or perpendicular to the fiber orientation (Fig. 9) (Graeser et al., 2007). The most interesting characteristic of ferromagnetic nanofibers is their massively enhanced coercivity, up to 427 (Fe), 65 (Co), and 177 (Ni) times the value for bulk metals (Barakat et al., 2009; Wu et al., 2007) that makes them suitable for applications like electromagnetic interference shielding, high-density information storage, flexible magnets, magnetic sensors, and spintronic devices.

Future work in the field of magnetic metal nanofibers could include the ferromagnetic lanthanides and other alloys exhibiting temperature-dependent ferromagnetic behavior.

3.4 Catalytically Active Metal Nanofibers

Catalytic properties of noble metals like gold (Chen et al., 2016a; Xue et al., 2017), silver (Halperin et al., 2015), rhodium (Xue et al., 2017), palladium (Zhang et al., 2015a), or platinum (Formo et al., 2008; Zhang et al., 2015a) have significant relevance in energy storage systems based on hydrogenation–dehydrogenation reactions and have therefore been intensively studied.

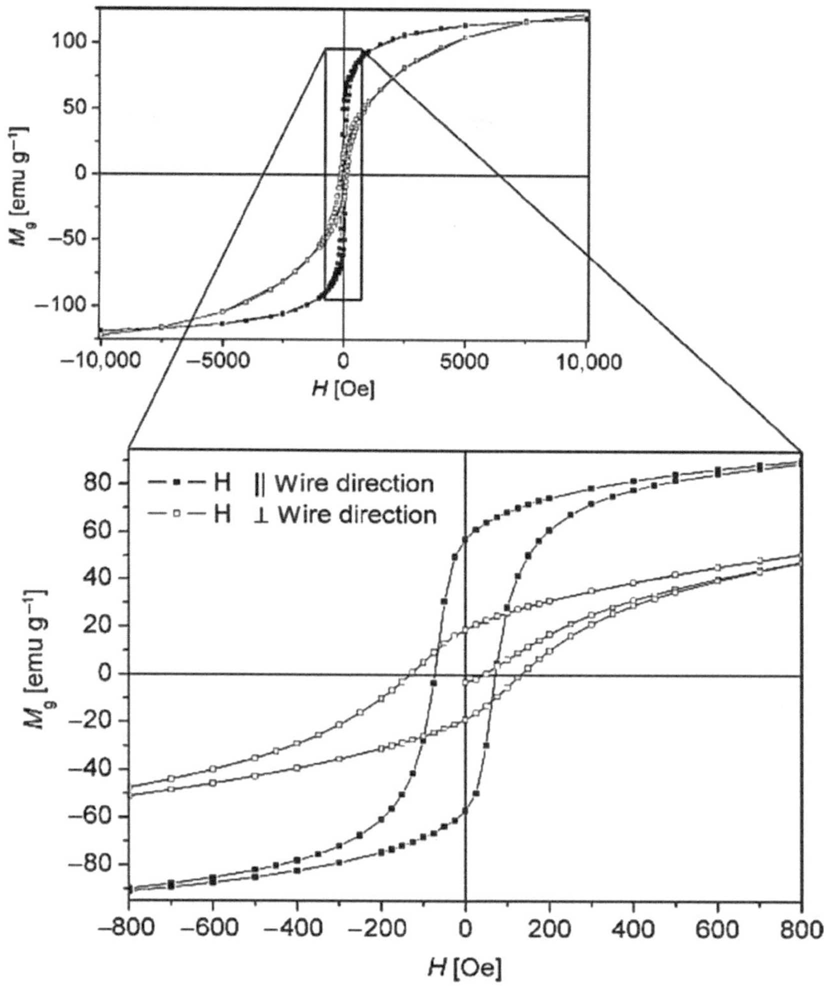

Fig. 9 Hysteresis curves of parallel aligned Fe nanofibers with field applied parallel and perpendicular to fiber alignment. *Reprinted with permission from Graeser, M., Bognitzki, M., Massa, W., Pietzonka, C., Greiner, A., Wendorff, J.H., 2007. Adv. Mater. 19, 4244–4247. Copyright 2007 John Wiley & Sons.*

Electrospun inorganic nanofibers decorated with metal nanoparticles constitute a suitable way to fix nanoparticles in a stable matrix, preventing loss of expensive catalyst material during recovering it from a reaction solution (Wen et al., 2015). Integration of nanoparticles can either be achieved by attaching them after the electrospinning process (Formo et al., 2008; Zhang et al., 2015a) or by adding small amounts of a metal salt to the precursor solution (Halperin et al., 2015).

Zhang and coworkers decorated PAN nanofibers with Pd/Pt alloy particles and tested them in hydrogen generation from ammonia borane (NH_3BH_3) (Zhang et al., 2015a), which, due to its excellent stability and high hydrogen capacity of 19.6% (Marder, 2007), is one of the favored hydrogen storage materials for being used in fuel cells for portable devices. One of the main challenges is to reduce the thermolysis temperature at high turnover rates. The synthesized fibers exhibited an excellent turnover frequency of 51.9 mol/min H_2 at ambient conditions and room temperature (Zhang et al., 2015a).

Also improving catalytic activity or reduction of processing temperatures by new structural design of catalysts can help to decrease energy consumption and production costs in large-scale industrial application. For example, the standard process for oxidation of methanol to formaldehyde (annual demand in 2016: 30 megatons (Bahmanpour et al., 2016)) takes place in a temperature range of 600–720°C (Reuss et al., 2000). Halperin et al. synthesized silver nanofibers, which reached a weight hourly space velocity (WHSV) of 140/h with 95% conversion rate at 500°C with a respective space-time yield (STY) of 97.5/h (Halperin et al., 2015). At 400°C WHSV of 370/h with 70% conversion rate and at 550°C WHSV of 540/h with 95% conversion rate were obtained. In these cases, STY was 126/h and 386/h, respectively.

A typical problem in catalysis is the surface passivation of the active material. Some metals like silver or gold feature plasmonic activity under illumination, which is beneficial in electrocatalytic alcohol oxidation reactions. Chen et al. demonstrated with their indirectly synthesized gold nanofibers that surface passivation was mainly occurring in the dark and could be removed by inducing surface plasmon interaction with blocking molecules at the gold/electrolyte interface by illumination (Chen et al., 2016a).

Other emerging fields of interest for electrospun catalytic active metal nanofibers are lithium-air batteries (Gittleson et al., 2014) and filter systems for water treatment (Choi et al., 2016a).

4. ELECTROSPUN INORGANIC OXIDE NANOFIBERS FOR ENERGY APPLICATIONS

Metal oxide nanofibers are suitable candidates for multiple applications due to their wide range of properties such as ferromagnetism, superconductivity, ionic and/or electrical conductivity, ferroelectricity, and photocatalytic activity, which can further be modified by controlling their morphology to fabricate 1D nanostructures and their random networks. Hence, electrospinning appears as an easy way for the integration of metal oxides in order to combine the unique properties of nanomaterials with a smart engineering approach (Ding and Yu, 2014; Odling and Robertson, 2015).

Electrospinning of metal oxides involves the electroforming of a solution of metalorganic precursor (alkoxides, acetylacetonates, chlorides, nitrates, or acetates) into a volatile solvent (water, alcohol acetone, DMF) (Luo et al., 2010). In addition, organic acids such as acetic or citric acid and polymers (PVA, PVP, PANI) are commonly used as catalysts and spinning aids, respectively. The spinning of a homogeneous precursor solution, consisting of a mixture of organic polymer and inorganic precursor sol, provides the as-spun nanofibers, which are calcined at elevated temperatures to obtain the desired ceramic phase under the removal of the organic components, and in case of salt-derived spinning solutions the removal of the undesired elements originating from counter ions. Calcination under inert conditions leads to the formation of ceramic/carbon composite fibers where carbon is in situ formed under carbonization of the polymer used in the electrospinning process.

4.1 Binary Metal Oxide Nanofibers

Binary metal oxides such as ZnO, TiO_2, CeO_x, Fe_2O_3 and WO_3 are promising in energy applications since they possess optical and electrical properties that make them ideal for sunlight harvesting, energy conversion and storage, or photocatalytic degradation of water pollutants. In addition, they are very easy to produce via electrospinning, since their fabrication only involves a single metalorganic precursor. The following sections describe a few representative examples of electrospun binary metal oxides applied in energy applications.

4.1.1 TiO_2 Nanofibers

TiO_2 is one of the most investigated materials for applications in photocatalysis, solar cells, and photoelectrochemical water splitting (Chuangchote et al., 2009; Kim et al., 2006b; Lepcha et al., 2015; Mali et al., 2015; Song et al., 2004; Zhu et al., 2008), mainly because it is an inexpensive and environmentally friendly wide bandgap semiconductor ($E_g \sim 3.0$–$3.4\,eV$ depending on the crystal structure, i.e., anatase, rutile, or brookite (Reyes-Coronado et al., 2008)) with high chemical and physical stability. In addition, the use of 1D structures provides an increased surface area for light absorption and faster collection and dispersion of charge carriers. For example, TiO_2 nanofibers have been investigated for the photocatalytic production of H_2 (Chuangchote et al., 2009) or for PEC water splitting by electrospinning fiber networks on conductive substrates such as FTO and ITO (Mali et al., 2015). Fig. 10A shows electrospun TiO_2 nanofibers after calcination at 500°C and the temperature-dependent PEC behavior of nanofibers calcined at different temperatures. The variation of electrochemical performance was attributed to a high degree of crystallinity, large surface area, and the predominant presence of anatase phase (Fig. 10B)

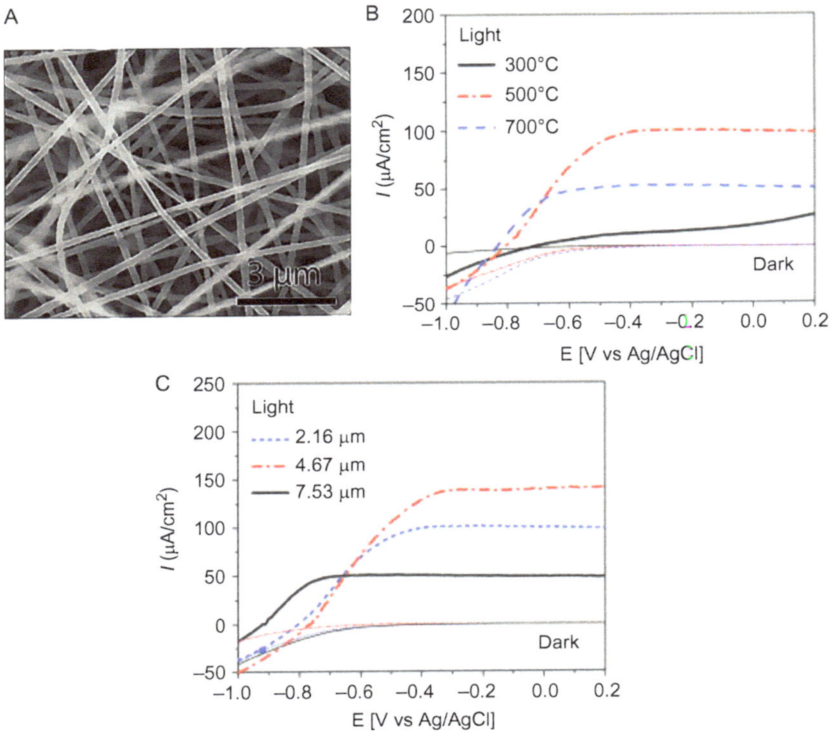

Fig. 10 (A) Morphology of TiO_2 nanofiber films (thickness = 4.67 μm) annealed at 500°C. Photocurrent measurements for TiO_2 nanofibers: (B) temperature dependence for films with a thickness of 2.16 μm, and (C) thickness dependence for films annealed at 500°C. *Reprinted from Mali, M.G., An, S., Al-Deyab, S.S., Yoon, S.S., 2015. Photoelectrochemical solar water splitting using electrospun TiO2 nanofibers. Appl. Surf. Sci. 328, 109–114. Copyright 2014, with permission from Elsevier.*

that possesses a broader bandgap, longer charge-carrier lifetime and mobility. In addition, it has been demonstrated that anatase TiO_2 has higher activity in producing free hydroxyl radicals as compared to other TiO_2 structures, which is a key factor for enhanced photocatalytic activity (Odling and Robertson, 2015). Fig. 10C shows that the photocurrent density increased from 100 to 150 μA/cm^2 by enhancing the film thickness from 2.16 to 4.67 μm, while a further growth to 7.53 μm results in the reduction of the photocurrent density. Mali et al. suggested that an increased thickness provides an enhanced surface area, which allows formation of more charge carriers. However, when the nanofiber film thickness becomes too large, the photogenerated electron/hole pairs need to travel farther and recombination processes become more relevant (Mali et al., 2015). TiO_2 nanofibers have also been employed as porous electrodes for DSSC (Song et al., 2004; Zhang et al., 2010). Song et al. treated calcinated anatase TiO_2 nanofibers with $TiCl_4$ in order to induce the formation of rutile nanocrystals on their surface (Song et al., 2004). Irrespective of the postsynthesis treatment, the 1D morphology arranged in a porous mesh led to photocurrent values between 8 and 11 mA/cm^2.

4.1.2 Fe_2O_3 and WO_3 Nanofibers

Although TiO_2 and equivalent binary oxides possess very interesting properties for numerous applications (Table 6), their usage for photoactive devices is restricted mainly to the ultraviolet (UV) range due to the large bandgap of 3.5 eV. Therefore, more suitable alternatives have been investigated in order to enhance the absorption of the entire solar spectrum. Two salient examples include α-Fe_2O_3 and WO_3 with bandgaps of 2.0–2.2 eV (Nalbandian et al., 2016) and 2.7 eV, respectively (Sui et al., 2011) that are suited for photocatalytic activities in the visible range of the solar spectrum. Hematite nanofibers have been evaluated for their capability to photodegrade different compounds. Sundaramurthy et al. tested the photocatalytic decomposition of Congo Red for different α-Fe_2O_3 nanofibers (Sundaramurthy et al., 2012). They were able to produce two types of nanofibers by changing the solution concentration and electrospinning parameters. The Congo Red degradation was significantly fast with a decomposition above 90% after 140 min, whereas other reports in the literature showed photodegradation values of approximately 30% for Fe_2O_3 nanorods and 60% when doped with platinum after 240 min. The superior photocatalytic activity was attributed to the smaller crystallite size, as well as the larger surface area of nanofibers. Einert et al. compared the PEC

Table 6 Summary of Metal Oxide Nanofibers Used for Energy Applications

Oxide	Bandgap (eV)	Precursors	Solvents and Catalysts	Polymers	Application	Important Parameters	References
TiO_2	3.0–3.4	Ti(IV) isopropoxide	DMF, EtOH, acetic acid	PVA, PVP	Photocatalytic H_2 production, PEC, DSSC	Evolved $H_2 = 270\,\mu mol/g$, $J = 150\,\mu A/cm^2$ (PEC), $J_{sc} = 8.07–11.24\,mA/cm^2$, FF = 0.56–0.60, $\eta = 3.80\%–5.02\%$ (DSSC)	Song et al. (2004), Chuangchote et al. (2009), Mali et al. (2015), and Zhang et al. (2010)
ZnO	3.4	$Zn(ac)_2$	DMF, acetic acid, H_2O	PVA, PAN	DSSC, photodegradation of MB, DE, and PAH	$J_{sc} = 3.58\,mA/cm^2$, FF = 0.62, $\eta = 1.34\%$ (DSSC)	Kim et al. (2007), Kanjwal et al. (2015), Singh et al. (2013)
NiO	3.7	$Ni(ac)_2$	H_2O	PVA	Photodegradation of MB and DE	—	Kanjwal et al. (2015)
Fe_2O_3	2.0–2.2	$FeCl_3$, $Fe(acac)_3$	DMF, EtOH	PVP	PEC, photodegradation of CR	$J = 2.1\,\mu A/cm^2$ (PEC)	Sundaramurthy et al. (2012) and Einert et al. (2016)
WO_3	2.7	WO_3 powder, tungstic acid	Ammonia, citric acid, H_2O	PVP, PVA	Photodegradation of MB and RhB	—	Sui et al. (2011) and Ofori et al. (2015)

performance of dense and hollow α-Fe_2O_3 nanofibers with other composite materials to demonstrate different photocurrent densities of 2.1 and 3.4 μA/cm^2, respectively (Einert et al., 2016). These values are rather low compared to TiO_2 nanofibers, which could be caused by the very short diffusion length of minority charge carriers in α-Fe_2O_3.

Although α-Fe_2O_3 nanofibers are very promising for solar-derived applications, they are still highly limited by their intrinsic material properties such as indirect bandgap, low electron mobility, short hole diffusion length, and slow oxidation kinetics, which leads to the fast recombination of photogenerated electron–hole pairs and a low photoresponse (Einert et al., 2016). Therefore, hematite nanofibers require further engineering through the combination of other materials, doping, or interface improvements, which can increase the charge carrier separation and, thus, achieve higher photocurrents. In this context, WO_3 offers promising potential due to its high photocatalytic and photoelectrocatalytic activities in the visible range of the solar spectrum, as well as high stability in most aqueous media. Ofori et al. have shown that electrospun WO_3 nanofibers fabricated from inexpensive precursor solutions containing PVP, citric acid, and tungstic acid can be employed to degrade organic water pollutants such as methylene blue (Ofori et al., 2015). The photodegradation of rhodamine B by WO_3 nanofibers has also been investigated under UV light (Sui et al., 2011). Moreover, WO_3 nanostructures have also been employed to degrade pharmaceutical compounds such as lidocaine, which makes tungsten trioxide nanofibers even more appealing for their use in wastewater treatment (Fakhri and Behrouz, 2015).

Table 6 shows a list of many other binary oxides synthesized by electrospinning. The precursors, solvents, and polymers used are compiled together with their properties and important parameters.

4.2 Ternary Oxide Nanofibers

This section is focused on ternary oxides that represent an attractive class of functional materials due to fundamentally new properties emerging from electronic and magnetic coupling among the constituting atoms. Although the bonding in oxides is mostly ionic in nature, a certain amount of covalent character occurs due to significant overlap between p-orbitals of the filled O^{2-} anions and the unfilled d-orbitals of the transition metal cations that makes electronic effects more pronounced. The functional properties are augmented due to possible multiple oxidation states and possible dynamics of vacancies driven by charge neutrality. Nevertheless, the use of ternary or more complex oxide single nanofibers for energy applications is still in its infancy and only few reports are available on their use in photocatalysis, photoelectrochemical water splitting, or photovoltaics.

4.2.1 Perovskite Oxides and Other Complex Structures

Perovskite (ABO_3) oxides are one of the most investigated structures for complex oxides. Their properties highly depend on the oxygen arrangement in their structure, since even a tilt of the octahedron in the oxygen sublattice significantly alters their electronic behavior so that their functional properties are coupled with the metal–oxygen–metal interactions. Although many perovskite oxides have been examined for their use in solar applications (Zhang et al., 2016d), perovskite nanofibers are relatively less investigated.

Titanium-based perovskite oxides have been mainly used as photocatalysts in the UV range of the solar spectrum due to their large bandgaps (Zhang et al., 2016d). Hassan et al. investigated the photodegradation of rhodamine 6G under UV irradiation by $CdTiO_3$ nanofibers annealed at 600°C and 800°C (Shamshi Hassan et al., 2014). $CdTiO_3$ nanofibers annealed at 800°C exhibited a high photocatalytic activity probably attributed to the higher crystallinity of nanofibers and shift in the absorption due to a quantum size effect resulted in a bandgap decrease from 3.3 eV at 600°C to 2.8 eV at 800°C, reflected in the increase in the photocatalytic performance from 73% to 97%. Huan et al. have also compared the photocatalytic activity of Nb-based perovskite oxide nanofibers, namely, ferroelectric $KNbO_3$ ($E_g \sim 3.6$ eV) and antiferroelectric $NaNbO_3$ ($E_g \sim 3.8$ eV) (Huan et al., 2015). Both compounds were used for the degradation of RhB under UV light illumination as depicted in Fig. 11A. Despite the lower surface area of $KNbO_3$ nanofibers, the slightly higher decomposition rate as compared to $NaNbO_3$ was attributed to the spontaneously generated internal electric field in $KNbO_3$, which promoted the separation of charge carriers, as schematically illustrated in Fig. 11B and C.

The promising photocatalytic properties of perovskite oxides in the UV region can be augmented through judicious combination of cations. In this context, Fe-based perovskite oxides have great potential for solar applications in the visible range of the solar spectrum due to the reduced bandgap, stable structure, and the high abundance of iron compounds on Earth. For instance, $BiFeO_3$ is a *p*-type semiconductor and multiferroic material that is particularly interesting for photocatalytic applications due to its suitable bandgap of $E_g \sim 2.2$ eV. $BiFeO_3$ nanofibers were synthesized in a

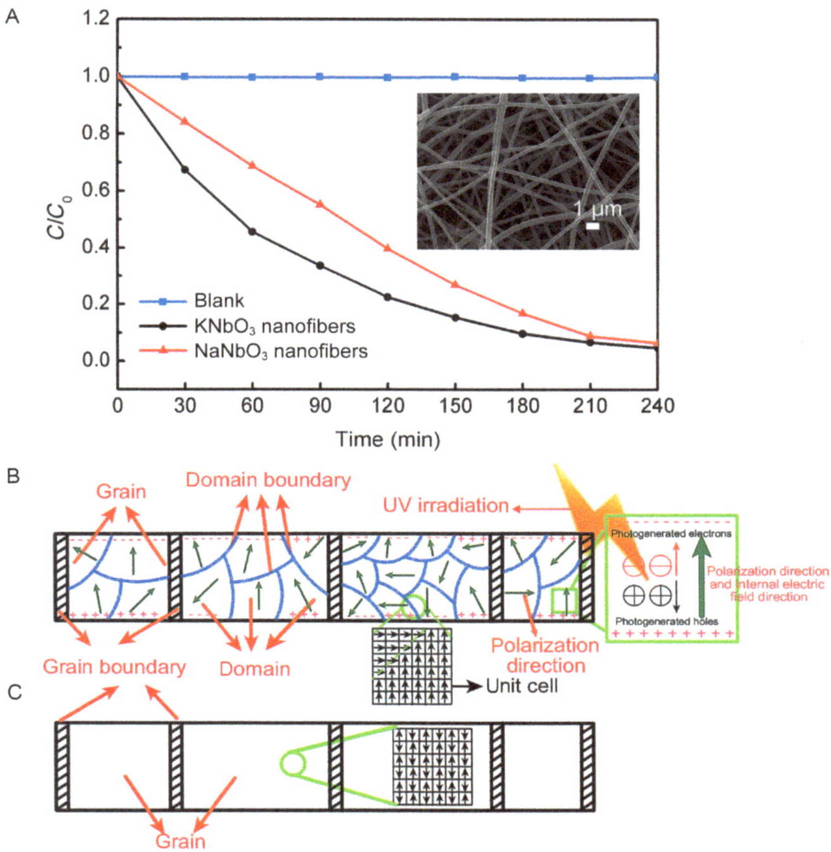

Fig. 11 (A) Photodecomposition curves for $KNbO_3$ and $NaNbO_3$ nanofibers of RhB in aqueous solution. *Inset*: SEM image depicting the calcined nanofibers. Domain structure of (B) $KNbO_3$ and (C) $NaNbO_3$ nanofibers and the influence of spontaneous polarization on charge carrier separation. *Adapted from Huan, Y., Wang, X., Hao, W., Li, L., 2015. Enhanced photocatalysis activity of ferroelectric KNbO3 nanofibers compared with anti-ferroelectric NaNbO3 nanofibers synthesized by electrospinning. RSC Adv. 5, 72410–72415. Copyright 2015, with permission from The Royal Society of Chemistry.*

photochemical reactor and were recovered through an external magnetic field (Wang et al., 2013). Wang et al. showed that $BiFeO_3$ nanofibers were able to photodecompose organic pollutants such as RhB under UV, as well as visible light irradiation. Interestingly, the addition of H_2O_2 during photodegradation under a visible light source enhanced the decomposition of RhB by 80%. As a proposed mechanism, OH radicals are released into the solution due to the interaction between H_2O_2 and $BiFeO_3$ nanofibers, aiding the photodegradation of RhB when irradiated with a visible light source. Fei et al. designed a photovoltaic device by depositing $BiFeO_3$ nanofibers on symmetric gold electrodes, sputtered on transparent glass substrates (Fig. 12) (Fei et al., 2015). The photocurrent–photovoltage response measurements showed a remarkably high current density of $1\,mA/cm^2$ under AM 1.5G illumination for a single nanofiber which was around 10 times larger than values reported for thin films in literature. Nevertheless, achieving device current densities higher than film-based devices would require the deposition of a large number of nanofibers.

Besides solar applications, other perovskite oxide nanofibers have been investigated for their implementation in nanogenerator devices such as the one schematically represented in Fig. 13A. For instance, Chen et al. fabricated $Pb[Zr_xTi_{1-x}]O_3$ (PZT) nanofibers as mechanical energy harvesters by depositing PZT nanofibers on interdigitated platinum electrodes (Fig. 13A and B) (Chen et al., 2010). The whole device was then embedded into a conductive PDMS polymer matrix (Fig. 13C). Despite the flexibility of the nanogenerator, a silicon substrate was used as a rigid mechanical support in order to reduce the probability of damaging the electrical connections and to prevent excessive stress on the nanofibers. The working principle of the nanogenerator is the following: (i) pressure is applied on the surface of the device, (ii) the mechanical energy is transferred through the PDMS matrix into the PZT nanofibers, (iii) an electric dipole is produced in the nanofibers due to the piezoelectric properties of PZT, and (iv) the electric energy is collected by the electrodes. Fig. 13D illustrates the voltage outputs after applying a dynamic load. Positive voltages were caused by the appearance of a potential after pressure was applied on the device, whereas the negative values were related to the reverse flow of the mechanically generated carriers once the pressure was released.

Beyond the conventional perovskite structure, there is also a broad range of complex oxides that have been synthesized as nanofibers for their use in energy applications. For example, layered perovskite oxides are polymorphs of the typical perovskite structure ABO_3 due to the presence of oxygen vacancies or the inclusion of interstitials, which results in more complex structures. Hildebrandt et al. successfully fabricated layered perovskite oxide nanofibers, i.e., $Ba_5Ta_4O_{15}$, $Ba_5Nb_4O_{15}$, and $Ba_5Ta_2Nb_2O_{15}$ ($E_g = 4.5$, 4.2, and 3.9 eV), for overall water splitting (Hildebrandt et al., 2015). The experiments were performed by preparing a suspension of nanofibers in an aqueous methanol solution, which was then irradiated within a homemade system using a 500 W Hg lamp. Interestingly, the nanofibers are able

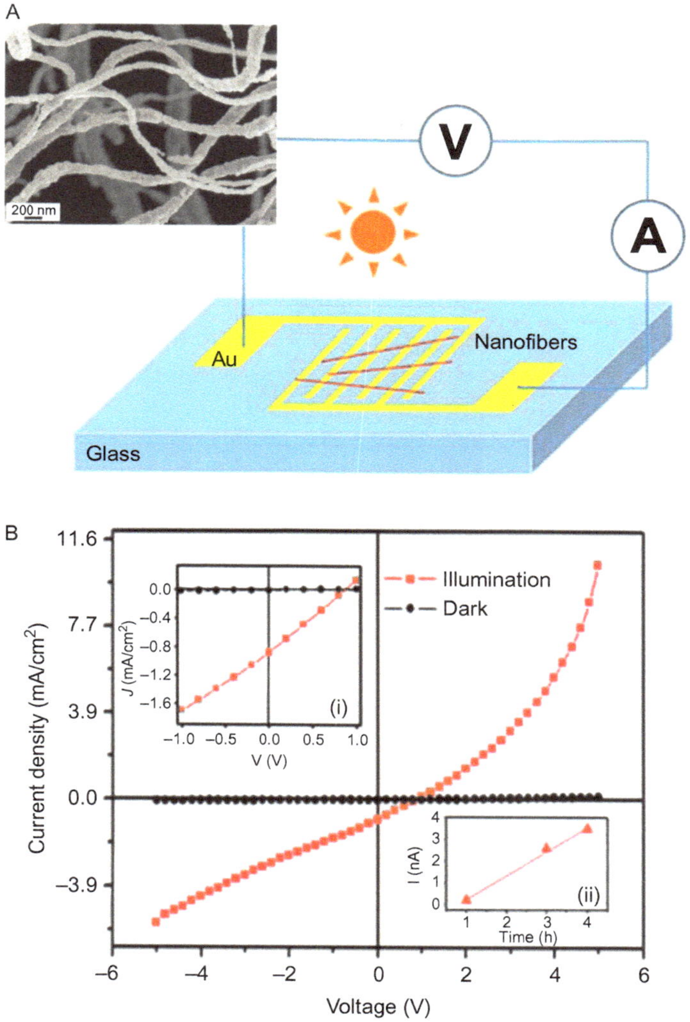

Fig. 12 (A) Schematic representation of the photovoltaic device fabricated from $BiFeO_3$ nanofibers, and (B) the dark/light photoresponse (current density vs voltage) for the same device. *Insets* depict (i) a zoom-in of the current density for the range of −1 to 1 V, and (ii) the average values of the photocurrent for deposition times between 1 and 4 h. *Adapted from Fei, L., Hu, Y., Li, X., Song, R., Sun, L., Huang, H., Gu, H., Chan, H.L., Wang, Y. 2015. Electrospun bismuth ferrite nanofibers for potential applications in ferroelectric photovoltaic devices. ACS Appl. Mater. Interfaces 7, 3665–3670. Copyright 2015, with permission from American Chemical Society.*

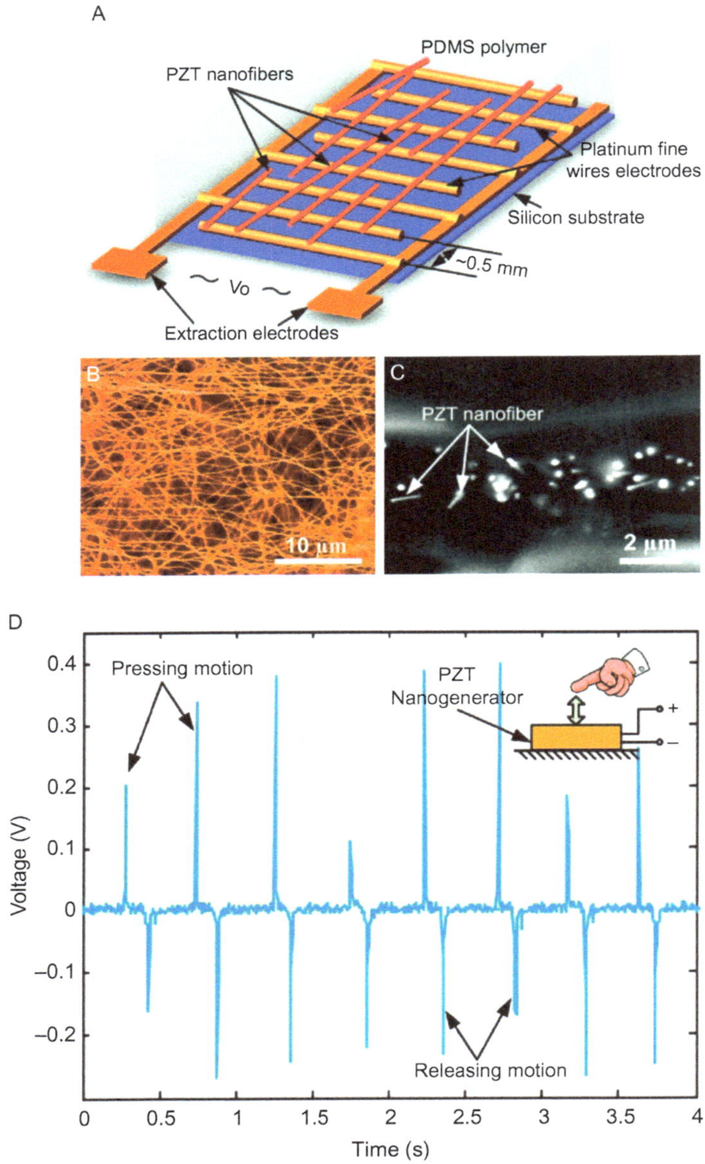

Fig. 13 (A) Sketch depicting the mechanical energy harvester made from PZT nanofibers. (B) SEM image of the electrospun PZT nanofibers. (C) SEM cross section of the PZT nanofibers embedded in the PDMS matrix. (D) Measurement of the voltage output from the harvesting device after applying a dynamic load with a fingertip. *Adapted from Chen, X., Xu, S., Yao, N., Shi, Y., 2010. 1.6 V nanogenerator for mechanical energy harvesting using PZT nanofibers. Nano Lett. 10, 2133–2137. Copyright 2010, with permission from American Chemical Society.*

to induce hydrogen evolution reactions. The nanofibers with a higher hydrogen production rate were made from $Ba_5Ta_4O_{15}$, whereas the lowest activity was found for $Ba_5Ta_2Nb_2O_{15}$ nanofibers. Although the photoactivity was expected to decrease with the amount of Nb due to a higher flatband potential, it was intriguing that the mixed Ta/Nb oxide phase was the one with the lowest hydrogen evolution rates. The H_2/O_2 production took around 2h to reach a steady state. It is worth mentioning that $Ba_5Nb_4O_{15}$ nanofibers were the only ones able to achieve the stoichiometric generation of H_2 and O_2, although overall water splitting was possible in all cases. Additional complex oxide nanofibers and parameters of interest used for multiple energy applications are listed in Table 7.

4.3 Challenges and Perspectives

Oxide materials can be successfully electrospun into nanofibers showing good properties for their use in a broad range of energy applications such as photocatalysis, nanogenerators, or water splitting. The advantages of nanofibers are evident since they are interconnected with each other allowing the measurement of properties in different areas of the sample. Moreover, their large surface area allows a high degree of interaction between the nanofibers and different media such as sunlight, electrolytes, dyes, etc., being important for solar applications among others. Nevertheless, there are some engineering aspects that still need to be further developed to have a technological impact, especially when materials are to be deposited on complex substrate architectures. The need for the transfer of charge carriers in most of energy applications makes it critical to achieve a good nanofiber/substrate adhesion. The brittleness of ceramic oxide nanofibers after polymer removal during calcination is another issue of concern due to their usage limitation for flexible or freestanding devices. As demonstrated very recently, the smart fabrication of nanofiber architectures such as sponges could improve flexibility (Wang et al., 2017). Preparation of the spin solution is another key issue in the fabrication of ceramic nanofibers since identifying appropriate precursors for mixed-cation oxides with similar solution behavior is a chemical and synthetic challenge. In this context, the use of new metalorganic single-source precursors, in which the right stoichiometry is already determined during precursor synthesis, is gaining increased importance (Jamil et al., 2016).

Table 7 Summary of Ternary Metal Oxide Nanofibers Used for Energy Applications

Oxide	Bandgap (eV)	Precursors	Solvents and Catalysts	Polymers	Application	Important Parameters	References
$BiFeO_3$	2.2	$Bi(NO_3)_3$, $Fe(NO_3)_3$	DMF, acetone, H_2O, nitric acid, 2-MOE, EtOH, acetylacetone	PVP	Photodegradation of RhB, photovoltaics	$J_{sc} = 1\,mA/cm^2$ (single nanofiber), $3.1 \times 10^{-4}\,mA/cm^2$ (device)	Wang et al. (2013) and Fei et al. (2015)
$BaTiO_3$	3.2	$Ba(ac)_2$, $Ti(O^tBu)_4$	Ethyl acetate, acetic acid, acetylacetone	PVP	Nanogenerators	Output power = 0.0086–0.1841 μW, output voltage = 0.56–2.67 V, output current = 57.78–261.4 nA	Yan and Jeong (2016)
$Pb[Zr_{0.52}Ti_{0.48}]O_3$	3.4–3.6	$Pb[Zr_{0.52}Ti_{0.48}]O_3$ sol–gel	EtOH, acetic acid	PVP	Nanogenerators	Output power = 0.03 μW, output voltage = 1.63 V, output current = 272 nA	Chen et al. (2010)
$Bi_4Ti_3O_{12}$	3.0	$Bi(NO_3)_3$, tetrabutyl-titanate	DMF	PVP	Photodegradation of RhB	—	Hou et al. (2013)
$Bi_2Fe_4O_9$	2.1	$Bi(NO_3)_3$, $Fe(NO_3)_3$	2-MOE, ethanolamine, glacial acetic acid, DMF, EtOH	PVP	Photodegradation of MO	—	Qi et al. (2013)
$BiVO_4$	2.4–2.5	$Bi(NO_3)_3$, vanadyl acetylacetonate, ammonium metavanadate	Acetic acid, EtOH, DMF, nitric acid, citric acid, H_2O	PVP	Photodegradation of RhB, PEC	$J = 0.05\,mA/cm^2$ (at 1.2 V vs RHE, 0.1 M K_2HPO_4) $J - 0.6\,mA/cm^2$ (at 1.2 V vs RHE, 0.1 M K_2HPO_4 and 1 M Na_2SO_3) $J = 0.36\,mA/cm^2$ (at 1.2 V vs RHE, 0.5 M Na_2SO_4), $J = 0.6\,mA/cm^2$ (at 1.2 V vs RHE, 0.5 M Na_2SO_4 and 0.5 M Na_2SO_3)	Cheng et al. (2015b), Jung et al. (2016), and Antony et al. (2016)

$MgFe_2O_4$	2.0	$Mg(NO_3)_2$, $Fe(NO_3)_3$	$EtOH/H_2O$	PVP	Photodegradation of tetracycline	—	Jiang et al. (2016)
$InVO_4$	3.2	Ammonium metavanadate, $In(NO_3)_3$	EtOH, H_2O, citric acid	PVP	Photodegradation of nitrobenzene	—	Song et al. (2012)
$MNbO_3$ (M = K, Na)	3.2, 3.4	K acetate, Na acetate, Nb ethoxide	2-MOE, acetylacetone	PVP	Photodegradation of RhB	—	Huan et al. (2015)
YSZ	—	Zr acetate hydroxide, $Y(NO_3)_3$	DMF	PAN	SOFC (interlayer)	ASR = 186 Ωcm^2 at 450°C	Koo et al. (2017)
$La_{0.58}Sr_{0.4}Co_{0.2}Fe_{0.8}O_3$	—	$La(ac)_3$, $Sr(ac)_2$	DMF	PAN	SOFC (cathode)	ASR = 0.474 Ωcm^2 at 750°C	Zhi et al. (2012)
$Sm_{0.5}Sr_{0.5}CoO_{3-\delta}$	—	$Sm(NO_3)_3$, $Sr(NO_3)_2$, $Co(NO_3)_2$	EtOH, H_2O, acetic acid	PVP	SOFC (cathode)	ASR = 0.024 Ωcm^2 at 650°C	Ahn et al. (2017)
$Ba_{0.5}Sr_{0.5}Co_{0.8}Fe_{0.2}O_{3-\delta}$	—	$Ba(NO_3)_2$, $Sr(NO_3)_2$, $Co(NO_3)_2$, $Fe(NO_3)_3$	H_2O, EDTA, citric acid, ammonium hydroxide	PVP	SOFC (cathode)	ASR = 0.094 Ωcm^2 at 600°C	Hieu et al. (2012)

5. NANOFIBER ARCHITECTURE: FROM CORE–SHELL FIBERS TO YARNS

Besides the challenge of controlling parameters to obtain desired material compositions and phases, various possibilities to control their shape, morphology, and structure are known, ranging from controlled molecular diffusion by precise annealing procedures to design of special electrospinning (ES) setups or needles (Fig. 14). The starting point for structural diversity in the field of electrospinning is marked by Loscertales and coworkers, who performed electrospraying experiments with a core–shell needle to form corresponding nanoparticles (Loscertales et al., 2002). Following this, core–shell (Sun et al., 2003) and hollow (Li and Xia, 2004) structures were published, which are already comprehensively reviewed in the literature (Zander, 2013). Nanofiber collector design in form of drum collectors, parallelly oriented collectors, or disc collectors (Teo and Ramakrishna, 2006), as well as experimental setups for the fabrication of electrospun yarns (Wei and Qin, 2016), have been discussed elsewhere and are not part of this chapter. In the following sections, most recent results on especially inorganic fibers including new net-shaping techniques, as well as new aspects of multiple composition fibers and electrospun yarns will be discussed.

5.1 Core–Shell and Hollow Structures

The formation of core–shell or hollow structures can be achieved by different approaches. Most established methods are electrospinning setups, which utilize multiwalled needles. In this case, two different precursor solutions are fed into different chambers of the needle, leading to the core–shell geometry at its tip (Fig. 15).

In the present example, a cobalt salt/PVP mixture was loaded as the shell solution, while a nickel salt/PVP mixture acted as the core solution. The geometry was maintained throughout the electrospinning process due to a jointly formed Taylor cone, while the fast process prevents intermixing of the two solutions. To obtain an all inorganic core–shell fiber, the as-spun fibers are calcined at high temperatures to remove the scaffold polymer and form an oxide phase. Hollow fiber structures can be easily obtained by replacing the inner solution with mineral oil, used as a sacrificial scaffold (Hu et al., 2013; Li and Xia, 2004). Variations and combinations of these methods possess unexplored potential to fabricate complex structures (Fig. 16). Especially, the silicon-mediated wire-in-tube structure showed

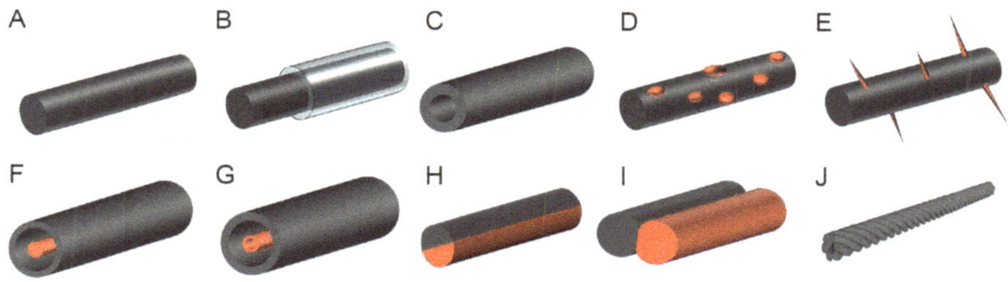

Fig. 14 General morphologies of nanofibers. (A) Single fibers, (B) core–shell fibers, (C) tubes or hollow fibers, (D) particle-decorated fibers, (E) fiber/spike-decorated fibers, (F) wire-in-tube, (G) tube-in-tube, (H) monoaxial Janus fiber, (I) side-by-side Janus fibers, (J) yarns.

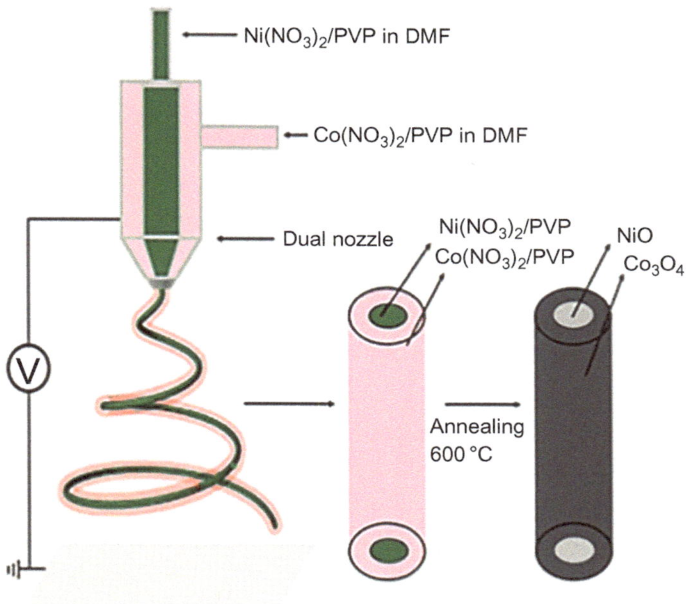

Fig. 15 Schematic formation of core–shell structures using a core–shell needle. *Reprinted from Jiang, P., Wang, Q., Dai, J., Li, W., Wei, Z., 2017. Mater. Lett. 188, 69–72. Copyright 2017 Elsevier.*

a high specific capacity, stable cycling behavior, and excellent rate capability leading to promising applications in Li-ion batteries (Lee et al., 2014).

It is noteworthy, that core solutions do not necessarily need polymeric content. Gebeyehu and coworkers, for example, were able to form Ag/PVP nanocables with a simple aqueous $AgNO_3$ solution as core compound (Gebeyehu et al., 2016). Other upcoming approaches use single-walled

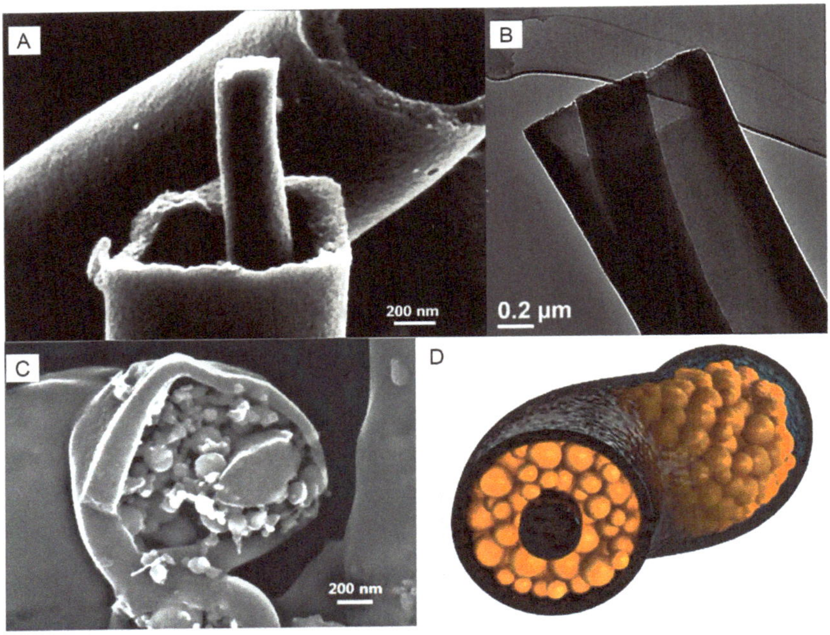

Fig. 16 (A and B) SEM and TEM images of a wire-in-tube structure, (C and D) SEM image and model of a wire in silica filled tube structure. *Reprinted from Lee, B.S., Yang, H.S., Jung, H., Jeon, S.Y., Jung, C., Kim, S.W., Bae, J., Choong, C.L., Im, J., Chung, U.I., et al., 2014. Nanoscale 6, 5989–5998. Copyright 2014 Royal Society of Chemistry.*

needles. This is possible due to careful diffusion control, either by polymer blends/gradients, precise annealing procedures, or mixtures of both. Possible structures by controlling the postannealing steps are core–shell (Yang et al., 2013), hollow (Han et al., 2015; Huang et al., 2014; Yang et al., 2014), or wire/fiber-in-tube arrangements (Fu et al., 2013; Wang et al., 2014a). The proposed formation mechanism, described in several publications, includes the development of a polymer gel layer followed by particle or ion generation, as well as diffusion due to solvent or ion gradients during electrospinning and calcination steps. Fig. 17, for example, shows the formation of a gel layer due to positively charged polymer moving to the surface and fast solvent evaporation during electrospinning. Upon annealing, a temperature gradient is formed in the nanofiber with a higher core temperature, resulting in preferred decomposition of manganese acetate into Mn_3O_4 in the core region. By increasing the temperature, the polymer starts to decompose at the surface, turning the gel layer into an inorganic, rigid shell. The exposure to air leads to formation of Mn_2O_3 at the surface,

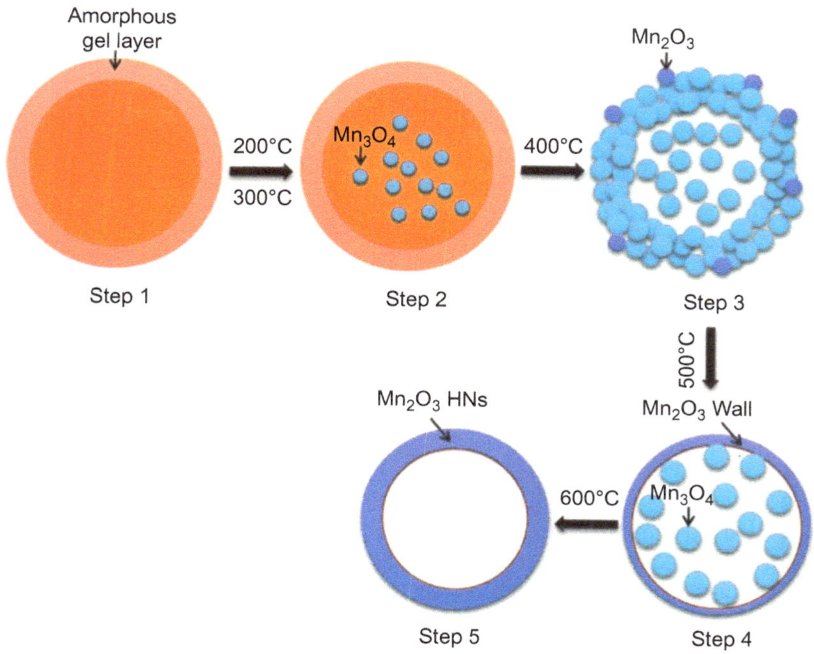

Fig. 17 Proposed formation mechanism of hollow fibers. *Reprinted from Yang, G., Yan, W., Wang, J., Yang, H., 2014. CrystEngComm 16, 6907–6913. Copyright 2014 Royal Society of Chemistry.*

creating a diffusion gradient between core and shell, resulting in particle migration toward the nanofibers surface.

These diffusion phenomena can be used to create the above-mentioned wire-in-tube or core–shell structures. If different metal salts are present in the precursor solution, the solvation of metal ions may differ, resulting in intrinsic gradients. Zinc ions are likely bound to PVP chains, while cadmium ions exhibit a higher affinity toward sulfur and are therefore preferentially coordinated by thiourea (Fig. 18). Upon charging during the electrospinning process, PVP with attached zinc ions move to the surface and form the gel layer. During calcination, polymer, solvents, and byproducts evaporate resulting in a core–shell structure.

In this case, CdS and ZnO both have similar hexagonal crystal structures (Xu et al., 2012), which is responsible for the core–shell formation. If crystal structures do not match, it is likely that fibers become either a wire-in-tube-like structure or particle-decorated fibers at more elevated temperatures (Fig. 19).

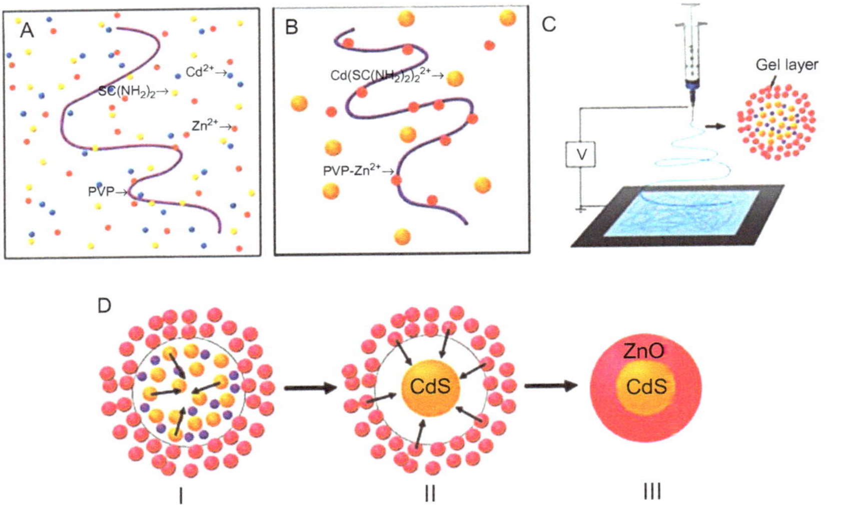

Fig. 18 Formation mechanisms of core–shell structures by single nozzle approach. *Reprinted from Yang, G., Yan, W., Zhang, Q., Shen, S., Ding, S., 2013. Nanoscale 5, 12432–12439. Copyright 2013 Royal Society of Chemistry.*

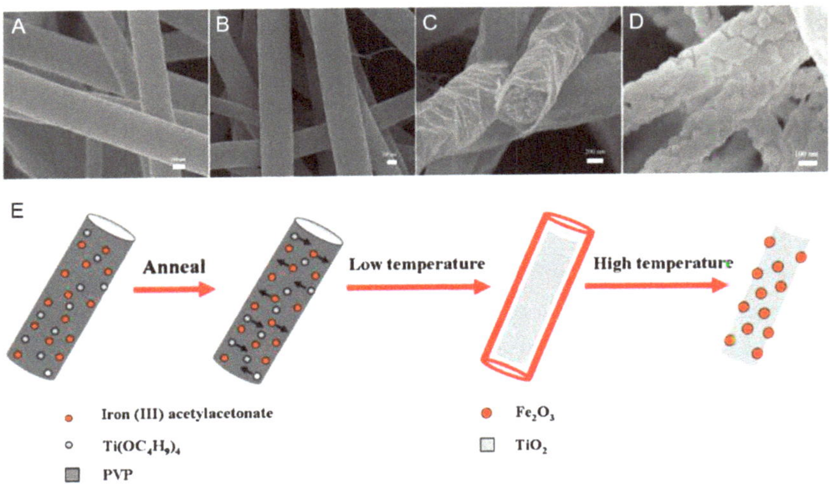

Fig. 19 Proposed formation mechanism of wire-in-tube structures. *Reprinted from Wang, H., Li, Y., Liu, W., Wan, Y., Li, Y., Duan, Q., 2014. RSC Adv. 4, 23125–23130. Copyright 2014 Royal Society of Chemistry.*

Nanotubes (Qin et al., 2017) or core–shell fibers (Samuel et al., 2017) can be fabricated using polymer blends as well. Due to different physical properties of polymers, some will tend to form emulsions (Fig. 20), leading to phase separation during the electrospinning process (e.g., PMMA/PAN in DMF).

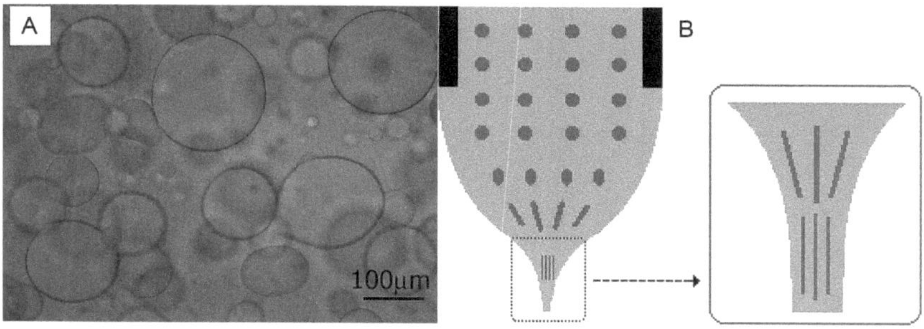

Fig. 20 (A) PMMA/PAN emulsion in DMF. (B) Taylor cone with polymer gradient. *Reprinted from Bazilevsky, A.V., Yarin, A.L., Megaridis, C.M., 2007. Langmuir 23, 2311–2314. Copyright 2007 American Chemical Society.*

As a variation of this process, Niu et al. used PVA with distinct molecular weights, resulting in a polymer gradient inside the as-spun fibers due to viscosity differences. PVA decomposes at different temperatures depending on its molecular weight, allowing the control of the architecture by post-processing (Fig. 21). Slow and controlled annealing causes the inner core to evaporate first, giving the solved metal ion core time to move to outer layers. Depending on atmospheric conditions in the last annealing step, completely inorganic, hollow fibers, or carbon-supported hollow fibers can be obtained. If the as-spun fibers are placed into a preheated furnace, polymer decomposition is too fast for metal salts to follow, leading toward pea-like structures formed due to local segregation of inorganic phases in the core.

A combination of a polymer blend and annealing control showed the power of these techniques, by demonstrating the easy architecture control and leading to either solid, hollow, or tube-in-tube structures (Fig. 22) (Peng et al., 2015). In a first step, PAN-core-PVP-shell fibers were generated by polymer blending. If the heating rate during the annealing procedure was kept low, a rigid, tubular structure was formed. At high heating rates, the decomposition was too fast, leading to a collapse of the structure and ultimately to a solid fiber. At a medium heating rate, a tube-in-tube structure was achieved due to opposing forces (contracting and adhesive) being in equilibrium (Zhang et al., 2012).

5.2 Janus-Type Structures

Even though first approaches toward Janus-type fibers have been conducted several years ago using a side-by-side spinneret, publications in this field are rather scant (Gupta and Wilkes, 2003). The report on Janus-type belts demonstrated the problems when trying to spin compositionally heterogeneous fibers. Due to the fact, that each solution is positively charged on its surface

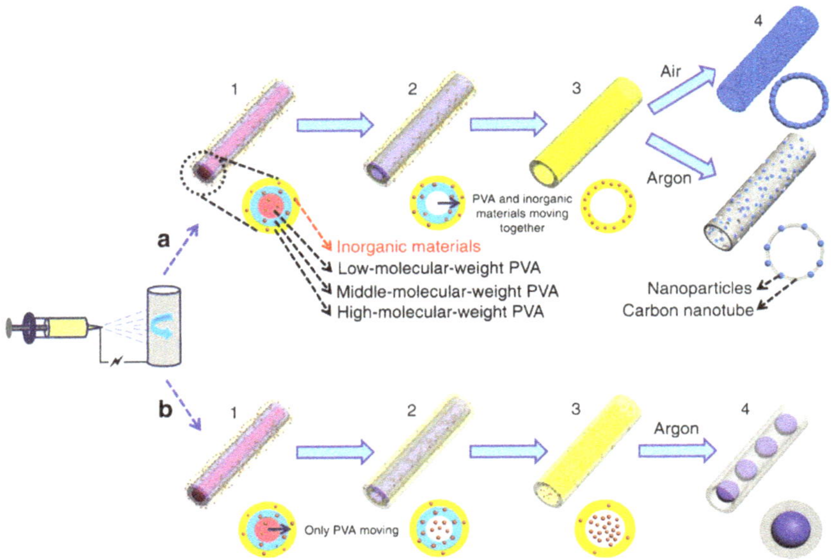

Fig. 21 Gradient electrospinning can lead to pea-like or hollow structures. *Reprinted from Niu, C., Meng, J., Wang, X., Han, C., Yan, M., Zhao, K., Xu, X., Ren, W., Zhao, Y., Xu, L., et al., 2015. Nat. Commun. 6, 7402. Copyright 2015 Springer Nature.*

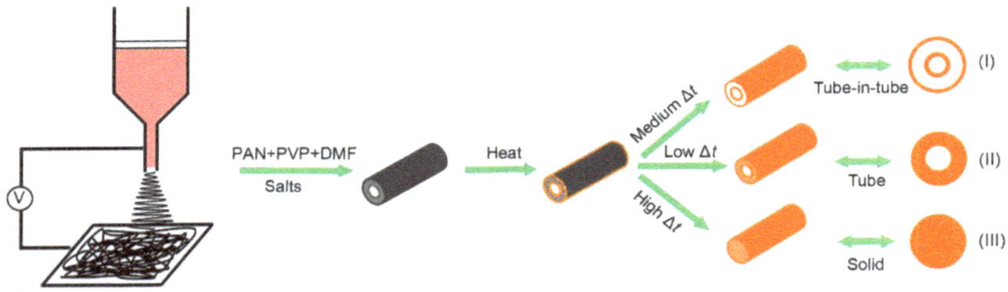

Fig. 22 Combining polymer blending and temperature control, leading to different structures. *Reprinted from Peng, S., Li, L., Hu, Y., Srinivasan, M., Cheng, F., Chen, J., Ramakrishna, S., 2015. ACS Nano 9, 1945–1954. Copyright 2015 American Chemical Society.*

individually, repulsion occurs in most cases, leading to separated or roughly connected fibers (Starr and Andrew, 2013; Starr et al., 2015). Additionally, both solutions solidify at different rates, causing a separation due to stress and strain (Deng-Guang Yu et al., 2017). To overcome these problems, different approaches have been chosen.

Ma et al. used a plastic nozzle, which prevented the two fluids from separation (Fig. 23) and led to an asymmetric Taylor cone (Ma et al., 2014b). The resulting nanobelts had very distinct domains and proved their simple setup to be efficient.

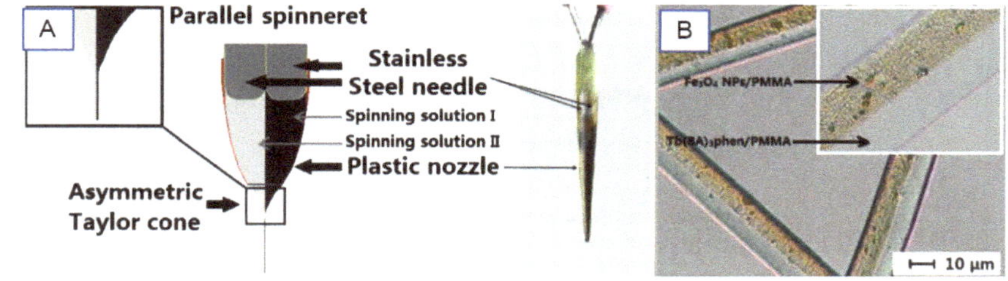

Fig. 23 (A) Plastic nozzle spinneret. (B) Janus nanobelts. *Reprinted from Ma, Q., Yu, W., Dong, X., Wang, J., Liu, G., 2014. Nanoscale 6, 2945–2952. Copyright 2014 Royal Society of Chemistry.*

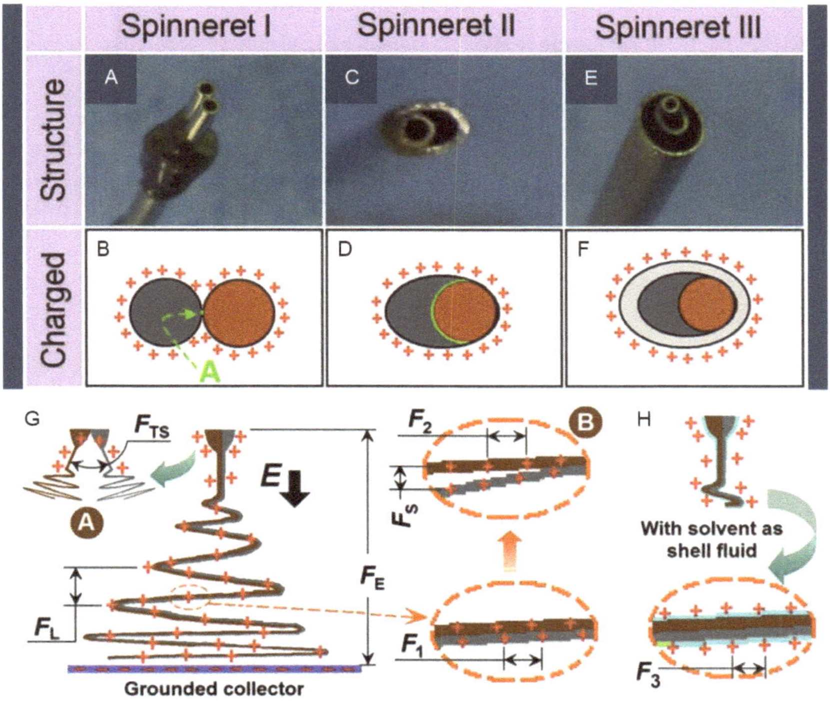

Fig. 24 Different spinnerets for Janus fibers and their influence on fiber formation. *Reprinted from Deng-Guang Yu, R., Li, J.-J., Zhang, M., Williams, G.R., 2017. Chem. Commun. 53, 4542–4545. Copyright 2017 Royal Society of Chemistry.*

Yu et al. tested different spinneret designs (Fig. 24) and confirmed the problems occurring with "classical" side-by-side spinnerets. They presented a new design, which led to an equally distributed positive charge around both fluids. The setup was further improved by setting this needle into a core–shell setup and using a solvent as outer fluid, thus providing a more simultaneous fiber solidification, leading to a reduced fiber separation. This technique has recently been developed for conventional electrospinning, helping to prevent clogging and to enable an overall more stable process (Wen et al., 2016; Yang et al., 2016, 2017a).

5.3 Nanofiber Yarns

Yarns can be fabricated from nanofibers in a single-step electrospinning method. This is achieved by replacing the conventional collector with a rotating metal funnel or cone. The as-spun fibers form a net over the funnel and a yarn, produced through the spinning of the funnel, is drawn and collected on a winder (Fig. 25). This became a prominent method for direct yarn production from electrospun nanofibers since it was first presented by Afifi et al. (2010).

Most publications deal with only polymer containing yarns, sometimes incorporating organic or bioorganic chemistry, and target biomedical applications. Yarns containing inorganic or functional materials are still quite rare. Very recently, Fan et al. realized heterogeneous yarns by loading different precursor solutions in two syringes, spinning simultaneously on a funnel

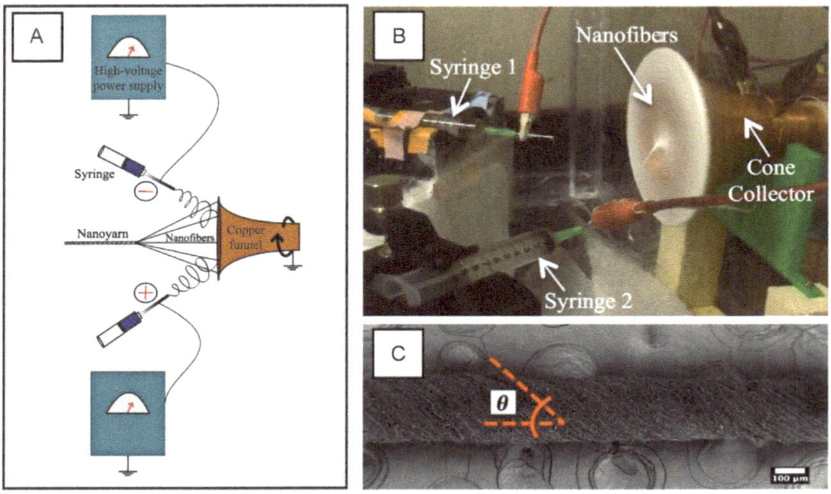

Fig. 25 (A) Scheme of a yarn production setup, (B) image of the setup during process, (C) produced yarn. *Reprinted from Levitt, A.S., Knittel, C.E., Vallett, R., Koerner, M., Dion, G., Schauer, C.L., 2017. J. Appl. Polym. Sci. 134, 1–7. Copyright 2017 John Wiley & Sons.*

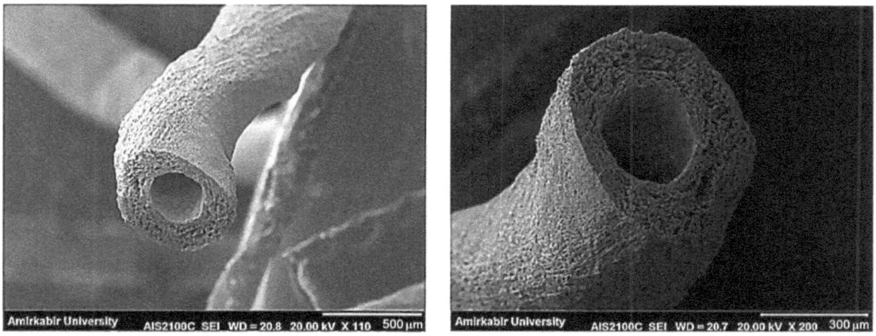

Fig. 26 Hollow yarn structure. *Reprinted from Najafi, S.J., Gharehaghaji, A.A., Etrati, S.M., 2016. Mater. Des. 99, 328–334. Copyright 2016 Elsevier.*

collector. These trifunctionalized yarns, formed from [Fe_3O_4/PANI/PAN] and [$Eu(BA)_3phen$/PAN] fibers, show excellent fluorescent, superparamagnetic, and conductive properties (Fan et al., 2017). Targeting smart textiles, Yang and coworkers presented piezoelectric P(VDF-TrFE) yarns. They tested different plied yarns and incorporated them into woven fabrics (Yang et al., 2017b). A new variation in yarn morphology has been achieved by Najafi et al., who managed to produce hollow yarns (Fig. 26). A core template of eightfolded PVA multifilament yarn was used and surrounded with polyurethane sheath yarns. Afterward, the PVA component was dissolved in hot water.

A triaxial nanofiber sourced yarn has been presented by Zhou et al., featuring a PMMA/copper acetate core/outer shell and a PAN interlayer. Calcination transforms the structure into a Cu/C composite yarn with a conductivity as high as 470 S/cm (Zhou et al., 2016). One of the few reports for inorganic yarns has been contributed by Memarian et al., who managed to realize pure TiO_2 yarns by annealing a TTIP/PVAc yarn. These yarns showed good results in photocatalysis for water purification with the advantage of a very easy removal of the catalytic yarn (Memarian et al., 2014, 2015).

Even though some research from the last decade present impressive geometries with tailored functionality, investigations toward more complex architectures produced by electrospinning are required. While core–shell fibers have become more prominent in literature, yarns are still in their beginning and Janus-type fibers seem to be relegated to a niche existence, despite their great potential (Table 8). For example, the setup used by

Table 8 Overview of Published Articles in the Past 5 Years, Concerning Core–Shell/Hollow Fibers, Janus Fibers and Yarns, Excluding Biomedical-Related Publications

Materials	Applications	Content	References
Core–shell and hollow fibers, one step			
SnO_x/C	Supercapacitors	Single-source core–shell, capacitance 289 F/g, 86% capacitance after 5000 cycles at 1 A/g	Samuel et al. (2017)
NiO/Co_3O_4	Supercapacitors	Double-source core–shell, capacitance 437 F/g, 83% capacitance after 1000 cycles at 5 A/g	Jiang et al. (2017)
Hollow/$Co_3V_2O_8$	Li-ion battery	Single-source hollow, 1040 mAh/g for 350 cycles at 1 A/g, 630 mAh/g for 1100 cycles at 5 A/g	Qin et al. (2017)
Hollow/SnO_2/C	Li-ion battery	Double-source structure, 833 mAh/g for 500 cycles at 0.6 A/g	Zhou et al. (2015a)
C/Si/C	Li-ion battery	Triple-source core–shell, 820 mAh/g at 12 A/g	Lee et al. (2014)
Sn/C	Li-ion battery	Double-source core–shell, 626 mAh/g, 73% capacity after 50 cycles	Xia et al. (2014)
TiO_2/hollow/Fe_2O_3	Li-ion battery	Single-source wire-in-tube, 987 mAh/g, 65% capacity after 240 cycles	Wang et al. (2014a)
Hollow/: $M_3V_2(PO_4)_3$ (M = Li, Na), $Na_{0.7}Fe_{0.7}Mn_{0.3}O_2$, LiV_3O_8, CuO, $LiNi_{1/3}Co_{1/3}Mn_{1/3}O_2$, SnO_2, LiM_2O_4 (M = Mn, Co), $LiMO_2$ (M = Mn, Co), Co_3O_4 Pea-like: Co/C, $LiCoO_2$/C, $Na_{0.7}Fe_{0.7}Mn_{0.3}O_2$/C, $Li_3V_2(PO_4)_3$/C	Exemplary: supercapacitors, Li-ion battery	Single-source hollow or pea-like structures by molecular weight gradient polymer	Niu et al. (2015)

SnO_2/In_2O_3:Au	Gas sensing	Double-source core–shell, acetone sensor—good selectivity, fast response	Li et al. (2017a)
SiO_2/SnO_2, hollow/SnO_2:Pt	Gas sensing	Single-source core–shell, etching of core leads to hollow structure, H_2S sensor, enhanced selectivity, fast response	Bulemo et al. (2017)
In_2O_3/SnO_2	Gas sensing	Double-source core–shell, triethylamine sensor, high selectivity, rapid response/recovery	Li et al. (2016)
Hollow/WO_3/Pd	Gas sensing	Double-source hollow, porous, Pd-decorated nanotubes, porosity by incorporation of sacrificial PS particles, H_2 sensor, high selectivity, high and fast response	Choi et al. (2016b)
Co_3O_4/Fe_2O_3	Gas sensing	Double-source core–shell, acetone sensor, high selectivity, fast response/recovery	Cao et al. (2015a)
Fe_2O_3/hollow/NiO	Gas sensing	Double-source core–shell, wire-in-tube by following annealing process, formaldehyde sensing, good selectivity, fast response/recovery, low-detection limit	Cao et al. (2015b)
$La_{0.8}Sr_{0.2}MnO_3/Y_2O_3$, $Y_2O_3/La_{0.8}Sr_{0.2}MnO_3$	Solid oxide fuel cells	Double-source corn-cob-like core–shell structure, maximum power density $1.15\,W/cm^2$ at 800°C	Jeon et al. (2017)
Hollow/Fe_2O_3, ITO/Fe_2O_3	Solar water oxidation	Single-source hollow/core–shell, differentiation between dense and hollow structures by humidity control	Einert et al. (2016)

Continued

Table 8 Overview of Published Articles in the Past 5 Years, Concerning Core–Shell/Hollow Fibers, Janus Fibers and Yarns, Excluding Biomedical-Related Publications—continued

Materials	Applications	Content	References
$BaTiO_3/Al_2O_3$	Expected: dielectric capacitors	Double-source core–shell, discharge energy density 12.18 J/cm^3 at a field of 400 MV/m	Pan et al. (2016)
Ag/PVP	Expected: electronic circuits, optoelectronics, sensing	Double-source core–shell, core solution containing no polymer	Gebeyehu et al. (2016)
$BaTiO_3/TiO_2$	Expected: thin film capacitors	Double-source core–shell, 19.94 J/cm^3 energy density in a field of 360 kV/mm	Lin et al. (2016)
Hollow/Y_2O_3:Yb^{3+},Er^{3+}	Upconversion luminescence	Single-source hollow, emits strong green and weak red upconversion emission bands centered at 526, 546, and 667 nm	Han et al. (2015)
γ-$Fe_2O_3/Ti_{0.9}Si_{0.1}O_2$	Wastewater purification	Double-source core–shell, magnetic mesoporous fibers	Zhan et al. (2014)
C/TiO_2	Dye-sensitized solar cell	Double-source core–shell, used as anode with 7.5% solar cell efficiency	Hieu et al. (2014)
$NiFe_2O_4/PbZr_{0.52}Ti_{0.48}O_3$	Expected: magnetic sensors, high-frequency devices	Double-source core–shell, magnetic fibers	Sreenivasulu et al. (2014)
Hollow/PtCo	Oxygen reduction reaction	Single-source hollow, very thin walled tubes (2–4 nm thick wall), well-preserved catalytic activities after 10,000 cycles	Huang et al. (2014)

SiO_2/Al_2O_3		Wastewater purification	Double-source core–shell, mesoporous shell ensures high adsorption capacity	Wang et al. (2014b)
Hollow/Mn_2O_3		Expected: catalysts, sensors, electrodes in batteries	Single-source hollow	Yang et al. (2014)
CdS/ZnO		Photocatalysis	Single-source core–shell, used for catalytic hydrogen production	Yang et al. (2013)
$Ni_{0.5}Zn_{0.5}Fe_2O_4$/hollow/$Ni_{0.5}Zn_{0.5}Fe_2O_4$		Expected: magnetic sensors and devices	Single-source wire-in-tube	Fu et al. (2013)
Materials	**Cotechnique**	**Applications**	**Content**	**References**
Core–shell and hollow fibers with combined techniques, multistep				
TiO_2/ZnO, ZnO/TiO_2	ALD	Gas sensing, photocatalysis	Single-source ES coupled with ALD for a thin outer shell	Boyadjiev et al. (2017)
C/Au	Electroplating	Flexible transparent electrode	Fiber mats transferred to flexible, transparent substrates, transparency ~91%, sheet resistance ~2,7 Ω/sq., mechanically stable over 10,000 cycles, fast joule heating up to 280°C with 3 V	Huh et al. (2017)
Ta_3N_5/Bi_2MoO_6	Solvothermal	Photocatalysis	High photodegradation efficiency	Li et al. (2017b)
Hollow/Fe_2O_3/MnO_2	Solvent-assisted synthesis	Supercapacitors	Flower-like structured shell, 290 F/g, 85% capacitance after 1200 cycles at 1 A/g	Nie et al. (2017)
IrO_2/NiO	Aqueous synthesis	Glucose sensor	At 0.35 V, sensitivity 1439.4 μA/(mM cm) (one order higher than pure NiO) with a wider linear range from 0.5 μM to 2.5 mM, a low-detection limit of 0.31 μM	Wang et al. (2016c)

Continued

Table 8 Overview of Published Articles in the Past 5 Years, Concerning Core–Shell/Hollow Fibers, Janus Fibers and Yarns, Excluding Biomedical-Related Publications—continued

Materials	Cotechnique	Applications	Content	References
Zno/MnO_2	Hydrothermal	Supercapacitors	907 F/g, 86% at 0.6 A/g, maximum energy density 17 Wh/Kg, power density 6.5 KW/Kg, capacitance retention 94%, coulombic efficiency 100%	Radhamani et al. (2016)
(Ag-GO-C composite)/ $ZnCo_2O_4$	Solvothermal	Electrode	Urchin-like surface structure, specific capacity of 460 mAh/g at 20 mV/s, 96% retention after 9000 cycles, flexible electrode	Chen et al. (2016b)
PVDF/Ni/ MnO_2	Electroless deposition, solvothermal	Li-ion battery	Specific reversible capacity of 1031 mAh/g after 70 cycles	Zhang et al. (2015b)
Hollow/ZnO	Atomic layer deposition	Photocatalysis	Comparison between 80 and 650 nm inner diameter tubes	Kayaci et al. (2015)
Hollow/C/ MnO_2	Dip coating	Capacitors	Double-source core–shell PMMA/PAN fibers, hollow/C by calcination, specific capacitance of 237 F/g	Hong et al. (2014)
ZnO/TiO_2, TiO_2/ZnO	Atomic layer deposition	Photocatalysis	Determination of the individual roles of electrons and holes in photocatalysis	Kayaci et al. (2014)
Hollow/SiO_2	Atomic layer deposition	Catalyst support	Thermal removal of electrospun PVA to obtain hollow silica fibers	Müller et al. (2014)
α-Fe_2O_3/γ-Al_2O_3	Vapor deposition	Expected: water purification	Water purification by adsorption of heavy metals, e.g., Cr(III) expected	Li et al. (2014c)

SnO_2/ZnO	Atomic layer deposition	Gas sensing	CO sensor, influence of shell thickness on sensing properties	Katoch et al. (2013)
C/Au	Solvent-assisted synthesis	Catalysis	Reduction of 4-nitrophenol	Zhang et al. (2013)

Materials	**Applications**	**Content**	**References**
Janus-type fibers			
PVP/shellac	—	Different nozzle designs tested and presented, solvent shell for less separation of fibers	Deng-Guang Yu et al. (2017)
[Eu(BA)$_3$phen + Tb(BA)$_3$phen]-PMMA)/(PANI-Fe_3O_4-PMMA)	Expected: full-color display, electromagnetic shield	Ribbon structure, excellent luminescent performance, electrical conduction, and magnetic properties, color tunable by variation of dopants	Tian et al. (2016)
(Fe_3O_4-PMMA)/($NaYF_4$:Yb^{3+}, Er^{3+}-PMMA)	Upconversion	Ribbon structure, magnetic-upconversion luminescent bifunctionality	Ma et al. (2015a)
(Fe_3O_4-PVP)/(Dy(BA)$_3$phen-PVP + Eu(BA)$_3$phen-PVP)	Expected: magnetic–luminescent devices, full-color displays	Janus-type dual fibers, color-tunable by dopant variation, first white light emitting magnetic–luminescent bifunctional 1D nanomaterial	Zhou et al. (2015b)
PS(Core)/Au(one side)/Pt-Pd(other side)	Self-propulsion	ES combined with sputtering on both sides, self-propulsion perpendicular to the long axis of the fiber when placed in a H_2O_2 solution	Reddy et al. (2013)
$CoFe_2O_4$/$BaTiO_3$	Expected: multiferroics	Janus-type dual fibers	Starr and Andrew (2013) and Wei and Qin (2016)

Continued

Table 8 Overview of Published Articles in the Past 5 Years, Concerning Core–Shell/Hollow Fibers, Janus Fibers and Yarns, Excluding Biomedical-Related Publications—continued

Materials	Applications	Content	References
($NaYF_4$:Tb^{3+}-PVP)/(Fe_3O_4-PVP)	Expected: medical diagnostics, drug target delivery, optical imaging, anticounterfeiting technology, nanomechanics	Janus-type dual fibers, magnetic–luminescent bifunctionality	Xi et al. (2014)
(Fe_3O_4-PMMA)/(Tb(BA)$_3$phen-PMMA)	Expected: biomedical, nanodevices	Ribbon structure, magnetic-fluorescent bifunctionality	Ma et al. (2014b)
(Fe_3O_4-PANI-PMMA)/(Tb(BA)$_3$phen-PMMA)	Anisotropic conductive film, expected: biomedical, nanodevices	Ribbon structure, strongly anisotropic conductivity, magnetism, and photoluminescence	Ma et al. (2015b)

Materials	Applications	Content	References
Yarns			
(Fe_3O_4-PANI-PAN)/(Eu(BA)$_3$phen-PAN)	Expected: electromagnetic shielding, gas sensors, drug targeting, flexible nanodevices	Fluorescent electrical-superparamagnetic, trifunctional heterogeneous nanofiber yarns	Fan et al. (2017)
PCL	Biomedical	Yarn woven into fabric	Wu et al. (2017)
PAN/LCGO	Expected: supercapacitors, biomedical, microfiltration	Enhanced mechanical properties	Mehrpouya et al. (2017)
P(VDF-TrFE)	Expected: piezoelectric fabrics	Comparison between single, 2-ply and 3-ply yarns, incorporation into fabrics	Yang et al. (2017b)
PLLA	—	Variation of twist rates	Maleki et al. (2017)
PAN	Biomedical	Woven fabric composites as 3D fibrous scaffold for tissue engineering and regenerative medicine applications	Wu et al. (2016)

Hollow/PU	—	First hollow yarn structures by using a sacrificial inner yarn	Najafi et al. (2016)
Cu/C/Cu	—	Yarns made of triaxial nanofibers, triple-source core–shell, highly conductive (600 S/cm)	Zhou et al. (2016)
PAN	—	Cone collector in comparison with spiral collectors	Lee et al. (2016c)
PVDF-HFP	—	Online stretching enables production rates up to 10 m/min, different stretching ratios have been tested	Ali et al. (2016)
TiO_2	Photocatalysis	High photocatalytic activity, easily removable after process	Memarian et al. (2014) and Memarian et al. (2015)
PVDF-HFP	—	Combination of needleless and conventional spinning technique for high-throughput fiber generation	Shuakat and Lin (2015)
C	Expected: high-performance carbon fibers	Carbonized nanofiber yarns show comparable tensile properties to the commercial carbon fibers	Xie et al. (2015)
CNT/graphene	Expected: supercapacitors, batteries, high current capable cables, artificial muscles	Highly conductive (900 S/cm), good capacitance (111 F/g)	Foroughi et al. (2014)
PCL, P(VDF-TrFE), PAN	Expected: wearables	High strain-to-failure (up to 300% elongation)	Levitt et al. (2017)

Materials in order: core/(interlayer)/shell.

Yu et al. (Fig. 24) could be extended by a combination of core–shell and Janus-type fibers by replacing the stabilizing outer fluid by a polymeric solution. This could lead to exciting new materials with high potential in applications.

6. SUMMARY AND OUTLOOK

Electrospinning offers a versatile platform for flexible net-shape forming of ceramics and composites to provide access to individual as well as integrated nanoarchitectures (e.g., webs and yarns) through the principles of electrohydrodynamics. The possibility of spinning single- and multi-phased materials in a single process provides opportunities to investigate multimaterial interfaces and phenomena occurring at artificially constructed heterojunctions. In addition, both surface states and the axially oriented voids can be used to explore new properties useful for a number of applications spanning superior transport of charge carriers to controlled delivery of medicinally active molecules (Maleki et al., 2014). As demonstrated in this account, the electrospinning process can be scaled up to construct 3D structures and scaffolds that create a bridge to principles of additive manufacturing and can thus lead to new synergies with 3D printing techniques, for instance, by chemically triggered cross-linking of starting materials required for densification of the final material (Fig. 27). A peculiar advantage of electrospinning is that it is independent to a larger extent of the chemistry of spinning sol or solution that allows by combining different spinneret designs to fabricate material that will not occur naturally due to thermodynamic impediments. In this context, the fabrication of metallic

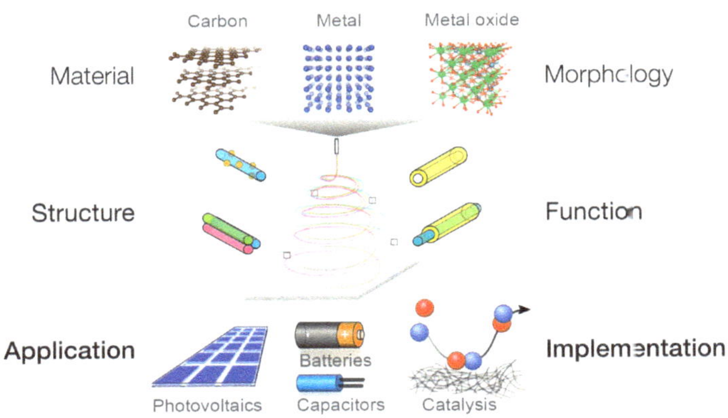

Fig. 27 Schematic illustration of electrospinning materials, structures, and applications.

nanomeshes that show high optical transmittance and low sheet resistance illustrates the potential of electrospinning in transparent conductive electrodes and heaters required for flexible and wearable electronics. Moreover, the emerging field of multimetallic ceramics is promising to demonstrate the strength of chemical processing of new materials from precursor-based routes. One of the major limitations of electrospinning lies in the processing of air-sensitive materials that has been circumvented in few studies by using inert electrospinning setup (Wilfert et al., 2012).

Electrospinning provides access to nanomaterials that can be obtained without any influence of substrate as light weight and flexible predevice structures. Such self-supporting and substrate-free electrospun webs have been fabricated for a large number of materials compositions such as carbon, metals, metal oxides, and composites thereof useful for different applications in the energy sector. In particular, electrode designs for Li-ion batteries and supercapacitors benefit strongly from the inherent pores and high-surface area of electrospun meshes that allow optimal wetting with the electrolyte and enhance the charge transport. Nevertheless, the high porosity remains a limitation that can be addressed by the infiltration of a secondary phase (von Hagen et al., 2012). Despite its long history and wide-ranging applications, possibilities of electrospinning are far from being exhausted and manifest high technological potential for future devices and materials not only related to energy but also in other fields like biomedical applications or filtration systems.

ACKNOWLEDGMENTS

All authors kindly acknowledge the financial support of the University of Cologne. In particular, the financial support of the ERA.Net RUS Plus project FONSENS (Project-ID: 01DJ16017) and the IGF funding scheme of the Arbeitsgemeinschaft industrieller Forschungsvereinigungen (AiF) are highly acknowledged. C.B. thanks the Evonik Foundation for financial support, and A.Q. appreciates the support of the Alexander von Humboldt foundation and the German Federal Ministry for Education and Research, supporting his postdoctoral research.

REFERENCES

Afifi, A.M., Nakano, S., Yamane, H., Kimura, Y., 2010. Macromol. Mater. Eng. 295, 660–665.

Ahn, M., Lee, J., Lee, W., 2017. J. Power Sources 353, 176–182.

Ali, U., Niu, H., Abbas, A., Shao, H., Lin, T., 2016. RSC Adv. 6, 30564–30569.

An, S., Jo, H.S., Kim, D.-Y., Lee, H.J., Ju, B.-K., Al-Deyab, S., Ahn, J.-H., Qin, Y., Swihart, M.T., Yarin, A.L., Yoon, S.S., 2016. Adv. Mater. 28, 7149–7154.

Antony, R.P., Bassi, P.S., Abdi, F.F., Chiam, S.Y., Ren, Y., Barber, J., Loo, J.S.C., Wong, L.H., 2016. Electrochim. Acta 211, 173–182.

Bahmanpour, A.M., Hoadley, A., Mushrif, S.H., Tanksale, A., 2016. ACS Sustain. Chem. Eng. 4, 3970–3977.
Bailie, C.D., Christoforo, M.G., Mailoa, J.P., Bowring, A.R., Unger, E.L., Nguyen, W.H., Burschka, J., Pellet, N., Lee, J.Z., Grätzel, M., Noufi, R., Buonassisi, T., Salleo, A., McGehee, M.D., 2015. Energ. Environ. Sci. 8, 956–963.
Barakat, N.A.M., Kim, B., Kim, H.Y., 2009. J. Phys. Chem. C 113, 531–536.
Bognitzki, M., Becker, M., Graeser, M., Massa, W., Wendorff, J.H., Schaper, A., Weber, D., Beyer, A., Gölzhäuser, A., Greiner, A., 2006. Adv. Mater. 18, 2384–2386.
Boyadjiev, S.I., Kéri, O., Bárdos, P., Firkala, T., Gáber, F., Nagy, Z.K., Baji, Z., Takács, M., Szilágyi, I.M., 2017. Appl. Surf. Sci. 424, 190–197.
Brown, T.D., Daltona, P.D., Hutmacher, D.W., 2016. Prog. Polym. Sci. 56, 116–166.
Bulemo, P.M., Cho, H.-J., Kim, N.-H., Kim, I.-D., 2017. ACS Appl. Mater. Interfaces 9, 26304–26313.
Cai, J., Niu, H., Wang, H., Shao, H., Fang, J., He, J., Xiong, H., Ma, C., Lin, T., 2016. J. Power Sources 324, 302–308.
Cao, J., Wang, Z., Wang, R., Liu, S., Fei, T., Wang, L., Zhang, T., 2015a. RSC Adv. 5, 36340–36346.
Cao, J., Wang, Z., Wang, R., Liu, S., Fei, T., Wang, L., Zhang, T., 2015b. J. Mater. Chem. A 3, 5635–5641.
Chen, Z., Cotterell, B., Wang, W., Guenther, E., Chua, S.-J., 2001. Thin Solid Films 394, 202–206.
Chen, X., Xu, S., Yao, N., Shi, Y., 2010. Nano Lett. 10, 2133–2137.
Chen, Z., Ye, S., Stewart, I.E., Wiley, B.J., 2014. ACS Nano 9, 9673–9679.
Chen, D., Zhang, R., Wang, R., Dal Negro, L., Minteer, S.D., 2016a. J. Electrochem. Soc. 163, H1132–H1135.
Chen, H., Jiang, G., Yu, W., Liu, D., Liu, Y., Li, L., Huang, Q., Tong, Z., 2016b. J. Mater. Chem. A 4, 5958–5964.
Chen, R., Hu, Y., Shen, Z., Pan, P., He, X., Wu, K., Zhang, X., Cheng, Z., 2017. J. Mater. Chem. A 5, 12914–12921.
Cheng, Y., Huang, L., Xiao, X., Yao, B., Yuan, L., Li, T., Hu, Z., Wang, B., Wan, J., Zhou, J., 2015a. Nano Energy 15, 66–74.
Cheng, J., Feng, J., Pan, W., 2015b. ACS Appl. Mater. Interfaces 7, 9638–9644.
Cheng, Y., Huang, L., Xiao, X., Yao, B., Hu, Z., Li, T., Liu, K., Zhou, J., 2016. J. Power Sources 327, 488–494.
Choi, J., Chan, S., Yip, G., Joo, H., Yang, H., Ko, F.K., 2016a. Water Res. 101, 46–54.
Choi, S.-J., Chattopadhyay, S., Kim, J.J., Kim, S.-J., Tuller, H.L., Rutledge, G.C., Kim, I.-D., 2016b. Nanoscale 8, 9159–9166.
Choudhury, A., Kim, J.H., Mahapatra, S.S., Yang, K.S., Yang, D.J., 2017. ACS Sustain. Chem. Eng. 5, 2109–2118.
Chuangchote, S., Jitputti, J., Sagawa, T., Yoshikawa, S., 2009. ACS Appl. Mater. Interfaces 1, 1140–1143.
Conway, B.E., 1999. Electrochemical Supercapacitors: Scientific Fundamentals and Technological Applications. Springer US, New York.
Cooley, J.F., 1902. Apparatus for Electrically Dispersing Fluids. US Patent No. 692, 631.
Deng, B., Hsu, P.-C., Chen, G., Chandrashekar, B.N., Liao, L., Ayitimuda, Z., Wu, J., Guo, Y., Lin, L., Zhou, Y., Aisijiang, M., Xie, Q., Cui, Y., Liu, Z., Peng, H., 2015. Nano Lett. 15, 4206–4213.
Deng-Guang Yu, R., Li, J.-J., Zhang, M., Williams, G.R., 2017. Chem. Commun. 53, 4542–4545.
Ding, B., Yu, J., 2014. Electrospun Nanofibers for Energy and Environmental Applications. Springer, Berlin Heidelberg.
Dong, Z., Kennedy, S.J., Wu, Y., 2011. J. Power Sources 196, 4886–4904.

Dong, Y., Lin, H., Zhou, D., Niu, H., Jin, Q., Qu, F., 2015. Electrochim. Acta 159, 116–123.
Dufficy, M.K., Khan, S.A., Fedkiw, P.S., 2016. ACS Appl. Mater. Interfaces 8, 1327–1336.
Einert, M., Ostermann, R., Weller, T., Zellmer, S., Garnweitner, G., Smarsly, B.M., Marschall, R., 2016. J. Mater. Chem. A 4, 18444–18456.
Fakhri, A., Behrouz, S., 2015. Sol. Energy 112, 163–168.
Fan, L., Yang, L., Ni, X., Han, J., Guo, R., Zhang, C., 2016. Carbon 107, 629–637.
Fan, L., Ma, Q., Tian, J., Li, D., Xi, X., Dong, X., Yu, W., Wang, J., Liu, G., 2017. RSC Adv. 7, 48702–48711.
Fei, L., Hu, Y., Li, X., Song, R., Sun, L., Huang, H., Gu, H., Chan, H.L., Wang, Y., 2015. ACS Appl. Mater. Interfaces 7, 3665–3670.
Formo, E., Peng, Z., Lee, E., Lu, X., Yang, H., Xia, Y., 2008. J. Phys. Chem. C 112, 9970–9975.
Foroughi, J., Spinks, G.M., Antiohos, D., Mirabedini, A., Gambhir, S., Wallace, G.G., Ghorbani, S.R., Peleckis, G., Kozlov, M.E., Lima, M.D., Baughman, R.H., 2014. Adv. Funct. Mater. 24, 5859–5865.
Fu, J., Zhang, J., Peng, Y., Zhao, C., He, Y., Zhang, Z., Pan, X., Mellors, N.J., Xie, E., 2013. Nanoscale 5, 12551–12557.
Gebeyehu, M.B., Chang, Y.-H., Abay, A.K., Chang, S.-Y., Lee, J.-Y., Wu, C.-M., Chiang, T.-C., Murakami, R.-I., 2016. RSC Adv. 6, 54162–54168.
Gittleson, F.S., Sekol, R.C., Doubek, G., Linardi, M., Taylor, A.D., 2014. Phys. Chem. Chem. Phys. 16, 3230–3237.
Go, D., Lott, P., Stollenwerk, J., Thomas, H., Möller, M., Kuehne, A.J.C., 2016. ACS Appl. Mater. Interfaces 8, 28412–28417.
González, A., Goikolea, E., Barrena, J.A., Mysyk, R., 2016. Renew. Sustain. Energy Rev. 58, 1189–1206.
Graeser, M., Bognitzki, M., Massa, W., Pietzonka, C., Greiner, A., Wendorff, J.H., 2007. Adv. Mater. 19, 4244–4247.
Greiner, A., Wendorff, J.H., 2007. Angew. Chem. Int. Ed. 46, 5670–5703.
Gupta, P., Wilkes, G.L., 2003. Polymer 44, 6353–6359.
Halperin, V., Shter, G.E., Gelman, V., Peselev, D.M., Mann-Lahav, M., Grader, G.S., 2015. Cat. Sci. Technol. 5, 1153–1162.
Han, L., Hu, Y., Pan, M., Xie, Y., Liu, Y., Li, D., Dong, X., 2015. CrstEngComm 17, 2529–2535.
Hansen, N.S., Cho, D., Joo, Y.L., 2012. Small 8, 1510–1514.
Hasan, A., Memic, A., Annabi, N., Hossain, M., Paul, A., Dokmeci, M.R., Dehghani, F., Khademhosseini, A., 2014. Acta Biomater. 10, 11–25.
Hieu, N.T., Park, J., Tae, B., 2012. Mater. Sci. Eng. B 177, 205–209.
Hieu, N.T., Baik, S.J., Jun, Y., Lee, M., Chung, O.H., Park, J.S., 2014. Electrochim. Acta 142, 144–151.
Hildebrandt, N.C., Soldat, J., Marschall, R., 2015. Small 11, 2051–2057.
Hong, S., Lee, S., Paik, U., 2014. Electrochim. Acta 141, 39–44.
Hou, D., Luo, W., Huang, Y., Yu, J.C., Hu, X., 2013. Nanoscale 5, 2028–2035.
Hsu, P.-C., Wu, H., Carney, T.J., McDowell, M.T., Yang, Y., Garnett, E.C., Li, M., Hu, L., Cui, Y., 2012. ACS Nano 6, 5150–5156.
Hsu, P.-C., Wang, S., Wu, H., Narasimhan, V.K., Kong, D., Lee, H.R., Cui, Y., 2013. Nat. Commun. 4, 2522.
Hsu, P.-C., Kong, D., Wang, S., Welch, A.J., Wu, H., Cui, Y., 2014. J. Am. Chem. Soc. 136, 10593–10596.
Hu, L., Wu, H., Cui, Y., 2011. MRS Bull. 36, 760–765.
Hu, W., Antoine, D., Yu, X., 2013. J. Compos. Mater. 48, 3019–3024.
Huan, Y., Wang, X., Hao, W., Li, L., 2015. RSC Adv. 5, 72410–72415.

Huang, Z.M., Zhang, Y.Z., Kotaki, M., Ramakrishna, S., 2003. Composite Sci. Tech. 63, 2223–2253.
Huang, Y., Garcia, M., Habib, S., Shui, J., Wagner, F.T., Zhang, J., Jorné, J., Li, J.C.M., 2014. J. Mater. Chem. A 2, 16175–16180.
Huh, J.W., Jeon, H.J., Ahn, C.W., 2017. Curr. Appl. Phys. 17, 1401–1408.
Jamil, A., Schläfer, J., Gönüllü, Y., Lepcha, A., Mathur, S., 2016. Cryst. Growth Des. 16, 5260–5267.
Jang, J., Hyun, B.G., Ji, S., Cho, E., An, B.W., Cheong, W.H., Park, J.-U., 2017. NPG Asia Mater. 9, e432.
Jeon, Y., Myung, J., Hyun, S., Shul, Y., Irvine, J.T.S., 2017. J. Mater. Chem. A 5, 3966–3973.
Jiang, J., Fan, W., Zhang, X., Bai, H., Liu, Y., Huang, S., Mao, B., Yuan, S., Liu, C., Shi, W., 2016. New J. Chem. 40, 538–544.
Jiang, P., Wang, Q., Dai, J., Li, W., Wei, Z., 2017. Mater. Lett. 188, 69–72.
Jing, P., Liu, M., Pu, Y., Cui, Y., Wang, Z., Wang, J., Liu, Q., 2016. Sci. Rep. 6, 37701.
Jung, H., Chae, S.Y., Kim, H., Min, B.K., Hwang, Y.J., 2016. Catal. Commun. 75, 18–22.
Kanjwal, M.A., Chronakis, I.S., Barakat, N.A.M., 2015. Ceram. Int. 41, 12229–12236.
Katoch, A., Choi, S.-W., Sun, G.-J., Kim, S.S., 2013. J. Mater. Chem. A 1, 13588–13596.
Kayaci, F., Vempati, S., Ozgit-Akgun, C., Donmez, I., Biyikli, N., Uyar, T., 2014. Nanoscale 6, 5735.
Kayaci, F., Vempati, S., Ozgit-Akgun, C., Donmez, I., Biyikli, N., Uyar, T., 2015. Appl. Catal. Environ. 176–177, 646–653.
Kholmanov, I.N., Domingues, S.H., Chou, H., Wang, X., Tan, C., Kim, J.-Y., Li, H., Piner, R., Zarbin, A.J.G., Ruoff, R.S., 2013. ACS Nano (2), 1811–1816.
Kim, C., Yang, K.S., Kojima, M., Yoshida, K., Kim, Y.J., Kim, Y.A., Endo, M., 2006a. Adv. Funct. Mater. 16, 2393–2397.
Kim, I.D., Rothschild, A., Lee, B.H., Kim, D.Y., Jo, S.M., Tuller, H.L., 2006b. Nano Lett. 6, 2009–2013.
Kim, I.-D., Hong, J.-M., Lee, B.H., Kim, D.Y., Jeon, E.-K., Choi, D.-K., Yang, D.-J., 2007. Appl. Phys. Lett. 91, 163109.
Kim, S., Lee, H., Kim, D., Ko, D., Kim, D., Kim, H., Yoon, S.-G., 2015. J. Nanomater. 518589.
Kim, C.H., Wee, J.-H., Kim, Y.A., Yang, K.S., Yang, C.-M., 2016. J. Mater. Chem. A 4, 4763–4770.
Koo, J.Y., Lim, Y., Kim, Y.B., Byun, D., Lee, W., 2017. Int. J. Hydrogen Energy 42, 15903–15907.
Kumar, P.S., Sundaramurthy, J., Sundarrajan, S., Babu, V.J., Singh, G., Allakhverdiev, S.I., Ramakrishna, S., 2014. Energ. Environ. Sci. 7, 3192–3222.
Lai, C.C., Lo, C.T., 2015. Electrochim. Acta 183, 85–93.
Lange, N.A., 2005. Lange's Handbook of Chemistry, Sixteen Ed. McGraw-Hill Companies, Inc., pp. 1128–1130.
Lawrence, D.W., Tran, C., Mallajoysula, A.T., Doorn, S.K., Mohite, A., Gupta, G., Kalra, V., 2016. J. Mater. Chem. A 4, 160–166.
Le, T., Yang, Y., Huang, Z., Kang, F., 2015. J. Power Sources 278, 683–692.
Lee, Y.-I., Choa, Y.-H., 2012. J. Korean Cryst. Growth Cryst. Technol. 22, 218–222.
Lee, B.S., Yang, H.S., Jung, H., Jeon, S.Y., Jung, C., Kim, S.W., Bae, J., Choong, C.L., Im, J., Chung, U.I., Park, J.J., Yu, W.R., 2014. Nanoscale 6, 5989–5998.
Lee, K.M., Kim, M., Lee, E., Baeck, S.H., Shim, S.E., 2016a. Electrochim. Acta 213, 124–131.
Lee, H., Kim, M., Kim, I., Lee, H., 2016b. Adv. Mater. 28, 4541–4548.
Lee, J.H., Shin, D.W., Nam, K.B., Gim, Y.H., Ko, H.S., Seo, D.K., Hui Lee, G., Kim, Y.H., Kim, S.W., Oh, T.S., Yoo, J.B., 2016c. Polymer 84, 52–58.

Lepcha, A., Maccato, C., Mettenbörger, A., Andreu, T., Mayrhofer, L., Walter, M., Olthof, S., Ruoko, T.P., Klein, A., Moseler, M., Meerholz, K., Morante, J.R., Barreca, D., Mathur, S., 2015. J. Phys. Chem. C 119, 18835–18842.
Levitt, A.S., Knittel, C.E., Vallett, R., Koerner, M., Dion, G., Schauer, C.L., 2017. J. Appl. Polym. Sci. 134, 1–7.
Li, D., Xia, Y.N., 2004. Nano Lett. 4, 933–938.
Li, W., Zeng, L., Yang, Z., Gu, L., Wang, J., Liu, X., Cheng, J., Yu, Y., 2014a. Nanoscale 6, 693–698.
Li, L., Wu, Z., Yuan, S., Zhang, X.-B., 2014b. Energ. Environ. Sci. 7, 2101–2122.
Li, X., Zhao, R., Sun, B., Lu, X., Zhang, C., Wang, Z., Wang, C., 2014c. RSC Adv. 4, 42376–42382.
Li, W., Li, M., Wang, M., Zeng, L., Yu, Y., 2015. Nano Energy 13, 693–701.
Li, F., Gao, X., Wang, R., Zhang, T., Lu, G., Barsan, N., 2016. ACS Appl. Mater. Interfaces 8, 19799–19806.
Li, F., Zhang, T., Gao, X., Wang, R., Li, B., 2017a. Sens. Actuators B Chem. 252, 822–830.
Li, S., Shen, X., Liu, J., Zhang, L., 2017b. Environ. Sci. Nano 4, 1155–1167.
Liao, Y., Loh, C.-H., Tian, M., Wang, R., Fane, A.G., 2018. Prog. Polym. Sci. 77, 69–94.
Lin, X., Hu, P., Jia, Z., Gao, S., 2016. J. Mater. Chem. A 4, 2314–2320.
Liu, Y., Zhou, J., Chen, L., Zhang, P., Fu, W., Zhao, H., Ma, Y., Pan, X., Zhang, Z., Han, W., Xie, E., 2015. ACS Appl. Mater. Interfaces 7, 23515–23520.
Liu, Y., Fan, L.-Z., Jiao, L., 2017. J. Mater. Chem. A 5, 1698–1705.
Loscertales, I.G., Barrero, A., Guerrero, I., Cortijo, R., Marquez, M., Gañán-Calvo, A.M., 2002. Science 295, 1695–1698.
Lu, X., Wang, C., Wie, Y., 2009. Small 5, 2349–2370.
Lu, X., Wang, C., Favier, F., Pinna, N., 2017. Adv. Energy Mater. 7, 1–43.
Luo, C.J., Nangrejo, M., Edirisinghe, M., 2010. Polymer 51, 1654–1662.
Ma, C., Li, Y., Shi, J., Song, Y., Liu, L., 2014a. Chem. Eng. J. 249, 216–225.
Ma, Q., Yu, W., Dong, X., Wang, J., Liu, G., 2014b. Nanoscale 6, 2945–2952.
Ma, Q., Wang, J., Dong, X., Yu, W., Liu, G., 2015a. Chem. Eng. J. 260, 222–230.
Ma, Q., Wang, J., Dong, X., Yu, W., Liu, G., 2015b. Adv. Funct. Mater. 25, 2436–2443.
Ma, C., Sheng, J., Ma, C., Wang, R., Liu, J., Xie, Z., Shi, J., 2016. Chem. Eng. J. 304, 587–593.
Maleki, M., Amani-Tehran, M., Latifi, M., Mathur, S., 2014. Comput. Methods Programs Biomed. 113, 92–100.
Maleki, H., Gharehaghaji, A.A., Dijkstra, P.J., 2017. J. Mech. Behav. Biomed. Mater. 71, 231–237.
Mali, M.G., An, S., Liou, M., Al-Deyab, S.S., Yoon, S.S., 2015. Appl. Surf. Sci. 328, 109–114.
Marder, T.B., 2007. Angew. Chem. Int. Ed. 46, 8116–8118.
Martin-Gullon, I., Vera, J., Conesa, J.A., González, J.L., Merino, C., 2006. Carbon 44, 1572–1580.
Mehrpouya, F., Foroughi, J., Naficy, S., Razal, J., Naebe, M., 2017. Nanomaterials 7, 293.
Memarian, F., Latifi, M., Amani-Tehran, M., 2014. J. Ind. Eng. Chem. 20, 1886–1891.
Memarian, F., Tehran, M.A., Latifi, M., 2015. J. Ind. Eng. Chem. 23, 182–187.
Miller, J.R., Simon, P., Patrice, S., 2008. Science 321, 651–652.
Mirjalili, M., Zohoori, S., 2016. J. Nanostruct. Chem. 6, 207–213.
Müller, G.F.J., Stürzel, M., Mülhaupt, R., 2014. Polymer 55, 465–470.
Najafi, S.J., Gharehaghaji, A.A., Etrati, S.M., 2016. Mater. Des. 99, 328–334.
Nalbandian, M.J., Zhang, M., Sanchez, J., Choa, Y.H., Nam, J., Cwiertny, D.M., Myung, N.V., 2016. Chemosphere 144, 975–981.
Nan, D., Huang, Z.-H., Lv, R., Yang, L., Wang, J.-G., Shen, W., Lin, Y., Yu, X., Ye, L., Sun, H., Kang, F., 2014. J. Mater. Chem. A 2, 19678–19684.

Nie, G., Lu, X., Chi, M., Zhu, Y., Yang, Z., Song, N., Wang, C., 2017. Electrochim. Acta 231, 36–43.

Niu, C., Meng, J., Wang, X., Han, C., Yan, M., Zhao, K., Xu, X., Ren, W., Zhao, Y., Xu, L., Zhang, Q., Zhao, D., Mai, L., 2015. Nat. Commun. 6, 7402.

Odling, G., Robertson, N., 2015. ChemSusChem 8, 1838–1840.

Ofori, F.A., Sheikh, F.A., Appiah-Ntiamoah, R., Yang, X., Kim, H., 2015. Nano-Micro Lett. 7, 291–297.

Pan, Z., Yao, L., Zhai, J., Shen, B., Liu, S., Wang, H., Liu, J., 2016. J. Mater. Chem. A 4, 13259–13264.

Park, S.W., Kim, J.C., Dar, M.A., Shim, H.W., Kim, D.W., 2017. Chem. Eng. J. 315, 1–9.

Patil, J.V., Mali, S.S., Kamble, A.S., Hong, C.K., Kim, J.H., Patil, P.S. 2017. Appl. Surf. Sci. 423, 641–674.

Peng, S., Li, L., Hu, Y., Srinivasan, M., Cheng, F., Chen, J., Ramakrishna, S., 2015. ACS Nano 9, 1945–1954.

Perananthan, S., Bonso, J.S., Ferraris, J.P., 2016. Carbon 106, 20–27.

Persano, L., Camposeo, A., Pisignano, D., 2015. Prog. Polym. Sci. 43, 48–95.

Qi, S., Zuo, R., Wang, Y., Chan, H.W.L.-W., 2013. J. Mater. Sci. 48, 4143–4150.

Qin, Z., Pei, J., Chen, G., Chen, D., Hu, Y., Lv, C., Bie, C., 2017. New J. Chem. 41, 5974–5980.

Radhamani, A.V., Shareef, K.M., Rao, M.S.R., 2016. ACS Appl. Mater. Interfaces 8, 30531–30542.

Ramakrishnan, P., Shanmugam, S., 2016. ACS Sustain. Chem. Eng. 4, 2439–2448.

Ramakrishnan, P., Park, S.-G., Shanmugam, S., 2015. J. Mater. Chem. A 3, 16242–16250.

Rathmell, A.R., Wiley, B.J., 2011. Adv. Mater. 23, 4798–4803.

Reddy, N.K., Palangetic, L., Stappers, L., Buitenhuis, J., Fransaer, J., Clasen, C., 2013. J. Mater. Chem. C 1, 3646–3650.

Reuss, G., Disteldorf, W., Gamer, A.O., Hilt, A., 2000. Ullmann's Encyclopedia of Industrial Chemistry. Wiley-VCH Verlag GmbH & Co. KGaA.

Reyes-Coronado, D., Rodriguez-Gattorno, G., Espinosa-Pesqueira, M.E., Cab, C., de Coss, R., Oskam, G., 2008. Nanotechnology 19, 145605.

Samuel, E., Joshi, B., Jo, H.S., Il Kim, Y., Swihart, M.T., Yun, J.M., Kim, K.H., Yoon, S.S., 2017. J. Alloys Compd. 728, 1362–1371.

Shamshi Hassan, M., Amna, T., Khil, M.-S., 2014. Ceram. Int. 40, 423–427.

Shao, H., Zhang, X., Liu, S., Chen, F., Xu, J., Feng, Y., 2011. Mater. Lett. 65, 1775–1777.

Shi, Z., Jin, G., Wang, J., Zhang, J., 2017. J. Electroanal. Chem. 795, 26–31.

Shuakat, M.N., Lin, T., 2015. RSC Adv. 5, 33930–33937.

Sierros, K.A., Morris, N.J., Ramji, K., Cairns, D.R., 2009. Thin Solid Films 517, 2590–2595.

Simon, P., Gogotsi, Y., 2008. Nat. Mater. 7, 845–854.

Singh, P., Mondal, K., Sharma, A., 2013. J. Colloid Interface Sci. 394, 208–215.

Soltanian, S., Rahmanian, R., Gholamkhass, B., Kiasari, N.M., Ko, F., Servati, P., 2013. Adv. Energy Mater. 3, 1332–1337.

Song, M.Y., Kim, D.K., Ihn, K.J., Jo, S.M., Kim, D.Y., 2004. Nanotechnology 15, 1861–1865.

Song, L., Liu, S., Lu, Q., Zhao, G., 2012. Appl. Surf. Sci. 258, 3789–3794.

Sreenivasulu, G., Popov, M., Zhang, R., Sharma, K., Janes, C., Mukundan, A., Srinivasan, G., 2014. Appl. Phys. Lett. 104, 52910.

Starr, J.D., Andrew, J.S., 2013. Chem. Commun. 49, 4151–4153.

Starr, J.D., Budi, M.A.K., Andrew, J.S., 2015. J. Am. Ceram. Soc. 98, 12–19.

Stewart, I.E., Ye, S., Chen, Z., Flowers, P.F., Wiley, B.J., 2015. Chem. Mater. 27, 7788–7794.

Sui, C., Gong, J., Cheng, T., Zhou, G., Dong, S., 2011. Appl. Surf. Sci. 257, 8600–8604.

Sun, Z., Zussman, E., Yarin, A.L., Wendorff, J.H., Greiner, A., 2003. Adv. Mater. 15, 1929–1932.

Sun, Y., Sills, R.B., Hu, X., Seh, Z.W., Xiao, X., Xu, H., Luo, W., Jin, H., Xin, Y., Li, T., Zhang, Z., Zhou, J., Cai, W., Huang, Y., Cui, Y., 2015. Nano Lett. 15, 3899–3906.

Sundaramurthy, J., Kumar, P.S., Kalaivani, M., Thavasi, V., Mhaisalkar, S.G., Ramakrishna, S., 2012. RSC Adv. 2, 8201–8208.

Tan, J., Han, Y., He, L., Dong, Y., Xu, X., Liu, D., Yan, H., Yu, Q., Huang, C., Mai, L., 2017. J. Mater. Chem. A 5, 23620–23627.

Teo, W.E., Ramakrishna, S., 2006. Nanotechnology 17, R89–R106.

Theron, S.A., Yarin, A.L., Zussman, E., Kroll, E., 2005. Polymer 46, 2889–2899.

Tian, X., Zhao, N., Song, Y., Wang, K., Xu, D., Li, X., Guo, Q., Liu, L., 2015. Electrochim. Acta 185, 40–51.

Tian, J., Ma, Q., Dong, X., Yu, W., Yang, M., Yang, Y., Wang, J., Liu, G., 2016. RSC Adv. 6, 36180–36191.

von Hagen, R., Lorrmann, H., Möller, K.C., Mathur, S., 2012. Adv. Energy Mater. 2, 553–559.

Wang, W., Li, N., Chi, Y., Li, Y., Yan, W., Li, X., Shao, C., 2013. Ceram. Int. 39, 3511–3518.

Wang, H., Li, Y., Liu, W., Wan, Y., Li, Y., Duan, Q., 2014a. RSC Adv. 4, 23125–23130.

Wang, Y., Ding, W., Jiao, X., Chen, D., 2014b. RSC Adv. 4, 30790–30797.

Wang, H.-G., Yuan, S., Ma, D.-L., Zhang, X.-B., Yan, J.-M., 2015. Energ. Environ. Sci. 8, 1660–1681.

Wang, S., Cui, Z., Qin, J., Cao, M., 2016a. Nano Res. 9, 2270–2283.

Wang, J., Tang, J., Xu, Y., Ding, B., Chang, Z., Wang, Y., Hao, X., Dou, H., Kim, J.H., Zhang, X., Yamauchi, Y., 2016b. Nano Energy 28, 232–240.

Wang, J., Xu, L., Lu, Y., Sheng, K., Liu, W., Chen, C., Li, Y., Dong, B., Song, H., 2016c. Anal. Chem. 88, 12346–12353.

Wang, H., Zhang, X., Wang, N., Li, Y., Feng, X., Huang, Y., Zhao, C., Liu, Z., Fang, M., Ou, G., Gao, H., Li, X., Wu, H., 2017. Sci. Adv. 3, 1603170.

Wei, L., Qin, X., 2016. Text. Res. J. 86, 1885–1898.

Wen, S., Liang, M., Zou, R., Wang, Z., Yue, D., Liu, L., 2015. RSC Adv. 5, 41513–41519.

Wen, H.F., Yang, C., Yu, D.G., Li, X.Y., Zhang, D.F., 2016. Chem. Eng. J. 290, 263–272.

Wilfert, J., von Hagen, R., Fitz, R., Jansen, M., Mathur, S., 2012. J. Mater. Chem. 22, 2099–2104.

Wu, H., Zhang, R., Liu, X., Lin, D., Pan, W., 2007. Chem. Mater. 19, 3506–3511.

Wu, H., Hu, L., Rowell, M.W., Kong, D., Cha, J.J., McDonough, J.R., Zhu, J., Yang, Y., McGehee, M.D., Cui, Y., 2010. Nano Lett. 10, 4242–4248.

Wu, H., Pan, W., Lin, D., Li, H., 2012. J. Adv. Ceramics 1, 2–23.

Wu, H., Kong, D., Ruan, Z., Hsu, P.-C., Wang, S., Yu, Z., Carney, T.J., Hu, L., Fan, S., Cui, Y., 2013. Nat. Nanotechnol. 8, 421–425.

Wu, S., Duan, B., Liu, P., Zhang, C., Qin, X., Butcher, J.T., 2016. ACS Appl. Mater. Interfaces 8, 16950–16960.

Wu, S., Wang, Y., Streubel, P.N., Duan, B., 2017. Acta Biomater. 62, 102–115.

Xi, X., Wang, J., Dong, X., Ma, Q., Yu, W., Liu, G., 2014. Chem. Eng. J. 254, 259–267.

Xia, X., Wang, X., Zhou, H., Niu, X., Xue, L., Zhang, X., Wei, Q., 2014. Electrochim. Acta 121, 345–351.

Xie, Z., Niu, H., Lin, T., 2015. RSC Adv. 5, 15147–15153.

Xu, F., Yuan, Y., Han, H., Wu, D., Gao, Z., Jiang, K., 2012. CrstEngComm 14, 3615–3622.

Xue, J., Xie, J., Liu, W., Xia, Y., 2017. Acc. Chem. Res. 50, 1976–1987.

Yan, J., Jeong, Y.G., 2016. ACS Appl. Mater. Interfaces 8, 15700–15709.

Yan, X., Liu, Y., Fan, X., Jia, X., Yu, Y., Yang, X., 2014. J. Power Sources 248, 745–751.

Yang, G., Yan, W., Zhang, Q., Shen, S., Ding, S., 2013. Nanoscale 5, 12432–12439.
Yang, G., Yan, W., Wang, J., Yang, H., 2014. CrstEngComm 16, 6907–6913.
Yang, C., Yu, D.G., Pan, D., Liu, X.K., Wang, X., Bligh, S.W.A., Williams, G.R., 2016. Acta Biomater. 35, 77–86.
Yang, G.Z., Li, J.J., Yu, D.G., He, M.F., Yang, J.H., Williams, G.R., 2017a. Acta Biomater. 53, 233–241.
Yang, E., Xu, Z., Chur, L.K., Behroozfar, A., Baniasadi, M., Moreno, S., Huang, J., Gilligan, J., Minary-Jolandan, M., 2017b. ACS Appl. Mater. Interfaces 9, 24220–24229.
Yao, Y., Wu, H., Huang, L., Li, X., Yu, L., Zeng, S., Zeng, X., Yang, J., Zou, J., 2017. Electrochim. Acta 246, 606–614.
Yarin, A.L., 2011. Polymer Adv. Tech. 22, 310–317.
Zander, N.E., 2013. Polymer 5, 19–44.
Zhan, S., Zhu, D., Ren, G., Shen, Z., Qiu, M., Yang, S., Yu, H., Li, Y., 2014. ACS Appl. Mater. Interfaces 6, 16841–16850.
Zhang, W., Zhu, R., Ke, L., Liu, X., Liu, B., Ramakrishna, S., 2010. Small 6, 2176–2182.
Zhang, G., Yu, L., Bin Wu, H., Hoster, H.E., Lou, X.W., 2012. Adv. Mater. 24, 4609–4613.
Zhang, P., Shao, C., Li, X., Zhang, M., Zhang, X., Su, C., Lu, N., Wang, K., Liu, Y., 2013. Phys. Chem. Chem. Phys. 15, 10453.
Zhang, B., Yu, Y., Xu, Z.L., Abouali, S., Akbari, M., He, Y.B., Kang, F., Kim, J.K., 2014. Adv. Energy Mater. 4, 1–9.
Zhang, Z., Jiang, Y., Chi, M., Yang, Z., Wang, C., Lu, X., 2015a. RSC Adv. 5, 94456–94461.
Zhang, Y., Luo, Z., Xiao, Q., Sun, T., Lei, G., Li, Z., Li, X., 2015b. J. Power Sources 297, 442–448.
Zhang, B., Kang, F., Tarascon, J.M., Kim, J.K., 2016a. Prog. Mater. Sci. 76, 319–380.
Zhang, L., Jiang, Y., Wang, L., Zhang, C., Liu, S., 2016b. Electrochim. Acta 196, 189–196.
Zhang, L., Xia, G., Guo, Z., Sun, D., Li, X., Yu, X., 2016c. J. Power Sources 324, 294–301.
Zhang, G., Liu, G., Wang, L., Irvine, J.T., 2016d. Chem. Soc. Rev. 45, 5951–5984.
Zhang, M., Zhao, X., Zhang, G., Wie, G., Su, Z., 2017. J. Mater. Chem. B 5, 1699–1711.
Zhi, M., Lee, S., Miller, N., Menzler, N.H., Wu, N., 2012. Energ. Environ. Sci. 5, 7066–7071.
Zhou, G., Li, F., Cheng, H.-M., 2014. Energ. Environ. Sci. 7, 1307–1338.
Zhou, D., Song, W.L., Fan, L.Z., 2015a. ACS Appl. Mater. Interfaces 7, 21472–21478.
Zhou, X., Ma, Q., Dong, X., Wang, J., Yu, W., Liu, G., 2015b. RSC Adv. 5, 35948–35957.
Zhou, Y., He, J., Wang, H., Qi, K., Ding, B., Cui, S., 2016. Mater. Des. 95, 591–598.
Zhu, R., Jiang, C.-Y., Liu, X.-Z., Liu, B., Kumar, A., Ramakrishna, S., 2008. Appl. Phys. Lett. 93, 013102.
Zhu, J., Chen, C., Lu, Y., Ge, Y., Jiang, H., Fu, K., Zhang, X., 2015. Carbon 94, 189–195.
Zhu, S., Sun, J., Wu, T., Su, X., Su, H., Qu, S., Xie, Y., Chen, M., Diao, G., 2016. RSC Adv. 6, 83185–83195.

CHAPTER TWO

Top-Down Etching of Si Nanowires

Amit Solanki[1], Handon Um[1]
Harvard John A. Paulson School of Engineering and Applied Sciences, Harvard University, Cambridge, MA, United States
[1]Corresponding authors: e-mail address: asolanki@seas.harvard.edu; handon@seas.harvard.edu

Contents

1. INTRODUCTION

Over the last few decades, Si nanowires have been intensively studied due to their novel chemical, mechanical, optical, electronic, and thermal properties. Si nanowires have many potential applications including nanoelectronics, biosensors, energy conversion/storage systems, and nanophotonic devices. Numerous techniques have been developed to make vertically aligned Si nanowire arrays, which can be classified into two categories: bottom-up and top-down. For bottom-up processes, Si atoms are

Semiconductors and Semimetals, Volume 98
ISSN 0080-8784
https://doi.org/10.1016/bs.semsem.2018.04.001

assembled piece-by-piece and scaled up to produce nanowire structures on a substrate. Si nanowires have been synthesized by the chemical vapor deposition, physical vapor deposition (PVD), and laser ablation method. Among them, the vapor–liquid–solid (VLS) method is one of the most common methods. Gas-phase precursors, such as SiH_4, $SiCl_4$ for silicon, dissolve into a liquid–metal catalyst, and then crystalline Si nanowire forms from the supersaturated metal–silicon catalyst. VLS method provides the opportunity for nanowire composition control during growth, enabling the fabrication of complex superlattice nanowire for the application of quantum well devices. However, high process temperature ($\geq$400°C) is required to decompose source precursor. In addition, a noble metal catalyst is used to reduce the process temperature and selectively grow the nanowire, leading to metal impurity contamination of the Si nanowire. Another main challenge of VLS method is to organize perfectly aligned arrays of vertically oriented nanowires in a large scale. Because Si nanowires grow along the preferred crystalline orientation, multiple growth directions are available due to crystal symmetry, resulting in growth at multiple angles.

Compared to the bottom-up method, top-down methods etch nanowire structures from a bulk Si substrate in a subtractive fashion. Fabrication of nanostructures by top-down method provides a promising path forward for making complex device geometries which otherwise would be difficult to achieve by bottom-up growth methods. Top-down methods can be categorized into one of the two paradigms, dry or wet etch processes. Nanowire fabrication is particularly challenging when the requirements include high aspect ratios, small diameters, and patterned arrays. In such situations, dry etching processes are useful as they do not have issues related to structural collapse due to van der Waals forces, or surface tension as seen in wet etching methods (Chang et al., 2009; Dawood et al., 2012; Yoon and Khang, 2012). Furthermore, dry etching processes prove to be advantageous over dry chemical growth processes such as chemical vapor deposition and molecular beam epitaxy as they permit patterned fabrication of horizontal and vertical nanowires in an array via lithography.

As a cost-effective approach, wet etch processes have been used to fabricate Si nanowires. Usually, wet etching processes are isotropic, i.e., vertical and lateral etch rates are equal. Metal-assisted chemical etching (MacEtch), first proposed by Li and Bohn (2000), is fundamentally a wet etching process, but etches anisotropically. A metal catalyst is deposited on a semiconductor, and then the metal-coated semiconductor is dipped into an etching solution consisting of an acid and an oxidant. The metal catalyst reduces the activation energy required for the chemical etch reaction, leading to an enhanced etching rate at the metal/semiconductor contact. Owing to its anisotropic etching behavior, MacEtch allows fabricating high aspect ratio nano/microstructures, especially vertically aligned nanowires. The MacEtch method has been utilized to produce various kinds of nanowires based on Si (Um et al., 2014) and Ge (Rezvani et al., 2016) as well as compound semiconductors such as GaAs (Dejarld et al., 2011), GaN (Geng et al., 2012), and InP (Kim et al., 2015).

This chapter will first discuss strategies for making arrays of nanowires with different dry etching processes, while highlighting the benefits and drawbacks of these dry etching processes. Then, we will present an overview of the fabrication of the Si nanostructures based on MacEtch methods. The influences of metal catalyst, substrate, and etchant concentration on MacEtch are also considered. Experimental results will be discussed to provide guidance for accurate control over Si nanowire morphologies via both dry etching and MacEtch processes.

2. NANOWIRES BY DRY ETCHING

2.1 Introduction to Dry Etching

The term "dry etching" refers to processes where a surface is etched by exposure to gaseous phase molecules either in the form of neutral, ionic, or reactive species. Dry etching can be divided into physical and chemical etching processes. An etching process is referred to as "physical etching" if it relies exclusively on high kinetic energy molecules or ions to physically bombard the etching surface. A variation of the physical etching process commonly employed in microfabrication research and the semiconductor industry is ion beam etching (IBE). A typical IBE machine employs high energy Ar^+ ions which bombard the substrate and knock the atoms out of the substrate leading to ablation. IBE is inherently anisotropic due to the highly directional ions that have a relatively low selectivity between different materials. It has thus been widely used in literature to make nanowires and anisotropic nanostructures from materials which are hard to etch by wet/dry chemical etching (Cybart et al., 2015; He et al., 2008; Reyntjens and Puers, 2001; Tettamanzi et al., 2009). In comparison, a "chemical etching" process relies solely on the chemical reaction with the substrate for etching. The process is facilitated by either a neutral or radical species that reacts with the substrate to form volatile byproducts that can be pumped out. A barrel etcher, used most commonly for O_2 plasma etching of resist and organics, is an example of a chemical dry etching process where highly reactive oxygen

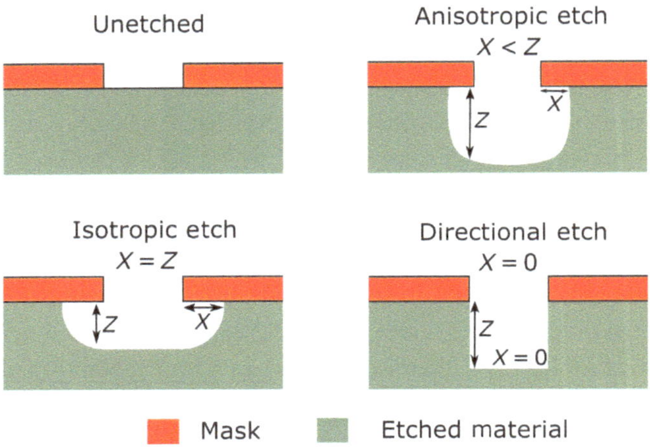

Fig. 1 Isotropic vs anisotropic etch.

ions are produced by ionization of the O_2 gas. Since the reactive ions produced have random kinetic energy in all the directions, a pure chemical dry etching process shows an isotropic (equal etch rate in all directions) etching behavior. Such isotropic etching does not allow for etching high aspect ratio structures as shown in Fig. 1. We define the lateral etch ratio as $L=$(horizontal etch rate/vertical etch rate) and degree of anisotropy as $A=(1-L)$. For isotropic etching, $L=1$. For most anisotropic etching, the etch ratio is $0<L<1$ and for a purely anisotropic or directional etching $L=0$.

An improved process can be achieved by combining the physical and chemical dry etching processes. Such a process is referred to as reactive ion etching (RIE) where the ionized species are accelerated toward the substrate by applying a bias on the substrate holder. RIE is the most commonly used dry etching process due to the ease of tuning the etching profile and the versatility of etching various materials including semiconductors, metals, and oxides. The process can be improved to further increase the directional etching by adding a noble gas like Ar in the reaction mixture. Directed etching by highly energetic reactive species can result in anisotropic etching, but for high aspect ratio and dense features, reducing the sidewall etching and sidewall redeposition becomes crucial. Cryogenic RIE, a variation of standard RIE process, addresses this issue and has been used for anisotropic Si etching since its introduction in 1988 (Tachi et al., 1988). In cryoetching RIE, the substrate is cooled between −80 and −140°C to reduce surface reactivity and to provide more directional etch. Since the advent of cryogenic RIE, many research efforts have been dedicated to obtaining anisotropic structures like mesa's, holes, and nanowires in Si using the cryogenic etching process (Aachboun et al., 2000; Jansen et al., 1996; Sökmen et al., 2009). For aspect ratios greater than 30, a process famously known as the "Bosch process" was developed by Laermer and Schilp at Bosch in 1994 (Laermer and Schilp, 1994). The Bosch process creates a temporal separation between the etching and the passivation processes by time multiplexing. This method has heralded a new era of extreme anisotropic etching of microstructures with aspect ratios >1:100 with the only limitation being mask erosion.

2.2 Plasma Etching Process

Dry chemical etching of group IV semiconductors like Si (or Ge) can be done by exposing the material to gaseous phase reactants like XeF_2 which spontaneously react with Si to form SiF_4 gas and Xe gas. Such a process does not require plasma. For faster etching, a plasma etching process is needed. A typical plasma chamber is shown in Fig. 2. The working principle of a plasma etcher is quite simple. The sample is placed in a closed chamber and an etching gas is introduced. A radio frequency (RF) field is fed into the chamber from the top electrode. The RF causes the gas molecules to vibrate at high frequency. When the energy supplied is high enough, the gas becomes ionized, resulting in plasma. The dissociated gas molecules in the plasma generate ionized gas species which are then used to etch the substrate.

It is imperative to choose a gas which creates radicals that react with the sample to form volatile compounds which readily desorb from the

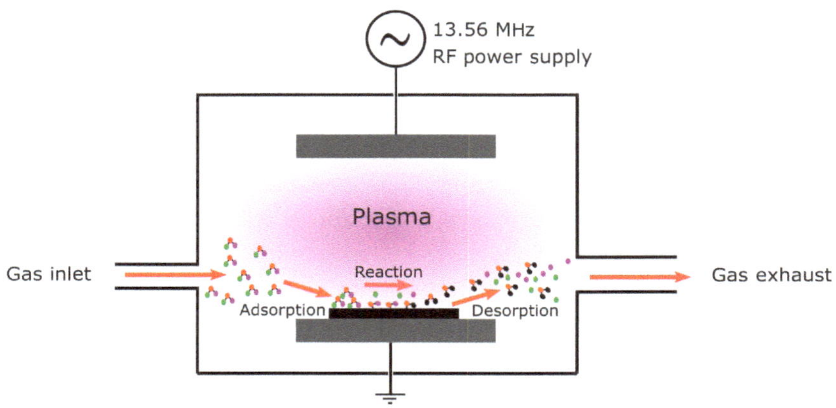

Fig. 2 Working principle of dry etching with plasma.

surface. The volatile compounds are then purged from the chamber, and the reactant gas is continuously introduced into the chamber to keep the etching process going. The etching process can be divided into six basic steps:

1. Generation

   The energy from the RF field dissociates the gas and generates a plasma which turns into a soup of electrons, photons, neutrals, positive, and negative ions. The neutrals and radicals (positive and negative ions) are the reactive species that participate in the etching.

2. Diffusion

   The radicals subsequently diffuse from the generation region to the surface of the substrate. If a plate bias is applied to the sample holder, then a drift field is also present which accelerates the ionized species toward the sample, giving a more directional etching behavior. In most RIE systems, a self-bias is also generated due to the separation of charge between the plasma and the substrate (Coburn and Winters, 1979a; Nojiri, 2015). The intensity of the self-bias depends on the plasma RF power and has the same influence as DC bias on etching.

3. Adsorption

   Next, the reactive species get adsorbed on the sample surface. In most anisotropic RIE processes, the sample surface is passivated by nonvolatile products. This passivation layer inhibits further adsorption of the reactive species on the sample surface. Therefore, a DC bias is usually required to sputter the nonvolatile layer and for the reaction to proceed.

4. Reaction

   Following adsorption, a reaction occurs between the adsorbed molecules and the substrate. For Si etching in SiF_6 plasma, a chemical reaction produces highly volatile SiF_x ($x < 5$) byproducts.

5. Desorption

   After the reaction, the etch products need to desorb from the surface. For desorption to occur, the etch byproducts should be volatile (i.e., they should have a relatively high vapor pressure to vaporize at the etching temperature). The high volatility is an essential feature required for fast mass transfer of the byproducts, and it also makes the etch faster.

6. Pump out

   The chamber needs to be continuously pumped to remove the etch byproducts quickly. If the desorbed species are not removed quickly, they can enter the plasma and cause undesirable effects like redeposition.

2.3 RIE Process

Since the advent of plasma etching, various modifications have been made to address more stringent process demands with the miniaturization in microelectronics industry following the Moore's law. For the RIE process, electron cyclotron resonance (ECR-RIE) was developed in 1984 to decouple the ion density and the ion energy, thus giving more control over the etching process (Suzuki et al., 1984). Inductively coupled plasma (ICP-RIE) was developed around the same time which overcomes the limitations of pressure, tuning, and scaling imposed by ECR-RIE (Perry and Boswell, 1989; Reinberg et al., 1984). Implementation of the ICP design is widespread in mainstream RIE etchers available today. A schematic of the Oxford Instruments ICP-RIE is shown in Fig. 3.

In an ICP-RIE, the plasma is generated by an RF source which is supplied by inductively coupling the copper coils surrounding the ICP generation chamber. The ICP power controls the degree of ionization of the gas in the plasma and thus the density of the plasma. A charge-coupled plasma (CCP) generator supplied through the chuck provides the so-called plate bias for vertical acceleration of the ions and radicals in the plasma. The above design allows for almost independent control of energy and directional momentum of species in the plasma, which translates to independent control of chemical etching rate with ICP and ion milling rate with CCP. Increasing the ICP power results in an increase in isotropic etching, whereas an increase in plate power produces an increase in directional etching.

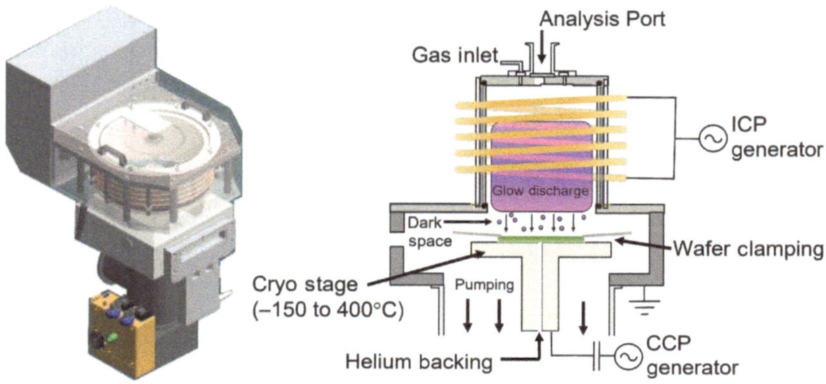

Fig. 3 A 3D view and a cross-sectional schematic of Oxford Instruments ICP-RIE. *Reproduced with permission from Henry, M.D., 2010. ICP Etching of Silicon for Micro and Nanoscale Devices. California Institute of Technology.*

The earlier process can be used to etch a multitude of semiconductor materials. Since the etching process and chemistry are different for different materials, it is beyond the scope of this chapter to cover all of them. So, we will limit our discussion to the most widely used semiconductors in group IV, namely Si. The process optimizations and techniques covered in later sections are core to the RIE process in general and are mostly independent of the material or the gases used to etch. As a result, the concepts discussed can very well be applied toward etching other materials. We will also limit the content to understanding the challenging task of etching vertically oriented nanowires which have a more profound application in energy harvesting and optoelectronics research. The etching of horizontal nanowires on a substrate is trivial once a vertical nanowire etching process is realized.

2.4 RIE Etching Chemistry

Silicon is the most extensively used material for microelectronics and the reigning champion of the global solar market (Green et al., 2015; Thompson and Parthasarathy, 2006). Germanium, on the other hand, is used in niche applications including solid-state electronics and infrared detectors (Colace et al., 2007; Soref, 2010; Tang et al., 2008). In the research literature, both Si and Ge have been etched using similar chemistries usually involving either hydrogen or halogen plasma with either Cl, Br, or F. Since the reaction probability depends on the bond strengths, it is essential to study the bond dissociation energies to predict the etching behavior. Table 1 shows various bond dissociation energy of Si and Ge products. The Si–Si or Ge–Ge bond energy is much less than halogen and hydrogen byproducts, and thus Si and Ge should be easily etched in halogen and hydrogen chemistry (Luo, 2007). Si and Ge have similar etching behavior, and most Si etching characteristics have a parallel in Ge. Furthermore, it is important to note

Table 1 Si and Ge Bond Dissociation Energies (kJ/mol) at 298 K

Silicon		Germanium	
Si–Si	310	Ge–Ge	264.4 ± 6.8
Si–Br	358.2 ± 8.4	Ge–Br	347 ± 8
Si–Cl	416.7 ± 6.3	Ge–Cl	390.8 ± 9.6
Si–F	576.4 ± 17	Ge–F	523 ± 13
H_3Si–H	383.7 ± 2.1	H_3Ge–H	348.9
H_3Si–H_3Si	313 ± 8		

however that most of the nanowire etching results in Si can be translated to Ge etching with little effort (Solanki et al., 2018; Solanki and Crozier, 2014).

Table 2 shows essential gases used for Si etching and their key plasma radicals, products, and passivation compounds. The products are the volatile compounds that are produced as a byproduct of the etch and pumped away. The passivation is composed of low- or nonreactive, nonvolatile compounds produced during etching which coat the surface of Si and prevent the radicals from reaching the Si surface. They can either be produced by the reaction of the etching gas with Si or deliberately by addition of a gas like C_xF_y which forms a fluorocarbon polymer in plasma.

Pure hydrogen-based etching of Si is uncommon. Most of the studies on hydrogen plasma etching of Si are directed toward understanding the surface roughness and the anisotropic etching that arises due to the preferential etching of different lattice planes (Ishii et al., 1992). The etch rate for hydrogen plasma etching is much lower than halogen plasma etching due to lower reactivity. Moreover, difficulties with surface roughness and anisotropic etching make hydrogen plasma etching less suitable for etching high aspect ratio structures compared to halogen-based etching.

The reaction with Si starts by adsorption of the F, Cl, or Br atoms on the silicon surface. The reaction of each halogen atom with Si is characterized by its reaction probabilities. The reaction probabilities are defined by reaction coefficients which are plotted in Fig. 4 (left). The plot is reproduced from Jansen et al. (1996) where they assume the etch byproducts to be SiF_4 for F, $SiCl_2/SiCl_4$ for Cl, and $SiBr_2/SiBr_4$ for Br etching. The reaction coefficients for F is much higher than Cl and Br for most of the temperature range. This is as expected from the bond ionization energies in Table 1 where Si–F > Si–Cl > Si–Br. The etch rate depends on reaction coefficients and thus we can deduce a much higher etch rate for F etching of Si compared to Cl and Br. We also see a strong dependence of reaction coefficients on doping for Cl and Br. This can be attributed to the strong electronegativity of the halogens which control the rate of chemisorption by transfer of electron from Si to physically adsorbed radicals. Baldi et al. postulated a two-step process of electron transfer through surface states (Baldi and Beardo, 1985). Owing to the high etch rate of Si in halogen plasma, an additional inhibitor or passivation gas is needed with Cl, Br, and especially F which has the highest reactivity.

The dependence of etching yield on halogen ion energy, reproduced from Donnelly and Kornblit (2013) is shown in the right side of Fig. 4. The energy of ions in the plasma plays an essential role in the etching yield

Table 2 Gases Used for Si Etching and Respective Radicals, Products, and Passivation Compounds

Gas	Radicals	Products	Passivation	Gas	Radicals	Products	Passivation
H_2		$HSiH_4$	Si_xH_y [a]	CHF_3	CF_2	HF, SiF_4	$Si_xC_yF_z$
CH_4	H, CH_3, CH_2	SiH_4, H_2	$Si_xC_yH_z$	CH_2F_2	CFH, C	HF	$Si_wC_xF_yH_z$
F_2	F	SiF_4	Si_xF_y [a]	CH_3F	CH_2, CFH	HF, H_2	$Si_wC_xF_yH_z$
NF_3	F, NF_2	SiF_4	$Si_xN_yF_z$ [a]	CF_4/O_2	CF_3, F, O	SiF_4, F_2, OF, O_2F, COF_2	$Si_xO_yF_z$
SiF_4	F, SiF_3	SiF_4	Si_xF_y [a]	CF_4/H_2	CF_3, F, H	SiF_4, HF, CHF_3	$Si_xC_yF_z$
CF_4	F, CF_3	SiF_4	$Si_xC_yF_z$	SF_6/O_2	SF_5, F, O	SiF_4, SOF_4	$Si_xO_yF_z$
SF_6	F, SF_5	SiF_4	$Si_xS_yF_z$ [a]	SF_6/H_2	SF_5, F, H	SiF_4, HF	$Si_xS_yF_z$ [a]
S_2F_2	F, S_2F	SiF_4	$Si_xS_yF_z$ [a]	SF_6/N_2	SF_5, F, N_2	SiF_4,	$Si_xS_yF_z$ [a]
Cl_2	Cl	$SiCl_4$	Cl	SF_6/CHF_3	SF_5, F, CF_2	SiF_4, HF	$Si_xC_yF_z$
Br_2	Br	$SiBr_4$	Br	$CBrF_3$	Forbidden	Greenhouse	Ozone
CBr_4	Br, CBr_3	$SiBr_4$	Br, $Si_xC_yBr_z$	CCl_4	Forbidden	Greenhouse	Ozone

[a]Only with cryogenic cooling.

Reproduced with permission from Jansen, H., Gardeniers, H., Boer, M. de, Elwenspoek, M., Fluitman, J., 1996. A survey on the reactive ion etching of silicon in microtechnology. J. Micromech. Microeng. 6, 14–28. https://doi.org/10.1088/0960-1317/6/1/002.

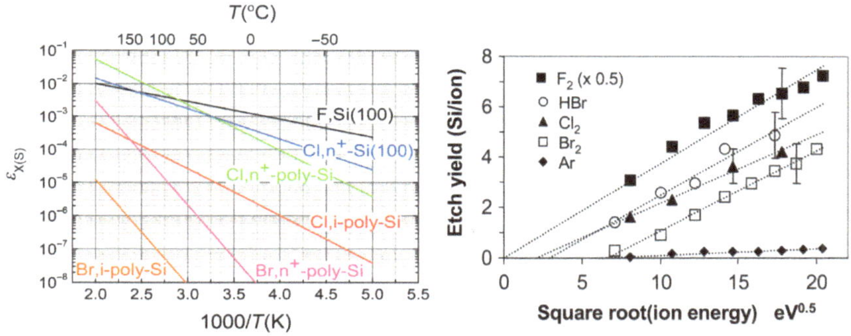

Fig. 4 Temperature dependence of reaction coefficients for F, Cl, and Br atoms with Si (*left*). Dependence of etching yield on ion energy (*right*). *Reproduced with permission from Donnelly, V.M., Kornblit, A., 2013. Plasma etching: yesterday, today, and tomorrow. J. Vac. Sci. Technol. A. 31, 50825. https://doi.org/10.1116/1.4819316.*

for all plasma etching processes. This was first shown conclusively by Coburn and Winters by three separate Si etching experiments with XeF_2, Ar^+ ions, and the combination $XeF_2 + Ar^+$. The combination of XeF_2 and Ar^+ ions provides an etch rate greater than sum of the individual XeF_2 and Ar^+ etching processes (Coburn and Winters, 1979b). The etching yield is inearly proportional to the square root of ion energy above a threshold energy, *E*th. This *E*th is denoted by the intersection of the dotted extrapolated curves with the *x*-axis which depicts the ion energy. The threshold energy for *F* is near zero, which suggests that F_2 can rapidly etch Si even in the absence of ion bombardment. For most other gases, *E*th greater than zero suggests that ion bombardment is needed to initiate the etch. As indicated in Fig. 4, if etch yield is the main criterion for etching, then the recommended choice of etching gases would be $F_2 > HBr > Cl > Br_2 > Ar$. Furthermore, the substrate can be cooled or the pressure reduced to increase the ratio of ion-assisted etching vs the thermally activated isotropic etching. Thus by combining chemical and physical etching parts, any RIE process can be divided into three mechanisms based on (Hirobe et al., 1987):

a. Etch rate from physical sputtering of the material by ion bombardment—ER_p

The sputter etch rate ER_p is given by,

$$ER_p = \Gamma_{ion}\left(Y_p/N\right) \tag{1}$$

where Γ_{ion}, Y_p, and N are the radical flux, the sputter yield, and the atomic density of Si, respectively.

b. Etch rate from chemical etching by free radicals—ER_c

The sputter etch rate ER_c is given by,

$$ER_c = \Gamma_n(Y_n/N) \tag{2}$$

where Γ_n and Y_n are the ion flux and the reaction probability with Si.

c. Etch rate from ion-assisted chemical reaction due to ion bombardment on surface adsorbed with radicals—ER_i

$$ER_i = n\Gamma_{\text{ion}}(Y_{\text{ion}}/N) \quad \text{for } n\Gamma_{\text{ion}} \leq \Gamma_n \tag{3}$$

Or

$$ER_i = \Gamma_n(Y_n/N) \quad \text{for } n\Gamma_{\text{ion}} \geq \Gamma_n \tag{4}$$

where n is the number of adsorbed radical atoms activated by an incident ion and Y_n is the chemical sputter yield of the ion at the ion energy under investigation. The total etch rate for an RIE process is then determined by the sum of ER_p, ER_c, and ER_i etch rates. For anisotropic etching, the ER_c and ER_i should be maximized to increase the etch rate, and ER_p should be minimized to reduce the physical damage to mask and Si.

Having discussed the factors determining the etch rate, we will briefly look at the etching mechanism for Si etching in most commonly used Cl and F plasma. The Cl etching of Si proceeds by Cl covering the Si surface and reaction at the Si surface to form $SiCl_{(surf)}$. Cl further reacts with $SiCl_{(surf)}$ to form di- and trihalides to give $SiCl_{x(surf)}$, where $x < 4$ (Cheng et al., 1994; Layadi et al., 1998; Radny et al., 1996). The surface reactions with chlorination and ion bombardment-assisted etching can be summarized below:

1. Dissociation

$$Cl_2 + e \rightarrow 2Cl + e \tag{5}$$

2. Surface adsorption

$$Cl_2 \text{ or } Cl + Si_{(surf)} \rightarrow Si_{(surf)} - nCl \tag{6}$$

3. Chemical reaction

$$Cl_{2(g)} + 2Si_{(surf)} \rightarrow 2SiCl_{(ads)} \tag{7}$$

$$Cl_{(g)} + 2Si_{(surf,\, bulk)} \rightarrow SiCl_{(ads)} \tag{8}$$

$$Cl_2^+ + nSi_{(surf,\, bulk)} + (2-n)SiCl_{(ads)} \rightarrow SiCl_{(ads)} + (2-n)SiCl_{2(ads)} \tag{9}$$

where $n = 0$, 1, or 2 and the subscripts surf and bulk represent the surface and subsurface Si atoms, respectively.

4. Lastly, ion bombardment-assisted desorption as

$$Cl_2^+ + Si_{(bulk)} + SiCl_{x(ads)} \rightarrow SiCl_{y(g)} + SiCl_{z(ads)} \tag{10}$$

For fluorine etching, SF_6 source is commonly used. In rare cases, gases like CF_4 or NF_3 are also used for Si etching. The reaction in SF_6 plasma proceeds as below:

$$SF_6 + e^- \rightarrow SF_5 + F + e^- \tag{11}$$

$$SF_5 + e^- \rightarrow SF_4 + F + e^- \tag{12}$$

The F then gets adsorbed on the Si surface and forms SiF_x byproducts as below:

$$F \text{ or } F_2 + Si_{(surf)} \rightarrow Si - F_{(ads)} \tag{13}$$

$$F_2 + Si_{(surf)} \rightarrow SiF_{(ads)} + F \tag{14}$$

$$\begin{aligned} F + SiF_{x(surf)} &\rightarrow SiF_{x+1(surf/gas)} \\ F_2 + SiF_{x(surf)} &\rightarrow SiF_{x+1(surf/gas)} + F \end{aligned} \tag{15}$$

where x is between 0 and 3. The etching proceeds when Si is removed by formation of volatile SiF_4 or desorption of SiF_2 (Coburn and Winters, 1979b; Donnelly, 2017). A detailed discussion on the etching mechanisms can be found in Cheng et al. (1994), Layadi et al. (1998), and Radny et al. (1996) for Cl and Coburn and Winters (1979b) and Donnelly (2017) for F.

2.5 RIE Etching Processes for Si Nanowire Etching

The RIE process can be divided into three main categories for Si etching: (a) room temperature RIE (RTRIE), also known as the pseudo-Bosch process, (b) cryogenic RIE (CRIE), where the Si substrate is cooled to less than −90°C, and (c) time-multiplexed RIE, also known as the Bosch or DRIE process.

2.5.1 Near Room Temperature RIE

To etch any material in a patterned fashion, a mask is required to protect the nonetching region from ions and radicals in the plasma. This is usually accomplished by some sort of lithographic patterning which forms a masking layer on the Si surface. This mask can be made by various processes such as optical lithography (Mack, 2007), electron beam lithography (Chen, 2015; Fischer, 1993; Tseng et al., 2003), nanosphere lithography (Haynes and Van Duyne, 2001), and nanoimprint lithography (Traub et al., 2016). The choice of the lithography technique is important as it sets the critical dimension. Electron beam lithography and nanoimprint lithography provide the highest resolution and are most commonly used. Nanosphere lithography is another

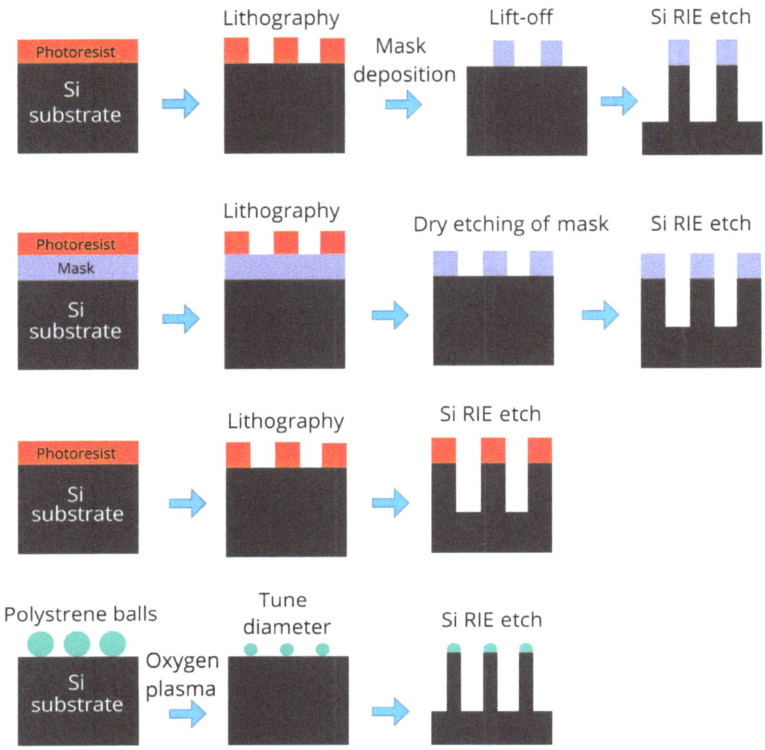

Fig. 5 Schematic of different lithography processes for nanowire etching.

alternative for nanoscale patterning. It uses a self-assembled monolayer of polystyrene nanospheres as a mask. It is unique in that the critical dimension can be tuned by performing an oxygen plasma etching step which reduces the nanosphere diameter thereby delivering a smaller feature size.

Fig. 5 shows the various options available for fabrication of mask on Si. The most straightforward case is using resist as a mask but as we will see later in the chapter, the resist selectivity (resist vs Si etch rate) is the modest at best. It works well for low aspect ratio structures, but for aspect ratios above 10, using a hard mask becomes crucial. In such cases, SiO_2 and Si_3N_4 are routinely used as hard masks. For aspect ratios greater than 30, the mask required should have excellent halogen etch and sputtering resistance. Al, Cr, Ni, and alumina fit these requirements and are therefore perfect mask candidates for nanowire etching. We will discuss selectivity in more detail in Section 2.5.1.3.

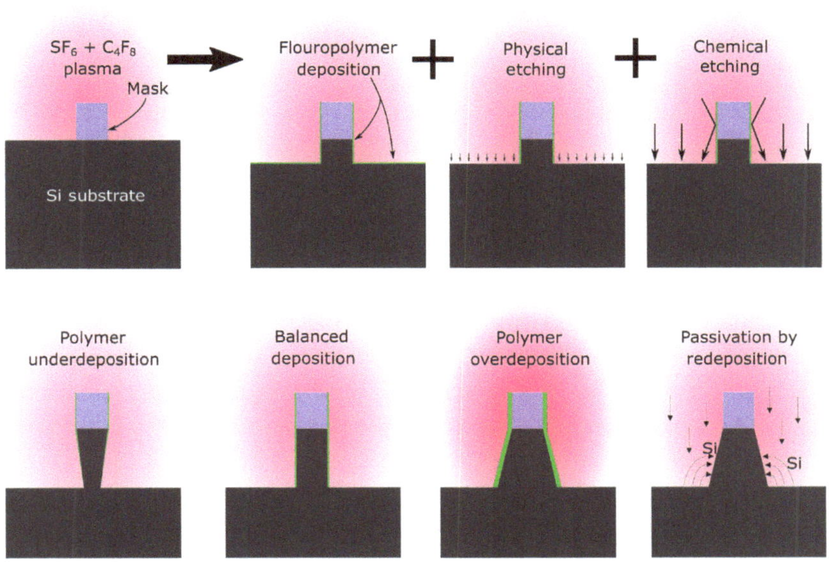

Fig. 6 Schematic of RTRIE etching process (*top*). Influence of over/under passivation and redeposition on etch profile (*bottom*).

Once the Si surface is patterned, it can be etched in an RTRIE process to give nanowires. Although, a delicate balance between physical and chemical etching is required for etching nanowires with vertical sidewalls. Some anisotropy can be introduced in the etching by increasing the plate power or by reducing the pressure or temperature. Even so, achieving nanowires with a sidewall angle greater than 80 degree can be challenging. Thus, most of the studies on RIE, Bosch, and even cryogenic RIE processes have a passivation gas to protect the nanowire sidewalls from etching. A schematic with different mechanism involved in anisotropic Si etching is shown in the top row of Fig. 6.

In an RTRIE process, both etching (SF_6) and passivation (C_4F_8), gases are present simultaneously. The passivation, physical etching, and chemical etching occur concurrently. The fluorocarbon polymer acts as a passivation layer and is deposited on all the surfaces. The polymer layer inhibits further chemical reaction with Si and stops the etching process. Thus, to avoid a complete etch stop, the process includes a physical etching component via plate bias which increases the energy anisotropy of the ions in the direction toward the substrate. The accelerated species then bombard the bottom surface of the sample and remove the protective polymer layer. Ar being a heavy atom and a noble gas is sometimes added in the etching mixture to increase the physical sputtering without significantly influencing the chemical etching. The etching proceeds again with the reaction of the exposed bottom surface with SF_6.

The RTRIE process requires a significant effort to achieve the right balance between the chemical etching rate, physical etching rate, and polymer deposition rate since they all occur simultaneously. The ratio of SF_6/C_4F_8 directly controls the chemical etching vs polymer deposition rate. The schematic in the bottom row of Fig. 6 shows the effect of under deposition, balanced deposition, over deposition, and excess ion bombardment. Under deposition causes the recently etched sidewalls to be etched further due to insufficient protection from radicals. This gives a tapered nanowire with inverted tapering. On the other hand, with over deposition, the polymer layer thickness increases during etching. This results in an increased diameter due to excess polymer shell which shadows the region below and marginally increases the diameter of the freshly etched region below. The process continues over the course of etching and forms a tapered nanowire (Dussart et al., 2014). It should be noted that although we are only describing the etching behavior in relation to the C_4F_8 passivation gas flow, it is the percentage change in C_4F_8 flow with respect to the total $C_4F_8 + SF_6$ mixture that matters and not the individual flow rate. Lastly, having excessive physical etching can lead to Si sputtering from the substrate and redepositing. The thickness of this redeposited layer increases with further etching and leads to a tapered nanowire.

The Bosch process is the most widely used for high aspect ratio etching, but RTRIE has its advantages for nanowire etching. Hung et al. provide a very compelling argument for etching nanowires with RTRIE instead of Bosch process (Hung et al., 2011). Their tests show that Bosch process provides a higher etch rate of 2.3 μm/min and a higher selectivity (to the resist mask) of ~88.46 compared to ~16.52 for the RTRIE process. But then, they also observe a scalloping width of 265 nm which is unacceptable for nanowire etching and is a central issue in most Bosch processes. The scalloping process will be explained in detail in Section 2.5.3. They observe etch rates of 300–400 nm/min for RTRIE process which are reasonable considering the length of most dry etched nanowires are <10 μm. We will next discuss the influence of gas mixture, pressure, RF, and plate power, and a linearly ramped gas flow on the nanowire etching rate and sidewall angle.

2.5.1.1 Effect of Gas Mixture

The effect of gas flow on the etching was studied by Hung et al. as reproduced in Fig. 7. Tapering can be studied by measuring the sidewall angle of the etched nanowires as shown in the right side of Fig. 7 (Hung et al., 2011). They

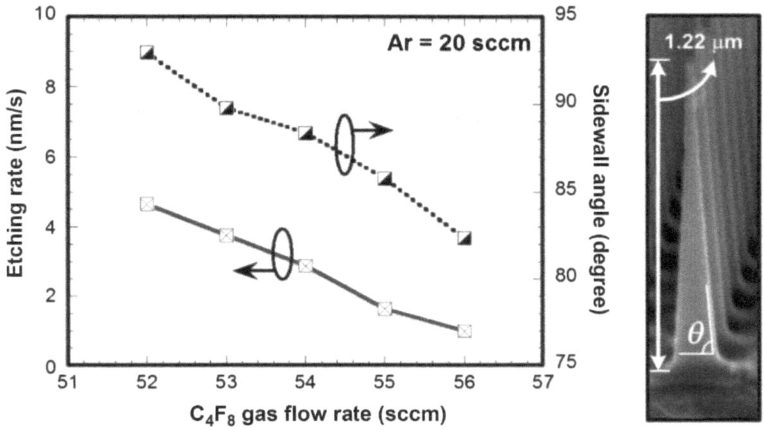

Fig. 7 Etching rate and sidewall angle on C4F8 in the gas mixture (*left*). Definition of sidewall angle (*right*). *Reproduced with permission from Hung, Y.-J., Lee, S.-L., Thibeault, B.J., Coldren, L.A., 2011. Fabrication of highly ordered silicon nanowire arrays with controllable sidewall profiles for achieving low-surface reflection. IEEE J. Sel. Top. Quantum Electron. 17, 869–877. https://doi.org/10.1109/JSTQE.2010.2068540.*

fix the total flow of $SF_6 + C_4F_8$ to 80 sccm and use Ar to stabilize the plasma. The plot shows a linear reduction in etch rate as the C_4F_8 flow is increased from 52 to 56 sccm. This is expected due to increase in passivation thickness with increasing C_4F_8 flow.

The sidewall angle plot shows the tuning of the etching regime from under to over passivation as highlighted in the bottom row of Fig. 6. For low C_4F_8 flows, the etching dominates over passivation and leads to an inverted tapered nanowire with sidewall angles greater than 90 degree. For etching high aspect ratio nanowires with a small diameter, such an inverted tapering can be very detrimental as it causes pinching of the nanowire base and results in fallen nanowires. On the other hand, C_4F_8 flows greater than 53 sccm leads to passivation dominating over etching and gives a more tapered etch geometry with increasing flow rates. The C_4F_8 flow rate of 53, seems to have the right balance between etching/passivation and gives a sidewall angle near 90 degree. Lastly, it should be noted that although the SF_6:C_4F_8 flow ratio defines the etching:passivation balance, increasing the individual flow rates to higher (or lower) value with same flow ratio will not give the same etching behavior. This is due to ion scattering and ionization rate which depend on the density of the gas in the chamber which in turn changes with total flow rate. For example, although SF_6 flow increases the etching rate, an increase in the SF_6 flow beyond a critical point for a fixed ICP power will reduce the etch rate due to increased scattering of ions.

2.5.1.2 Effect of Pressure

The pressure also has an important effect on the number of species or the density of the plasma. With higher pressure, the chance of collisions between species increases. Thus, pressure affects the mean free path of the gas. Mean free path is an important parameter in anisotropic etching. It is defined as the average distance a particle travels between successive collisions. The mean free path (λ) of the molecules in a gas is inversely related to pressure (p) as in the relation below,

$$\lambda = \frac{k_b T}{\sqrt{2}\pi d^2 p}$$

where k_b is the Boltzmann constant and d is the molecule diameter. Thus, the pressure can have a significant impact on the directionality of the etch.

As we see in Fig. 8, high pressure reduces the etch anisotropy and increases sidewall damage. This can result in an inverted taper in nanowires

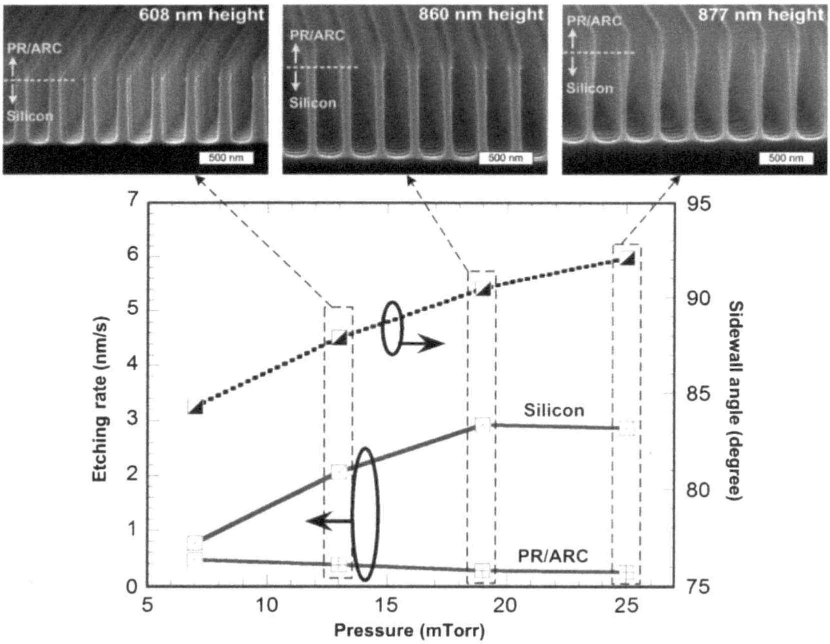

Fig. 8 Etching rate and sidewall angle vs pressure for Si and photoresist. *Reproduced with permission from Hung, Y.-J., Lee, S.-L., Thibeault, B.J., Coldren, L.A., 2011. Fabrication of highly ordered silicon nanowire arrays with controllable sidewall profiles for achieving low-surface reflection. IEEE J. Sel. Top. Quantum Electron. 17, 869–877. https://doi.org/10.1109/JSTQE.2010.2068540.*

due to more isotropic etching behavior. Lower pressure on the other hand means that ions and radicals have a larger mean free path and thus higher probability of following their vertical trajectory resulting in a more anisotropic etch. If the bias is very high, it can also induce substantial physical etching due to ion bombardment. As expected the etch rate of photoresist which does not chemically react with SF_6 plasma sees an increased etch rate for lower pressure. The etch rate of Si usually increases with pressure due to increase in F concentration. Very high pressures however can cause instability in plasma and the etch rate saturates or starts to decrease due to a decline in ion energy and radical flux (Wu et al., 2010). As seen in Fig. 8, an etch rate saturation is seen for high pressures.

2.5.1.3 Effect of Plasma Energy and Direction

The etching rate and passivation rate are directly correlated to the radical density and ionization rate of input gases. The RF source power and the bias directly influence the plasma density, momentum, and energy. An ICP reactor gives higher densities and plasma ionization rate compared to other plasma generation techniques (Donnelly and Kornblit, 2013). As a rule, increasing the RF power increases the ionization rate of the plasma and the energy of ions in the plasma. In an ICP-RIE, the plasma is generated away from the sample, and at low source powers, the process can be dominated by neutral etching. In contrast, for high powers the ion-driven etching plays a leading part in the etching.

The plate bias and self-bias in a RIE process generate a downward-oriented electric field. The field accelerates the ionized species toward the substrate. This imparts a considerable downward momentum to the charged species in the plasma. A strategy of using high bias along with low pressure gives ions a ~90 degree impact angle on the Si surface. The ions have very low momentum perpendicular to the nanowire sidewalls, and this makes the etch highly anisotropic. The high downward momentum also helps in stripping the passivation layer and increases the etch rate by ion-assisted chemical etching.

Fig. 9 shows the influence of bias power on etching rate and selectivity as studied by Hung et al. The etching time is fixed to 3 min so the scanning electron microscope (SEM) images in the top row give a qualitative idea of the increase in Si etch rate with increasing bias. On careful examination, we can see the resist etch rate also increases with increasing bias. The results are plotted in the bottom row of Fig. 9. As expected, a positive correlation is observed between the vertical component of Si etching rate and bias. The lateral component of Si etch rate reduces with increasing bias as the ion impact angle

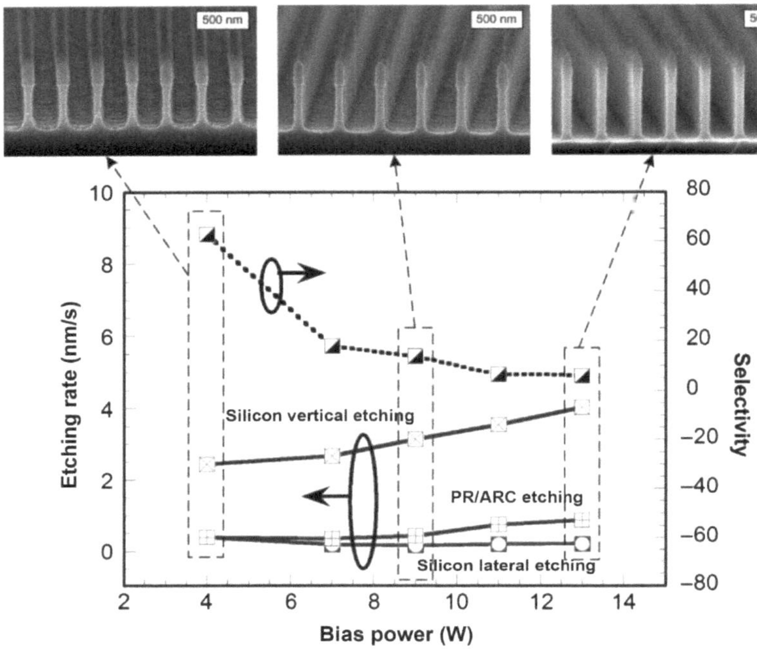

Fig. 9 Etching rate and selectivity vs bias power for Si and photoresist etching. *Reproduced with permission from Hung, Y.-J., Lee, S.-L., Thibeault, B.J., Coldren, L.A., 2011. Fabrication of highly ordered silicon nanowire arrays with controllable sidewall profiles for achieving low-surface reflection. IEEE J. Sel. Top. Quantum Electron. 17, 869–877. https://doi.org/10.1109/JSTQE.2010.2068540.*

becomes progressively parallel to the nanowire sidewalls. Reducing the lateral etch is crucial as it leads to pattern shrinkage and collapsed nanowires in extreme cases.

The ratio of Si etching rate/resist etching rate gives the etching resistance of the mask associated with Si and is known as "selectivity." The selectivity should be maximized for any etch and is a critical parameter when aiming for high aspect ratio structures. The selectivity inversely follows the mask etch rate. As seen in Fig. 9, it is ~70 for low plate bias but is lowered to ~5 at high plate bias as the etch rate of Si is modestly higher than the resist. A compromise is achieved around a bias of 9–11 W, where we see smooth sidewalls and selectivity of ~16 with a reasonable etching rate of ~3 nm/s.

The etching selectivity can be broken down into chemical and physical etching selectivities. The chemical etching selectivity can be increased by using a mask material which makes highly nonvolatile compounds with the etching gas. Table 3 shows halogen byproducts of various metals and the

Table 3 Temperature at Which the Various Chemical Compounds Achieve a Vapor Pressure of 10 mTorr

Chemical Compound	Temperature (°C)	Chemical Compound	Temperature (°C)	Chemical Compound	Temperature (°C)	Chemical Compound	Temperature (°C)
$AgCl$	1074	CdI_2	512	HgI_2	204.5	$SbCl_3$	85.2
AgI	983	CrO_2Cl_2	13.8	KBr	982	SbI_3	223.5
$AlBr_3$	118.0	Cu_2Br_2	718	KC1	968	$SiCl_4$	−34.7
$AlCl_3$	123.8	Cu_2Cl_2	702	KF	1039	$SiClF_3$	−127.0
AlF_3	1324	Cu_2I_2	656	KI	887	$SiCl_2F_2$	−102.9
AlI_3	225.8	$FeCl_2$	700	$LiBr$	888	$SiCl_3F$	−68.3
As	437	$FeCl_3$	235.5	$MgCl_2$	930	SiF_4	−130.4
AsH_3	−124.7	$GeBr_4$	56.8	$NaBr$	952	$SnBr_4$	72.7
BBr_3	−10.1	$GeCl_4$	−15.0	NaF	1240	$SnCl_2$	391
BCl_3	−66.9	H_2S	−116.40	NaI	903	$SnCl_4$	10.0
BF_3	−141.3	H_2S_2	−19	$NiCl_2$	759	SnH_4	−118.5
$CdCl_2$	656	H_2Se	−100	PbI_2	571	SnI_4	175.8
CdF_2	1559	$HgBr_2$	179.8	PbF_2	904	$ZnCl_2$	508
		$HgCl_2$	180.2	$SbBr_3$	142.7	ZnF_2	1359

Reproduced with permission from Nojiri, K., 2015. Mechanism of dry etching. In Dry Etching Technology for Semiconductors. Springer International Publishing, Cham, pp. 11–30. https://doi.org/10.1007/978-3-319-10295-5_2.

temperature needed to reach a vapor pressure of 10 mTorr. As seen in the table, Al or its oxide Alumina is an ideal mask candidate. The Al and F byproduct AlF_3 has a very high vapor temperature of 1324°C, thus at room temperature Si etching will dominate Al etching by several orders of magnitude over Al etching in F even at low etching pressures. Next, to increase the physical etching selectivity we can choose a material which has a lower physical etching or ion milling rate compared to Si. Table 4 shows ion milling rates of various metals and oxides. Al and especially its oxide has a small milling rate, and thus are an exceptional masking material for any physical etching. SiO_2 and resist are other useful mask materials for physical etching. Comparing vapor pressure temperature and ion milling rate for etching/masking materials provides an easy path to find the right mask irrespective of the etching chemistry.

We discussed the independent influence of bias and the gas flow rates earlier. In some cases where the process is in an over passivation regime or if the plate power is the too low, micrograss can form on the Si surface. This occurs due to selective micromasking of the surface by a thin residual layer of passivation polymer which acts as tiny masks that facilitate the etching of micro or nanoscale columns of Si. An obvious way to eliminate micrograss is by increasing the bias power to remove the passivation layer from the bottom surface, but the down side is that it also reduces selectivity. An alternative strategy is to reduce the C_4F_8 flow until the micromasking stops. The temperature also plays an important role. With low temperatures the etch rate of the polymer is low and passivation rate is high. Thus, increasing the temperature in 5°C steps until micromasking is resolved is an option. Other parameters like ICP power and pressure can also bring about micromasking by changing the plasma conditions but even in these cases reducing the passivation gas flow provides a much faster solution to micromasking. Lastly, micrograss also depends on the critical dimensions of the etch pattern. It is more likely to occur on wide open areas where passivation rate is higher than on high aspect ratio channels, holes, or dense nanowire arrays where the rate is low due to slower diffusion.

2.5.1.4 Optimization With Linearly Graded Gas Flow

An ideal etching system should allow etching of vertical nanowire arrays in different densities and aspect ratios and give same sidewall angles on all structures. For nanowire etching, the nanoscale dimensions, high aspect ratios, and nanowire density play a vital role in detuning or unbalancing the etch process. Interestingly, the etch rate depends on the feature size and density

Table 4 Ion Beam Estimate of Etch Rates Obtained From INTLVAC Nanoquest Ion Beam Etcher Datasheet (Beam Voltage = 200 V, Beam Current = 56 mA) at Center of Nanoscale Science, Harvard University

Chemical Compound	Etch Rate (A/min)	Selectivity to Si	Chemical Compound	Etch Rate (A/min)	Selectivity to Si	Chemical Compound	Etch Rate (A/min)	Selectivity to Si
Al	28	1.649	Fe	16	0.946	Ta	19	1.135
Be	6	23.74	Pb	118	7.027	Sn	54	3.243
Bi	400	0.297	Mg	10	0.608	Ti	15	0.892
C	5	1.432	Mn	40	2.351	W	15	0.919
Cr	24	1.216	Mo	19	1.108	Al_2O_3	4	0.224
Co	20	2.378	Ni	24	1.432	SiC	16	0.946
Cu	40	2.703	Nb	21	1.270	SiO_2	15	0.892
Gd	50	2.973	Pd	50	2.973	TaC	7	0.405
Ge	42	2.486	Pt	28	1.676	Ta_2O_6	27	1.622
Au	49	2.919	Si	17	1	Soda glass	9	0.541
Hf	30	1.784	Ag	82	4.865	Shipley AZ 1350	9	0.541

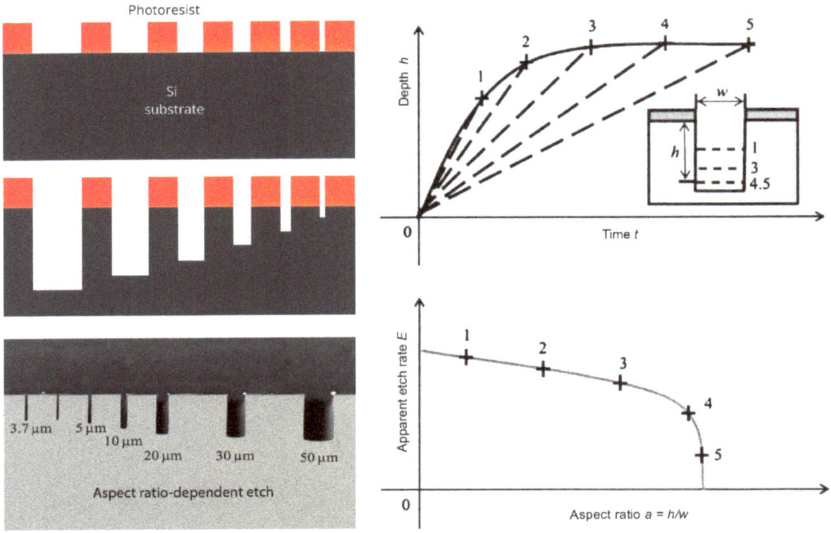

Fig. 10 Schematic highlighting aspect ratio-dependent etching in dry etching (*left*). A conceptual plot for change in etch depth vs time and etch rate vs aspect ratio (*right*). *Reproduced with permission from Wu, B., Kumar, A., Pamarthy, S., 2010. High aspect ratio silicon etch: a review. J. Appl. Phys. 108, 51101. https://doi.org/10.1063/1.3474652.*

as seen in Fig. 10. This phenomenon is known as aspect ratio-dependent etch (ARDE) and is very common when fabricating high aspect ratio structures where the etching conditions evolve over the duration of the etch (Jansen et al., 1996; Wu et al., 2010). The change in the etching condition is mainly due to depletion of fluorine (Blauw et al., 2000). Consequently, the main factors are SF_6 flow rate and the dissociation probability. For high aspect ratio structures, as the etch progresses, the mass transfer rate can reduce dramatically. This is because the reactants have difficulty reaching in the narrow regions and the etch products need more time to be pumped out. The etching kinetics becomes transport limited and are governed by Knudsen transport (Blauw et al., 2000). The results are highlighted in the schematic and SEM in the left half of Fig. 10. The top right of Fig. 10 shows a conceptual plot of reduction in etch depth with increasingly deep etching on the top right. Fig. 10 when translated to etch rate vs aspect ratio is plotted in the bottom right. The etch rate plot indicates two regimes (a) with a gradual reduction in etch rate and (b) with a sharp decline in etching rate where the mass transfer is highly limited. The effect of ARDE can be reduced by working at low temperature and low pressure. Thus, CRIE outperforms RTRIE in this regard (Liu et al., 2013).

When etching high-density nanowire arrays, the ARDE causes the nanowires to have a broader base. A more complicated situation occurs when trying to etch a silicon sample to obtain two arrays with nanowires of different diameters (e.g., 50 and 150 nm) on a constant pitch (e.g., 200 nm). The 50 nm diameter on a 200 nm spacing has a nanowire to nanowire gap of 300 nm whereas the 150 nm nanowire has a gap of 100 nm. This difference in the nanowire gaps results in etched nanowires with different heights and tapering angles. Therefore, an RIE etching recipe optimized for etching 100 nm diameter nanowires with a sidewall angle of 90 degree would give a sidewall angle <90 degree for diameter >100 nm and a sidewall angle >90 degree for diameters <100 nm. A narrow range of diameter around 100 nm, however, will exhibit a complete anisotropic etch behavior. By pushing the boundaries of this diameter range with process optimizations, we can achieve complete directional etching for a wide range of diameters and pitch.

Since ARDE is primarily caused by modification of the etching chemistry due to reduced mass transfer, one way to solve this issue is by gradually changing the etch parameters during the process to offset the imbalance in etching vs passivation rates in deep regions. Increasing the bias power is another option (Jo et al., 2005). After the initial stage where the ARDE sets in, the bias power can be increased gradually to provide ions with more directional energy to reach the base of nanowire forest. Lowering the pressure can have a similar effect due to the reduction of the angular spread of ion energy (Juan and Pang, 1995). However, the plasma conditions strongly depend on the pressure and bias and thus changing them during a running process could make the plasma unstable and increase the reflected RF power at the source. A better alternative is to change the gas etch ratio (Hung et al., 2011). We have seen in Fig. 7 that a small change in C_4F_8 is required to go from inverted taper to tapered nanowire. This small change in gas ratios has a negligible change in plasma stability but has a significant impact on the etching behavior.

Such a graded process with linearly varying gas ratios was tested by Hung et al. and compared with the standard RTRIE process. Their results are summarized in Table 5. They start with a standard process and achieve a very high sidewall angle of 90.6 degree but with a low filling factor of 0.04 due to sidewall etching. The same process with increased etching time to obtain taller nanowires (process B) gives a filling fraction of 0.02 due to more prolonged sidewall etching. This lateral undercut is a critical issue in etching of fine features which can be seen from SEMs on 800 nm tall nanowires

Table 5 A Comparison of Etching Results Between Standard RTRIE Process and RTRIE Process With a Graded Gas Flow

Etching Process		RTRIE Process Conditions				Etching Results		
		Time (min)	C_4F_8 (sccm)	SF_6 (sccm)	Bias Power (W)	Height/ Width (nm)	Filling Factor	Sidewall Angle (degree)
RTRIE A		5	54	26	9	987/30.2	0.04	90.6
RTRIE B		7	54	26	9	1390/24	0.03	90.3
RTRIE C		5	55	25	13	906/93.5	0.13	90
RTRIE with graded flow D	Step 2	1.5 per step	55 ↓ 55.8	25 ↓ 24.2	13	632/50	0.21	90.5
	Step 3					849/139	0.20	89.6
	Step 4					1110/130	0.19	89.1
	Step 5					1340/123	0.18	88.4

Adapted with permission from Hung, Y.-J., Lee, S.-L., Thibeault, B.J., Coldren, L.A., 2011. Fabrication of highly ordered silicon nanowire arrays with controllable sidewall profiles for achieving low-surface reflection. IEEE J. Sel. Top. Quantum Electron. 17, 869–877. https://doi.org/10.1109/JSTQE.2010.2068540.

in Fig. 8. The issue seems to arise from etching balance tipping in favor of excess SF_6 flow during the latter part of etching. The direction in which the balance tips can depend on a multitude of factors ranging from plasma conditions to reactor geometry. The next intuitive step after understanding the etch behavior in Fig. 9 is to reduce SF_6 to reduce the under etching and increase the plate bias for more directional etch. This is shown by process C in Table 5. The high bias process has a significantly higher filling factor of 0.13 with nanowire height like process A. The sidewall angle is improved to 90 degree giving highly vertical structures but while retaining some lateral etching.

Finally, Hung et al. moved to a linearly graded gas flow recipe where the C_4F_8/SF_6 ratios were changed from 55 sccm/25 sccm to 55.8 sccm/24.2 sccm in five 1.5 min steps over the course of the etch. The etching results (process D) now show a much higher filling factor. It is important to note that they switched to SiN_x mask for process D instead for photoresist which was used in processes A–C which biases the selectivity in favor of process D. Nevertheless, they obtained much larger filling factors of 0.21 with small lateral etching. The SEM images of the nanowires after different etch steps are shown in Fig. 11. The nanowire width reduces from 150 to 123 nm from steps 2 to 5. The ion bombardment-assisted etching of the mask is seen for the longest nanowires which show a tapering angle of 88.4 degree.

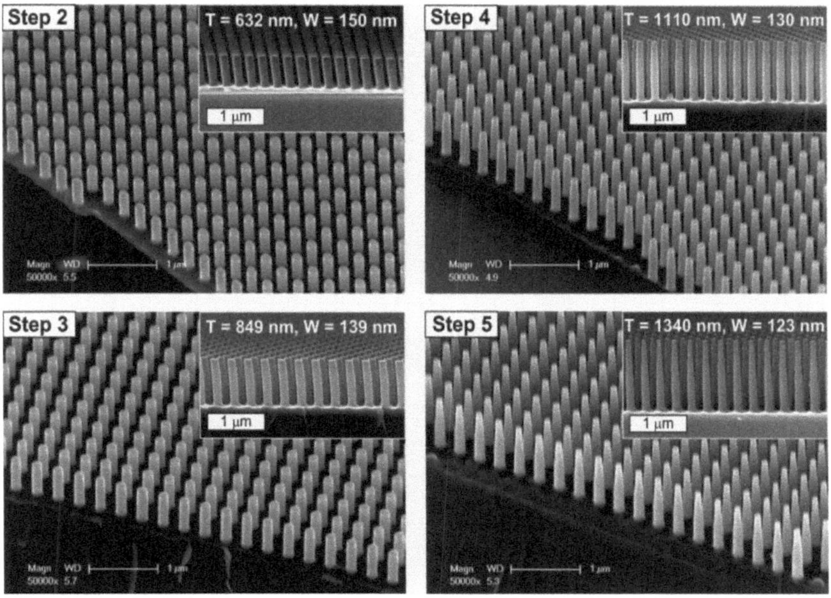

Fig. 11 SEM views of a resulting profile in an RTRIE process with graded flow rate at different steps during the etching. *Reproduced with permission from Hung, Y.-J., Lee, S.-L., Thibeault, B.J., Coldren, L.A., 2011. Fabrication of highly ordered silicon nanowire arrays with controllable sidewall profiles for achieving low-surface reflection. IEEE J. Sel. Top. Quantum Electron. 17, 869–877. https://doi.org/10.1109/JSTQE.2010.2068540.*

ICP or ECR reactors are widely used in literature for Si nanowires etching. In this section, we have summarized the different etching efforts on Si nanowires by RTRIE in Table 6. Table 6 along with the discussion in the sections earlier can serve as a starting point for any reader interested in etching nanowires or any high aspect ratio structures in Si on an RIE tool. Fig. 12 shows some of the state-of-the-art nanowires with high-density (top left), high aspect ratio (top right), multidiameter (bottom left), and extremely small diameter nanowires (bottom right) that have been reported in literature. The readers are encouraged to read the articles by Donnelly and Kornblit (2013), Jansen et al. (2009), Karouta (2014), and Wu et al. (2010) for detailed information on RIE etching processes, the issues, and solutions. Most of the applications with nanowires require clean, freestanding nanowires without any masking layer. Therefore, frequently after a RIE process some wet or dry chemical etching is performed to clean the polymer, oxide, or metal masking layer.

Table 6 Literature Review on the RTRIE Etching Process

References	RF Power/Plate Power (W)	Gases	Gas Flow (sccm)	Pressure (mTorr)	Mask
Lin et al. (2007)	150–50/50	CH_4/Ar	40/40	5	Ni
	AR = 5–8. ICP 50 W → high selectivity + low etch rate. ICP 100 W → low selectivity + high etch rate. Self-assembled Ni dots used as mask				
Hung et al. (2011)	850/09	$C_4F_8/SF_6/Ar$	52/28/20	19	Resist
	Selectivity = 16.52 and etch rate = 300–400 nm/min				
	800/9–13	$C_4F_8/SF_6/Ar$	54–55/25–26/20	19	Resist
	Sidewall angle = 90–90.6				
	800/13	$C_4F_8/SF_6/Ar$	55–55.8/25–24.2	19	SiN_x
	Graded gas flow used. Sidewall angle reduces with increasing height from 90.5 to 88.4				
Kuo et al. (2003a, b)	70–80	CHF_3/O_2	20/2	25	Cr
	70	$SF_6/CHF_3/Ar/O_2$	2/16/2/2	20	
	100 nm short nanowires obtained				
Chang et al. (2006)	150	SF_6/CHF_3	26/25–50	10	Ni
	120 nm diameter and 2.8 μm tall nanowires				
Park et al. (2012, 2014), Park and Crozier (2013), and Seo et al. (2011)	1200/10–20	SF_6/C_4F_8	60/160–162	20	Al
	80–280 nm diameter and 2.7 μm tall on 1 μm pitch. Photodetectors and color imaging done using Si nanowires				

Henry (2010) and Henry et al. (2009)	1200/20	SF_6/C_4F_8	33/57	—	Al_2O_3
	40 nm minimum diameter, 1.2 μm long, and 200 nm spacing				
Mirza et al. (2012)	600/12	SF_6/C_4F_8	20–40/90	15/10/8.5	HSQ
	Sub-10 nm diameter and 50:1 aspect ratio. Studied pressure and bias effect on selectivity				
Khorasaninejad et al. (2012a, b, c)	1200/20	SF_6/C_4F_8	20/40	10	Al
	30 nm minimum diameter and 210 nm long				
	1200/20	SF_6/C_4F_8	30/50	10	Al
	105–346 nm and 1.5 μm tall nanowires				
Martin et al. (2013)	1500/200	$SF_6/O_2/HBr/SiF_4$	28–32/45/ 30–60/15–30	25	SiO_2
	65°C process temperature, 200 nm diameter, and 12 μm tall. Sidewall angle 87.5–90. SiO_xBr_y or SiO_xF_y passivation				
Park et al. (2010)	500/20	SF_6/C_4F_8	20:30	10	NiCr
	25 nm diameter and 500 nm tall. 220 nm/min etch rate				
Kanechika et al. (2002)	110	$HBr/Cl/O_2$	30/15/4	120	SiO_2
	Magnetic enhanced RIE process. Selectivity = 2.1 and optimized for nanoneedles				
Ayari-Kanoun et al. (2016)	1200/45–120	C_4F_8/SF_6	40/25–40	25	Cr
	Sidewall angle = 95–80 degree. Studied effect of RF power on etching				

Continued

Table 6 Literature Review on the RTRIE Etching Process—cont'd

References	RF Power/Plate Power (W)	Gases	Gas Flow (sccm)	Pressure (mTorr)	Mask
Han et al. (2010, 2011)	40	Cl_2	30	10	HSQ
	No ICP process, clean surface after etch, 86% anisotropy				
	40	Cl_2/BCl_3	15/15	10	HSQ
	No ICP process, clean surface after etch, 82% anisotropy				
	40	$Cl_2/SiCl_4$	15/15	10	HSQ
	No ICP process, clean surface after etch, 80% anisotropy				
	70	$SF_6/O_2/N_2$	10/10/10	10	HSQ
	No ICP process, rough surface after etch, 88% anisotropy				
	40	CHF_3/SF_6	20/10	10	HSQ
	No ICP process, rough surface after etch, 15% anisotropy				

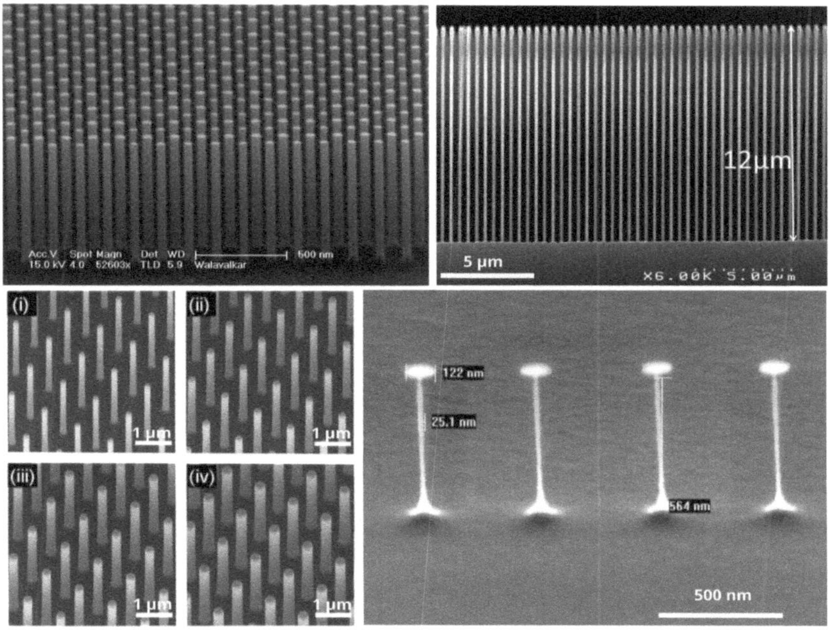

Fig. 12 State-of-the-art silicon nanowire etched by RTRIE process in literature. *Reproduced with permission from Henry, M.D., 2010. ICP Etching of Silicon for Micro and Nanoscale Devices. California Institute of Technology; Martin, M., Avertin, S., Chevolleau, T., Dhalluin, F., Ollivier, M., Baron, T., Joubert, O., Hartmann, J.M., 2013. Fabrication of high-density Si and Si_xGe_{1-x} nanowire arrays based on the single step plasma etching process. J. Vac. Sci. Technol. B Nanotechnol. Microelectron. Mater. Process. Meas. Phenom. 31, 41806. https://doi.org/10.1116/1.4812792; Park, H., Dan, Y., Seo, K., Yu, Y.J., Duane, P.K., Wober, M., Crozier, K.B., 2014. Filter-free image sensor pixels comprising silicon nanowires with selective color absorption. Nano Lett. 14, 1804–1809. https://doi.org/10.1021/nl404379w; Park, S.-Y., Di Giacomo, S.J., Anisha, R., Berger, P.R., Thompson, P.E., Adesida, I., 2010. Fabrication of nanowires with high aspect ratios utilized by dry etching with SF6:C4F8 and self-limiting thermal oxidation on Si substrate. J. Vac. Sci. Technol. B. 28, 763–768. https://doi.org/10.1116/1.3455498.*

2.5.2 Cryogenic RIE for Si Nanowire Etching

The RTRIE or Bosch process is a relatively quick and easy way to etch Si nanowires. However there are concerns regarding the passivation gas octafluorocyclobutane C_4F_8 leading to carbon contamination. The process forces polymer deposition not only on the sidewalls as a means to achieve anisotropy but also does so on the chamber surface. This needs to be cleaned periodically to have a consistent etching recipe. A cryogenic etch process alleviates those concerns. This section will detail the cryogenic etching process, chemistry, optimizations, and issues. Furthermore, a summary of processes used for fabricating high aspect ratio nanowires will be provided.

A low-temperature process such as CRIE has benefits of being extremely clean. The process replaces C_4F_8 with O_2 as the passivation gas. In cryogenic conditions, the SiF_x radicals on the surface react with O_2 to form a SiO_xF_y passivation layer at temperatures below –75°C (Pereira et al., 2009). This passivation layer desorbs when the sample is removed while unloading the sample leaves behind a clean surface without residues or scalloping. The process has been actively studied for making dense high aspect ratio structures. The substrate is cryogenically cooled usually with liquid N_2 between −75 and −130°C. At cryogenic temperatures, even SF_6 has been shown to have an anisotropic etching behavior. Tachi et al. first proposed the process in 1988 and it was the first technique to obtain aspect ratios up to 30:1 (Tachi et al., 1988).

The etching primarily relies on the adsorption of the reactive species, usually from SF_6 and O_2 gas, to form SiO_xF_y compounds which form a protective layer on the sample surface. The surface reactions hypothesized for the formation of SiO_xF_y passivation layer are as below from (d'Agostino and Flamm, 1981):

$$O + SF_5 \rightarrow SOF_4 + F \tag{16}$$

$$\begin{aligned} O + SF_4 &\rightarrow SOF_4 \\ &\rightarrow SOF_2 + F_2 \end{aligned} \tag{17}$$

$$O + SF_3 \rightarrow SOF_2 + F \tag{18}$$

$$(F \text{ or } F_2) + SOF_2 \rightarrow SOF_3(+F) \tag{19}$$

$$(F \text{ or } F_2) + SOF_3 \rightarrow SOF_4(+F) \tag{20}$$

$$O + SOF_3 \rightarrow SO_2F_2 + F \tag{21}$$

$$O + SOF_3 \rightarrow SO_2F_2 + F \tag{22}$$

$$O + SOF_2 \rightarrow SO_2F_2 \tag{23}$$

$$e^- + SOF_4 \rightarrow SOF_3 + F^- \tag{24}$$

The reaction of SF_6 with Si is spontaneous, but models have suggested a difference in etching products depending on the etch temperature as seen below:

$$Si + 2SF_{6(g)} \rightarrow SiF_{4(g)} + 2SF_{2(g)} \quad T > 100^\circ C \tag{25}$$

$$2Si + 2SF_{6(g)} \rightarrow 2SiF_{4(g)} + S_2F_{2(g)} \quad T < -30^\circ C \tag{26}$$

The reactions suggest that even at cryogenic temperatures the etch products are volatile. When the temperature is above 100°C SF_4 is the sulfur

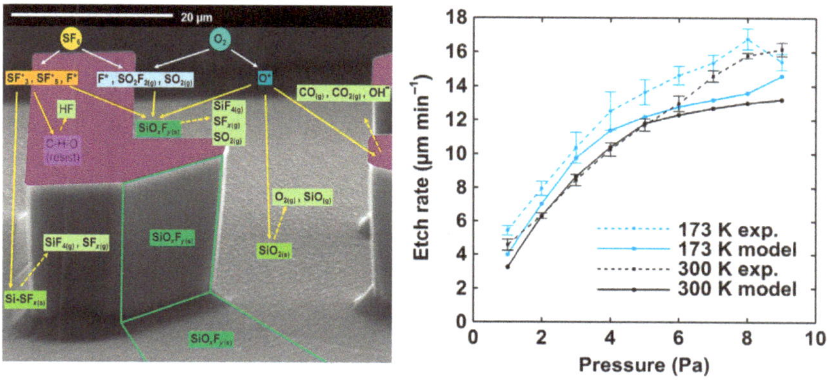

Fig. 13 A schematic of etching reactions on a Si sample with a resist mask in CRIE etching (*left*). The plot of etch rate vs pressure for Si etching at 173 and 300 K (*right*). *Reproduced with permission from Merzsch, S., Steib, F., Wasisto, H.S., Stranz, A., Hinze, P., Weimann, T., Peiner, E., Waag, A., 2014. Production of vertical nanowire resonators by cryogenic-ICP–DRIE. Microsyst. Technol. 20, 759–767. https://doi.org/10.1007/s00542-013-2032-4; Tinck, S., Tillocher, T., Dussart, R., Bogaerts, A., 2015. Cryogenic etching of silicon with SF 6 inductively coupled plasmas: a combined modelling and experimental study. J. Phys. D. Appl. Phys. 48, 155204. https://doi.org/10.1088/0022-3727/48/15/155204.*

byproduct and below −30°C S_2F_2 is the stable product. At room temperature, a mixture of both SF_4 and S_2F_2 exist. A detailed discussion on etching and passivation chemistry can be found in d'Agostino and Flamm (1981), Dussart et al. (2004, 2014), Knizikevičius (2009), and Tinck et al. (2015).

An overview of different species present in a cryogenic etch using Si with resist mask is shown in Fig. 13 (left). In an ICP-RIE chamber since the plasma generation is away from wafer, the low temperature of the wafer does not significantly affect the plasma properties. However, the flux of neutral and ion species near the substrate is locally affected by the low sample temperature (Tinck et al., 2015). The small temperature drop in species near the wafer causes a small increase in the density of neutral species such as F and F_2 compared to ions. Surprisingly, this implies that the etch rate of Si which is primarily mediated by neutral F species in plasma should increase at cryogenic temperatures (Tinck et al., 2015). The right plot in Fig. 13 shows simulated and experimental SF_6 etching of Si for a room temperature vs a cryogenic etching. Indeed, an increase in the etch rate is observed with a switch from 300 to 173 K substrate temperatures. Tinck et al. conclude that local reduction of temperature in the region above the sample yields higher densities of neutral flux on the sample which encourages higher etching rate in cryogenic etching.

Like RTRIE, the CRIE process also requires plate bias-assisted ion bombardment to remove the passivation layer on the bottom surface. The thickness of SiO_xF_y layer is thinner than the —CF2— layer formed in RTRIE or Bosch process. Hence, the required ion bombardment energy is much smaller. This is confirmed by Li et al. where they used a low-frequency power source at 25% duty cycle instead of high-frequency source commonly used in RIE (Li et al., 2016). Their studies demonstrate the ability to etch extremely high aspect ratio structures with higher mask etching selectivity compared to a high-frequency plate source. The directional ion bombardment due to plate power results in desorption of the SiO_xF_y layer on the bottom surface. Once the passivation layer is desorbed, the F radicals react with the exposed Si surface. As the etch progresses, the sidewall is coated with SiO_xF_y when Si is exposed. It is important to note that the passivation process is different from RTRIE, where a —CF2— layer is deposited, and so the sidewall passivation thickness can easily be >5 nm depending on the C_4F_8 flow and etching time. In contrast, only a few nanometer of SiO_xF_y is formed on the surface in cryogenic etching and it gets replenished only when depleted by bombardment. If the O_2 amount is too high the rate of SiO_xF_y growth will exceed the etching rate resulting in either micromasking or termination of etching. Hence, the ratio of SF_6:O_2 is critical in obtaining vertical etching profiles. Readers interested in a more detailed understanding of the cryoetching process and ways to optimize the etching process should refer to the work of Jansen et al. (1996). We will next provide a qualitative discussion on the anisotropic etch control by tuning various etch parameters.

A study by Merzsch et al. on cryogenic etching of nanowires provides an excellent outline on influence of temperature, oxygen, and HF plate power on NW etch profile (Merzsch et al., 2014). The results from their study are plotted in Fig. 14. The top row shows the influence of temperature. They start with −75°C which is the minimum temperature reported for formation of SiO_xF_y passivation layer. The temperature of −75°C is not low enough to allow formation of necessary passivation layer and thus isotropic etching is obtained. The anisotropy and etch rate data for Fig. 14 are tabulated in Table 7. The etch rate at −75°C is measured to be 114 nm/min but no nanowires are formed due to excessive sidewall etching. The temperature of −80°C also results in a small undercut but the pattern is maintained and tapered nanowires with significant sidewall etching are observed. As the temperature is increased to −95 and −105°C the anisotropy increases to 0.9 and 0.92, respectively. Moreover, the sidewall undercut reduces, suggesting the formation of good passivation layer.

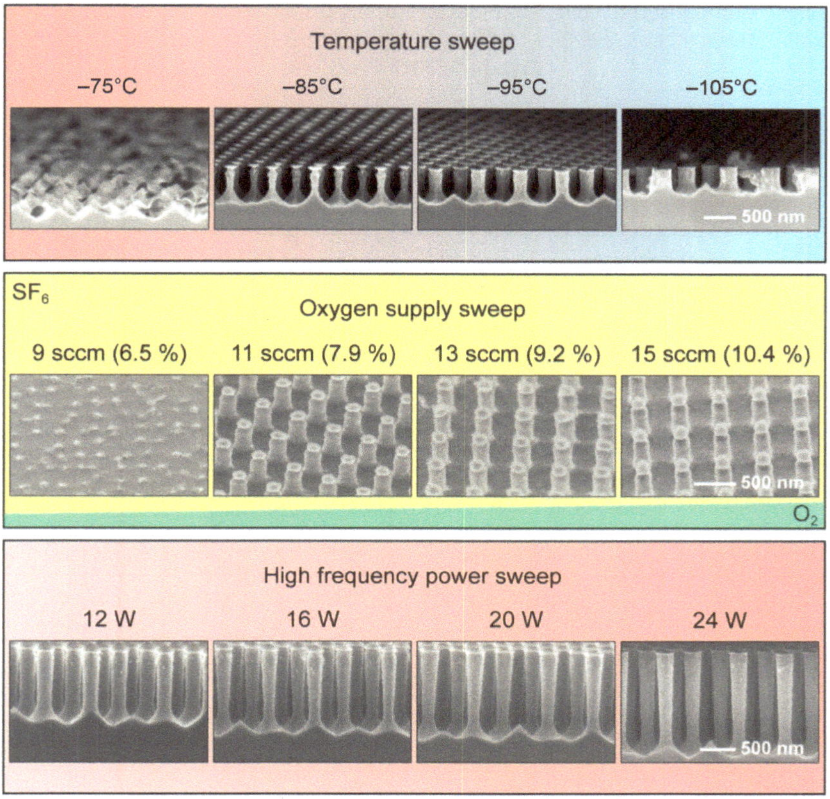

Fig. 14 Overview of etching profile for change in temperature, oxygen content, and plate bias. *Reproduced with permission from Merzsch, S., Steib, F., Wasisto, H.S., Stranz, A., Hinze, P., Weimann, T., Peiner, E., Waag, A., 2014. Production of vertical nanowire resonators by cryogenic-ICP–DRIE. Microsyst. Technol. 20, 759–767. https://doi.org/10.1007/s00542-013-2032-4.*

Next, we examine the influence of O_2 flow. Since the oxygen layer controls the formation of SiO_xF_y, the amount of oxygen available is critical toward optimizing the nanowire etching process. Merzsch et al. fix the temperature to an optimized value of −95°C and vary the O_2 flow rate from 6.5% to 10.4% of gas flow mixture. As expected, the lowest flow rate of 6.5% O_2 results in the removal of the mask pattern due to insufficient passivation and undercut.

Interestingly, for increasing O_2 flows they observe an increase in etch rate along with high anisotropy above 0.95. This suggests that their process is deficient in O_2 and thus the best process is obtained with the highest value

Table 7 Comparison of Etch Parameter for Tuning of Temperature, O_2 Flow, and Plate Power (Merzsch et al., 2014)

Cryogenic Etch Processes

		Process A	Process B	Process C	Process D	Parameters (°C, sccm, W, Pa, min)
Temperature (°C)		−75	−80	−95	−100	$SF_6=129$ $O_2=15$ ICP = 140 Plate = 20 Pressure = 1 Time = 5
	Etch rate (nm/min)	114	112.4	96	72.2	
	Anisotropy		0.8	0.9	0.92	
Oxygen flow (sccm)		9	11	13	15	$T=-95$ $SF_6=129$ ICP = 100 Plate = 2 Pressure = 1.5 Time = 20
	Etch rate (nm/min)		21.3	26.2	27.5	
	Anisotropy		0.99	0.98	0.97	
Plate power (W)		12	15	20	24	$T=-95$ $SF_6=129$ $O_2=15$ ICP = 100 Pressure = 1.5 Time = 20 min
	Etch rate	51.7	58	58.8	70	
	Anisotropy	0.98	0.96	0.95	0.97	

of O_2 −15 sccm in their tests. Smyrnakis et al. did a similar study on cryogenic etching of Si nanowires patterned using nanosphere lithography (Smyrnakis et al., 2015). They see an opposite trend of reduction in etch rate with increased O_2 low but with a higher O_2 flow percentage starting from 12% to 26% of total flow rate. Although their process conditions are very different, we can still draw a general inference that over passivation by excess O_2 flow inhibits the F-based etching of Si at high O_2 percentages which is not evident from the work of Merzsch et al. due to insufficient data.

Lastly, Merzsch et al. study the influence of HF plate power to improve both etch anisotropy and etch rate. Increasing the etch power from 12 to 24 W gradually increases the etch rate due to faster depassivation of the bottom surface and ion bombardment-assisted etching. The anisotropy is relatively unaffected and remains at a high value ranging between 0.95 and 0.98 for all the plate powers tested. Smyrnakis et al. show a similar cryogenic etch behavior. They push the etching recipe from low to high bias regime by increasing the bias by ~66%. They see a shift of their nominally tapered nanowires with etch rates of ~1.8 μm/min which increase to ~2.6 μm/min with no signs of tapering and an almost zero undercut. Interestingly, the base of their nanowires show pyramidal structures which can be attributed to the crystal orientation-dependent etching at cryogenic temperatures. This has been reported in Craciun et al. (2002) and Zijlstra et al. (1999), where (111) planes appear in the etching as the etch rate of (111) planes is slower than (100).

The etching selectivity with etching mask is much higher in CRIE compared to the RTRIE process but less than Bosch process. Indeed, the overall process conditions play a bigger role in selectivity. Merzsch et al. see a maximum selectivity of 10, whereas Smyrnakis et al. measure much higher selectivities of 85 for AZ photoresist, 45 for PMMA, 140 for polystyrene, and over 300 for SiO_2 for their cryogenic process. Fig. 15 shows

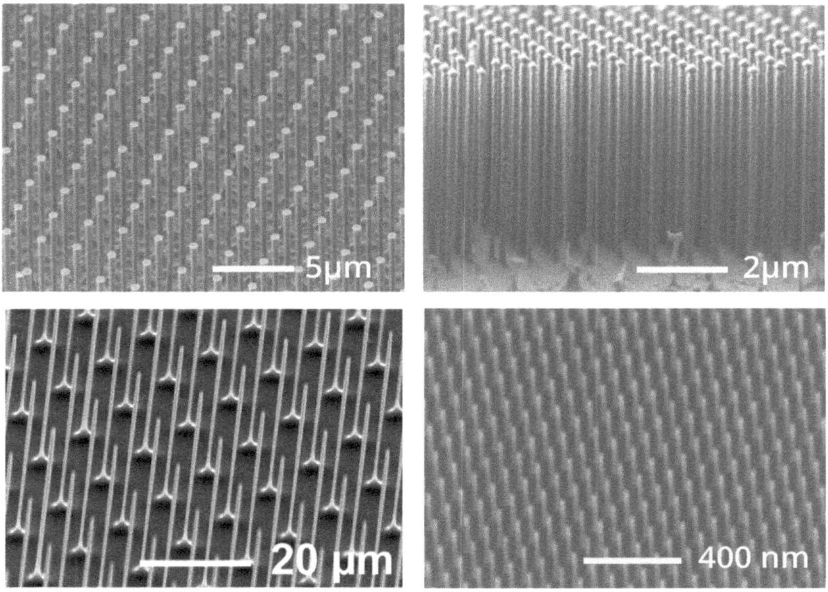

Fig. 15 State-of-the-art silicon nanowire etched by CRIE process in literature. *Reproduced with permission from Li, Z., Chen, Y., Zhu, X., Zheng, M., Dong, F., Chen, P., Xu, L., Chu, W., Duan, H., 2016. Fabrication of single-crystal silicon nanotubes with sub-10 nm walls using cryogenic inductively coupled plasma reactive ion etching. Nanotechnology 27, 365302. https://doi.org/10.1088/0957-4484/27/36/365302; Merzsch, S., Steib, F., Wasisto, H.S., Stranz, A., Hinze, P., Weimann, T., Peiner, E., Waag, A., 2014. Production of vertical nanowire resonators by cryogenic-ICP–DRIE. Microsyst. Technol. 20, 759–767. https://doi.org/10.1007/s00542-013-2032-4; Smyrnakis, A., Almpanis, E., Constantoudis, V., Papanikolaou, N., Gogolides, E., 2015. Optical properties of high aspect ratio plasma etched silicon nanowires: fabrication-induced variability dramatically reduces reflectance. Nanotechnology 26, 85301. https://doi.org/10.1088/0957-4484/26/8/085301.*

some high aspect ratio nanowires etched by cryogenic process. The top row shows nanowire etched by Smyrnakis et al. by e-beam lithography on the top right and nanosphere lithography on the top left. The nanosphere lithography shows higher aspect ratio of 26 compared to 20 for e-beam lithography due to different mask selectivities. The bottom left shows nanowires etched using the optimized process described in Table 7 by Merzsch et al. The bottom right figure shows high-density ultrafine nanowires on a 100 nm pitch fabricated by Li et al. In their excellent work, they push the boundaries of dry etching by their CRIE of nanotubes and ridges with sub-10 nm width. Fabrication of such high-density features <10 nm is enormously difficult with RTRIE and even more so with the Bosch process as the fluorocarbon passivation layer itself can be >10 nm. A literature summary of different nanowire cryoetching conditions is provided in Table 8 that can be used as a guide for readers interested in etching high-density nanowires with clean sidewalls and minimal damage.

2.5.3 Time-Multiplexed RIE for Si Nanowires

The time-multiplexed RIE process was first invented by Laemer and Schilip and patented by their employer Robert Bosch GmbH (Laermer and Schilp, 1994). Thus it is also called as Bosch process. The process consists of a gas switching recipe that utilizes alternating steps of etching gas and passivation gas in repeated cycles. The process is called time-multiplexed RIE due to the temporal separation between the etching and passivation steps. The etching and passivation gases are SF_6 and C_4F_8. The switching process requires a modified RIE tool with fast-acting mass flow controllers to allow for switching speed in seconds. An ICP source is typically used and the ICP power, plate power, and occasionally pressure are switched to appropriate values during the gas switching. The tools are also called deep RIE (DRIE) or Bosch etchers.

The process is ingenious as the temporal separation of etching and passivation subunits provides better control over the etching profile and passivation layer thickness. The result is a higher degree of anisotropy compared to other RIE processes. A drawback of the sequential etching and deposition steps is that it gives rise to undercuts during each etching step and shows a scalloped surface. The phenomenon is aptly called scalloping. It occurs as there is no passivation gas during the etching step which causes an isotropic etching and leads to an undercut. This is shown in Fig. 16. The scalloping makes the Bosch process unsuitable for etching nanowires for applications which require smooth surfaces. The scalloping usually results in surface roughness >100 nm and limits the achievable NW diameter to >200 nm

Table 8 Literature Review on the CRIE Processes for Etching Nanowires

References	RF Power/Plate Power (W)/Bias(V)	Gases	Gas Flow (sccm)	Pressure and Temperature (mTorr/°C)	Mask
Henry et al. (2009)	900/3–9	SF_6/O_2	70/7.2	10/−140	Al_2O_3
	5 μm diameter, 83 μm long micropillars etched. Selectivity = 755 and aspect ratio = 25 with sidewall angle of 88–91 degree				
Stranz et al. (2012)	—	SF_6/O_2	—	11.3/−75	Resist
	Etching rate = 2–4 μm/min, aspect ratio = 20, and diameter > 200 nm				
Sökmen et al. (2010)	500/−12 V	SF_6/O_2	159/9	11.3/−75	Resist
	Etch rate = 2.8–4 μm/min. Studied control of tapering angle				
Sökmen et al. (2009)	500/6	SF_6/O_2	129/9	11.3/−75	Resist
	Etch rate = 2.4–4 μm/min, etch depth = 7.3 μm, diameter (the tip) = 200 nm, and pitch = 2 μm				
Henry et al. (2010)	900/10 W/54 V	SF_6/C_4F_8	70/6.5	10/−130	Ga^+ implantation
	CRIE nanowires with Ga ion implantation				
Liu et al. (2013)	1000/2	SF_6/O_2	34/16	6/−120	Resist
	40 nm feature size and selectivity of 10:1. ARDE study with CRIE				
Tada (1998)	250/30 V	SF_6	3	0.075/−135	Au, Cu, Cr
	ECR-RIE used. Au, Ag, and Cu used as masks. Diameter = 20 nm and height = 280 nm				

Continued

Table 8 Literature Review on the CRIE Processes for Etching Nanowires—cont'd

References	RF Power/Plate Power (W)/Bias(V)	Gases	Gas Flow (sccm)	Pressure and Temperature (mTorr/°C)	Mask
Ellinas et al. (2011)	800/−30 V	SF_6/O_2	200/85	10/−105	Polystyrene
	Aspect ratio = 18, diameter = 200 nm, and selectivity = 25				
Hamdana et al. (2017)	140/8	SF_6/O_2	129/15	−95	Polystyrene
	Aspect ratio = 8, diameter = 146 nm, and pitch = 500 nm				
Smyrnakis et al. (2015)	1200/30–55 V	SF_6/O_2	200/30–70	10/−105	Polystyrene
	Aspect ratio = 37, diameter < 200 nm, and etch rate > 3 μm/min. They study the effect of O_2 flow and bias on etch rate and obtain clean and vertical sidewalls				
Merzsch et al. (2014)	100–140/12–24	SF_6/O_2	129/9–15	11.3/ −75 to −95	Resist
	Studied influence of O_2 flow, temperature, and HF on nanowire profile. Etch rate = 21–114 nm/min, anisotropy = 0.95–0.99, diameter = 130 nm, height = 4.8 μm, and aspect ratio > 60				
Li et al. (2016)	350/15	SF_6/O_2	30/20	7.5/−110	HSQ Resist
	Ultrahigh aspect ratio etching. Diameter < 20 nm, pitch = 100 nm, and height = 390 nm				

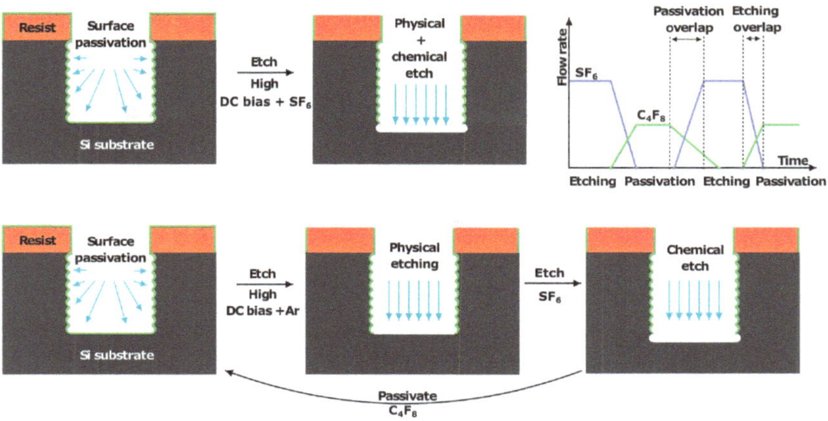

Fig. 16 Schematic of a two-step (*top*) and three-step Bosch etching process (*bottom*).

Table 9 Comparison of Etching Results Between Bosch and RTRIE Process

Dry Etching	Time (min)	$C_4F_8/SF_6/Ar$ (sccm)	Pressure (mTorr)	ICP/Bias (W)	Height (μm)	Scalloping (nm)	Selectivity Over Resist
Bosch process	2	0/100/40 70/0/40	23	825/13	4.6	265.5	88.46
RTRIE process	10	52/28/20	19	850/9	3.15	0	16.52

Adapted with permission from Hung, Y.-J., Lee, S.-L., Thibeault, B.J., Coldren, L.A., 2011. Fabrication of highly ordered silicon nanowire arrays with controllable sidewall profiles for achieving low-surface reflection. IEEE J. Sel. Top. Quantum Electron. 17, 869–877. https://doi.org/10.1109/JSTQE.2010.2068540.

in a standard Bosch process. Table 9 shows the etch comparison between RIE and Bosch process by Hung et al. For process with similar conditions, they observe a 265.5-nm scallop width for the Bosch process compared to zero scalloping with RIE process. As a result, reducing the scalloping width is an important prerequisite for generating nanowires with smooth sidewalls. Since the scalloping width is dictated by the cycle time, a DRIE tool with a shorter cycle and higher switching speed is desirable for nanowire etching. Nonetheless, there are numerous scientific reports of innovative processes that have been designed to alleviate the scalloping issue, such as nanowire oxidation and oxide removal.

The typical Bosch process is shown in the top row of Fig. 16. The top right shows the SF_6 and C_4F_8 flow rates during different etching and passivation cycles. Owing to the need for fast switching, there is no delay between the start of etching gas flow and the end of passivation gas flow which causes the cycles to slightly overlap giving a gradual change at the interface. This is not a cause for concern in Bosch process. An even more segmented process is shown in the bottom row of Fig. 16. Here the etching process is divided into three steps of passivation, physical etching, and chemical etching. Such a process is advantageous due to separate control of physical etching. The physical etching rate can be fine-tuned to adjust the amount needed to remove the passivation layer from the bottom surface, leading to a lower mask etching rate, and a substantial improvement in selectivity.

Since we have already discussed the most common challenges associated with the RIE process in previous sections, we will limit ourselves to issues specific to the Bosch process namely—scalloping. The Bosch process in general has a very high mask selectivity with values over 100:1 for resist mask and greater than 1000 for SiO_2 or A_2O_3 masks. This provides a unique opportunity for nanowires etching where high selectivity is required for etching long nanowires. Although extremely high aspect ratios are possible with the Bosch process, the minimum feature size is severely limited by scalloping. The schematic on the top right of Fig. 17 highlights the influence of scallop width during the etching cycle on the morphology of the nanowire finally obtained. The scalloping can be routinely few 100 nm, which limits the diameter to around 100 nm and results in extremely rough sidewalls. Zeniou et al. studied the scallop depth for a process with standard DRIE process with a cycle time of 7 s:3 s for SF_6:C_4F_8 and observed a scallop depth of 500 nm (Zeniou et al., 2014). Thus, optimization must be made to standard Bosch process such as reducing the etching and deposition time to reduce the scalloping width. Zeniou et al. subsequently reduced the timing to 5 s:2 s and observed that the scallop depth reduced to 150 nm. They could etch nanowires, albeit with large roughness as shown in the bottom half of Fig. 17. Interestingly, they found that DRIE initialization with passivation (bottom right) is crucial for obtaining the patterned diameter nanowires. Starting the DRIE with an etching cycle (bottom left) first caused an initial undercutting of the mask, which gives rise to a final diameter much smaller than the patterned diameter. The top right image in Fig. 17 shows an optimized DRIE process by Nakamura et al., where they observe a scalloping of ~20 nm on a 100 nm nanowire. Another optimal DRIE etch recipe can also be obtained from the work by Morton et al., where in addition to reducing the cycle time (7 s:5 s) they also reduce the aggressiveness of SF_6 etch by having ~16% O_2 and ~5% Ar in the etching gas mixture (Morton et al., 2008). The resulting process, a combination of RTRIE and Bosch process,

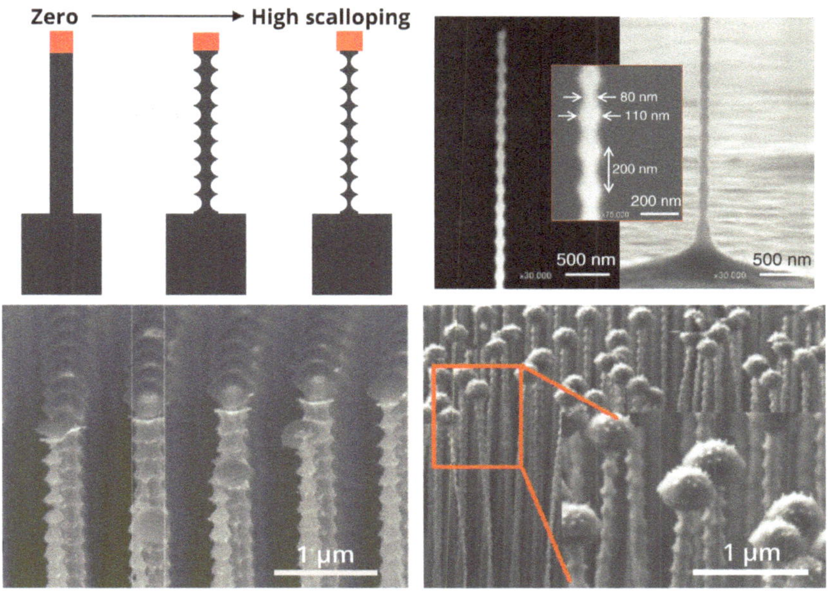

Fig. 17 Scalloping in DRIE process and its influence on nanowire roughness. *Reproduced with permission from Nakamura, J., Higuchi, K., Maenaka, K., 2013. Vertical Si nanowire with ultra-high-aspect-ratio by combined top-down processing technique. Microsyst. Technol. 19, 433–438. https://doi.org/10.1007/s00542-012-1662-2; Zeniou, A., Ellinas, K., Olziersky, A., Gogolides, E., 2014. Ultra-high aspect ratio Si nanowires fabricated with plasma etching: plasma processing, mechanical stability analysis against adhesion and capillary forces and oleophobicity. Nanotechnology 25, 35302. https://doi.org/10.1088/0957-4484/25/3/035302.*

leads to reduced scalloping with surface roughness of ~10 nm. They achieved nanowires with aspect ratios as high as 48 with nanowire diameter of 50 nm on a pitch of 235 nm.

An alternative strategy for achieving ultrahigh aspect ratio nanowires with small diameters is by narrowing the nanowire diameter by postetch oxidation. Initially, a large diameter nanowire greater than 500 nm is etched by deep Bosch process. The nanowires are then oxidized by thermal oxidation which reduces the nanowire diameter. The final silicon core diameter can then be precisely controlled as the oxide growth kinetics on silicon is very well studied. The oxide shell surrounding the nanowire can then be removed in dilute hydrofluoric acid solution. Nakamura et al. were successfully able to obtain ultrahigh aspect ratio ~100 nanowires fabricated by this process (Nakamura et al., 2013). The schematic in the top row of Fig. 18 summarizes the fabrication process. The initial nanowires obtained by

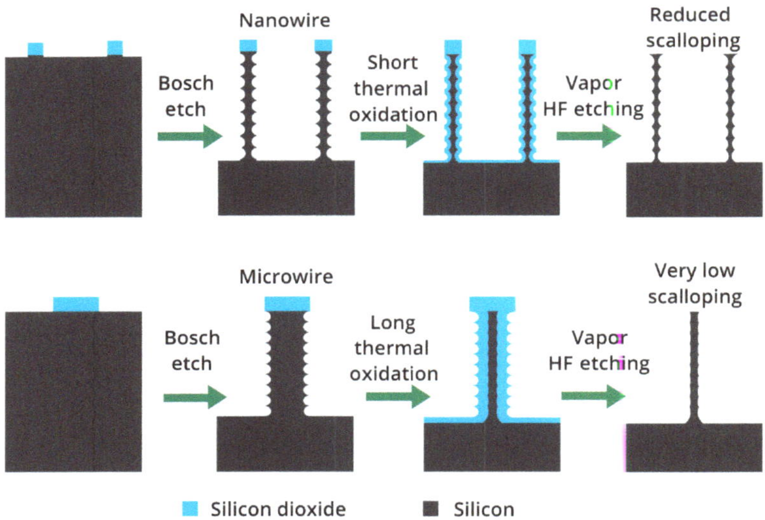

Fig. 18 Fabrication scheme for reducing scalloping in Bosch etched nanowires by oxidation and HF etching.

Nakamura et al. are 11 μm long with a diameter of 550 nm. They subsequently performed a wet oxidation step for 15 min at 1300 K. The oxide growth radially consumed 220 nm of Si, resulting in a smaller final nanowire diameter of ~110 nm.

In an isotropic etching process like oxidation, the sharp features provide a larger surface area for reactions that are the first to be smoothed out. A similar influence of reduction scallop width is seen during oxidation. The result is a much smaller scallop width of 20 nm in oxidized nanowire compared to original nanowires as observed by Nakamura et al. Following the same logic, a thicker oxidation can result in lesser scalloping as highlighted in the schematic in Fig. 18 (bottom row). Nakamura et al. tested this hypothesis with a Bosch etched microwire instead of nanowires and increased the oxidation time to 14 h at 1300 K to allow for growth of thicker oxide. After oxide removal they obtained nanowires with diameters varying from 160 to 260 nm. The variation in diameter seen along the nanowire is a result of the tapering in the original nanowire after Bosch etch. The final nanowire showed aspect ratio of 100 and observed a completely scallop free surface. This fabrication concept is not only limited to Bosch process but can also be applied to cryogenic or pseudo-Bosch nanowires as an easy means to increase the anisotropy. It should be noted that the earlier process of oxidation and etching imposes a restraint on the smallest pitch, which is now limited to the diameter of oxidized nanowire.

Nevertheless, Bosch process is an attractive method since it offers aspect ratios >100 which can be greatly increased by oxidation. Thermal oxidation when combined with Bosch etching presents a viable route to achieving ultrahigh aspect ratio nanowire which were only possible by bottom-up processes before.

Fig. 19 shows the state-of-the-art vertically aligned Si nanowire array etched by Bosch process. The top left image shows the oxidation-assisted extreme aspect ratio achieved by Nakamura et al. Their results show that with multiple oxidation and oxide etching steps, it is possible to further

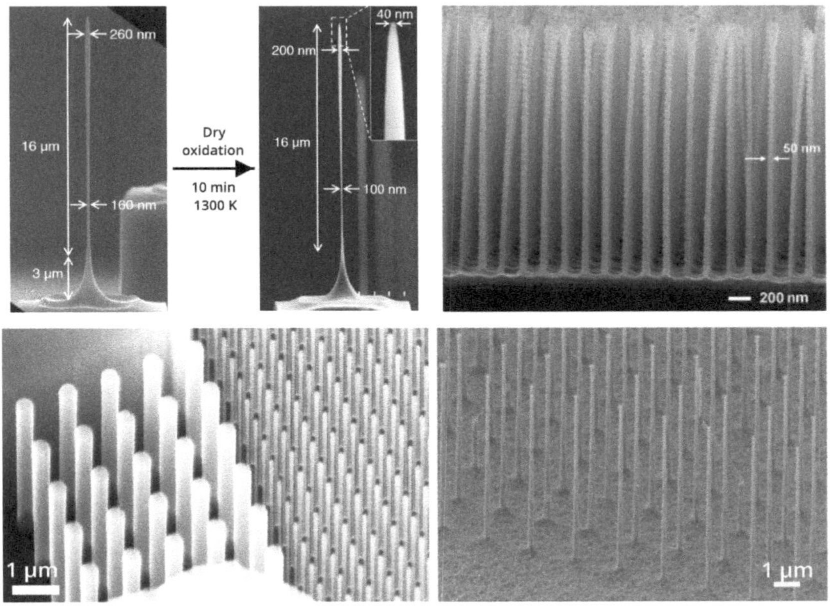

Fig. 19 State-of-the-art silicon nanowire etched by Bosch process in literature. *Reproduced with permission from Engstrom, D.S., Soh, Y.-A., 2013. Controlling the silicon nanowire tapering angle in dense arrays of silicon nanowires using deep reactive ion etching. J. Vac. Sci. Technol. B Nanotechnol. Microelectron. Mater. Process. Meas. Phenom. 31, 21806. https://doi.org/10.1116/1.4793500; Morton, K.J., Nieberg, G., Bai, S., Chou, S.Y., 2008. Wafer-scale patterning of sub-40 nm diameter and high aspect ratio (>50:1) silicon pillar arrays by nanoimprint and etching. Nanotechnology 19, 345301. https://doi.org/10.1088/0957-4484/19/34/345301; Nakamura, J., Higuchi, K., Maenaka, K., 2013. Vertical Si nanowire with ultra-high-aspect-ratio by combined top-down processing technique. Microsyst. Technol. 19, 433–438. https://doi.org/10.1007/s00542-012-1662-2; Zeniou, A., Ellinas, K., Olziersky, A., Gogolides, E., 2014. Ultra-high aspect ratio Si nanowires fabricated with plasma etching: plasma processing, mechanical stability analysis against adhesion and capillary forces and oleophobicity. Nanotechnology 25, 35302. https://doi.org/10.1088/0957-4484/25/3/035302.*

reduce the nanowire dimensions and obtain a scallop free surface with 100 nm diameter. The top right image shows high aspect ratio nanowires with an optimized Bosch process by Morton et al., which allows for etching of 50 nm nanowires with reduced scalloping of just 15 nm. The bottom left image shows 30:1 aspect ratio nanowires by Engstorm et al. where they optimize the process to control the tapering angle in dense arrays and obtain sidewall angles close to 90 degree. Lastly, the bottom right image shows the 50:1 aspect ratio nanowires optimized by Zeniou et al on a sparse pitch of 2 μm (Engstrom and Soh, 2013; Morton et al., 2008; Nakamura et al., 2013; Zeniou et al., 2014).

To conclude this section, we provide a summary of various nanowire etching recipes with Bosch process from literature in Table 10.

3. NANOWIRES BY MacEtch

3.1 Etching Mechanism and Chemical Reactions

First, we will focus on the MacEtch of Si to discuss the basic phenomenon and mechanism of the process as it has been intensively studied with Si. A microscale electrochemical cell consisting of a metal catalyst, Si substrate, and etchant solution is designed as shown in Fig. 20A. For simplicity, the catalyst in this section is an isolated metal particle. The influence of metal materials and morphology, etchant, and the crystalline and electrical properties of Si substrate will be discussed in the following sections. According to the localized galvanic cell model described in Peng et al. (2006b) and Zhang et al. (2008), the MacEtch mechanism involves a localized microscopic electrochemical process at the metal catalyst and the Si interface. The possible chemical reactions during the etching process are listed as the following:

Reaction at metal (cathode)

$$H_2O_2 + 2H^+ \rightarrow 2H_2O + 2h^+ \quad (27)$$

Reaction of Si etching (anode)

$$Si + 2H_2O + 4h^+ \rightarrow SiO_2 + 4H^+ \quad (28)$$

$$SiO_2 + 6HF \rightarrow [SiF_6]^{2-} + 2H_2O + 2H^+ \quad (29)$$

Overall reaction:

$$Si + 2H_2O_2 + 6F^- + 4H^+ \rightarrow [SiF_6]^{2-} + 4H_2O \quad (30)$$

In the MacEtch reaction, a critical feature is the generation of a hole (i.e., positive charge carrier) from the oxidant (e.g., H_2O_2) since charge

Table 10 Literature Review on the Nanowire Etching With Bosch Process

References	RF Power/Plate Power (W)	Etch/Passivation Gases	Etch/Passivation Gas Flow (sccm)	Etch/Passivation Timings (s)	Etch/Passivation Pressure (mTorr)
Logeeswaran et al. (2010)	1800/20	SF_6/C_4F_8	300/150	6/3	
	Etching process for micropillars. Diameter = 1.4–5 μm and height = 20–25 μm				
Chang et al. (2005)	600/15–0	SF_6/C_4F_8	130/80	9/8	33/18–19
	Optimized for etching >2 μm features. Sidewall roughness = 185–245 nm				
	600/50–0	SF_6+O_2/C_4F_8	20 + 15/20	4.4/4	1.7/8.9
	Optimized for etching antireflection surfaces. Sidewall roughness = 30–55 nm				
	350/11–0	SF_6+O_2/C_4F_8	50 + 5/80	3/3	13/18
	Optimized for etching submicron features. Sidewall roughness < 10 nm				
Fu et al. (2009)	500–800/50	$SF_6/C_4F_8/O_2$	250/200/100	1/1/0.5	
	Microwires to nanowires etched. Aspect ratio < 60. Nanowires are inverted tapered				
Qian et al. (2008)	800/10	SF_6+O_2/C_4F_8	30 + 13/100	8/5	15
	Ga ion lithography. Diameter = 1 μm, height = 810 nm, and etch rate = 1.85 μm/min				
Bui et al. (2015)	500–100/50–200 V	SF_6/C_4F_8	100/150	5–10/10	
	Investigation on influence of RF power and bias voltage on nanowire morphology				

Continued

Table 10 Literature Review on the Nanowire Etching With Bosch Process—cont'd

References	RF Power/Plate Power (W)	Etch/Passivation Gases	Etch/Passivation Gas Flow (sccm)	Etch/Passivation Timings (s)	Etch/Passivation Pressure (mTorr)
Miller et al. (2013)	400–800/6–14	SF_6/C_4F_8	130/13	5.5–10/4–7	
	Investigation on reduction of scalloping in nanowire etching				
Kang et al. (2008)	600/0–12	$SF_6 + O_2/C_4F_8$	130 + 13/100	7/6	
	Process optimized for fabrication of nanotips				
Morton et al. (2008)	600	$SF_6 + O_2 +$ Ar/C_4F_8	30 + 6 + 2/85	7/5	20
	Cycle time ramped by 0.1 s in each cycle. Aspect ratio > 60 and diameter = 50 nm				
Zeniou et al. (2014)	1800/70 V	SF_6/C_4F_8	171/110	7/3–1	37
	Etch rate = 2.8–3.6 μm/min, scalloping < 30 nm, aspect ratio > 100, and diameter = 200 nm				
Engstrom and Soh (2013)	600–800/12–100	$SF_6 + O_2/C_4F_8$	(130–15) + (6–3)/85	6/3.5–4	9–15/15
	Optimization for smooth and vertical sidewalls for dense nanowire arrays. Aspect ratio > 30 and diameter = 90 nm				

The spaces are left blank were parameter value was no available.

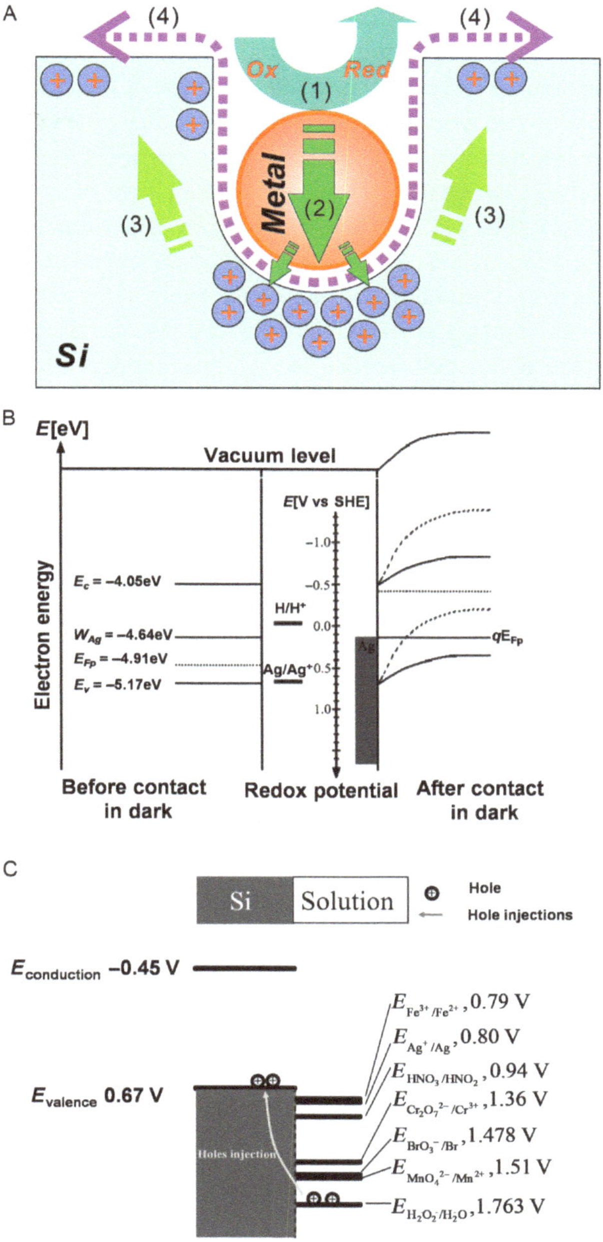

Fig. 20 (A) Schematic of MacEtch process: (1) reduction of oxidant, (2) hole injection into Si, (3) hole migration to surface, and (4) removal of oxidized Si by HF. (B) Energy band diagrams for p-type Si in HF/$AgNO_3$ solution: before contact (*left*) and after contact

(Continued)

transfer is necessary for the oxidation and dissolution of Si. When a Si substrate is brought into contact with a metal catalyst (e.g., Ag), band bending occurs on the Si surface, resulting from equilibration of the Fermi energy level in the Si side with the work function of the metal catalyst (Fig. 20B). The net result of such equilibration is the formation of an energy barrier (i.e., Schottky barrier) that inhibits charge transfer at the conduction band of Si. Since the redox potential of H_2O_2/H_2O (1.76 V, cathode reaction 27) is much more positive than the valence band of Si as shown in Fig. 20C, holes injected into the valence band (i.e., electrons released from the valence band) of Si will overcome the Schottky barrier, causing it to oxidize and dissolve the Si in contact with HF–H_2O_2 solution (anode reactions 28 and 29). Etching of Si by HF–H_2O_2 alone does occur, but the etch rate is lower than 10 nm/min when the etchant has a concentration of H_2O_2 that is much higher than that used in MacEtch (Chartier et al., 2008; Lehmann, 2002). In practice, the presence of metal particles increases the reduction of H_2O_2 (hole injection) due to the catalytic activity and electrostatic force. This dramatically enhances the etch rate at the Si underneath the metal particles. By preferentially catalyzing these reactions near metal particles, Si beneath the metal catalyst is dissolved. In the case of the isolated metal particle, a Si hole is formed as shown in Fig. 20A. If the catalyst is a metal film with a hole (rather than a metal particle), the Si surface covered by the metal film would be etched, resulting in the formation of Si nanowires. Therefore, the structure of the Si etched by MacEtch strongly depends on the morphology of the metal catalyst.

3.2 Effect of Metal Catalyst

As discussed earlier, the morphologies of the etched structures are normally defined by the morphological structure of the metal catalyst thanks to the

Fig. 20—Cont'd (*right*). (C) Potential relationship between Si band and standard potentials of oxidants. *Panel A: Reproduced with permission from Liu, X., Coxon, P.R., Peters, M., Hoex, B., Cole, J.M., Fray, D.J., 2014. Black silicon: fabrication methods, properties and solar energy applications. Energy Environ. Sci. 7, 3223–3263. https://doi.org/10.1039/C4EE01152J. Copyright 2014, The Royal Society of Chemistry; Panel B: Reproduced with permission from Peng, K., Fang, H., Hu, J., Wu, Y., Zhu, J., Yan, Y., Lee, S.T., 2006. Metal-particle-induced, highly localized site-specific etching of Si and formation of single-crystalline Si nanowires in aqueous fluoride solution. Chem. A Eur. J. 12, 7942–7947. https://doi.org/10.1002/chem.200600032. Copyright 2006, Wiley-VCH; Panel C: Reproduced with permission from Huang, Z., Geyer, N., Werner, P., de Boor, J., Gösele, U., 2011. Metal-assisted chemical etching of silicon: a review. Adv. Mater. 23, 285–308. https://doi.org/10.1002/adma.201001784. Copyright 2011, Wiley-VCH.*

selective etching of the semiconductor materials under the metal catalyst. The schematics of possible etched structures corresponding to the structures of the metal catalysts are shown in Fig. 21. The perfectly isolated metal particles (Fig. 21A) would produce hole structures (Fig. 21B) because the isolated metal particles are not able to interact with each other. Once the density of the metal particles significantly increases, the metal particles would cover most of the substrate surface, resulting in the formation of a network of metal particles (Fig. 21C and E). As the morphology of the metal catalyst processes from the isolated particles to a particle network, the shape of the etched structure changes from holes to a wall-like (Fig. 21D) or wire-like structure (Fig. 21F). If the metal catalyst has film structures with small holes as shown in Fig. 21G, the MacEtch will form the isolated nanowires (Fig. 21H). If the demand is for the periodic nanowires, periodically patterned metal catalyst can be used as shown in Fig. 21I and J. The formation of the periodic nanowires will be discussed in more detail later.

In MacEtch, noble metal catalysts such as Ag and Au have been mainly used (Chartier et al., 2008; Choi et al., 2017; Yae et al., 2012; Zhang et al., 2008). Besides these metals, others such as platinum (Li et al., 2013), copper (Lee et al., 2011a), palladium (Cichoszewski et al., 2013), and rhodium (Yae et al., 2012), have also been used in MacEtch. The deposition methods for metal catalysts are also very diverse. Among the various deposition methods, electroless galvanic deposition is one of the simple methods to form random metal particles, with which MacEtch will produce the wall-like or wire-like structures of random sizes. These metal particles are formed via electroless

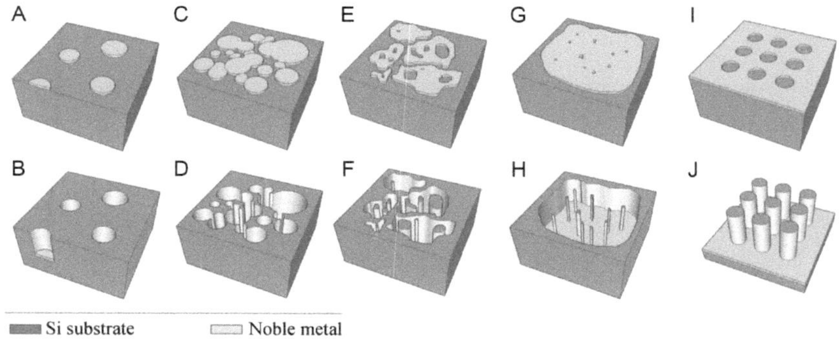

Fig. 21 Schematics of morphologies of etched structures depending on various shapes of metal catalysts. *Adapted with permission from Huang, Z., Geyer, N., Werner, P., de Boor, J., Gösele, U., 2011. Metal-assisted chemical etching of silicon: a review. Adv. Mater. 23, 285–308. https://doi.org/10.1002/adma.201001784. Copyright 2011, Wiley-VCH.*

galvanic displacement by injecting hole carriers from metal catalyst into the valance band of the Si. This results in the reduction of the metal ions as well as the oxidation of Si. In this section, we will focus on the electroless galvanic displacement solution with HF, enabling etching of the oxidized Si. In other words, the deposition of metal particles occurs with the MacEtch of Si, simultaneously. Experimentally, the Si substrate is dipped into a solution containing HF and metal salts such as $AgNO_3$, $KAuCl_4$, $HAuCl_4$, H_2PtCl_6, K_2PtCl_4, or $RhCl_3$. The metal ions inject the holes into Si spontaneously, resulting in the nucleation of the metal particles (Fig. 22). For example, the Fermi level of Si is lower than the electrochemical potential of Ag/Ag^+ for $HF/AgNO^3$, enabling spontaneous hole injection into the Si. While this occurs, the Si underneath the nucleated metal particles is locally oxidized by the hole injection and finally etched by HF. As a result, the metal particles sink into the Si as shown in Fig. 23A. Because the charge transfer of the hole injection is preferentially generated at the interface between the metal catalyst and the Si, the metal ions

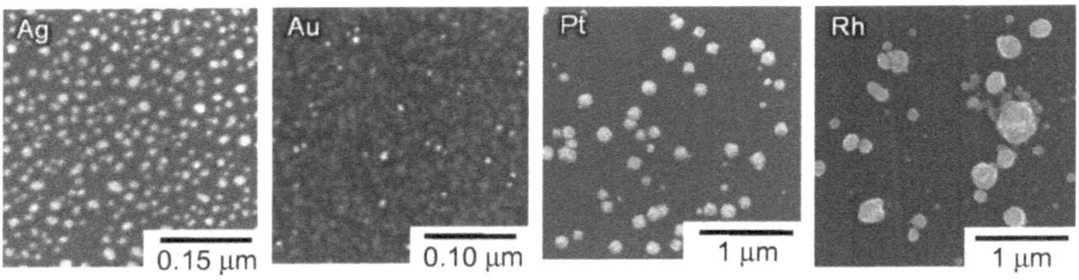

Fig. 22 SEM images of metal nanoparticles deposited via electroless galvanic displacement method. *Reproduced with permission from Yae, S., Morii, Y., Fukumuro, N., Matsuda, H., 2012. Catalytic activity of noble metals for metal-assisted chemical etching of silicon. Nanoscale Res. Lett. 7, 352. https://doi.org/10.1186/1556-276X-7-352. Copyright 2012, Springer.*

Fig. 23 Cross-sectional SEM images of etched Si structures in $HF/AgNO_3$ for (A) 5 min and (B and C) 30 min. *Reproduced with permission from Peng, K., Fang, H., Hu, J., Wu, Y., Zhu, J., Yan, Y., Lee, S.T., 2006. Metal-particle-induced, highly localized site-specific etching of Si and formation of single-crystalline Si nanowires in aqueous fluoride solution. Chem. A Eur. J. 12, 7942–7947. https://doi.org/10.1002/chem.200600032. Copyright 2006, Wiley-VCH.*

would be reduced onto predeposited metal particles at the bottom of the etched structure rather than the sidewall of the Si structures. With increasing etch time, the continuous metal deposition produces metal dendrites on the top of the Si structures (Fig. 23B). Owing to the highly dense metal particles, MacEtch using HF and metal salts etches the wall-like or the wire-like structures (Fig. 23C). Compared to Ag and Au salts, platinum and copper salts show different deposition phenomenon. A solution containing HF and platinum or copper salt forms a continuous film rather than dendrites. This continuous Pt or Cu film prevents the access of the HF solution to the oxidized substrate. This renders Pt and Cu catalysts unsuitable for MacEtch. In common practice, electroless galvanic displacement depositions of platinum and copper are performed to form continuous metal films, not to etch the semiconductor substrate.

If there are strict requirements on nanowire diameter and spacing, the metal catalyst should be patterned. As we saw, electroless galvanic displacement produces an undesirable random particle network. Thus, for continuous and thin metal films, vacuum-based metal depositions are used rather than the solution processes. Among the vacuum processes, PVD methods such as thermal evaporation, e-beam evaporation, and sputter methods are widely used to deposit the patterned metal catalyst. Because the PVD methods deposit a metal film onto the whole surface of the substrate, the size and position of the metal holes which determine the diameter and the position of the nanowires, should be defined before depositing the metal film. Lithography techniques such as photo-, e-beam, and laser interference lithography are commonly used to define the size and position of the metal hole arrays. First, the substrate is spin coated with a resist film (Fig. 24). The cured resist layer is selectively exposed by ultraviolet light, e-beam, or laser interference. The unexposed resist was then removed by developer chemistry, with the unexposed resist remaining on the substrate. After forming the resist patterns, a thin metal film is deposited via the PVD method onto the substrate. The patterned metal film then gives rise to patterned nanowire arrays during MacEtch. The size and position of the nanowires can be easily controlled through the design of the resist mask for lithography. For photolithography, however, nanoscale dimensions, smaller than few 100 nm, is a big challenge. Although e-beam lithography has the capability to produce a nanoscale pattern, the throughput is extremely low because of its direct writing method, limiting the fabrication of the large-scale nanowire arrays.

An alternative for large-scale nanofabrication, nanosphere lithography has been used. Huang et al. (2007) demonstrated ordered Si nanowires using

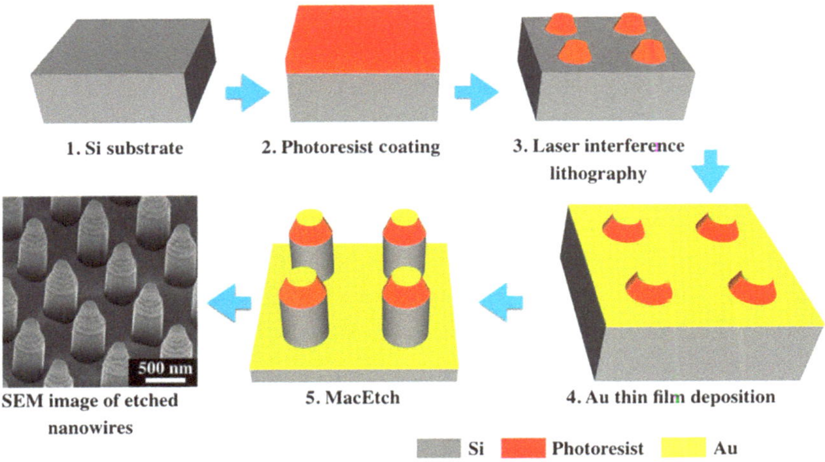

Fig. 24 Schematic showing the process steps in the fabrication of Si nanowire using laser interference lithography and MacEtch. *Reproduced with permission from Kong, L., Dasgupta, B., Ren, Y., Mohseni, P.K., Hong, M., Li, X., Chim, W.K., Chiam, S.Y., 2016. Evidences for redox reaction driven charge transfer and mass transport in metal-assisted chemical etching of silicon. Sci. Rep. 6, 36582. https://doi.org/10.1038/srep36582. Copyright 2016, Springer Nature.*

a self-assembled monolayer of polystyrene nanospheres (Fig. 25A). Thanks to close-packed nanosphere arrays, the pitch (i.e., the distance, center-to-center, of neighboring nanowires) of the nanowires array was determined by the diameter of the original nanospheres. For the control of the nanowire diameter, the diameter of the polystyrene nanospheres was adjusted with RIE time, transferring close-packed to nonclose-packed arrays. After forming the polystyrene mask arrays, the substrate was coated by the metal film and etched via the MacEtch. The average diameter of etched nanowires matches with the diameter of the remaining polystyrene nanospheres as shown in SEM images of Figs. 25B and 26C. Because the technique of nanosphere lithography has been intensively developed, the monolayer of nanospheres with the diameter in a range from 100 nm to a few micrometers can be easily assembled by Langmuir–Blodgett (Gwinner et al., 2009), using Dip coating (Nagao et al., 2007), or spin coating methods (Chen et al., 2013) on a wafer scale. In addition, with the state-of-the-art control of RIE processes, the diameter of nanospheres can be precisely controlled with RIE etch time. However, if another array pattern (e.g., square or mix patterns) is required, nanosphere lithography is not suitable. Nanosphere lithography typically produces hexagonal nanowire arrays because the nanosphere monolayer starts as a close-packed hexagonal array.

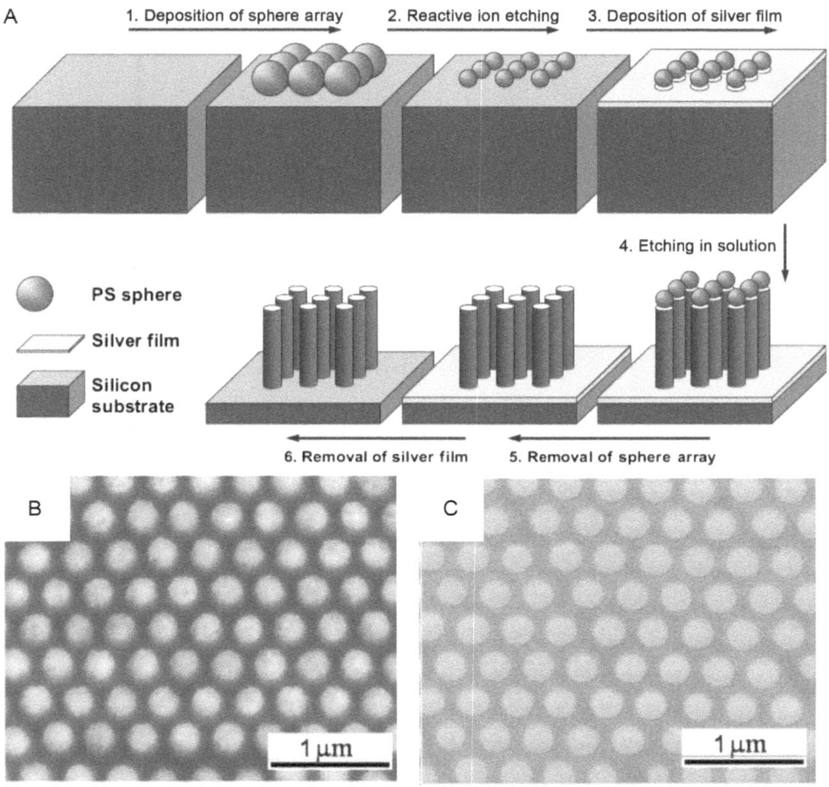

Fig. 25 (A) Schematic showing the process steps in the fabrication of Si nanowire using polystyrene nanosphere lithography and MacEtch. Top view SEM images of (B) etched nanowires and (C) polystyrene nanosphere mask. *Adapted with permission from Huang, Z., Fang, H., Zhu, J., 2007. Fabrication of silicon nanowire arrays with controlled diameter, length, and density. Adv. Mater. 19, 744–748. https://doi.org/10.1002/adma.200600892. Copyright 2007, Wiley-VCH.*

3.3 Mass Transfer of Chemical Reactants

As discussed earlier, redox reactions of MacEtch (i.e., the reduction of the oxidant and the oxidation of the semiconductor) occur at the interface between the metal catalyst and the substrate. In MacEtch, the reactants need to diffuse into the metal/semiconductor interface, and the byproducts should diffuse out to the etchant solution. For this mass transport of reactants and byproducts, two possible models could be considered as shown in Fig. 26A. One of the models is that the Si atoms diffuse through the metal catalyst, and then react with the chemicals at the metal/etchant interface (Model 1, left side in Fig. 26A). Another model is that the reactants and byproducts diffuse

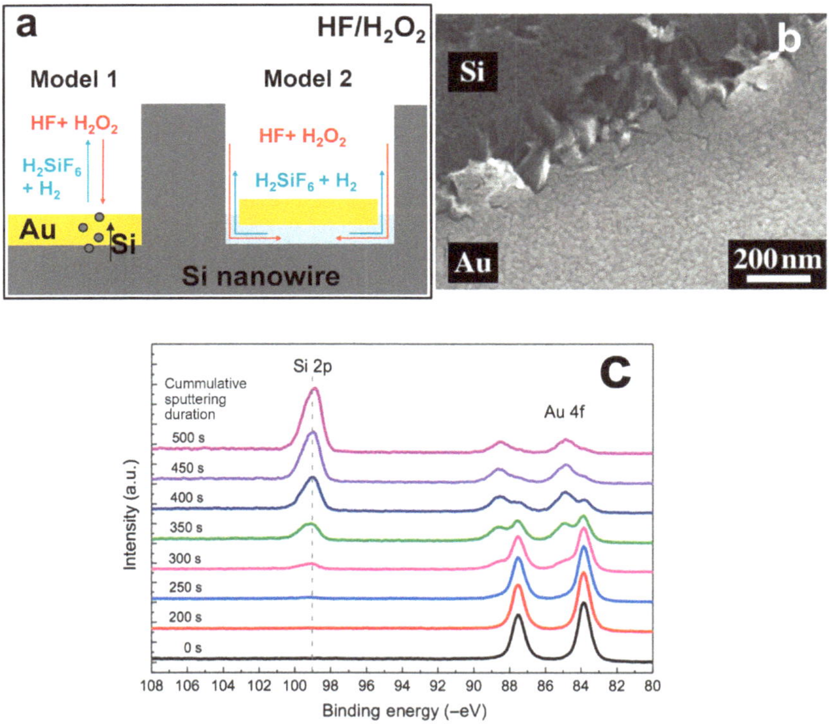

Fig. 26 (A) Schematic showing the possible models of mass transport in MacEtch. (B) Tilted-view SEM images of etched sample. (C) Chemical depth profile for the Au catalyst of etched sample. *Adapted with permission from Kong, L., Dasgupta, B., Ren, Y., Mohseni, P.K., Hong, M., Li, X., Chim, W.K., Chiam, S.Y., 2016. Evidences for redox reaction driven charge transfer and mass transport in metal-assisted chemical etching of silicon. Sci. Rep. 6, 36582. https://doi.org/10.1038/srep36582. Copyright 2016, Springer Nature.*

through the channel formed at the interface between the metal film and the substrate (Model 2, right side in Fig. 26B). Kong et al. analyzed the chemical composition of the metal catalyst for MacEtched samples (Fig. 26B) using X-ray photoelectron spectroscopy (XPS) (Kong et al., 2016). If the Si atom diffuses through the metal catalyst from the Si surface to the metal/etchant interface, elemental Si peak would be detected in the chemical depth profile. As shown in XPS depth profile of Fig. 26C, the response of Si 2p core-level binding energy in a range from ~98 to 104 eV was not observed while sputtering the Au catalyst up to 250 s. Even though the peak at ~99 eV started to show after 300 s of sputtering, however, this Si response came from the Si/Au interface, not the embedded Si atoms into the Au film. Based on this XPS observation, the MacEtch process does not follow Model 1 mechanism.

In contrast, various experiments have been shown to support Model 2. Geyer et al. studied a series of experiments on the etch rates as function of the thickness and the width of metal stripes (Geyer et al., 2012). The etch rate decreased with increasing the width of the metal stripes, indicating that the mass transfer suffered from the long diffusion path and became the rate-determining parameter in MacEtch. In addition, for the identical width of the metal stripes, the etch rate was consistent regardless of the metal thickness. This is also supported by the observation of other experiments that the etch rates of a 112-nm thick Au nanocube was higher than that of a 6.9-nm thick Au nanoprism (Li et al., 2014). Although both the experiments support Model 2, the authors have different views on the transport path of the reactants and byproducts. Geyer et al. claim that the mass transport is conducted along a porous Si which is formed between the metal and the Si substrate. However, Liu et al., who reported the nanocube and nanoprism experiments, explained that the porous Si was not necessary for transporting the chemicals. In practice, longer spacing between nanowires would decrease etch rate of MacEtch due to longer diffusion path of the chemicals according to Model 1.

To overcome long diffusion length issue, Choi et al. used the porous metal films with small holes as the metal catalyst (Choi et al., 2015; Um et al., 2015). By diffusing out the chemicals through the small holes in the metal film, the diffusion length was reduced compared to the continuous metal film. This leads to the significant increase in the etch rate even in the wider spacings.

3.4 Effect of Substrate Properties

In MacEtch, the semiconductor underneath the metal catalyst is selectively oxidized and then etched by HF. Subsequently, the metal catalyst sinks into the etched semiconductor in the vertical direction relative to the substrate surface. However, slanted nanowires have been fabricated along the ⟨100⟩ direction when (110) and (111) substrates were etched (Peng et al., 2005, 2006a). The nonvertical etch in a certain crystallographic direction can be explained by the different backbone strength depending on the crystalline planes of the substrate. This is a well-known concept used to explain the anisotropic etching in wet chemical reactions such as electrochemical etching (i.e., anodic oxidation) of Si (Xia et al., 2001) or Si dissolution in alkaline solution (Lehmann, 1993). To oxidize or dissolve the Si, the bond between the surface atoms and the underneath atoms should be broken. The stronger backbone strength needs more energy to break than the bonds of the surface atoms. Since the backbone strength is determined by the number of the atom on the surface of the crystalline substrate, the backbone strength of (100) surface having two subsurface bonds would be weaker than that of (110) or (111) surfaces with three subsurface bonds. As a result, the atoms on (100) surface are easily removed in comparison to the (110) or (111) surfaces. Therefore, the etch prefers to occur along (100) direction rather than any other direction.

In the literature, the other etch directions such as ⟨110⟩, ⟨111⟩, ⟨521⟩, etc., were also observed in (110) and (111) substrates. Although the MacEtch intrinsically prefers (100) direction, the etch direction can be also affected by the concentration of the oxidant in the etchant. If the backbone strength become weaker, the anisotropic etch behavior is reduced or eliminated, leading to the isotropic etch. For example, when an oxidant is added into the alkaline solution, the etched substrates show a smooth surface rather than the pyramidal structures usually formed by (100) substrate in alkaline solution. When an oxidant is added, the dangling bond of the Si atoms are modified from —H bonding in the alkaline solution to —OH bonding with the oxidant, leading to the decrease in the backbone strength, and the formation of a flat surface (Bressers et al., 1996). In other words, the addition of the oxidant reduces the anisotropy of an alkaline solution. This approach could be also applied to MacEtch of (110) or (111) substrates. Huang et al. etched the (111) Si substrates coated with the isolated Ag particles in HF/H_2O_2 solutions with different concentrations (Huang et al., 2010). For the low concentration (2 mM) of H_2O_2, MacEtch occurred along the ⟨100⟩ direction which was crystallographically preferred as shown in Fig. 27A. With increasing H_2O_2 concentration up to 100 mM, the etch direction changed from ⟨100⟩ directions to the vertical [111] direction (Fig. 27B). In the case of the HF/Fe$(NO_3)_3$ solutions, the etch direction of a (111) substrate with Ag particles was also affected by the concentration of $Fe(NO_3)_3$ (Huang et al., 2010). The etch directions for low (13.5 mM) and high $Fe(NO_3)_3$ concentration (135 mM) were ⟨100⟩ directions and vertical [111] direction, respectively (Fig. 27C and D). MacEtch of a (110) substrate showed similar phenomena in a HF/H_2O_2 solution with different concentrations (Huang et al., 2010). MacEtch proceeded along a [100] direction for low H_2O_2 concentration (400 mM) while the etch direction changed to the vertical [110] direction for high H_2O_2 concentration (1 M) as shown in Fig. 27E and F. These results indicate that the etch direction also depends on the concentration of the etchant as well as the crystalline orientation of the substrate. The switch of the etch direction during MacEtch is also reported in literature (Fang et al., 2008;

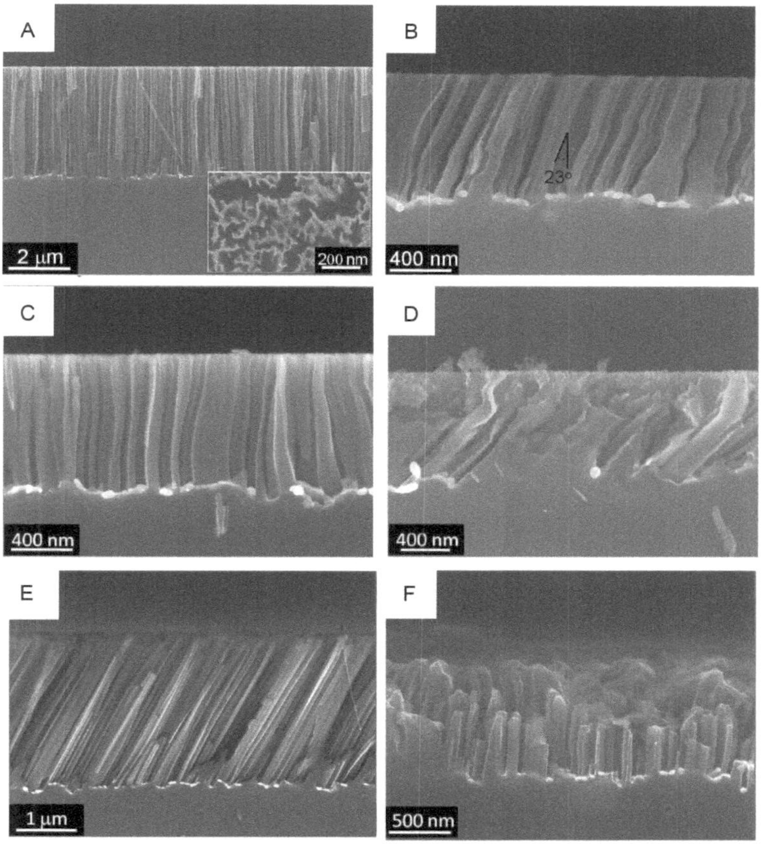

Fig. 27 Cross-sectional SEM images of p-type (111) substrates etched in etchant solutions with (A) 100 mM H_2O_2, 4.6 M HF, (B) 20 mM H_2O_2, 4.6 M HF, (C) 135 mM Fe^{3+} and 4.6 M HF, and (D) 13.5 mM Fe^{3+} and 4.6 M HF. Cross-sectional SEM images of p-type (110) substrates etched in etchant solutions with (E) 400 mM H_2O_2, 4.6 M HF and (F) 1 M H_2O_2, 4.6 M HF. *Adapted with permission from Huang, Z., Shimizu, T., Senz, S., Zhang, Z., Geyer, N., Gösele, U., 2010. Oxidation rate effect on the direction of metal-assisted chemical and electrochemical etching of silicon. J. Phys. Chem. C 114, 10683–10690. https://doi.org/10.1021/jp911121q. Copyright 2010, Wiley-VCH.*

Peng et al., 2005, 2006a, 2007, 2008). Since the metal catalyst goes deep into the substrate with increasing etch time, the oxidant is not able to diffuse into the metal/substrate interface properly due to long diffusion path of the chemicals. If the diffusion rate of the oxidant from the etchant to the metal catalyst is smaller than the consumption rate of the oxidant at the metal/substrate interface, the concentration of oxidant would be locally reduced near the metal catalyst. This causes a change of the etch direction from the initial vertical [111] direction to crystallographically preferred ⟨100⟩ directions.

In addition to the influence of the oxidant concentration, the morphology of the metal catalyst also affects the etch direction of non-(100) substrates. Chang et al. designed the experiments with three different morphologies of Au catalyst: isolated Au particles, Au mesh film with small hole pitch, and Au mesh film with relatively large hole pitch (Chang et al., 2010). MacEtch occurred along with the ⟨100⟩ directions for the isolated particles and mesh film with small hole pitch as shown in Fig. 28A because the Au could move in both the lateral and vertical directions. However, the mesh film with relatively large hole pitch has a restricted horizontal movement.

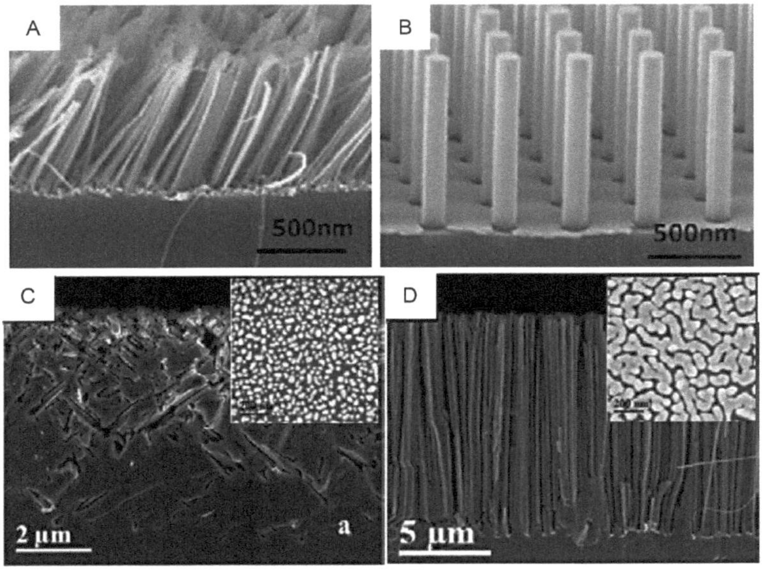

Fig. 28 Cross-sectional SEM images of (110) Si substrates etched using Au mesh film with (A) small holes and (B) large spacing. Cross-sectional SEM images of (111) Si substrates etched using Ag nanoparticles which were electroless galvanic deposited for (C) 5 s and (D) 120 s. *Panels A and B: Adapted with permission from Chang, S., Chuang, V.P., Boles, S.T., Thompson, C. V., 2010. Metal-catalyzed etching of vertically aligned polysilicon and amorphous silicon nanowire arrays by etching direction confinement. Adv. Funct. Mater. 20, 4364–4370. https://doi.org/10.1002/adfm.201000437. Copyright 2010, Wiley-VCH. Panels C and D: Adapted with permission from Milazzo, R.G., D'Arrigo, G., Spinella, C., Grimaldi, M.G., Rimini, E., 2012. Ag-assisted chemical etching of (100) and (111) n-type silicon substrates by varying the amount of deposited metal. J. Electrochem. Soc. 159, D521–D525. https://doi.org/10.1149/2.008209jes. Copyright 2012, The Electrochemical Society.*

This leads to etching the direction perpendicular to the substrate (Fig. 28B). Similar result was also observed for metal catalysts formed via electroless galvanic displacement (Milazzo et al., 2012). The isolated Ag particles with tiny size and low coverage produced irregular etch path along all equivalent ⟨100⟩ directions on (111) substrate (Fig. 28C). With the increase in density and the coverage of Ag particles, the surface morphology of the metal catalyst became interconnected particles in contrast to isolated particles obtained with low coverage. The interconnected network of Ag particles etches a (111) substrate along the [111] direction (Fig. 28D). This is mainly due to the fact that the etch direction is restricted due to limited lateral movement of the metal catalyst. Therefore, a continuous film could suppress crystal orientation preferred directional MacEtch, enabling the vertical etch direction regardless of the orientation of the substrate.

The effect of dopant type and dopant concentration of the substrate was investigated. Zhang et al. studied MacEtch of Si nanowires using various substrates with different orientations, doping types, and dopant concentrations (Zhang et al., 2008). For the same resistivity (i.e., same dopant concentration), they observed the different etch rate depending on the doping type. The etch rate of an n-type substrate with resistivity of $7-13\ \Omega$cm was faster than that of a p-type substrate with resistivity of $7-13\ \Omega$cm for both (100, 111) substrates. In addition, the morphology of etched Si nanowires is also affected by the dopant concentration. This can be explained by the fact that the dopant concentration also influences the hole injection over the Schottky barrier at the metal/Si interface. In Schottky diode, the current density contributed by hole injection is given by

$$J_{\mathrm{p}} = \frac{qD_{\mathrm{p}}}{W} p_{n0} \left[\exp\left(\frac{qV}{kT}\right) - 1 \right] \tag{31}$$

where W is the width of the surface charge region,

$$W = \left[\frac{2\varepsilon_0 \varepsilon}{qN_{\mathrm{d}}} (V_D - V) - 1 \right]^{1/2} \tag{32}$$

where q is the electronic charge, D_{p} is the diffusion constant of holes, p_{n0} is the equilibrium hole density in the neutral n region, V is the applied bias, k is the Boltzmann constant, T is the absolute temperature, ε_0 is the permittivity of free space, ε is the dielectric constant of Si, and N_{d} is the dopant concentration (Rhoderick, 1982). With increasing dopant concentration, the width of surface charge region decreases, leading to the increase in the current flow contributed by hole injection. Compared to moderately doped Si ($10^{15}-10^{18}\,cm^{-3}$) with $W>200\,nm$, degenerately doped Si ($\geq 10^{18}\,cm^{-3}$) with the small W ($9-16\,nm$) leads to more efficient hole injection, accelerating etching Si atoms.

Geyer et al. conducted the epitaxial growth of the multiple boron-doped Si layers by gradually varying the dopant concentration in the range of 6×10^{17} to $6\times10^{18}\,cm^{-3}$ (Geyer et al., 2015). To separate the boron-doped Si layers, germanium layers were inserted between the epitaxially grown Si layers. Then, the epitaxial layers were etched using the Ag film in the HF/H_2O_2 solution. The etched nanowires showed the different porosity along the length due to the dopant distribution in Si layers. The porosity increased with increasing dopant concentration because of the enhanced hole injection though the thinner surface charge region. When the dopant concentration of the epitaxy layer was lower than $1\times10^{18}\,cm^{-3}$, solid Si nanowires were obtained. For degenerately doped Si ($\geq 3\times10^{18}\sim10^{20}\,cm^{-3}$), however, porous Si nanowires with pore size of 6.9 ± 2.8 nm were observed. In the moderately doped Si, the injected holes are mainly consumed very close to the metal catalyst due to the electrostatic force of the metal, causing the localized etching of Si at the metal/Si interface. In contrast, MacEtch of degenerately doped Si is no longer confined to the metal/Si interface. When more holes are injected due to thinner width of the space charge region, the consumption rate of the holes at the metal/Si interface is smaller than the generation rate. Accordingly, unconsumed excess holes diffuse away from the metal catalyst to the side wall of nanowires, leading to the formation of a mesoporous structure on the side walls.

3.5 Effect of Etchant Concentration

The concentration of the etchant is one of the critical parameters that determine the morphology of the etched structures as well as the etch rate. Kim et al. measured the etch rate, the etch direction, and porosity of Si nanowires etched using Au/Ag bilayer mesh film in ternary mixtures of HF, H_2O_2, and H_2O at room temperature (Fig. 29) (Kim et al., 2011). Because of (100) substrate, the vertical and slanted (i.e., nonvertical) etch directions were $\langle 100\rangle$ and $\langle 110\rangle$, respectively. They observed that the concentration ratio (ε) of HF/H_2O_2 affected the porosity of the etched nanowires. Solid (i.e., nonporous) nanowires were obtained at $\varepsilon\geq1.15$ while the porous nanowires were formed at $0.3<\varepsilon\leq1.15$. In MacEtch, the overall reaction consists of two processes: hole injection into the Si through the metal/Si Schottky contact and chemical etch of oxidized Si by HF. Since the holes

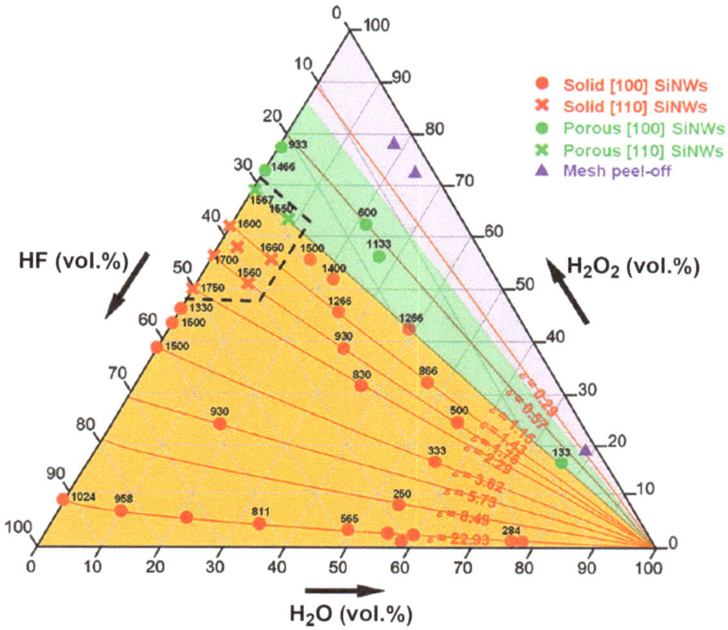

Fig. 29 Summary of etch rate, etch direction, and morphology of Si nanowires during MacEtch of (100) substrate in etchant solution with 48 wt% HF, 35 wt% H_2O_2 at room temperature. *Reproduced with permission from Kim, J., Han, H., Kim, Y.H., Choi, S.-H., Kim, J.-C., Lee, W., 2011. Au/Ag bilayered metal mesh as a Si etching catalyst for controlled fabrication of Si nanowires. ACS Nano 5, 3222–3229. https://doi.org/10.1021/nn2003458. Copyright 2011, The American Chemical Society.*

are generated by catalytic decomposition of H_2O_2, a higher concentration of H_2O_2 forms larger number of holes. At low ε condition (i.e., relatively low HF concentration), the etch rate is governed by the removal reaction of the oxidized Si. Extra holes, which are not consumed at the metal/Si interface, diffuse away to lattice defects or dopant sites on the surface of the as-etched nanowires (Chartier et al., 2008; Hochbaum et al., 2009; Li and Bohn, 2000; Zhu et al., 2015), leading to the formation of the new pores. As a result, porous Si nanowires were obtained at a low HF concentration (i.e., relatively high H_2O_2 concentration). In the chemical reactions (Eqs. 27–(30) of MacEtch, the actual potential at equilibrium could be calculated approximately by using the Nernst relationship as following (Qu et al., 2009):

$$\Delta E = \Delta E_0 + \frac{RT}{\mathrm{F}}\left[\log\frac{[\mathrm{SiF_2}^{2-}]}{[\mathrm{H_2O_2}]^2} - [\mathrm{H}^+]^4[\mathrm{F}^-]^6\right] \quad (33)$$

in which ΔE^0 is standard potential, R is the ideal-gas constant, T is the absolute temperature, F is the Faraday constant, $[SiF_6^{2-}]$, $[H_2O_2]$, $[H^+]$, and $[F^-]$ are the molarities of SiF_6^{2-}, H_2O_2, H^+, and, F^- ions in solution, respectively. As mentioned earlier, higher H_2O_2 concentration leads to a larger number of injected holes. Because of the increase in the driving force of hole injection, porous Si nanowires can be formed regardless of the dopant concentration (i.e., resistivity) of the substrate.

3.6 Fabrication of Vertically Aligned and Periodic Nanowires

We discussed the influence of MacEtch parameters on the morphology of the Si nanowires in previous sections. The etch rate, etch direction, and porosity of nanowires were strongly related to the MacEtch process conditions. For the applications in the field of photonic crystals, energy conversion, storage, and sensors, strict demands on the nanowire morphology are required. In this section, we will discuss the MacEtch process parameters which produced vertically aligned and periodic nanowires with smooth surfaces. Table 11 shows the summary of individually optimized MacEtch processes for obtaining vertically aligned Si nanowires. Fig. 30 shows the SEM images of optimized Si nanowires. As we have already seen degenerately doped substrate gives porous Si nanowires, moderately doped Si substrate is needed for the application interested in nonporous Si nanowires. Degenerately doped Si substrates form a shallow space charge region, the holes in the degenerately doped Si substrates can be easily injected into Si compared to moderately doped Si substrates. This leads to the formation of porous Si caused by excess holes. In addition, due to pore formation at dopants and defects, a larger number of pores are formed by a higher concentration of dopant.

For a metal catalyst, Au film was preferred rather than Ag or Pt. To fabricate Si nanowires with precise diameter and pitch, MacEtch needs metal catalyst in the form of a patterned metal film. The patterned film should keep the initial morphology during MacEtch to produce the uniform diameter and pitch of nanowires along the length. For Ag, however, deformation of fabricated metal pattern was observed in a relatively long MacEtch process with time (e.g., longer than 30 min). This is because H_2O_2 can oxidize Ag as well as inject hole into Si (Lu and Barron, 2013; Um et al., 2014). The deformed Ag pattern film produces an undesired morphology of the Si nanowires. In case of Pt catalyst, more holes are generated at the Pt/Si interface due to more efficient decomposition of H_2O_2 compared to Ag

... Process Conditions for Vertically Aligned Si Nanowires

Substrate	Resistivity (Ω cm)	Dopant Conc. (cm^{-3})	Etchant	Etchant Ratio $[H_2O_2]/[HF]$	Metal Catalyst	Temp. (°C)	Diameter, Length	Etch Rate (nm/min)	Porosity	References
p (100), n (100)	p-Si: 16.8, 0.005 n-Si: 2.3, 0.002	p-Si: $8e^{14}$, $2e^{19}$ n-Si: $2e^{15}$, $3e^{19}$	4.8 M HF, 0.4 M H_2O_2	8.33%	Ag 50 nm	n/a	~300 nm/~5 μm	n/a	Solid	Weisse et al. (2011)
n(100), n(110), n(111)	n/a	n/a	5.0 M HF, 0.4 M H_2O_2	8.8%	Au 12 nm	n/a	140 nm/~1 μm	66–200	Solid	Chang et al. (2010)
(100)	n/a	n/a	4.5 M HF, 0.9 M H_2O_2	20%	Ti/Au 3/20 nm	Room temp.	85 nm/~800 nm	~160	Solid	Chen and Liu (2014)
p(100)	6.8, 0.005	$2e^{15}$, $2e^{19}$	5.65 M HF, 0.1 M H_2O_2	1.7%	Ag 60 nm	Room temp.	~300 nm/4–7.5 μm	400–750	Solid, porous	Geyer et al. (2015)
p(100), n(100)	0.001–0.01	$5e^{18}$–$1e^{20}$	$HF:H_2O_2$: ethanol = 5:1:1/10:1:1	10%/20%	Au 20 nm	n/a	550, 800 nm/n/a	n/a	Porous	Balasundaram et al. (2012)
n(100)	2.2–2.6	$2e^{15}$	$HF:H_2O_2$:ethanol	12%–26%	Au 25 nm	30	500 nm/7 μm	40–100	Tapered, solid	Teng et al. (2017)
p(100)	n/a	n/a	4.6 M HF, 0.4 M H_2O_2	8.8%	Au 25 nm	Room temp.	150 nm/1.5 μm	250	Solid	Choi et al. (2008)
n(100)	1–10	$5e^{14}$–$3e^{15}$	7.8 M HF, 0.6 M H_2O_2, 0.85 M glycerol	7.0%	Ti/Au 3/30 nm	n/a	200 nm/4.9 μm	810 nm	Solid	Chen et al. (2017)
p(100)	10–20	$7e^{14}$–$1e^{15}$	1.9 M HF, 8.9 M H_2O_2	460%	Au/Fe/Au 10/10/10 nm	~18	1 μm/~200 μm	n/a	n/a	Oh et al. (2012)
p(100)	1–30	$4e^{14}$–$1e^{16}$	5.0 M HF, 0.2 M H_2O_2	4%	Au 20 nm	~25	~400 nm/n/a	n/a	Solid	Lee et al. (2011b)

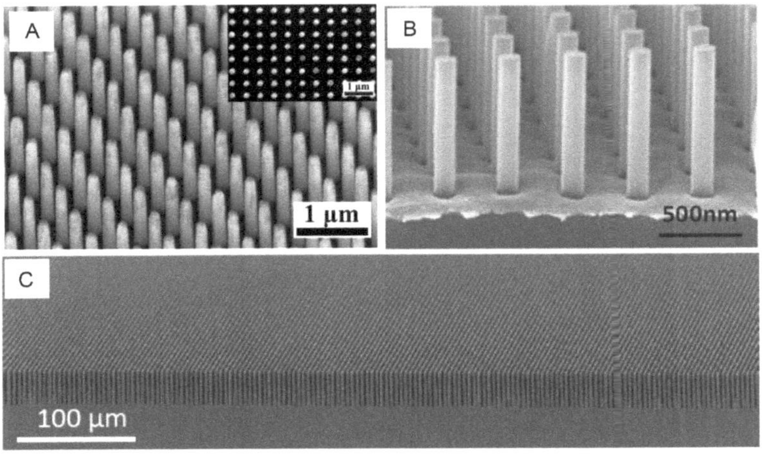

Fig. 30 Bird's eye view SEM images of vertically aligned Si nanowires etched (A) using 25-nm thick Au film in 4.6 M HF/0.4 M H_2O_2 solution, (B) using 12-nm thick Au film in 5 M HF/0.4 M H_2O_2 solution, and (C) using 10-nm thick Au film in HF/H_2O_2/IPA solution. *Panel A: Adapted with permission from Choi, W.K., Liew, T.H., Dawood, M.K., Smith, H.I., Thompson, C. V., Hong, M.H., 2008. Synthesis of silicon nanowires and nanofin arrays using interference lithography and catalytic etching. Nano Lett. 8, 3799–3802. https://doi.org/10.1021/nl802129f. Copyright 2018, The American Chemical Society; Panel B: Adapted with permission from Chang, S., Chuang, V.P., Boles, S.T., Thompson, C. V., 2010. Metal-catalyzed etching of vertically aligned polysilicon and amorphous silicon nanowire arrays by etching direction confinement. Adv. Funct. Mater. 20, 4364–4370. https://doi.org/10.1002/adfm.201000437. Copyright 2010, Wiley-VCH; Panel C: Adapted with permission from Romano, L., Vila-Comamala, J., Jefimovs, K., Stampanoni, M., 2017. Effect of isopropanol on gold assisted chemical etching of silicon microstructures. Microelectron. Eng. 177, 59–65. https://doi.org/10.1016/J.MEE.2017.02.008. Copyright 2017, Elsevier B.V.*

(Lee et al., 2008; Li et al., 2013). This results in the formation of the porous Si nanostructures rather than solid structures. In contrast, Au film is stable in HF/H_2O_2 solution. Therefore, the fabrication of the solid Si nanowires with high aspect ratio and periodic structure has been demonstrated by MacEtch based on the Au catalyst.

3.7 New Techniques for Controlling the Morphology of Nanowires

Fundamentally, the MacEtch of Si follows the crystallographic preferred directions of ⟨100⟩. Oh et al. demonstrated various Si structures such as Si sheets and zigzag Si nanowires by MacEtch based on a magnetically direction-induced metal catalyst (Fig. 31A–E) (Oh et al., 2012). To induce magnetic field, thin ferromagnetic metal film, iron, was inserted between the

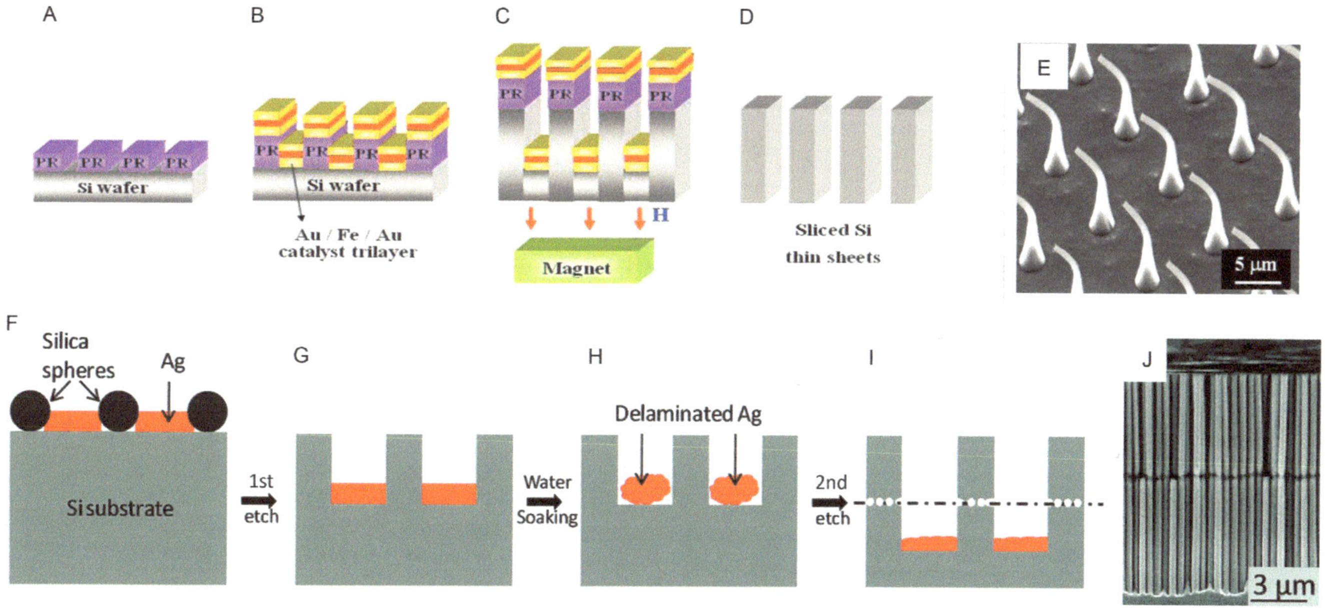

Fig. 31 (A)–(D) Schematic showing the MacEtch using magnetically direction-induced metal catalyst. (E) SEM image of bent Si nanowires etched with changing magnetic field direction. (F)–(I) Schematic showing the MacEtch processes in fabrication of cracked Si nanowires. (J) Cross-sectional SEM image of crack-inserted Si nanowires. *Adapted with permission from Oh, Y., Choi, C., Hong, D., Kong, S.D., Jin, S., 2012. Magnetically guided nano–micro shaping and slicing of silicon. Nano Lett. 12, 2045–2050. https://doi.org/10.1021/nl300141k. Copyright 2012, The American Chemical Society; Panel J: Adapted with permission from Weisse, J.M., Kim, D.R., Lee, C.H., Zheng, X., 2011. Vertical transfer of uniform silicon nanowire arrays via crack formation. Nano Lett. 11, 1300–1305. https://doi.org/10.1021/nl104362e. Copyright 2011, The American Chemical Society.*

thin Au films, forming 10 nm Au/10 nm Fe/10 nm Au thick film. The etch direction was controlled by the magnetically guiding the catalyst direction, enabling etching against the crystallographic preferred directions.

Vertical transfer of the Si nanowire onto non-Si substrates (e.g., glass and plastic substrates) has been investigated to study the optical and electrical properties of Si nanowires without the effect of bulk Si substrate. The Si substrate with as-etched Si nanowires was pressed against a poly-methyl methacrylate (PMMA)-coated receiver substrate at 200°C (over glass transition temperature of PMMA) (Shiu et al., 2009). After cooling down below the glass transition temperature of PMMA, the Si nanowires were fractured from the Si substrate by applying a shear force to the edge of the Si substrate, resulting in the transfer of vertically aligned Si nanowires to the receiver substrate. However, an extremely high insertion pressure could change the morphology of Si nanowire arrays such as bundling of nanowire tips. In other approaches, to easily fracture the Si nanowire from the substrate, the diameter at the root of Si nanowire was designed to be thinner than the original diameter of the nanowire. Shiu et al. first etched the Si nanowire in HF/H_2O_2 solution with low H_2O_2 concentration, then transferred into the other HF/H_2O_2 solution with relatively high H_2O_2 concentration (Shiu et al., 2011). Owing to isotropic etching at the high H_2O_2 concentration, the diameter near the bottom of nanowire shrinked. Instead of changing the etchant, the morphology of the metal catalyst was modified during MacEtch. This technique used water soaking process of the Ag film to create horizontal cracks into the Si nanowires (Weisse et al., 2011). During the water soaking step, some of the Ag film was delaminated and redistributed as particles onto the sidewall of as-etched nanowires close the Ag film (Fig. 31F–J). These Ag particles caused a localized MacEtch in the Si nanowires during the subsequent MacEtch step, resulting in the formation of pores and cracks perpendicular to the length.

Hu et al. demonstrated MacEtch of Si nanowires using aerated HF/H_2O vapor (Hu et al., 2014). Compared to conventional MacEtch based on aqueous HF solution, pseudogas phase MacEtch produced high aspect ratio nanowires due to replacement of the oxidant such as H_2O_2 and $AgNO_3$ by the oxygen in air. The aerated HF/H_2O vapor formed an invisibly thin electrolyte layer onto the substrate surface, leading to the high oxygen penetration rate. In this approach, HF was fully utilized, and the water consumption was also reduced, demonstrating environmentally friendly and scalable fabrication of Si nanowires.

4. SUMMARY

The RIE etching process has been extensively studied in last 30 years due to its immense potential in the fabrication of patterned high aspect ratio nanostructures. In this chapter, we show the fabrication of ultrahigh aspect ratio and high-density nanowires with RIE. The RIE process can be divided into three methods: (a) room temperature RIE, (b) cryogenic RIE, and (c) time-multiplexed RIE or Bosch process. The RTRIE is the most widely used for vertical nanowire etching as it does not need liquid nitrogen cooling as CRIE or an expensive RIE with fast switching mass flow controllers like Bosch process. All the RIE processes use halogen-based gas, usually SF_6 for etching Si. RTRIE allows anisotropic nanowire etching by protecting the nanowire sidewall by a —CF_2— passivation coating during etching. It provides high aspect ratio wires but has issues with dense pitch, and the process optimization can sometimes be challenging. The CRIE process uses O_2 gas for passivation instead C_4F_8 in RTRIE which forms a SiO_xF_y passivation layer. This SiO_xF_y passivation desorbs readily and thus provides CRIE an advantage of being very clean. Owing to the small thickness of passivation layer, it can be used to etch structures <10 nm and with nanowire aspect ratios >140. The Bosch process needs a more complicated setup but provides the flexibility of independent control of passivation, chemical and physical etching during the process. Thus aspect ratios >100 are routinely demonstrated with Bosch process. Unfortunately, the above independent control also gives rise to the serious issue of scalloping. Some relief from scalloping can be obtained by tuning the process for nanowire etching. Additional step like oxidation can reduce scalloping considerably and provide very high aspect ratio nanowires. An optimized Bosch process can provide aspect ratios >50 with a minimum diameter of 50 nm.

MacEtch is a simple and cost-effective method to fabricate high aspect ratio Si nanowires. Thanks to the selective etching of Si at the metal/Si contact, MacEtch can demonstrate anisotropic etching of Si even in a solution-based process. Positive holes are injected into the valence band of Si at the metal/Si interface and locally oxidize the Si. By removing the oxidized Si in the etchant, the metal catalyst sinks into the Si, resulting in the formation of remaining Si structures. As a result, the isolated metal particles produce the pore structure while the mesh metal film forms the nanowire structures. Since the MacEtch is a catalytic chemical etch process, the morphology of the etching results strongly depends on the metal catalyst, the crystalline orientation of the substrate, the dopant concentration of the substrate, and the etchant concentration as discussed in the previous sections. By optimizing these parameters, the accurate control of the orientation (e.g., vertical or slanted direction), porosity (e.g., solid or porous Si), diameter (nanoscale to microscale size), shape, density, and length of Si nanowires can be demonstrated at the wafer scale.

To conclude we discussed the fabrication of vertical Si nanowires top-down etching methods. Apart from the earlier advantages, the top-down process for etching nanowires provides a unique benefit of etching vertical nanowires directly in bulk semiconductors. Thus, they allow the option to retain the bulk semiconductor properties. Although the process optimization is remarkable, tricky results have been obtained with a minimum diameter of 20 nm achievable both with the MacEtch and the RIE processes. Unlike bottom-up growth methods, top-down etching techniques allow etching of vertical nanowires which can be obtained in a periodic, random, or any user-defined pattern. Furthermore, top-down etching techniques allow high aspect ratio etching with exceptional controllability. The versatility of top-down etching process along with the novel electrical and optical properties of Si nanowires sees immense potential in photovoltaics, thermoelectric energy storage system, and chemical/biological sensing applications.

ACKNOWLEDGMENTS

The authors thank the Center of Nanoscale Science and the School of Engineering and Applied Sciences at Harvard University for supporting the research and Jiangsu Industrial Technology Research Institute for the funding. The authors especially thank Nicholas Collins, Peter Duane, Sulagna Ghosh, Alex Ji, and Wenqi Zhu for their assistance in the preparation of this chapter. The authors would also like to thank Fawwaz Habbal for the understanding and support, without which this chapter would not been possible.

REFERENCES

Aachboun, S., Ranson, P., Hilbert, C., Boufnichel, M., 2000. Cryogenic etching of deep narrow trenches in silicon. J. Vac. Sci. Technol. A 18, 1848. https://doi.org/10.1116/1.582434.

Ayari-Kanoun, A., Aydinoglu, F., Cui, B., Saffih, F., 2016. Silicon nanostructures with very large negatively tapered profile by inductively coupled plasma-RIE. J. Vac. Sci. Technol. B 34, 06KD01. https://doi.org/10.1116/1.4964402.

Balasundaram, K., Sadhu, J.S., Shin, J.C., Azeredo, B., Chanda, D., Malik, M., Hsu, K., Rogers, J.A., Ferreira, P., Sinha, S., Li, X., 2012. Porosity control in metal-assisted chemical etching of degenerately doped silicon nanowires. Nanotechnology 23, 305304. https://doi.org/10.1088/0957-4484/23/30/305304.

Baldi, L., Beardo, D., 1985. Effects of doping on polysilicon etch rate in a fluorine-containing plasma. J. Appl. Phys. 57, 2221–2225. https://doi.org/10.1063/1.334366.

Blauw, M.A., Zijlstra, T., Bakker, R.A., van der Drift, E., 2000. Kinetics and crystal orientation dependence in high aspect ratio silicon dry etching. J. Vac. Sci. Technol. B 18, 3453. https://doi.org/10.1116/1.1313578.

Bressers, P.M.M.C., Kelly, J.J., Gardeniers, J.G.E., Elwenspoek, M., 1996. Surface morphology of p-type (100) silicon etched in aqueous alkaline solution. J. Electrochem. Soc. 143, 1744. https://doi.org/10.1149/1.1836710.

Bui, T.T., Tu, H.P., Dang, M.C., 2015. DRIE process optimization to fabricate vertical silicon nanowires using gold nanoparticles as masks. Adv. Nat. Sci. Nanosci. Nanotechnol. 6, 45016. https://doi.org/10.1088/2043-6262/6/4/045016.

Chang, C., Wang, Y.-F., Kanamori, Y., Shih, J.-J., Kawai, Y., Lee, C.-K., Wu, K.-C., Esashi, M., 2005. Etching submicrometer trenches by using the Bosch process and its application to the fabrication of antireflection structures. J. Micromech. Microeng. 15, 580–585. https://doi.org/10.1088/0960-1317/15/3/020.

Chang, Y.-F., Chou, Q.-R., Lin, J.-Y., Lee, C.-H., 2006. Fabrication of high-aspect-ratio silicon nanopillar arrays with the conventional reactive ion etching technique. Appl. Phys. A 86, 193–196. https://doi.org/10.1007/s00339-006-3748-0.

Chang, S.-W., Chuang, V.P., Boles, S.T., Ross, C.A., Thompson, C.V., 2009. Densely packed arrays of ultra-high-aspect-ratio silicon nanowires fabricated using block-copolymer lithography and metal-assisted etching. Adv. Funct. Mater. 19, 2495–2500. https://doi.org/10.1002/adfm.200900181.

Chang, S., Chuang, V.P., Boles, S.T., Thompson, C.V., 2010. Metal-catalyzed etching of vertically aligned polysilicon and amorphous silicon nanowire arrays by etching direction confinement. Adv. Funct. Mater. 20, 4364–4370. https://doi.org/10.1002/adfm.201000437.

Chartier, C., Bastide, S., Lévy-Clément, C., 2008. Metal-assisted chemical etching of silicon in HF-H2O2. Electrochim. Acta 53, 5509–5516. https://doi.org/10.1016/j.electacta.2008.03.009.

Chen, Y., 2015. Nanofabrication by electron beam lithography and its applications: a review. Microelectron. Eng. 135, 57–72. https://doi.org/10.1016/J.MEE.2015.02.042.

Chen, C.-Y., Liu, Y.-R., 2014. Exploring the kinetics of ordered silicon nanowires with the formation of nanogaps using metal-assisted chemical etching. Phys. Chem. Chem. Phys. 16, 26711–26714. https://doi.org/10.1039/C4CP04237A.

Chen, J., Dong, P., Di, D., Wang, C., Wang, H., Wang, J., Wu, X., 2013. Controllable fabrication of 2D colloidal-crystal films with polystyrene nanospheres of various diameters by spin-coating. Appl. Surf. Sci. 270, 6–15. https://doi.org/10.1016/J.APSUSC.2012.11.165.

Chen, Y., Zhang, C., Li, L., Tuan, C.C., Wu, F., Chen, X., Gao, J., Ding, Y., Wong, C.P., 2017. Fabricating and controlling silicon zigzag nanowires by diffusion-controlled metal-assisted chemical etching method. Nano Lett. 17 (7), 4304–4310. https://doi.org/10.1021/acs.nanolett.7b01320.

Cheng, C.C., Guinn, K.V., Donnelly, V.M., Herman, I.P., 1994. In situ pulsed laser-induced thermal desorption studies of the silicon chloride surface layer during silicon etching in high density plasmas of Cl_2 and Cl_2/O_2 mixtures. J. Vac. Sci. Technol. A 12, 2630–2640. https://doi.org/10.1116/1.579082.

Choi, W.K., Liew, T.H., Dawood, M.K., Smith, H.I., Thompson, C.V., Hong, M.H., 2008. Synthesis of silicon nanowires and nanofin arrays using interference lithography and catalytic etching. Nano Lett. 8, 3799–3802. https://doi.org/10.1021/nl802129f.

Choi, K., Song, Y., Oh, I., Oh, J., 2015. Catalyst feature independent metal-assisted chemical etching of silicon. RSC Adv. 5, 76128–76132. https://doi.org/10.1039/C5RA15745E.

Choi, K., Song, Y., Ki, B., Oh, J., 2017. Nonlinear etch rate of Au-assisted chemical etching of silicon. ACS Omega 2, 2100–2105. https://doi.org/10.1021/acsomega.7b00232.

Cichoszewski, J., Reuter, M., Schwerdt, F., Werner, J.H., 2013. Role of catalyst concentration on metal assisted chemical etchingof silicon. Electrochim. Acta 109, 333–339. https://doi.org/10.1016/j.electacta.2013.07.079.
Coburn, J.W., Winters, H.F., 1979a. Plasma etching—a discussion of mechanisms. J. Vac. Sci. Technol. 16, 391–403. https://doi.org/10.1116/1.569958.
Coburn, J.W., Winters, H.F., 1979b. Ion- and electron-assisted gas-surface chemistry—an important effect in plasma etching. J. Appl. Phys. 50, 3189–3196. https://doi.org/10.1063/1.326355.
Colace, L., Ferrara, P., Assanto, G., Fulgoni, D., Nash, L., 2007. Low dark-current germanium-on-silicon near-infrared detectors. IEEE Photonics Technol. Lett. 19, 1813–1815. https://doi.org/10.1109/LPT.2007.907578.
Craciun, G., Blauw, M.A., van der Drift, E., Sarro, P.M., French, P.J., 2002. Temperature influence on etching deep holes with SF6/O2 cryogenic plasma. J. Micromech. Microeng. 12, 307. https://doi.org/10.1088/0960-1317/12/4/307.
Cybart, S.A., Cho, E.Y., Wong, T.J., Wehlin, B.H., Ma, M.K., Huynh, C., Dynes, R.C., 2015. Nano josephson superconducting tunnel junctions in YBa2Cu3O7–δ directly patterned with a focused helium ion beam. Nat. Nanotechnol. 10, 598–602. https://doi.org/10.1038/nnano.2015.76.
d'Agostino, R., Flamm, D.L., 1981. Plasma etching of Si and SiO_2 in SF_6–O_2 mixtures. J. Appl. Phys. 52, 162–167. https://doi.org/10.1063/1.328468.
Dawood, M.K., Zheng, H., Kurniawan, N.A., Leong, K.C., Foo, Y.L., Rajagopalan, R., Khan, S.A., Choi, W.K., 2012. Modulation of surface wettability of superhydrophobic substrates using Si nanowire arrays and capillary-force-induced nanocohesion. Soft Matter 8, 3549. https://doi.org/10.1039/c2sm07279c.
Dejarld, M., Shin, J.C., Chern, W., Chanda, D., Balasundaram, K., Rogers, J.A., Li, X., 2011. Formation of high aspect ratio GaAs nanostructures with metal-assisted chemical etching. Nano Lett. 12, 5259–5263. https://doi.org/10.1021/nl202708d.
Donnelly, V.M., 2017. Review article: reactions of fluorine atoms with silicon, revisited, again. J. Vac. Sci. Technol. A 35, 05C202. https://doi.org/10.1116/1.4983922.
Donnelly, V.M., Kornblit, A., 2013. Plasma etching: yesterday, today, and tomorrow. J. Vac. Sci. Technol. A 31, 50825. https://doi.org/10.1116/1.4819316.
Dussart, R., Boufnichel, M., Marcos, G., Lefaucheux, P., Basillais, A., Benoit, R., Tillocher, T., Mellhaoui, X., Estrade-Szwarckopf, H., Ranson, P., 2004. Passivation mechanisms in cryogenic SF_6/O_2 etching process. J. Micromech. Microeng. 14, 190–196. https://doi.org/10.1088/0960-1317/14/2/004.
Dussart, R., Tillocher, T., Lefaucheux, P., Boufnichel, M., 2014. Plasma cryogenic etching of silicon: from the early days to today's advanced technologies. J. Phys. D Appl. Phys. 47, 123001. https://doi.org/10.1088/0022-3727/47/12/123001.
Ellinas, K., Smyrnakis, A., Malainou, A., Tserepi, A., Gogolides, E., 2011. "Mesh-assisted" colloidal lithography and plasma etching: a route to large-area, uniform, ordered nanopillar and nanopost fabrication on versatile substrates. Microelectron. Eng. 88, 2547–2551. https://doi.org/10.1016/J.MEE.2010.12.073.
Engstrom, D.S., Soh, Y.-A., 2013. Controlling the silicon nanowire tapering angle in dense arrays of silicon nanowires using deep reactive ion etching. J. Vac. Sci. Technol. B 31, 21806. https://doi.org/10.1116/1.4793500.
Fang, H., Li, X., Song, S., Xu, Y., Zhu, J., 2008. Fabrication of slantingly-aligned silicon nanowire arrays for solar cell applications. Nanotechnology 19, 255703. https://doi.org/10.1088/0957-4484/19/25/255703.
Fischer, P.B., 1993. 10 nm Si pillars fabricated using electron-beam lithography, reactive ion etching, and HF etching. J. Vac. Sci. Technol. B 11, 2524. https://doi.org/10.1116/1.586659.

Fu, Y.Q., Colli, A., Fasoli, A., Luo, J.K., Flewitt, A.J., Ferrari, A.C., Milne, W.I., 2009. Deep reactive ion etching as a tool for nanostructure fabrication. J. Vac. Sci. Technol. B 27, 1520. https://doi.org/10.1116/1.3065991.

Geng, X., Duan, B.K., Grismer, D.A., Zhao, L., Bohn, P.W., 2012. Monodisperse GaN nanowires prepared by metal-assisted chemical etching with in situ catalyst deposition. Electrochem. Commun. 19, 39–42. https://doi.org/10.1016/J.ELECOM.2012.03.011.

Geyer, N., Fuhrmann, B., Huang, Z., de Boor, J., Leipner, H.S., Werner, P., 2012. Model for the mass transport during metal-assisted chemical etching with contiguous metal films as catalysts. J. Phys. Chem. C 116, 13446–13451. https://doi.org/10.1021/jp3034227.

Geyer, N., Wollschläger, N., Fuhrmann, B., Tonkikh, A., Berger, A., Werner, P., Jungmann, M., Krause-Rehberg, R., Leipner, H.S., 2015. Influence of the doping level on the porosity of silicon nanowires prepared by metal-assisted chemical etching. Nanotechnology 26, 245301. https://doi.org/10.1088/0957-4484/26/24/245301.

Green, M.A., Emery, K., Hishikawa, Y., Warta, W., Dunlop, E.D., 2015. Solar cell efficiency tables (Version 45). Prog. Photovoltaics Res. Appl. 23, 1–9. https://doi.org/10.1002/pip.2573.

Gwinner, M.C., Koroknay, E., Fu, L., Patoka, P., Kandulski, W., Giersig, M., Giessen, H., 2009. Periodic large-area metallic split-ring resonator metamaterial fabrication based on shadow nanosphere lithography. Small 5, 400–406. https://doi.org/10.1002/smll.200800923.

Hamdana, G., Südkamp, T., Descoins, M., Mangelinck, D., Caccamo, L., Bertke, M., Wasisto, H.S., Bracht, H., Peiner, E., 2017. Towards fabrication of 3D isotopically modulated vertical silicon nanowires in selective areas by nanosphere lithography. Microelectron. Eng. 179, 74–82. https://doi.org/10.1016/J.MEE.2017.04.030.

Han, X.-L., Larrieu, G., Dubois, E., 2010. Realization of vertical silicon nanowire networks with an ultra high density using a top–down approach. J. Nanosci. Nanotechnol. 10, 7423–7427. https://doi.org/10.1166/jnn.2010.2841.

Han, X.-L., Larrieu, G., Fazzini, P.-F., Dubois, E., 2011. Realization of ultra dense arrays of vertical silicon nanowires with defect free surface and perfect anisotropy using a top-down approach. Microelectron. Eng. 88, 2622–2624. https://doi.org/10.1016/J.MEE.2010.12.102.

Haynes, C.L., Van Duyne, R.P., 2001. Nanosphere lithography: a versatile nanofabrication tool for studies of size-dependent nanoparticle optics. J. Phys. Chem. B 105 (24), 5599–5611. https://doi.org/10.1021/JP010657M.

He, L., Johansson, J., Murayama, M., Hull, R., 2008. Focused ion beam fabrication of novel core-shell nanowire structures. Nanotechnology. 19 (44), 445610–445615. https://doi.org/10.1088/0957-4484/19/44/445610.

Henry, M.D., 2010. Icp Etching of Silicon for Micro and Nanoscale Devices. California Institute of Technology.

Henry, M.D., Walavalkar, S., Homyk, A., Scherer, A., 2009. Alumina etch masks for fabrication of high-aspect-ratio silicon micropillars and nanopillars. Nanotechnology 20, 255305. https://doi.org/10.1088/0957-4484/20/25/255305.

Henry, M.D., Shearn, M.J., Chhim, B., Scherer, A., 2010. Ga^+ beam lithography for nanoscale silicon reactive ion etching. Nanotechnology 21, 245303. https://doi.org/10.1088/0957-4484/21/24/245303.

Hirobe, K., Kawamura, K., Nojiri, K., 1987. Formation of deep holes in silicon by reactive ion etching. J. Vac. Sci. Technol. B 5, 594. https://doi.org/10.1116/1.583955.

Hochbaum, A.I., Gargas, D., Hwang, Y.J., Yang, P., 2009. Single crystalline mesoporous silicon nanowires. Nano Lett. 9, 3550–3554. https://doi.org/10.1021/nl9017594.

Hu, Y., Peng, K.Q., Qiao, Z., Huang, X., Zhang, F.Q., Sun, R.N., Meng, X.M., Lee, S.T., 2014. Metal-catalyzed electroless etching of silicon in aerated HF/H2O vapor for facile fabrication of silicon nanostructures. Nano Lett. 14 (8), 4212–4219. https://doi.org/10.1021/nl500361u.

Huang, Z., Fang, H., Zhu, J., 2007. Fabrication of silicon nanowire arrays with controlled diameter, length, and density. Adv. Mater. 19, 744–748. https://doi.org/10.1002/adma.200600892.

Huang, Z., Shimizu, T., Senz, S., Zhang, Z., Geyer, N., Gösele, U. 2010. Oxidation rate effect on the direction of metal-assisted chemical and electrochemical etching of silicon. J. Phys. Chem. C 114, 10683–10690. https://doi.org/10.1021/jp911121q.

Hung, Y.-J., Lee, S.-L., Thibeault, B.J., Coldren, L.A., 2011. Fabrication of highly ordered silicon nanowire arrays with controllable sidewall profiles for achieving low-surface reflection. IEEE J. Sel. Top. Quantum Electron. 17, 869–877. https://doi.org/10.1109/JSTQE.2010.2068540.

Ishii, M., Nakashima, K., Tajima, I., Yamamoto, M., 1992. Investigation of hydrogen plasma etched Si surfaces. Jpn. J. Appl. Phys. 31, 4422–4427. https://doi.org/10.1143/JJAP.31.4422.

Jansen, H., Gardeniers, H., de Boer, M., Elwenspoek, M., Fluitman, J., 1996. A survey on the reactive ion etching of silicon in microtechnology. J. Micromech. Microeng. 6, 14–28. https://doi.org/10.1088/0960-1317/6/1/002.

Jansen, H.V., de Boer, M.J., Unnikrishnan, S., Louwerse, M.C., Elwenspoek, M.C., 2009. Black silicon method: X. A review on high speed and selective plasma etching of silicon with profile control: an in-depth comparison between Bosch and cryostat DRIE processes as a roadmap to next generation equipment. J. Micromech. Microeng. 19, 33001. https://doi.org/10.1088/0960-1317/19/3/033001.

Jo, S.-B., Lee, M.-W., Lee, S.-G., Lee, E.-H., Park, S.-G. Characterization of a modified Bosch-type process for silicon mold fabrication. J. Vac. Sci. Technol. A 23, 905–910. https://doi.org/10.1116/1.1943467.

Juan, W.H., Pang, S.W., 1995. High-aspect-ratio Si etching for microsensor fabrication. J. Vac. Sci. Technol. A 13, 834–838. https://doi.org/10.1116/1.579837.

Kanechika, M., Sugimoto, N., Mitsushima, Y., 2002. Control of shape of silicon needles fabricated by highly selective anisotropic dry etching. J. Vac. Sci. Technol. B 20, 1298. https://doi.org/10.1116/1.1484100.

Kang, C.K., Lee, S.M., Jung, I.D., Jung, P.G., Hwang, S.J., Ko, J.S., 2008. The fabrication of patternable silicon nanotips using deep reactive ion etching. J. Micromech. Microeng. 18, 75007. https://doi.org/10.1088/0960-1317/18/7/075007.

Karouta, F., 2014. A practical approach to reactive ion etching. J. Phys. D Appl. Phys. 47, 233501. https://doi.org/10.1088/0022-3727/47/23/233501.

Khorasaninejad, M., Abedzadeh, N., Singh Jawanda, A., Anantram, M.P., Singh Saini, S., 2012a. Bunching characteristics of silicon nanowire arrays. J. Appl. Phys. 111, 44328. https://doi.org/10.1063/1.3688025.

Khorasaninejad, M., Abedzadeh, N., Walia, J., Patchett, S., Saini, S.S., 2012b. Color matrix refractive index sensors using coupled vertical silicon nanowire arrays. Nano Lett. 12, 4228–4234. https://doi.org/10.1021/nl301840y.

Khorasaninejad, M., Walia, J., Saini, S.S., 2012c. Enhanced first-order Raman scattering from arrays of vertical silicon nanowires. Nanotechnology 23, 275706. https://doi.org/10.1088/0957-4484/23/27/275706.

Kim, J., Han, H., Kim, Y.H., Choi, S.-H., Kim, J.-C., Lee, W., 2011. Au/Ag bilayered metal mesh as a Si etching catalyst for controlled fabrication of Si nanowires. ACS Nano 5, 3222–3229. https://doi.org/10.1021/nn2003458.

Kim, S.H., Mohseni, P.K., Song, Y., Ishihara, T., Li, X., 2015. Inverse metal-assisted chemical etching produces smooth high aspect ratio InP nanostructures. Nano Lett. 15, 641–648. https://doi.org/10.1021/nl504136c.

Knizikevičius, R., 2009. Simulations of Si and SiO2 etching in SF6 + O2 plasma. Vacuum 83, 953–957. https://doi.org/10.1016/J.VACUUM.2008.11.002.
Kong, L., Dasgupta, B., Ren, Y., Mohseni, P.K., Hong, M., Li, X., Chim, W.K., Chiam, S.Y., 2016. Evidences for redox reaction driven charge transfer and mass transport in metal-assisted chemical etching of silicon. Sci. Rep. 6, 36582. https://doi.org/10.1038/srep36582.
Kuo, C.-W., Shiu, J.-Y., Chen, P., Somorjai, G.A., 2003a. Fabrication of size-tunable large-area periodic silicon nanopillar arrays with sub-10-nm resolution. J. Phys. Chem. B 107 (37), 9950–9953. https://doi.org/10.1021/JP035468D.
Kuo, C.-W., Shiu, J.-Y., Chen, P., 2003b. Size- and shape-controlled fabrication of large-area periodic nanopillar arrays. Chem. Mater. 15, 2917–2920. https://doi.org/10.1021/CM0343249.
Laermer, F., Schilp, A., 1994. A method for anisotropic etching of silicon. Google Patents. Retrieved from http://www.google.com/patents/DE4241045C1?cl=en.
Layadi, N., Donnelly, V.M., Lee, J.T.C., 1998. Cl2 plasma etching of Si(100): nature of the chlorinated surface layer studied by angle-resolved x-ray photoelectron spectroscopy. J. Appl. Phys. 81, 6738. https://doi.org/10.1063/1.365216.
Lee, C.-L., Tsujino, K., Kanda, Y., Ikeda, S., Matsumura, M., 2008. Pore formation in silicon by wet etching using micrometre-sized metal particles as catalysts. J. Mater. Chem. 18, 1015. https://doi.org/10.1039/b715639a.
Lee, J.-P., Choi, S., Park, S., 2011a. Extremely superhydrophobic surfaces with micro- and nanostructures fabricated by copper catalytic etching. Langmuir 27, 809–814. https://doi.org/10.1021/la1045354.
Lee, D.H., Kim, Y., Doerk, G.S., Laboriante, I., Maboudian, R., 2011b. Strategies for controlling Si nanowire formation during Au-assisted electroless etching. J. Mater. Chem. 21, 10359. https://doi.org/10.1039/c1jm11164g.
Lehmann, V., 1993. The physics of macropore formation in low doped n-type silicon. J. Electrochem. Soc. 140, 2836. https://doi.org/10.1149/1.2220919.
Lehmann, V., 2002. Electrochemistry of Silicon. Wiley-VCH Verlag GmbH, Weinheim, FRG. https://doi.org/10.1002/3527600272.
Li, X., Bohn, P.W., 2000. Metal-assisted chemical etching in HF/H2O2 produces porous silicon. Appl. Phys. Lett. 77, 2572. https://doi.org/10.1063/1.1319191.
Li, X., Xiao, Y., Yan, C., Song, J.-W., Talalaev, V., Schweizer, S.L., Piekielska, K., Sprafke, A., Lee, J.-H., Wehrspohn, R.B., 2013. Fast electroless fabrication of uniform mesoporous silicon layers. Electrochim. Acta 94, 57–61. https://doi.org/10.1016/J.ELECTACTA.2013.01.136.
Li, L., Liu, Y., Zhao, X., Lin, Z., Wong, C.-P., 2014. Uniform vertical trench etching on silicon with high aspect ratio by metal-assisted chemical Etching using nanoporous catalysts. ACS Appl. Mater. Interfaces 6, 575–584. https://doi.org/10.1021/am4046519.
Li, Z., Chen, Y., Zhu, X., Zheng, M., Dong, F., Chen, P., Xu, L., Chu, W., Duan, H., 2016. Fabrication of single-crystal silicon nanotubes with sub-10 nm walls using cryogenic inductively coupled plasma reactive ion etching. Nanotechnology 27, 365302. https://doi.org/10.1088/0957-4484/27/36/365302.
Lin, G.-R., Chang, Y.-C., Liu, E.-S., Kuo, H.-C., Lin, H.-S., 2007. Low refractive index Si nanopillars on Si substrate. Appl. Phys. Lett. 90, 181923. https://doi.org/10.1063/1.2736281.
Liu, Z., Wu, Y., Harteneck, B., Olynick, D., 2013. Super-selective cryogenic etching for sub-10 nm features. Nanotechnology 24, 15305. https://doi.org/10.1088/0957-4484/24/1/015305.
Logeeswaran, V.J., Katzenmeyer, A.M., Islam, M.S., 2010. Harvesting and transferring vertical pillar arrays of single-crystal semiconductor devices to arbitrary substrates. IEEE Trans. Electron Devices 57, 1856–1864. https://doi.org/10.1109/TED.2010.2051195.

Lu, Y.-T., Barron, A.R., 2013. Nanopore-type black silicon anti-reflection layers fabricated by a one-step silver-assisted chemical etching. Phys. Chem. Chem. Phys. 15, 9862. https://doi.org/10.1039/c3cp51835c.

Luo, Y.-R., 2007. Comprehensive Handbook of Chemical Bond Energies. CRC Press.

Mack, C., 2007. Fundamental Principles of Optical Lithography. John Wiley & Sons, Ltd, Chichester, UK. https://doi.org/10.1002/9780470723876.

Martin, M., Avertin, S., Chevolleau, T., Dhalluin, F., Ollivier, M., Baron, T., Joubert, O., Hartmann, J.M., 2013. Fabrication of high-density Si and Si_xGe_{1-x} nanowire arrays based on the single step plasma etching process. J. Vac. Sci. Technol. B 31, 41806. https://doi.org/10.1116/1.4812792.

Merzsch, S., Steib, F., Wasisto, H.S., Stranz, A., Hinze, P., Weimann, T., Peiner, E., Waag, A., 2014. Production of vertical nanowire resonators by cryogenic-ICP–DRIE. Microsyst. Technol. 20, 759–767. https://doi.org/10.1007/s00542-013-2032-4.

Milazzo, R.G., D'Arrigo, G., Spinella, C., Grimaldi, M.G., Rimini, E., 2012. Ag-assisted chemical etching of (100) and (111) n-type silicon substrates by varying the amount of deposited metal. J. Electrochem. Soc. 159, D521–D525. https://doi.org/10.1149/2.008209jes.

Miller, K., Li, M., Walsh, K.M., Fu, X.-A., 2013. The effects of DRIE operational parameters on vertically aligned micropillar arrays. J. Micromech. Microeng. 23, 35039. https://doi.org/10.1088/0960-1317/23/3/035039.

Mirza, M.M., Zhou, H., Velha, P., Li, X., Docherty, K.E., Samarelli, A., Ternent, G., Paul, D.J., 2012. Nanofabrication of high aspect ratio (~50:1) sub-10 nm silicon nanowires using inductively coupled plasma etching. J. Vac. Sci. Technol. B 30, 06FF02. https://doi.org/10.1116/1.4755835.

Morton, K.J., Nieberg, G., Bai, S., Chou, S.Y., 2008. Wafer-scale patterning of sub-40 nm diameter and high aspect ratio (>50:1) silicon pillar arrays by nanoimprint and etching. Nanotechnology 19, 345301. https://doi.org/10.1088/0957-4484/19/34/345301.

Nagao, D., Kameyama, R., Kobayashi, Y., Konno, M., 2007. Multiformity of particle arrays assembled with a simple dip-coating. Colloids Surfaces A 311, 26–31. https://doi.org/10.1016/J.COLSURFA.2007.08.031.

Nakamura, J., Higuchi, K., Maenaka, K., 2013. Vertical Si nanowire with ultra-high-aspect-ratio by combined top-down processing technique. Microsyst. Technol. 19, 433–438. https://doi.org/10.1007/s00542-012-1662-2.

Nojiri, K., 2015. Mechanism of dry etching. In: Dry Etching Technology for Semiconductors. Springer International Publishing, Cham, pp. 11–30. https://doi.org/10.1007/978-3-319-10295-5_2.

Oh, Y., Choi, C., Hong, D., Kong, S.D., Jin, S., 2012. Magnetically guided nano–micro shaping and slicing of silicon. Nano Lett. 12, 2045–2050. https://doi.org/10.1021/nl300141k.

Park, H., Crozier, K.B., 2013. Multispectral imaging with vertical silicon nanowires. Sci. Rep. 3, 2460. https://doi.org/10.1038/srep02460.

Park, S.-Y., Di Giacomo, S.J., Anisha, R., Berger, P.R., Thompson, P.E., Adesida, I., 2010. Fabrication of nanowires with high aspect ratios utilized by dry etching with SF6:C4F8 and self-limiting thermal oxidation on Si substrate. J. Vac. Sci. Technol. B 28, 763–768. https://doi.org/10.1116/1.3455498.

Park, H., Seo, K., Crozier, K.B., 2012. Adding colors to polydimethylsiloxane by embedding vertical silicon nanowires. Appl. Phys. Lett. 101, 193107-1–193107-4. https://aip.scitation.org/doi/pdf/10.1063/1.4766944.

Park, H., Dan, Y., Seo, K., Yu, Y.J., Duane, P.K., Wober, M., Crozier, K.B., 2014. Filter-free image sensor pixels comprising silicon nanowires with selective color absorption. Nano Lett. 14, 1804–1809. https://doi.org/10.1021/nl404379w.

Peng, K., Wu, Y., Fang, H., Zhong, X., Xu, Y., Zhu, J., 2005. Uniform, axial-orientation alignment of one-dimensional single-crystal silicon nanostructure arrays. Angew. Chem., Int. Ed. 44, 2737–2742. https://doi.org/10.1002/anie.200462995.

Peng, K., Fang, H., Hu, J., Wu, Y., Zhu, J., Yan, Y., Lee, S.T., 2006a. Metal-particle-induced, highly localized site-specific etching of Si and formation of single-crystalline Si nanowires in aqueous fluoride solution. Chem. A Eur. J. 12, 7942–7947. https://doi.org/10.1002/chem.200600032.

Peng, K.Q., Hu, J.J., Yan, Y.J., Wu, Y., Fang, H., Xu, Y., Lee, S.T., Zhu, J., 2006b. Fabrication of single-crystalline silicon nanowires by scratching a silicon surface with catalytic metal particles. Adv. Funct. Mater. 16, 387–394. https://doi.org/10.1002/adfm.200500392.

Peng, K., Zhang, M., Lu, A., Wong, N.-B., Zhang, R., Lee, S.-T., 2007. Ordered silicon nanowire arrays via nanosphere lithography and metal-induced etching. Appl. Phys. Lett. 90, 163123. https://doi.org/10.1063/1.2724897.

Peng, K., Lu, A., Zhang, R., Lee, S.T., 2008. Motility of metal nanoparticles in silicon and induced anisotropic silicon etching. Adv. Funct. Mater. 18, 3026–3035. https://doi.org/10.1002/adfm.200800371.

Pereira, J., Pichon, L.E., Dussart, R., Cardinaud, C., Duluard, C.Y., Oubensaid, E.H., Lefaucheux, P., Boufnichel, M., Ranson, P., 2009. *In situ* x-ray photoelectron spectroscopy analysis of SiOxFy passivation layer obtained in a SF6/O2 cryoetching process. Appl. Phys. Lett. 94, 71501. https://doi.org/10.1063/1.3085957.

Perry, A.J., Boswell, R.W., 1989. Fast anisotropic etching of silicon in an inductively coupled plasma reactor. Appl. Phys. Lett. 55, 148–150. https://doi.org/10.1063/1.102127.

Qian, H.X., Zhou, W., Miao, J., Lim, L.E.N., Zeng, X.R., 2008. Fabrication of Si microstructures using focused ion beam implantation and reactive ion etching. J. Micromech. Microeng. 18, 35003. https://doi.org/10.1088/0960-1317/18/3/035003.

Qu, Y., Liao, L., Li, Y., Zhang, H., Huang, Y., Duan, X., 2009. Electrically conductive and optically active porous silicon nanowires. Nano Lett. 9, 4539–4543. https://doi.org/10.1021/nl903030h.

Radny, M.W., Smith, P.V., Cao, P.L., 1996. An ab initio hartree-fock/density functional study of the cluster simulated Si(111)7 × 7:Cl adsorption system. Surf. Sci. 365, 15–28. https://doi.org/10.1016/0039-6028(96)00696-6.

Reinberg, A.R., Stienberg, G.N., Zarowin, C.B., 1984. Inductively coupled discharge for plasma etching and resist stripping. Retrieved from https://patents.google.com/patent/US4431898A/en.

Reyntjens, S., Puers, R., 2001. A review of focused ion beam applications in microsystem technology. J. Micromech. Microeng. 11, 287–300. https://doi.org/10.1088/0960-1317/11/4/301.

Rezvani, S.J., Pinto, N., Boarino, L., 2016. Rapid formation of single crystalline Ge nanowires by anodic metal assisted etching. CrstEngComm 18, 7843–7848. https://doi.org/10.1039/C6CE01598K.

Rhoderick, E.H., 1982. Metal-semiconductor contacts. IEE Proc. I Solid State Electron Devices 129, 1. https://doi.org/10.1049/ip-i-1.1982.0001.

Seo, K., Wober, M., Steinvurzel, P., Schonbrun, E., Dan, Y., Ellenbogen, T., Crozier, K.B., 2011. Multicolored vertical silicon nanowires. Nano Lett. 11, 1851–1856. https://doi.org/10.1021/nl200201b.

Shiu, S.-C., Hung, S.-C., Chao, J.-J., Lin, C.-F., 2009. Massive transfer of vertically aligned Si nanowire array onto alien substrates and their characteristics. Appl. Surf. Sci. 255, 8566–8570. https://doi.org/10.1016/J.APSUSC.2009.06.021.

Shiu, S.-C., Lin, T.-C., Pun, K.-L., Syu, H.-J., Hung, S.-C., Chao, J.-J., Lin, C.-F., 2011. Recycling Si wafers to fabricate multiple Si nanohole thin films by metal-assisted etching. 2011 11th IEEE International Conference on Nanotechnology. IEEE, pp. 1703–1706. https://doi.org/10.1109/NANO.2011.6144571.

Smyrnakis, A., Almpanis, E., Constantoudis, V., Papanikolaou, N., Gogolides, E., 2015. Optical properties of high aspect ratio plasma etched silicon nanowires: fabrication-induced variability dramatically reduces reflectance. Nanotechnology 26, 85301. https://doi.org/10.1088/0957-4484/26/8/085301.

Sökmen, Ü., Stranz, A., Fündling, S., Wehmann, H.-H., Bandalo, V., Bora, A., Tornow, M., Waag, A., Peiner, E., 2009. Capabilities of ICP-RIE cryogenic dry etching of silicon: review of exemplary microstructures. J. Micromech. Microeng. 19, 105005. https://doi.org/10.1088/0960-1317/19/10/105005.

Sökmen, Ü., Stranz, A., Fündling, S., Merzsch, S., Neumann, R. Wehmann, H.-H., Peiner, E., Waag, A., 2010. Shallow and deep dry etching of silicon using ICP cryogenic reactive ion etching process. Microsyst. Technol. 16, 863–870. https://doi.org/10.1007/s00542-010-1035-7.

Solanki, A., Crozier, K., 2014. Vertical germanium nanowires as spectrally-selective absorbers across the visible-to-infrared. Appl. Phys. Lett. 105, 191115-1–191115-5. https://doi.org/10.1063/1.4901438.

Solanki, A., Li, S., Park, H., Crozier, K.B., 2018. Harnessing the interplay between photonic resonances and carrier extraction for narrowband germanium nanowire photodetectors spanning the visible to infrared. ACS Photonics 5, 520–527. https //doi.org/10.1021/acsphotonics. 7b01034.

Soref, R., 2010. Mid-infrared photonics in silicon and germanium. Nat. Photonics 4, 495–497. https://doi.org/10.1038/nphoton.2010.171.

Stranz, A., Kähler, J., Merzsch, S., Waag, A., Peiner, E., 2012. Nanowire silicon as a material for thermoelectric energy conversion. Microsyst. Technol. 18, 857–862. https://doi.org/10.1007/s00542-011-1390-z.

Suzuki, K., Ninomiya, K., Mishimatsu, S., 1984. Microwave plasma etching. Vacuum 34, 953–957. https://doi.org/10.1016/0042-207X(84)90177-5.

Tachi, S., Tsujimoto, K., Okudaira, S., 1988. Low-temperature reactive ion etching and microwave plasma etching of silicon. Appl. Phys. Lett. 52, 616–618. https://doi.org/10.1063/1.99382.

Tada, T., 1998. Regular array of Si nanopillars fabricated using metal clusters. J. Vac. Sci. Technol. B 16, 3934. https://doi.org/10.1116/1.590440.

Tang, L., Kocabas, S.E., Latif, S., Okyay, A.K., Ly-Gagnon, D.-S., Saraswat, K.C., Miller, D.A.B., 2008. Nanometre-scale germanium photodetector enhanced by a near-infrared dipole antenna. Nat. Photonics 2, 226–229. https://doi.org/10.1038/nphoton.2008.30.

Teng, F., Li, N., Xu, D., Xiao, D., Yang, X., Lu, N., 2017. Precise regulation of tilt angle of Si nanostructures via metal-assisted chemical etching. Nanoscale 9, 449–453. https://doi.org/10.1039/C6NR08384F.

Tettamanzi, G.C., Pakes, C.I., Potenza, A., Rubanov, S., Marrows, C.H., Prawer, S., 2009. Superconducting transition in Nb nanowires fabricated using focused ion beam. Nanotechnology 20, 465302. https://doi.org/10.1088/0957-4484/20/46/465302.

Thompson, S.E., Parthasarathy, S., 2006. Moore's law: the future of Si microelectronics. Mater. Today 9, 20–25. https://doi.org/10.1016/S1369-7021(06)71539-5.

Tinck, S., Tillocher, T., Dussart, R., Bogaerts, A., 2015. Cryogenic etching of silicon with SF_6 inductively coupled plasmas: a combined modelling and experimental study. J. Phys. D Appl. Phys. 48, 155204. https://doi.org/10.1088/0022-3727/48/15/155204.

Traub, M.C., Longsine, W., Truskett, V.N., 2016. Advances in nanoimprint lithography. Annu. Rev. Chem. Biomol. Eng. 7, 583–604. https://doi.org/10.1146/annurev-chembioeng-080615-034635.

Tseng, A.A., Chen, K., Chen, C.D., Ma, K.J., 2003. Electron beam lithography in nanoscale fabrication: recent development. IEEE Trans. Electron. Packag. Manuf. 26, 141–149. https://doi.org/10.1109/TEPM. 2003.817714.

Um, H.-D., Park, K.-T., Jung, J.-Y., Li, X., Zhou, K., Jee, S.-W., Lee, J.-H., 2014. Incorporation of a self-aligned selective emitter to realize highly efficient (12.8%) Si nanowire solar cells. Nanoscale 6, 5193–5199. https://doi.org/10.1039/C4NR00455H.

Um, H.-D., Kim, N., Lee, K., Hwang, I., Hoon Seo, J., Yu, Y.J., Duane, P., Wober, M., Seo, K., 2015. Versatile control of metal-assisted chemical etching for vertical silicon microwire arrays and their photovoltaic applications. Sci. Rep. 5, 11277. https://doi.org/10.1038/srep 11277.

Weisse, J.M., Kim, D.R., Lee, C.H., Zheng, X., 2011. Vertical transfer of uniform silicon nanowire arrays via crack formation. Nano Lett. 11, 1300–1305. https://doi.org/10.1021/nl104362e.

Wu, B., Kumar, A., Pamarthy, S., 2010. High aspect ratio silicon etch: a review. J. Appl. Phys. 108, 51101. https://doi.org/10.1063/1.3474652.

Xia, X., Ashruf, C.M.A., French, P.J., Rappich, J., Kelly, J.J., 2001. Etching and passivation of silicon in alkaline solution: a coupled chemical/electrochemical system. J. Phys. Chem. B 105 (24), 5722–5729. https://doi.org/10.1021/JP003208F.

Yae, S., Morii, Y., Fukumuro, N., Matsuda, H., 2012. Catalytic activity of noble metals for metal-assisted chemical etching of silicon. Nanoscale Res. Lett. 7, 352. https://doi.org/10.1186/1556-276X-7-352.

Yoon, S.-S., Khang, D.-Y., 2012. Switchable wettability of vertical Si nanowire array surface by simple contact-printing of siloxane oligomers and chemical washing. J. Mater. Chem. 22, 10625. https://doi.org/10.1039/c2jm30619k.

Zeniou, A., Ellinas, K., Olziersky, A., Gogolides, E., 2014. Ultra-high aspect ratio Si nanowires fabricated with plasma etching: plasma processing, mechanical stability analysis against adhesion and capillary forces and oleophobicity. Nanotechnology 25, 35302. https://doi.org/10.1088/0957-4484/25/3/035302.

Zhang, M.L., Peng, K.Q., Fan, X., Jie, J.S., Zhang, R.Q., Lee, S.T., Wong, N.B., 2008. Preparation of large-area uniform silicon nanowires arrays through metal-assisted chemical etching. J. Phys. Chem. C 112, 4444–4450. https://doi.org/10.1021/jp077053o.

Zhu, H., Fu, Y., Meng, F., Wu, X., Gong, Z., Ding, Q., Gustafsson, M.V., Trinh, M.T., Jin, S., Zhu, X.Y., 2015. Lead halide perovskite nanowire lasers with low lasing thresholds and high quality factors. Nat. Mater. 14, 636–642. https://doi.org/10.1038/nmat4271.

Zijlstra, T., Drift, E., van der Dood, M.J.A., de Snoeks, E., Polman, A., 1999. Fabrication of two-dimensional photonic crystal waveguides for 1.5 μm in silicon by deep anisotropic dry etching. J. Vac. Sci. Technol. B 17, 2734. https://doi.org/10.1116/1.591054.

CHAPTER THREE

Group IV Nanowires for Carbon-Free Energy Conversion

Samik Mukherjee, Simone Assali, Oussama Moutanabbir[1]
École Polytechnique de Montréal, Succursale Centre-Ville, Montréal, QC, Canada
[1]Corresponding author: e-mail address: oussama.moutanabbir@polymtl.ca

Contents

1. INTRODUCTION

Semiconductor nanowires are quasi-one-dimensional nanostructures that have sparked a surge of interest as powerful and versatile nanotechnological building blocks (Barth et al., 2010; Boukai et al., 2008; Chen et al., 2011a; Colinge et al., 2010; Ferry, 2008; Garnett et al., 2011; Hochbaum and Yang, 2010; Hochbaum et al., 2008; Hu et al., 2007; Kelzenberg et al., 2010; Lee et al., 2012; Palacios, 2012; Tarun et al., 2011; Tian et al., 2007; Wallentin et al., 2013; Xie et al., 2012; Yan et al., 2011; Yang, 2012). Particularly,

Semiconductors and Semimetals, Volume 98
ISSN 0080-8784
https://doi.org/10.1016/bs.semsem.2018.02.003

nanowires made of group IV semiconductors (Si, Ge, and their alloys) have been a rich playground to develop innovative nanoscale devices enabling a vast variety of applications in nanoelectronics (Colinge et al., 2010; Ferry, 2008; Tarun et al., 2011; Yan et al., 2011), biosensing (Chen et al., 2011a; Xie et al., 2012), quantum computing (Hu et al., 2007), and environment-friendly energy conversion such as thermoelectrics (Boukai et al., 2008; Hochbaum and Yang, 2010; Hochbaum et al., 2008; Lee et al., 2012) and solar energy (Kelzenberg et al., 2010; Tian et al., 2007; Wallentin et al., 2013). These novel device concepts exploit size-related effects, the flexibility in design and fabrication offered by the use of nanowires, and the concomitant progress in probing nanoscale properties. This ubiquitous family of semiconductors and its applications can be further extended by introducing additional degrees of freedom to manipulate their basic properties. In this regard, two main paradigms have been pursued by our group aiming at uncovering and harness new phenomena in innovative nanowire-based structures and devices. Phonon-engineered nanowires are the first paradigm. Herein, we present recent advances in this area and discuss the implications for the field of thermoelectrics. We show how the introduction of stable isotopes as an additional degree of complexity is effective to control subtle but important phenomena in nanowire devices. This approach exploits the difference in the properties between stable isotopes leading to important changes in lattice dynamics (Cardona and Thewalt, 2005). The mass fluctuation induced by isotope disorder acts as a substitutional defect in the crystal, thus affecting the phonon mean free path and consequently the phononic thermal conductivity (Bracht et al., 2014; Cahill et al., 2014). Our group explores and exploits these phenomena through an unprecedented control of the distribution of isotopes in individual nanowires of group IV semiconductors. These new concepts expand our fundamental understanding of these important materials and pave the way to develop innovative devices relevant for heat transport studies.

The second paradigm consists of the growth of nonequilibrium Sn-containing group IV nanowires aiming at exploring new frontiers in group IV semiconductors. The key strategy here is to develop quantum-sized structures based on group IV silicon–germanium–tin (SiGeSn) nanowires. This emerging family of semiconductors provides two degrees of freedom for band structure engineering, namely, alloying and strain (Attiaoui and Moutanabbir, 2014; Kouvetakis et al., 2006; Lin et al., 2012; Soref, 2010; Soref and Perry, 1991; Soref et al., 2016; Wirths et al., 2015, 2016). This ability to independently manipulate strain and lattice parameter is central to engineer low-dimensional systems and heterostructures in a similar fashion to the more mature III–V materials (Soref, 2010; Soref et al., 2016).

Moreover, unlike the indirect bandgap featured by Si and Ge, a direct bandgap semiconductor can be achieved using Sn-rich alloys (Sn content >10 at.%) (Attiaoui and Moutanabbir, 2014; Kouvetakis et al., 2006; Soref, 2010; Soref et al., 2016; Wirths et al., 2016), thus providing a promising path toward Si-compatible efficient devices for emission and conversion of light (Kouvetakis et al., 2006; Lin et al., 2012). Developing these direct bandgap semiconductors was proven to be very challenging from materials perspective due to the low solubility of Sn in Si and Ge (<1 at.%). Recent progress, made by several groups including ours, in the control of growth kinetics overcame the aforementioned thermodynamic constraints leading to the epitaxial growth of dislocation-free binary and ternary layers with high Sn content (>10 at.%). This progress paves the way for a range of exciting integrated photonic and optoelectronic devices (Soref, 2010). The ability to grow these nonequilibrium alloys also creates unique opportunities to develop new low-dimensional systems to study exciting fundamental phenomena. We present here recent advances in growth and functionalization of Sn-containing group IV nanowires.

The fundamental concept at the core of the paradigms addressed in this chapter is the capability to control the physical properties through a precise, atomistic-level control of structure and composition. This has been stimulating the development of new nanowires with tunable properties that are a consequence of nanoscale and quantum phenomena. Herein, we describe new approaches to study and control surface and interface processes leading to these new classes of nanowires and their use to address some of the most exciting problems in materials physics by providing a rich playground to investigate fundamental phenomena related to heat transport and quantum confinement and their implications in carbon-free energy conversion. More specifically, we demonstrate how stable isotope engineering is used to shed new light on the basic understanding of thermal transport and thermoelectrics and how bandgap engineering in SiGeSn-based nanowires can be a valuable approach to implement innovative photovoltaic cells. This chapter mainly focuses on the current understanding at the materials science level. When it is relevant, new device concepts are proposed and discussed.

2. PHONON-ENGINEERED GROUP IV NANOWIRE AND NANOWIRE-BASED THERMOELECTRICS

2.1 Basic Concepts of Thermoelectricity

Before we address the progress in group IV nanowire-based thermoelectrics and underlying phonon-engineering approaches, we first present the theoretical background and the broader context of this research area. In 1909, Altenkirch proposed that the goodness of a material in terms of thermoelectricity is given by the dimensionless parameter known as the thermoelectric figure of merit (ZT) at a temperature T defined as

$$ZT = \frac{S^2 \sigma}{\kappa} T \quad (1)$$

where S is the Seebeck coefficient, σ is the electrical conductivity, and $\kappa = \kappa_e + \kappa_L$ is the thermal conductivity having contributions from both electrons (κ_e) and the phonons (κ_L). Thomas Seebeck (1770–1831) observed a compass needle to deflect when a temperature gradient is maintained on two different electrically coupled conductors. Later, it was realized that the deflection of the needle was in response to a magnetic field created by the passage of an electrical current through the circuit. When a conductor is placed in a temperature gradient between hot and cold regions at temperatures T_H and T_C, respectively, charge carriers diffuse from the hot end to the cold end setting up a current, a phenomenon known as thermoelectricity. The charge separation establishes a potential difference V between the two ends of the conductor known as the Seebeck voltage. At equilibrium, the charge current flux is zero, meaning that the electric field ($\mathcal{E}$) established between the ends of the conductor is just enough to stop the flow of the charge carriers from the hot end to the cold end. The Seebeck coefficient is defined as the ratio of the electric field to the temperature gradient at equilibrium as

$$S = \frac{\mathcal{E}}{\nabla T} \quad (2)$$

In order to harness the potential difference into an external circuit, two materials with different S are required. By convention S of a p-type material (holes as majority carrier) is taken as positive. The basic block diagram of a thermoelectric generator is shown in the schematic in Fig. 1A. Let x-axis be the reference axis and let the cold end be at $x=0$. The temperature along the p-type leg of length L increases along positive x. The holes diffusing from T_H toward T_C accumulate near $x=0$ and are depleted from $x=L$. $\mathcal{E}$ is thus directed along positive x, making S positive according to Eq. (2). For the n-type leg (electrons as majority carrier), the electrons also accumulate near $x=0$ and get depleted from $x=L$, but $\mathcal{E}$ now points toward negative x, making S negative. It is easy to see that the block acts as thermal parallel circuit to the heat flow but an electrical series circuit for current flow. The heat flux is always directed from T_H toward T_C and so is the thermal diffusion of the

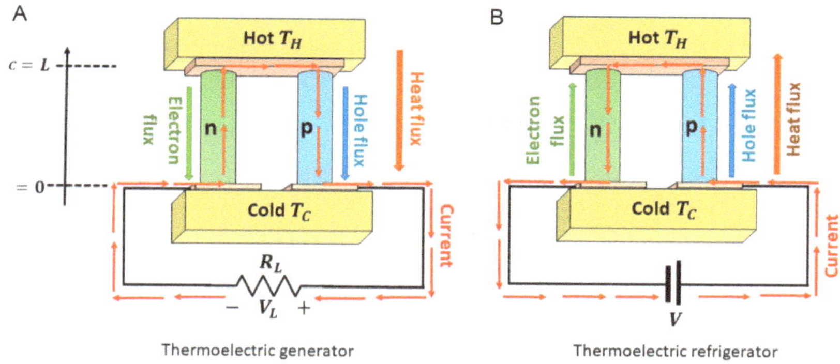

Fig. 1 Schematic illustrations of (A) a thermoelectric power generator and (B) a thermoelectric refrigerator. The p-type leg and the n-type leg are connected in a closed loop to form a thermoelectric module.

charge carriers. The current, however, flows from T_H to T_C in the p-type leg but from T_C to T_H in the n-type leg, thereby helping to maintain a potential difference V across the ends of the legs given by

$$V = \left(S_p + |S_n|\right)(T_H - T_C) \tag{3}$$

This can be connected to an external load to do some useful work. The charge carrier flow and hence the potential difference can be maintained as long as the temperature gradient is maintained. A thermoelectric device consists of many such alternating p-type and n-type legs. The reverse design, shown schematically in Fig. 1B, is that of a thermoelectric cooler. If the external load is replaced by a voltage source, then it forces the charge carriers to flow through the semiconductors following the externally imposed potential difference, thereby setting up currents in the p-type and n-type legs. As the carriers flow, they carry away thermal energy with them and build up a temperature gradient between the two ends. The temperature gradient is maintained as long as the potential difference is maintained by the external source. This effect, first observed by Jean C. A. Peltier in 1834, is known as the Peltier effect. At equilibrium, when the motion of carriers due to the external voltage source is equal and opposite to the thermal diffusion of carriers due to the established temperature gradient, the heat flux (Φ) and the current flux (J) are related by the equation $\Phi = \Pi J$, where Π is known as the Peltier coefficient.

Under nonequilibrium conditions, the current flux in response to a temperature gradient is given by

$$\text{Seebeck effect}: \; J = \sigma\mathcal{E} - S\sigma\nabla T \tag{4a}$$

Similarly, the temperature gradient that is generated in response to a current flux can be shown to be

$$\text{Peltier effect}: \; \nabla T = \frac{1}{\kappa}(\Pi J - \Phi) \tag{4b}$$

It is easy to see that S and Π are related to each other by the equation $\Pi = ST$. A material with high S would have a high Π and vice versa and therefore could be used either as a generator or as a refrigerator, depending upon the application in question.

2.2 Choice of Thermoelectric Materials

The figure of merit of a thermoelectric material is what determines the power conversion efficiency (Liu et al., 2015)

$$\eta = \left(\frac{T_H - T_C}{T_H}\right)\frac{\sqrt{ZT+1}-1}{\sqrt{ZT+1}+\frac{T_C}{T_H}} \tag{5}$$

where T is the average temperature given by $(T_H + T_C)/2$. For high values of ZT ($ZT \rightarrow \infty$), the maximum conversion efficiency (η_{max}) is fixed by the second law of thermodynamics. η_{max} is the prefactor $(T_H - T_C)/T_H$ or simply the efficiency of a Carnot engine connected between T_H and T_C and becomes independent of the material properties. $S^2\sigma$ appearing in the numerator of Eq. (1) is called the power factor. The power factor determines the amount of output power that can be delivered to the external load. Thus, at a first glance, it appears that a thermoelectric device should always be designed with the aim of maximizing the power factor. However, there is a strong interdependence between S and σ which puts a limit on how much the power factor can be increased. The interdependence is to be highlighted in the following where the transport properties are discussed. Graphene is a unique element with an extremely high intrinsic conductivity (Xiao et al., 2011) and S between 10 and 100 μV/K (Zuev et al., 2009) and can easily fit into the category of materials with a high power factor. However, such materials with a high thermoelectric power factor are suitable only for some special applications which demand high output power and require a large supply of thermal energy in the heat source to keep T_H as high as possible. The optimal design of a thermoelectric device should therefore maximize the conversion efficiency (η) of whatever thermal energy is available in the heat source rather than maximizing the output power.

Eq. (5) shows that η depends strongly on the properties of the thermoelectric material through the parameter Z or rather the dimensionless parameter ZT. A material with $ZT=1$ operating between room temperature ($T_C=300$ K) and the temperature of boiling water ($T_H=373$ K) would have η as low as 3.5% according to Eq. (5), while a material with $ZT=2$ operating between the same extremes would have η close to 5.5%. Needless to say, Eq. (5) does not take into account the parasitic losses, the electro-thermal contact resistances, or the Thompson effect on the conversion efficiency.

As far as maximizing the ZT is concerned, a material like graphene despite its high power factor is an inefficient thermoelectric material due to its high thermal conductivity (κ) of about 3000 W/m K (Balandin et al., 2008) and so are metals. In metals, the thermal conduction is almost entirely due to the motion of the free electrons. The electronic contribution to the thermal conductivity (κ_e) at a temperature T is directly proportional to the electrical conductivity (σ) by the well-known Wiedmann–Franz law

$$\kappa_e = \left(\frac{\pi}{3}\right)^2 \left(\frac{k_B}{q}\right)^2 \sigma T \tag{6}$$

where the prefactor equates to $(156\ \mu V/K)^2$ and is called the Lorentz number. While metals occupy one end of the spectrum with extremely high electrical and thermal conductivities due to the presence of large number of free carriers, glasses occupy the opposite end of the spectrum with extremely low thermal and electrical conductivity. The best thermoelectric materials should have a thermal conductivity like that of a glass and an electrical conductivity like that of a well-ordered crystalline material. The terminology "phonon glass, electron crystal" is often found in the literature to describe the properties of an ideal thermoelectric material. However, the interdependence of S, σ, and κ makes it extremely challenging to find or synthesize such "phonon glass, electron crystal"-type materials in the quest to maximize the ZT. Also, as stated earlier a thermoelectric module consists of a thermocouple made up of a p-type leg and an n-type leg in the form of a closed circuit and a thermoelectric device consists of many such thermocouples. The ZT of the thermocouple, taking into account the contributions from each leg, is given by (Ali et al., 2017)

$$ZT = \frac{\left(S_p + |S_n|\right)^2 T}{\left[\sqrt{\frac{\kappa_p}{\sigma_p}} + \sqrt{\frac{\kappa_n}{\sigma_n}}\right]^2} \tag{7}$$

where the suffix p/n denote the contributions from the p-type and the n-type legs, respectively. It is also important to maximize the ZT of the couple rather than that of the individual legs.

Historically, during the period of the discovery of Seebeck and Peltier effects, metals were heavily studied to explore the possibilities of thermoelectricity. Soon it was realized that metals were inefficient as thermoelectric generators or cooler due to their low figure of merit and the focus then shifted to low bandgap semiconductors like the tellurium-based compounds. Throughout the mid-1990s, Bi_2Te_3 was one of the most heavily investigated materials for thermoelectricity (Goldsmid and Douglas, 1954; Satterthwaite and Ure, 1957) and showed a Z factor of $\sim 2.5 \times 10^{-3}$ K^{-1} leading to a ZT of 0.7–0.8 at room temperature (Goldsmid et al., 1958). Its narrow bandgap, a low room temperature κ, and the possibility of doping Bi_2Te_3 with either n-type or p-type dopants aided to this end. Doped ternary alloys of bismuth telluride containing antimony and selenium showed further improvements in ZT to above unity within a small temperature window around 300 K (Yamashita and Tomiyoshi, 2004; Yamashita et al., 2003). For PbTe, the Z factor is smaller ($\sim 1.5 \times 10^{-3}$ K^{-1}) than that of Bi_2Te_3 and its alloys, but the Z factor stays fairly constant over a wider range of temperatures (Dughaish, 2002; Gelbstein et al., 2005). Nonetheless, one major drawback of these materials is the toxicity of tellurium- or lead-based compounds and the potential impacts on the environment that can be caused by their disposal. The lookout has therefore always been for new material families which are toxic free, abundant on Earth, and possess suitable properties for thermoelectric and energy conversion applications. Several routes were taken to this end like exploring heavy fermion materials with complex band structure (Witas et al., 2015), disorder engineering (Eilertsen et al., 2011; Snyder et al., 2004), and nanostructuring. Heavy fermion compounds exhibit a strong electronic correlation which leads to the formation of narrow bands near the Fermi level with very high electronic density of states. Since the Seebeck coefficient is proportional to the energy derivative of the density of states at the Fermi energy, it is several times greater in these compounds compared to ordinary metals, boosting up the thermoelectric power factor (Witas et al., 2015). Disorder engineering in doped alloyed semiconductors like SiGe also forms a class of efficient bulk thermoelectric materials. The disorder causes a decrease in κ, while S and σ can be optimized by controlling the doping concentration (Dismukes et al., 1964; Joshi et al., 2008). The main topic of discussion in the remainder of this section will be devoted to the third formalism, nanostructuring with special attention to transport measurements and thermoelectric properties of nanowires.

In the pioneering theoretical work by Hicks and Dresselhaus, a significant improvement was predicted for the thermoelectric figure of merit for low-dimensional systems like nanowires and quantum wells (Hicks and Dresselhaus, 1993a,b) which was further investigated by other authors (Liang et al., 2010; Pennelli and Macucci, 2013). These theoretical predictions together with the development of novel techniques to synthesize and characterize nanostructured materials caused a surge of experimental studies on the thermoelectric effects in nanoscale materials (Boukai et al., 2008; Harman et al., 2002; Kim et al., 2006; Venkatasubramanian et al., 2001; Yu et al., 2005). In the works by Hicks and Dresselhaus, the increase in ZT in nanoscale materials was predicted to be due to the increase in the power factor $S^2\sigma$ owing to quantum confinement effects. Experimentally, it has been shown that the dominant contribution to the increase in ZT in nanoscale semiconductor materials comes from the reduction in κ due to phonon scattering at the boundaries and the interfaces of such materials (Boukai et al., 2008; Kim et al., 2006; Majumdar, 2004; Poudel et al., 2008; Venkatasubramanian et al., 2001) as compared to quantum effects on the power factor, $S^2\sigma$. Nonetheless, the theoretical understanding of the quantum effects on the power factor is important since it leaves us with an additional parameter to tune the ZT, apart from the current efforts which are primarily directed toward reducing the κ in group IV-based nanowires.

2.3 Generalized Transport Model for Thermoelectric Materials

This section has been devoted to the key concepts of the generalized transport model developed from the Landauer formalism (Landauer, 1957, 1989) by Datta and Lundstrom, popularly known as Landauer–Datta–Lundstrom (LDL) transport model. It is a near-equilibrium model the detailed theoretical developments of which can be found in the books by Datta (2012) and Lundstrom and Jeong (2013). The model takes into account a material forming the channel, characterized by the electronic density of states $D^{el}(E)$, between two ideal nonreflecting contacts, as shown schematically in Fig. 2A. The material forming the channel could be a 3D bulk structure, a thin 2D film, a 1D nanowire, a 0D quantum dot, or even a single molecule. At equilibrium, electrons fill up the contacts following the Fermi–Dirac statistics given by

$$f_j(E) = \left[1 + \exp\left(\frac{E - E_{Fj}}{k_B T}\right)\right]^{-1} \tag{8}$$

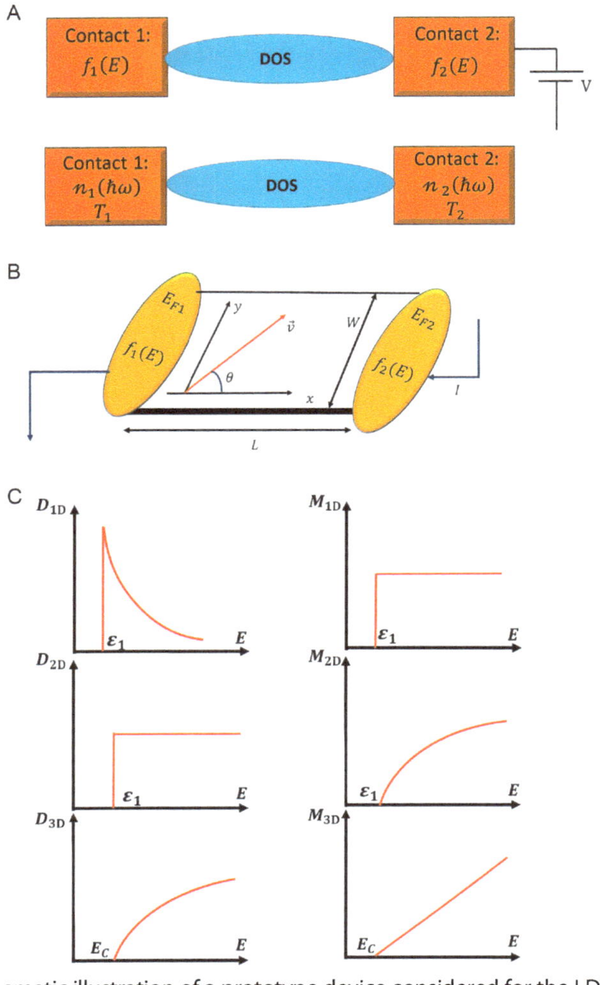

Fig. 2 (A) Schematic illustration of a prototype device considered for the LDL generalized transport model. The device consists of two identical contacts characterized by their Fermi or Bose functions and a channel characterized by its density of electronic or phononic states. The material forming the channel could be bulk 3D, 2D thin film, or 1D nanowire. A potential difference and/or a temperature difference could be applied between the two contacts. (B) Schematic illustration of a 2D thin film of width W and length L, as the channel material between the two contacts. The current flow is depicted along the $-x$-axis; hence the electron flow is along the $+x$-axis. The instantaneous velocity of an electron is depicted to be at an angle θ with the x-axis. (C) Sketches for the electronic density of states $D^{el}_{1D}(E)$, $D^{el}_{2D}(E)$, $D^{el}_{3D}(E)$ and the number of modes for electronic conduction $M^{el}_{1D}(E)$, $M^{el}_{2D}(E)$, $M^{el}_{3D}(E)$ or a 1D, 2D, and 3D channel, respectively. *Panels (A)–(C): Redrawn and adapted from Lundstrom, M., Jeong, C., 2013. Near-Equilibrium Transport: Fundamentals and Applications. World Scientific Publishing Company, New Jersey.*

where E_{Fj} is the Fermi energy level of the contact j ($j=1$ or 2). At equilibrium, it can be shown that the $D^{el}(E)$ of the channel are filled with electrons according to the Fermi level of the contacts. If the contacts are identical in all sense, half the available states are filled by contact 1 and the other half by contact 2. Thus, the total electron concentration in the channel is given by

$$N_t = \int n(E)dE = \int \left[\frac{D^{el}(E)f_1(E)}{2} + \frac{D^{el}(E)f_2(E)}{2}\right] dE \tag{9}$$

where $n(E)$ is the differential electron concentration. When a positive bias is applied to contact 2, its Fermi level is lowered relative to contact 1 such that E_{F2} is $E_{F1} - V$. The current that flows through the channel is given by (Lundstrom and Jeong, 2013)

$$I = \int i(E)dE = \int \frac{2q}{h}\frac{\gamma(E)}{2}\pi D^{el}(E)\{f_1(E) - f_2(E)\}dE \tag{10}$$

where $i(E)$ is the differential current, q is the magnitude of electronic charge, and $\gamma(E)$ is a broadening function related to some characteristic time of electrons $\tau_b^{el}(E)$. From Eqs. (9) and (10), $qn(E)dE/i(E)$ is given by

$$\frac{qn(E)}{i(E)} = \frac{[\text{Coulomb}]}{[\text{Ampere}]} = \frac{(h/2\pi)}{\gamma(E)}\frac{[f_1(E)+f_2(E)]}{[f_1(E)-f_2(E)]} = [\text{Time}] \tag{11}$$

where $qn(E)dE/i(E)$ is the differential charge over the differential current and therefore should have the units of time. To simplify things, a situation can be considered where the applied bias to contact 2 in Fig. 2A is large enough that the Fermi level of contact 2 is pulled down way below that of contact 1. Essentially, contact 2 is rendered ineffective in injecting any electron into the device which is now done entirely by contact 1. Under such a situation, the contributions from $f_2(E)$ are essentially zero. From Eq. (11), it is easy to see that $h/2\pi\gamma(E)$ should have the dimensions of time and this is the characteristic time $\tau_b^{el}(E)$. Thus, $\gamma(E) = \hbar/\tau_b^{el}(E)$.

To better understand the function $\gamma(E)\pi D^{el}(E)/2$ appearing in Eq. (10), a 2D thin film can be used for demonstration purpose, as shown in Fig. 2B and can be easily generalized to the 3D and 1D case by using appropriate transformation of the key parameters. $n(E)$, the differential electron concentration, can be written as $n_s(E)WL$, where $W(L)$ is the width (length) of the 2D film and $n_s(E)$ is the differential sheet electron concentration. If $\langle v_x^+(E)\rangle$ is the average electron velocity along the positive x-direction, then $i(E) = qWn_s(E)\langle v_x^+(E)\rangle$. From Eq. (11) the characteristic time $\tau_b^{el}(E)$ is $L/\langle v_x^+(E)\rangle$, which is the ballistic transit time of the electrons from one contact to the other across the channel (hence the suffix "b"). So,

$$\gamma(E)\pi\frac{D(E)}{2}=\frac{h}{2\pi\tau_b^{el}(E)}\pi\frac{D_{2D}^{el}(E)}{2}WL \tag{12}$$

where $D_{2D}(E)$ is the density of states per unit area. Putting the expression of $\tau_b^{el}(E)$ in Eq. (12)

$$\gamma(E)\pi\frac{D(E)}{2}=W\frac{h}{4}\langle v_x^+(E)\rangle D_{2D}^{el}(E) \tag{13}$$

The right-hand side of Eq. (13) is called the effective number of modes that are available for conduction at an electron energy E, $M^{el}(E)$. $M^{el}(E)=WM_{2D}^{el}(E)$, where $M_{2D}^{el}(E)$ is the number of modes per unit width at an electron energy E. For a 3D bulk semiconductor of volume $\boldsymbol{\Omega}$ and cross-section A, and 1D nanowire of length L, similar equations can be written. For the 3D bulk, $M^{el}(E)=AM_{3D}^{el}(E)=A(h/4)\langle v_x^+(E)\rangle D_{3D}^{el}(E)$, and for the 1D nanowire case, $M^{el}(E)=(h/4)\langle v_x^+(E)\rangle D_{1D}^{el}(E)$. During the ballistic transport, electrons that are injected into the channel through contact 1 in Fig. 2B transit the channel into contact 2 without suffering any scattering events. During diffusive transport however, all those electrons which are scattered at an angle θ between $-\pi/2$ and $+\pi/2$ add to the forward momentum and shall eventually reach contact 2. The remaining fraction of the electrons are back-scattered and shall return back to contact 1. Therefore, under diffusive transport, $\langle v_{x+}^{el}(E)\rangle$ in Fig. 2B is cosine component of $v^{el}(E)$ averaged over $-\pi/2$ to $+\pi/2$ given by

$$\langle v_x^+(E)\rangle=v^{el}(E)\frac{\int_{-\pi/2}^{+\pi/2}\cos\theta\,d\theta}{\int_{-\pi/2}^{+\pi/2}d\theta}=v^{el}(E)\frac{2}{\pi} \tag{14a}$$

where $v^{el}(E)$ is the electron velocity at an energy E. In thin-film semiconductors like the one depicted in Fig. 2B, if an electron inside the conduction band has an energy $v^{el}(E)$, then under the effective mass approximation $m^* v^{el}(E)^2/2=E-\varepsilon_1$, where ε_1 is the energy at the bottom of the first subband. This gives

$$\langle v_x^+(E)\rangle=\frac{2}{\pi}\sqrt{\frac{2(E-\varepsilon_1)}{m^*}} \tag{14b}$$

and with the expression of $D_{2D}^{el}(E) = m^* H(E - \varepsilon_1)/\pi\hbar^2$

$$M^{el}(E) = WM_{2D}^{el}(E) = W\frac{\sqrt{2m^*(E-\varepsilon_1)}}{\pi\hbar}H(E-\varepsilon_1) \tag{14c}$$

where H is a heavy-side step function. For the 3D and 1D cases

$$3\text{D}: \quad D^{el}(E) = D_{3D}^{el}(E)\Omega = \Omega\frac{m^*\sqrt{2m^*(E-E_C)}}{\pi^2\hbar^3}H(E-E_C) \tag{15a}$$

$$1\text{D}: \quad D^{el}(E) = D_{1D}^{el}(E)\text{L} = \frac{\text{L}}{\pi\hbar}\sqrt{\frac{2m^*}{(E-\varepsilon_1)}}H(E-\varepsilon_1) \tag{15b}$$

where E_C is the conduction band edge in a bulk semiconductor. The expression for the effective number of modes

$$3\text{D}: \quad M^{el}(E) = M_{3D}^{el}(E)\text{A} = \text{A}\frac{m^*(E-E_C)}{2\pi\hbar^2}H(E-E_C) \tag{16a}$$

$$1\text{D}: \quad M^{el}(E) = M_{1D}^{el}(E) = H(E-\varepsilon_1) \tag{16b}$$

Similar expressions can be obtained for the holes in the valence band. The evolution of $D_{1D}^{el}(E)$, $D_{2D}^{el}(E)$, $D_{3D}^{el}(E)$ and $M_{1D}^{el}(E)$, $M_{2D}^{el}(E)$, $M_{3D}^{el}(E)$ as a function of E is depicted in the schematic in Fig. 2C (Lundstrom and Jeong, 2013). Fig. 3A shows the evolution of $M^{el}(E)$ for bulk Ge as a function of energy (Jeong et al., 2010). The mid bandgap is located at zero energy. $M^{el}(E)$ was evaluated using three different approaches: the effective-mass parabolic-band approximation, from full band structure calculation using the nearest-neighbor tight-binding model, and by band counting. As Eq. (16a) reveals, the effective-mass parabolic-band approximation yields $M^{el}(E)$ proportional to E in 3D structures. Fig. 3A however shows that $M^{el}(E) \sim E$ is true only close to the band edge where the parabolic-band approximation describes the actual band structure fairly accurately. Far from the band edge, when the actual band structure deviates from the parabolic approximation, $M^{el}(E)$ calculated from the effective-mass parabolic-band approximation falls short of that calculated from the full band structure. However, under the assumption that the applied fields and/or temperature are not too high, electrons (holes) are located within a few $k_B T$ from the conduction (valence) band edge and the effective-mass parabolic-band approximation, which is far simpler than the full band calculation, still holds good.

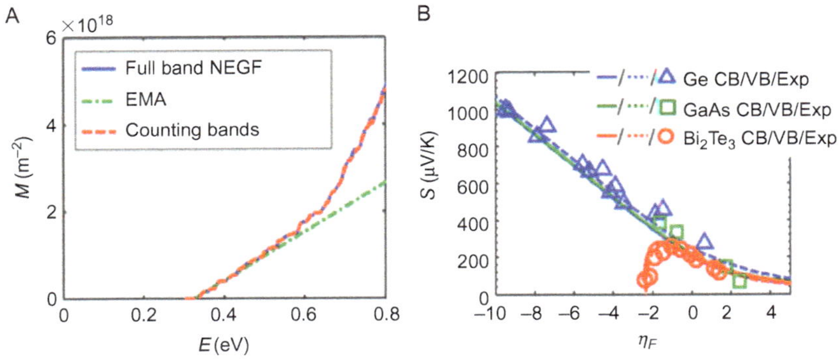

Fig. 3 (A) Comparison of the number of modes per unit area $M^{el}_{3D}(E)$ for bulk Ge computed by three different approaches: the effective-mass parabolic-band approximation (*greed dot-dash*), full-band-structure calculation (*blue*), and band counting technique (*red dash*). (B) The evolution of S as a function of the reduced Fermi level η_F. The plot shows the calculated value of S using the LDL approach for bulk Ge, GaAs, and Bi_2Te_3 both electrons (*continuous lines*) and holes (*dotted lines*) and experimentally measured S. *Panels (A) and (B): Adapted from Jeong, C., Kim, R., Luisier, M., Datta, S., Lundstrom, M., 2010. On Landauer versus Boltzmann and full band versus effective mass evaluation of thermoelectric transport coefficients. J. Appl. Phys. 107, 23707. https://doi.org/10.1063/1.3291120.*

$\tau_b^{el}(E) = L/\langle v_{x+}^{el}(E)\rangle$, the ballistic transit time of an electron through the channel, needs to be revised for the diffusive regime. The diffusive transit time can be defined as $\tau_d^{el}(E) = L^2/2\mathfrak{D}^{el}(E)$, the suffix "d" symbolizing the diffusive transport. $\mathfrak{D}^{el}(E)$ is the electron diffusion coefficient given by $\langle v_{x+}^{el}(E)\rangle\lambda^{el}(E)/2$, $\lambda^{el}(E)$ being the energy-dependent electron mean free path. $\lambda^{el}(E)$ is often expressed in a power law form as $\lambda_0(E_k/k_BT)^r$, where E_k is electron kinetic energy and r is a characteristic exponent. The LDL model takes into account the mean free path for back scattering. For an electron undergoing scattering in a 1D material, there are only two directions to scatter, either forward or backward. Thus, only one-half of all scattering events are back-scattering events. The back-scattering time constant is thus twice the scattering time constant and $\lambda^{el}_{1D,back}(E) = v^{el}(E)2\tau_d^{el}(E) = 2\lambda^{el}_{1D}(E)$. For the 2D and the 3D cases, it can be shown that $\lambda^{el}_{2D,back}(E) = \pi\lambda^{el}_{2D}(E)/2$ and $\lambda^{el}_{3D,back}(E) = 4\lambda^{el}_{3D}(E)/3$, respectively. Under the diffusive regime, the factor $\gamma(E)\pi D^{el}(E)/2$ can be written as

$$\gamma(E)\pi\frac{D^{el}(E)}{2} = \frac{\hbar}{\tau_d^{el}(E)}\pi\frac{D^{el}(E)}{2} = \frac{\hbar}{\tau_b^{el}(E)}\pi\frac{D^{el}(E)}{2}\times\frac{\tau_b^{el}(E)}{\tau_d^{el}(E)}$$
$$= M^{el}(E)\times T^{el}(E) \qquad (17)$$

where $T^{el}(E) = \tau_b^{el}(E)/\tau_d^{el}(E)$ is a transmittance function which gives the probability that a carrier injected at contact 1 with an energy E will come out of contact 2. From the expressions of $\tau_b^{el}(E)$ and $\tau_d^{el}(E)$, it is easy to see that $T^{el}(E) = \lambda^{el}(E)/L$, which is $\ll 1$ in the diffusive regime, <1 in the quasi-ballistic regime, and $\rightarrow 1$ in the ballistic regime. The generalized expression of current in Eq. (10) boils down to

$$I = \frac{2q}{h}\int M^{el}(E)\,T^{el}(E)(f_1 - f_2)\,dE \tag{18a}$$

For near-equilibrium transport or the low-field linear-response regime, the voltage difference (or the temperature difference) applied to the contacts is taken to be small. Near-equilibrium measurements are often used to characterize the transport properties of new materials and can easily be extended to treat more complicated situations. Under low fields, the Fermi level of contact 2 is pulled down only slightly relative to contact 1 such that $\Delta V = (E_{F1} - E_{F2})/q = \Delta E_F/q$ and $f_1(E) \approx f_0(E) > f_2(E)$, where $f_0(E)$ is the equilibrium Fermi function. For a small ΔE_F, $f_2(E)$ can be Taylor expanded under constant temperature condition as $f_2 = f_1 + (\partial f_1/\partial E_F)\Delta E_F = f_1 + (\partial f_1/\partial E_F)q\Delta V$ and using the property of Fermi function $(\partial f/\partial E_F) = -(\partial f/\partial E)$, the function $(f_1 - f_2)$ appearing in Eq. (18a) can be written as

$$f_1 - f_2 = \left(-\frac{\partial f_0}{\partial E}\right) q\Delta V \tag{18b}$$

Eq. (18a) thus reduces to

$$I = \left[\frac{2q^2}{h}\int M^{el}(E)\,T^{el}(E)\left(-\frac{\partial f_0}{\partial E}\right)dE\right]\Delta V = G\Delta V \tag{18c}$$

where G is the electrical conductance given by

$$G = \int g(E)\,dE = \frac{2q^2}{h}\int M^{el}(E)\,T^{el}(E)\left(-\frac{\partial f_0}{\partial E}\right)dE \tag{19}$$

where $g(E)$ is the differential electrical conductance. The parameter $(-\partial f_0/\partial E)$ is called the electron window function $\mathcal{W}^{el}(E)$. It is normalized since $\int_{-\infty}^{+\infty}(-\partial f_0/\partial E)\,dE = 1$. If in addition to the voltage difference (ΔV), a temperature difference of ΔT is also applied between the two contacts, then the difference in Fermi function between the two contacts will be partly due to ΔV and partly due to ΔT. The Taylor expansion of $f_1 - f_2$ in both voltage and temperature difference can be shown to be

$$f_1 - f_2 \approx \left(-\frac{\partial f_0}{\partial E}\right) q\Delta V - \left(-\frac{\partial f_0}{\partial E}\right) \frac{E - E_F}{T} \Delta T \tag{20a}$$

The electronic current through the channel under the presence of both voltage and temperature difference is given by

$$I = G\Delta V + S_T \Delta T \tag{20b}$$

where S_T is called the Soret coefficient given by

$$S_T = \int s_T(E) dE = -\frac{2q^2}{h} \int M^{el}(E) T^{el}(E) \left(\frac{E - E_F}{qT}\right) \left(-\frac{\partial f_0}{\partial E}\right) dE \tag{20c}$$

where $s_T(E)$ is the differential Soret coefficient. Under open-circuit condition ($I = 0$), Eq. (20b) gives $-G\Delta V = S_T \Delta T$. The ratio $(\Delta V/\Delta T) = -S_T/G$ is the Seebeck coefficient (S) given by

$$S = \frac{-\frac{2q^2}{h} \int M^{el}(E) T^{el}(E) \frac{(E - E_F)}{qT} \left(-\frac{\partial f_0}{\partial E}\right) dE}{\int g(E) dE} = \frac{-\frac{k_B}{q} \int \frac{(E - E_F)}{k_B T} g(E) dE}{\int g(E) dE} \tag{20d}$$

The interdependence of S and G can now be highlighted from Eqs. (19) and (20d). The electron window function $\mathcal{W}^{el}(E) = (-\partial f_0 / \partial E)$ picks up those modes which effectively takes part in conduction. It is nonzero only at $E = E_F$ and has a finite width of the order of few $k_B T$. On the other hand, all the available modes for conduction lie within the conduction (valence) bands and nothing actually lies within the bandgap. Thus, to ensure that as many as of the available modes take part in conduction (a condition necessary to have a high G), the Fermi level should be as close to the conduction (valence) band as possible. This can be achieved by doping. The function S on the other hand is proportional to $(E - E_F)$, meaning that the farther the Fermi level is from the band edge, higher is the Seebeck coefficient. So G is the highest for degenerately doped semiconductors and the lowest for intrinsic semiconductors, while S is the highest for intrinsic semiconductor and the lowest for degenerately doped semiconductors, making it difficult to optimize S^2G by doping or band structure engineering. Also, important to note that when a semiconductor is doped (degenerately or non-degenerately), the charge carriers may contribute to thermal conduction as well. From Eq. (18a), the expression for electronic current is given by $I=(2q/h)\int M^{el}(E)T^{el}(E)(f_1-f_2)dE$. Now, the electrons in either of the two contacts in Fig. 2A possess an energy which is close to E_F. To enter a mode $M^{el}(E)$ at an energy E within the channel, it has to either gain (if $E>E_F$) or lose (if $E<E_F$) an energy equal to $E-E_F$. To go to the expression of electronic heat current (I_Q), a factor $(E-E_F)/q$ therefore must be inserted into the expression for I in Eq. (18a). This gives the expression of electronic heat current as

$$I_Q=\frac{2}{h}\int(E-E_F)M^{el}(E)T^{el}(E)(f_1-f_2)dE \tag{21a}$$

Under the presence of both voltage and temperature difference between the contacts

$$I_Q=-TS_T\Delta V-K_0\Delta T \tag{21b}$$

where K_0 is the electronic thermal conductance. Under a short-circuit condition when $\Delta V=0$, Eq. (21b) reduces to the Fourier law of thermal conduction $I_Q=-K_0\Delta T$, where

$$K_0=T\left(\frac{k_B}{q}\right)^2\int\left(\frac{E-E_F}{k_BT}\right)^2 g(E)dE \tag{21c}$$

With $S=S_T/G$ and $\Pi=TS$, it is easy to see that I_Q in Eq. (21b) is $-\Pi I-K_0\Delta T$. Without the open circuit, when both ΔV and ΔT contribute to carrier flow, the net electronic to thermal conductance (K_{el}) is given by

$$K_{el}=K_0-\Pi SG \tag{21d}$$

Fig. 3B shows the evolution of S (calculated from Eq. 20d) for bulk Ge, GaAs, and Bi_2Te_3 as a function of the reduced Fermi function of η_F, where $\eta_F=E_F-\varepsilon_1/k_BT$ for 2D and 1D and $\eta_F=E_F-E_C/k_BT$ for 3D (Jeong et al., 2010). The calculations were done assuming the electron scattering rate to be proportional to the density of states. The plot demonstrates a good overlap of the calculated value (taking into account the contribution of both electrons in the conduction band and holes in the valence band) with the experimental data. The sudden reduction in S for Bi_2Te_3 around $\eta_F=-2$ was attributed to the bipolar conduction in Bi_2Te_3 due to its small bandgap. Fig. 4A shows

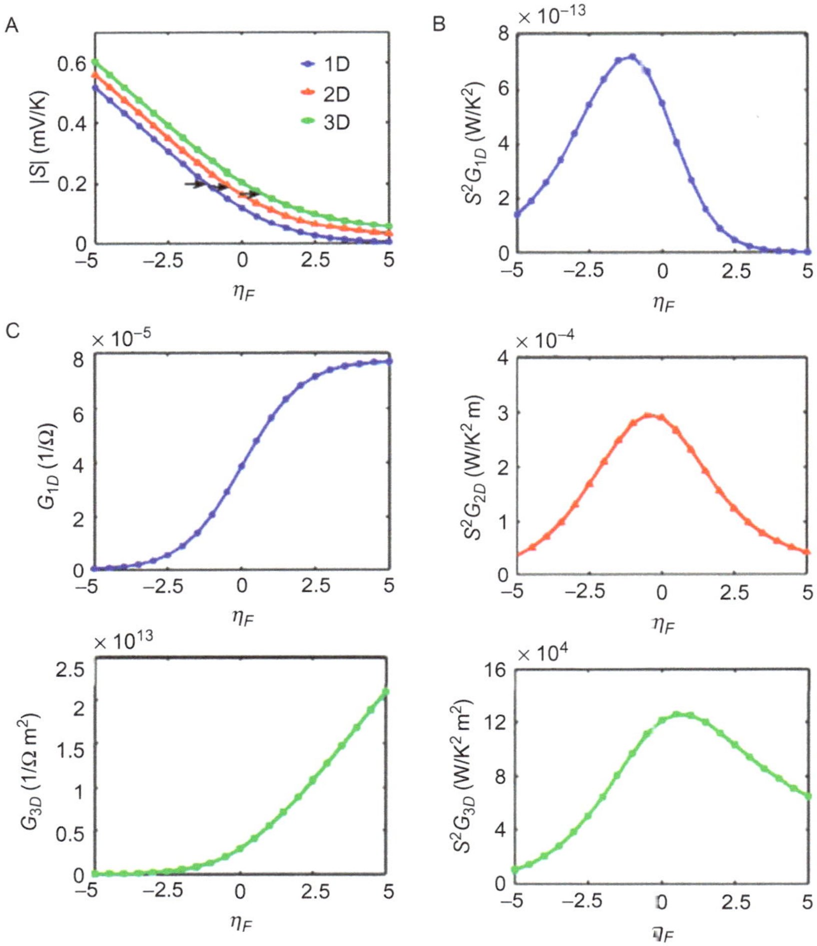

Fig. 4 (A) Calculated results showing the evolution of the Seebeck coefficient $|S|$ as a function of the reduced Fermi level η_F, for a 1D, 2D, and 3D ballistic conductor. For any η_F $|S_{3D}|>|S_{2D}|>|S_{1D}|$. (B) Calculated results showing the evolution of the power factor S^2G as a function of η_F, for a 1D, 2D, and 3D ballistic conductor. $\eta_{F,max}$, value of η_F at which the power factor reaches a maxima, is pointed out using the *black arrows* in (A). (C) Calculated results showing the evolution of the electrical conductance G as a function of η_F, for a 1D and 3D ballistic conductor. All the calculations are at 300 K and for $m^*=m_0$. *Panels (A)–(C): Adapted from Kim, R., Datta, S., Lundstrom, M.S., 2009. Influence of dimensionality on thermoelectric device performance. J. Appl. Phys. 105, 34506. https://doi.org/10.1063/1.3074347.*

the calculated value of S for 3D, 2D, and 1D ballistic conductors at $T=300$ K as a function of η_F (Kim et al., 2009). Fig. 4A shows that at any given η_F, $|S_{3D}|>|S_{2D}|>|S_{1D}|$ but that is not the end of the story. The evolution of power factor S^2G as a function of η_F is shown in Fig. 4B (Kim et al., 2009). For the 1D, 2D, and 3D cases, the power factor displays a maxima at $\eta_{F,max}$ of -1.14, -0.367, and 0.668, respectively. The value of S at $\eta_{F,max}$, that is, $\hat{S}=S(\eta_{F,\max})$, is however highest for 1D ($\hat{S}_{1D}=2.29k_B/q$) and lowest for 3D ($\hat{S}_{1D}=1.94k_B/q$), an improvement of about 18%. The reason that $|S_{3D}|>|S_{2D}|>|S_{1D}|$ is not too far to conceive. Eq. (20d) shows that the numerator in the expression of S contains the term $M^{el}(E)$ which depends on the dimension of the material forming the channel. In Fig. 2B, if the width of the film (W) is doubled, there will be twice as many modes available to the electrons for conduction. While comparing across dimension, it is therefore more suitable to compare the power factor per mode rather than the absolute value of the power factor. For ballistic conduction, $T^{el}(E)$ in Eq. (19) is close to unity. Thus, the expression of G is reduced to

$$G_{ballistic}=\frac{2q^2}{h}\int M^{el}(E)\left(-\frac{\partial f_0}{\partial E}\right)dE=\frac{2q^2}{h}M^{el}_{eff} \tag{22}$$

where the factor $2q^2/h$ is the quantum of electrical conductance (Landauer, 1957, 1989). $M^{el}_{eff}=\int M^{el}(E)\mathcal{W}^{el}(E)dE$ is an integral over the product of the total available number of modes and the electron window function and is the effective number of modes that actually take part in the conduction of electrons. It was found that $S^2G/M^{el}_{eff,1D}$ is $5.24\times 2k_B^2/h$, while $S^2G/M^{el}_{eff,2D}$ is $4.68\times 2k_B^2/h$ and $S^2G/M^{el}_{eff,3D}$ is $3.75\times 2k_B^2/h$, meaning that the available modes are used about 12% and 40% more effectively in 1D as compared to that in 2D and 3D, respectively. The reason for this can be found in Fig. 4C, which shows the evolution of G as a function of η_F for 1D and 3D (Kim et al., 2009). Clearly, the electrical conductance increases more rapidly with η_F for 1D as compared to 3D. Despite these benefits, it is still difficult for a 1D nanowire to compete the maximum power factor available for 3D bulk, simply due to the larger M^{el}_{eff}. Note that M^{el}_{eff} is proportional to the area of cross-section for a 3D material and to the width for a 2D material. It was shown that $S^2G_{3D,max}$ for a bulk device of cross-section $1\,cm^2$ is $12.6\,W/K^2$ which gives $M^{el}_{eff,3D}=5.84$ E12 (Kim et al., 2009). For a thin film of width 1 cm, $M^{el}_{eff,2D}=1.10$ E6, while for a

1D nanowire $M^{el}_{eff,1D}$= 0.242. So, if the 3D bulk material (of cross-section 1 cm^2) is to be replaced with 1D nanowires, then assuming a 100% packing fraction, the dimension of the nanowires must be $(M^{el}_{eff,1D}/M^{el}_{eff,3D})^{1/2} \times (M^{el}_{eff,1D}/M^{el}_{eff,3D})^{1/2}$ if the same effective number of modes as the bulk material is to be achieved. This means that the dimensions of the nanowires must be as small as ~2.03 nm × 2.03 nm. For 2D thin films, individual films stacked with 100% packing fraction must have a thickness of $M^{el}_{eff,2D}/M^{el}_{eff,3D}$~1.83 nm to achieve the same effective number of modes as the bulk material. Analogously it can be said that, to obtain the same maximum power factor as a 3D bulk material, the dimension of the individual 1D nanowires must be $(S^2G_{1D,max}/S^2G_{3D,max})^{1/2} \times (S^2G_{1D,max}/S^2G_{3D,max})^{1/2}$~2.40 nm × 2.40 nm. Not only are these extremely small dimensions which are very difficult to synthesize in practice, but also the requirement of barriers between quantum wells and nanowires means that the packing fraction in reality is never 100%. Hence, on a device scale, as far as the maximum achievable power factor or the total effective number of modes is concerned, it is still an open challenge for nanowires to compete with bulk materials. These conclusions, although obtained for ballistic transport, hold true for diffusive transport as well.

For thermal conductance calculation using the LDL model, the voltage difference between the two contacts is to be replaced by a temperature difference as shown in Fig. 2A. The material forming the channel needs to be characterized by its phonon density of states. Also, the available states in the channel shall be filled up by phonons from the contacts following the Bose–Einstein distribution. An equation similar to (18a) can be developed for the phonon transport as

$$Q=\frac{1}{h}\int(\hbar\omega)M^{ph}(\hbar\omega)T^{ph}(\hbar\omega)(n_1-n_2)d(\hbar\omega) \tag{23}$$

The current flow in the left-hand side is replaced by heat flow (Q). The charge (q) is replaced with energy flow ($\hbar\omega$). $M^{ph}(\hbar\omega)$ is the energy-dependent number of modes for the phonons and the n's are the Bose–Einstein functions for the two contacts. In the near-equilibrium regime when the difference in temperature (ΔT) between the two contacts is small, just like the electronic counterpart,

$$n_1-n_2\approx-\left(\frac{\partial n_o}{\partial T}\right)\Delta T \tag{24a}$$

where $n_o(\hbar\omega)$ is the equilibrium Bose–Einstein distribution given by

$$n_o(\hbar\omega) = \left[\exp\left(\frac{\hbar\omega}{k_B T}\right) - 1\right]^{-1} \quad (24b)$$

Also, $$\frac{\partial n_o}{\partial T} = \frac{\hbar\omega}{T}\left(-\frac{\partial n_o}{\partial(\hbar\omega)}\right) \text{ and } \frac{\partial n_o}{\partial(\hbar\omega)} = \left(-\frac{1}{k_B T}\right)\frac{e^{\hbar\omega/k_B T}}{\left[e^{\hbar\omega/k_B T} - 1\right]^2} \quad (24c)$$

Combining Eqs. (23) and (24a–24c) and using the relation $Q = -K_L \Delta T$, the expression of lattice thermal conductivity boils down to

$$K_L = \frac{k_B^2 T}{h}\int M^{ph}(\hbar\omega)\,T^{ph}(\hbar\omega)\left[\left(\frac{\hbar\omega}{k_B T}\right)^2\left(-\frac{\partial n_o}{\partial(\hbar\omega)}\right)\right]d(\hbar\omega) \quad (25a)$$

The term inside the square bracket serves as the phonon window function. But unlike the electron window function which is normalized, the phonon window function has a value of $\pi^2/3$. Thus, the term within the square bracket is multiplied and divided by a factor of $\pi^2/3$ to obtain the normalized phonon window function. With a little mathematical manipulation, the expression of K_L can be written as

$$K_L = \frac{\pi^2 k_B^2 T}{3h}\int M^{ph}(\hbar\omega)\,T^{ph}(\hbar\omega)\,\mathcal{W}^{ph}(\hbar\omega)\,d(\hbar\omega) \quad (25b)$$

with

$$\mathcal{W}^{ph}(\hbar\omega) = \frac{3}{\pi^2}\left(\frac{\hbar\omega}{k_B T}\right)\left(\frac{\partial n_o}{\partial(\hbar\omega)}\right) \quad (25c)$$

where the prefactor $\pi^2 k_B^2 T/3h$ is the quantum of thermal conductance (Schwab et al., 2000). $\mathcal{W}^{ph}(\hbar\omega)$ selects out only those phonon modes which take part in thermal conduction. Both the electrical and thermal conductances have a similar structure, a quantum of conductance times an integral over the product of the number of modes, transmittance, and the window function. For a ballistic phonon transport when $T^{ph}(\hbar\omega) = \lambda^{ph}(\hbar\omega)/L \approx 1$, the integral $\int M^{ph}(\hbar\omega)\mathcal{W}^{ph}(\hbar\omega)d(\hbar\omega)$ appearing in the right hand of Eq. (25b) gives M_{eff}^{ph}, that is, the effective number of phonon modes that take part in the thermal conduction.

Just like the effective-mass parabolic-band approximation for electrons, the Debye approximation for phonons helps bypass the need for a full phonon band structure calculation. In a solid with N atoms within a volume Ω, the total density of vibrational modes for each polarization is $3N/\Omega$, two transverse modes and one longitudinal mode per atom. For the acoustic modes, which are the principal contributors to thermal conduction (the low group velocity due to the flatness of the optical modes renders them ineffective in thermal conduction), the Debye approximation assumes a linear dispersion close to the center of the Brillouin zone as $\hbar\omega = \hbar v_D q$, as $q \to 0$, where q is the phonon wave vector and v_D is the Debye velocity. With the Debye approximation the phonon density of states can be written as (Kittel, 2005)

$$D^{ph}(\hbar\omega) = \frac{3(\hbar\omega)^2}{2\pi^2(\hbar v_D)^3} \tag{26a}$$

The width of the bands in a phonon dispersion relation is of the order of few hundred meV. At moderate temperatures, almost all the states across the Brillouin zone would be occupied by phonons. This is in contrast to the electronic dispersion wherein the width of the conduction and valence bands is of the order of eV. Thus, at moderate temperatures, the carriers are still located close to the band edge making the effective-mass parabolic-band approximation a good alternative to full-band-structure calculation. For the phonons, the dispersion of the acoustic modes tends to flatten out close to the edge of the Brillouin zone and deviate from the linear dispersion of the Debye approximation. So, the Debye approximation for the phonons holds good as long as q is not too far from the center of the Brillouin zone. Additionally, integrating $D^{ph}(\hbar\omega)$ over the entire Brillouin zone gives too many vibrational modes, way higher than the allowed $3N/\Omega$ modes in a solid with N atoms and volume Ω. Thus, the integration is performed up to a wave vector k_D (the Debye wave vector, the corresponding energy ω_D being the Debye energy) which gives the correct number of vibrational modes as

$$\int_0^{\hbar\omega_D} D^{ph}(\hbar\omega)d(\hbar\omega) = \frac{3N}{\Omega} \text{ gives } \hbar\omega_D = \hbar v_D\left(\frac{6\pi^2 N}{\Omega}\right)^{1/3} = k_B T_D \tag{26b}$$

where $T_D = \hbar\omega_D/k_B$ is the Debye temperature. When $T \ll T_D$, only those phonon states close to the center of the zone are occupied and the Debye approximation describes the actual phonon band structure quite accurately. A full phonon band structure calculation is required as T approaches or becomes higher than T_D.

Similar to that of the electrons, the number of modes for thermal conduction in a bulk material can be written as $M^{ph}(\hbar\omega) = AM^{ph}_{3D}(\hbar\omega) = A(h/4)$ $v^{ph}(\hbar\omega)D^{ph}(\hbar\omega)$. With the Debye approximation, this becomes

$$M^{ph}(\hbar\omega) = A\frac{3(\hbar\omega)^2}{4\pi(\hbar v_D)^2}; \quad \text{thus,} \quad M^{ph}_{3D}(\hbar\omega) = \frac{3(\hbar\omega)^2}{4\pi(\hbar v_D)^2} \tag{27a}$$

The phonon transmittance function has a similar form to that of the electrons, $T^{ph}(\hbar\omega) = \lambda^{ph}(\hbar\omega)/L$, and for phonon back scattering, $\lambda^{ph}_{3D,back}(\hbar\omega) = 4\lambda^{ph}_{3D}(\hbar\omega)/3$. Under the Debye approximation

$$\lambda^{ph}_{3D,\,back}(\hbar\omega) = \frac{4}{3}v_{ph}(\hbar\omega)\tau^{ph}(\hbar\omega) = \frac{4}{3}v_D\tau^{ph}(\hbar\omega) \tag{27b}$$

Similar equations for $M^{ph}_{2D}(\hbar\omega)$, $M^{ph}_{1D}(\hbar\omega)$ and $\lambda^{ph}_{2D,back}(\hbar\omega)$, $\lambda^{ph}_{1D,back}(\hbar\omega)$ can be developed for the 2D and 1D materials, respectively. Phonon engineering refers to the ways and means of controlling the phonon scattering time $\tau^{ph}(\hbar\omega)$ which directly affects the phonon mean free path and thus the transmittance function, $T^{ph}(\hbar\omega)$. Different phonon scattering mechanisms include the normal (τ_N^{-1}) and Umklapp anharmonic scattering (τ_U^{-1}), the scattering from point defects (τ_D^{-1}), boundary scattering (τ_B^{-1}) important in nanoscale materials like the nanowire. The ω dependence of different phonon scattering rates is as follows (Chen, 2005):

$$\tau_N^{-1} \propto \omega T^3 \tag{28a}$$

$$\tau_U^{-1} \propto \omega^2 T^3 \exp\left(-\frac{T_D}{bT}\right), \quad \text{where } b \text{ is a constant} \tag{28b}$$

$$\tau_D^{-1} \propto \omega^4 \tag{28c}$$

$$\tau_B^{-1} \propto \frac{v_D}{d}, \quad \text{where } d \text{ is the dimension} \tag{28d}$$

Additional phonon scattering could take place if homo- or heterointerfaces are present within the material. When several phonon scattering mechanisms act simultaneously and independently of each other, the scattering rates are additive and the total scattering rate is given by

$$\tau_{total}^{-1} = \tau_N^{-1} + \tau_U^{-1} + \tau_D^{-1} + \tau_B^{-1} + \cdots \tag{28e}$$

2.4 Phonon Engineering and Thermal Conductivity of Silicon-Based Nanowires

An overwhelming majority of the studies undertaken on thermal transport in nanowires are related to Si and Si-based nanowires, and this section has been devoted to highlight some of the key findings. There also exists several important theoretical and experimental investigations related to κ_L in III–V (Carrete et al., 2011; Mingo and Broido, 2004; Soini et al., 2010; Swinkels et al., 2015), IV–VI (Liang et al., 2009; Ma et al., 2017), and V–VI (Biswas et al., 2009; Park et al., 2016) nanowires.

Using molecular dynamic simulations, Volz and Chen showed that the κ_L in Si nanowires is smaller by over an order of magnitude compared to bulk Si (Volz and Chen, 1999). Apart from the contribution of boundary phonon scattering (τ_B^{-1}) not present in bulk, it was also discussed that the low-frequency acoustic phonons whose wavelengths are longer than the nanowire diameter cannot survive and die out. These low-frequency acoustic phonons contribute significantly to thermal conduction in bulk Si, imparting it with a high thermal conductivity of about 150 W/m K at room temperature. Experimentally the effect was first observed by Li et al. (2003b). Fig. 5A shows the evolution of κ_L as a function of temperature for Si nanowires of different diameters. Few things are worth noting in Fig. 5A. In bulk materials without point defects (isotopically pure bulk crystals for example (Asen-Palmer et al., 1997)), it is well known that the low-temperature regime is dominated by the frequency- and temperature-independent phonon boundary scattering, since the normal anharmonic scattering rate at low temperature is negligible. κ_L essentially follows a $\sim T^3$ dependence of the specific heat. At higher temperatures, when the normal anharmonic scattering rate starts to dominate, κ_L follows an $\sim T^{-1}$ dependence. The maxima of κ_L occur in between these two regimes, which is close to 25 K for bulk Si. For the Si nanowire of diameter 115 nm, κ_L has a bulk-like $\sim T^3$ dependence in the low-temperature regime, whereas for the nanowires with smaller diameter the deviation from the $\sim T^3$ dependence is prominent. This could be a result of the phonon confinement effects in nanowires with smaller diameter. It was known from theoretical works that confinement renormalizes the phonon dispersion and tends to flatten the acoustic dispersion relation especially close to the center of the Brillouin zone, reducing the phonon group velocity (Balandin and Wang, 1998; Zou and Balandin, 2001). Second, the peak of κ_L for the nanowires in Fig. 5A shifts toward higher temperature as the diameter

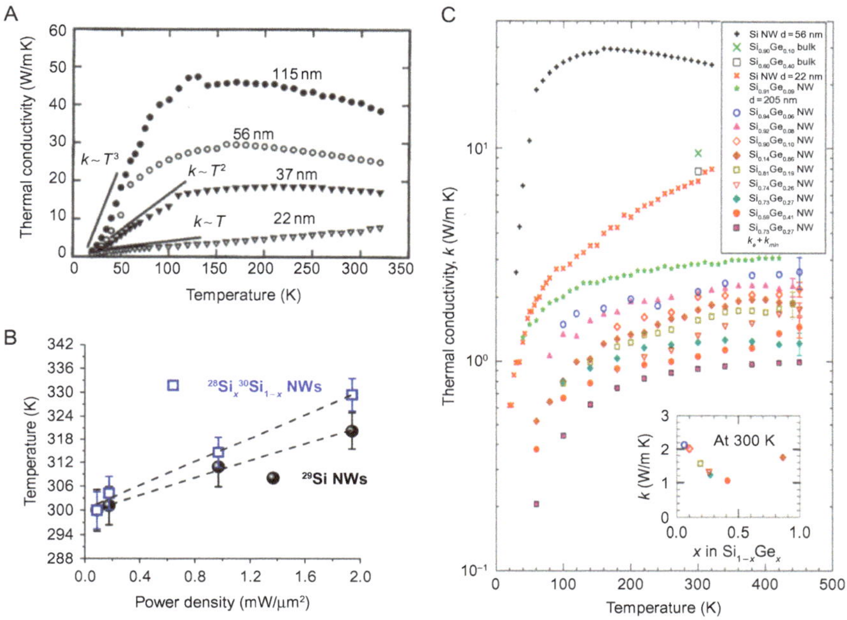

Fig. 5 (A) Evolution of κ_L as a function of temperature for Si nanowires of different diameters. (B) Raman nanothermometry measurement, depicting the evolution of the local temperature of isotopically mixed $^{28}Si_x^{30}Si_{1-x}$ nanowires and isotopically pure ^{29}Si nanowires as a function of incident laser power density. The isotopically mixed nanowires can be seen to get heated up faster as compared to the isotopically pure ones, depicting that the mixed nanowires have a lower κ_L. (C) Evolution of κ_L of Si nanowires of diameter 56 nm (+) and 22 nm (×) from the work of Li et al. (2003b), bulk SiGe alloys of different Ge contents (× and □) and SiGe nanowires of different diameters and Ge contents as a function of temperature. *Inset*: κ_L at 300 K of only the SiGe nanowires as a function of Ge content. *Panel (A): Adapted from Li, D., Wu, Y., Kim, P., Shi, L., Yang, P., Majumdar, A., 2003. Thermal conductivity of individual silicon nanowires. Appl. Phys. Lett. 83, 2934. https://doi.org/10.1063/1.1616981. Pane (B): Adapted from Mukherjee, S., Givan, U., Senz, S., Bergeron, A., Francoeur, S., de la Mata, M., Arbiol, J., Sekiguchi, T., Itoh, K.M., Isheim, D., Seidman, D.N., Moutanabbir, O., 2015. Phonon engineering in isotopically disordered silicon nanowires. Nano Lett. 15, 3885–3893. https://doi.org/10.1021/acs.nanolett.5b00708. Panel (C): Adapted from Lee, E.K., Yin, L., Lee, Y., Lee, J.W., Lee, S.J., Lee, J., Cha, S.N., Whang, D., Hwang, G.S., Hippalgaonkar, K., Majumdar, A., Yu, C., Choi, B.L., Kim, J.M., Kim, K., 2012. Large thermoelectric figure-of-merits from SiGe nanowires by simultaneously measuring electrical and thermal transport properties. Nano Lett. 12, 2918–23. https://doi.org/10.1021/nl300587u.*

gets smaller. In fact, for the nanowire with the smallest diameter of 22 nm, the peak in κ_L was never observed, demonstrating the dominance of the phonon boundary scattering over the anharmonic phonon scattering even at elevated temperatures.

Molecular dynamics-based simulations predicted the κ_L for Si nanowires to vary with the cross-sectional shape (Chen et al., 2011b). Si nanowires of rectangular, circular, and triangular cross-sections of the same surface-to-volume ratio were analyzed. The results showed that as the surface-to-volume ratio is increased at a fixed temperature of 300 K, κ_L of the nanowires with different cross-sections shows almost the same linear decrease all the way down to a critical cross-sectional area of about 20 nm^2. Below the critical cross-sectional area (above the critical surface-to-volume ratio of about 0.13), the decrease in κ_L with increasing surface-to-volume ratio not only deviates from linearity but also starts depending on the cross-sectional shape of the nanowire. Also for the same cross-sectional area, the surface-to-volume ratio of different nanowires is different, with the highest and lowest surface-to-volume ratio belonging to the triangular and the rectangular cross-sections, respectively. The surface-to-volume ratio directly affects the phonon surface scattering rate. As the temperature was varied at a fixed cross-sectional area of 33 nm^2, the κ_L of the nanowire with triangular cross-section was found to be suppressed by almost 20% relative to the nanowire with rectangular cross-section. The nanowire with circular cross-section was found to have an intermediate κ_L. Simulations also showed the κ_L of Si nanowire to vary with the growth direction (Markussen et al., 2008). $\langle 110 \rangle$-oriented ultrathin Si nanowires were found to possess a higher κ_L as compared to those oriented along the $\langle 100 \rangle$ and $\langle 111 \rangle$ directions. The reason for this behavior was attributed to the anisotropy in phonon dispersion along different crystal directions, causing the group velocity of the acoustic mode along the $\langle 110 \rangle$ direction to be the highest of the three well-known orientations of Si nanowires.

Naturally occurring isotopes and other point defects like vacancies and substitutional impurity atoms can act as phonon scattering centers and can be used to optimize the κ_L in nanowires. In as early as 1942, Pomeranchuk commented that isotopes disturb the periodicity of a lattice and contribute to thermal resistance (Pomeranchuk, 1942). Later, Klemens proposed the formula to calculate the rate of phonon scattering from the isotopes as $\tau_{iso}^{-1} = \pi V_0 g_2 \omega^2 N_{d1}(\omega)/6$, where g_2 is the degree of mass disorder, V_0 is the volume per atom, and $N_{d1}(\omega)$ is the one-phonon density of states (Klemens, 1955). With a Debye-like phonon density of states $N_{d1}(\omega) \sim \omega^2$, the expression for τ_{iso}^{-1} reduces to $V_0 g_2 \omega^4 / 4\pi v_D^3$, where g_2 is given by

$$g_2 = \sum_i c_i \left(\frac{m_i - \overline{m}}{\overline{m}} \right)^2 \tag{29}$$

where c_i is the fractional composition of the ith isotope of mass m_i and $\overline{m}$ is the average isotopic mass. Using the virtual crystal approximation, a real lattice can be replaced by a virtual lattice wherein every lattice site is occupied by a virtual atom of average isotopic mass $\overline{m} = \sum_i c_i m_i$. Natural Si ($^{Nat}Si$) is composed of three stable isotopes ^{28}Si, ^{29}Si, and ^{30}Si with natural abundances of 92.3%, 4.7%, and 3.0%, respectively. Although ^{Nat}Si is almost entirely ^{28}Si, the degree of mass disorder (g_2) in a ^{Nat}Si nanowire would still be higher than that in an isotopically pure Si nanowire. ^{Nat}Ge has five stable isotopes with a difference of 6 amu between the extreme isotopes. The relative abundances of ^{70}Ge, ^{72}Ge, and ^{74}Ge in ^{Nat}Ge are also comparable which makes the g_2 of ^{Nat}Ge higher than that of ^{Nat}Si. Molecular dynamics simulations showed that as the atoms within an isotopically pure ^{28}Si nanowire are substituted with a heavier isotope, κ_L starts to decrease and reaches a minimum when the fractional composition of the two isotopes is close to one another (Yang et al., 2008). The work used substitutional ^{29}Si and the unstable ^{42}Si, which is practically irrelevant. The vapor–liquid–solid (VLS) growth of isotopically pure Si nanowires (Moutanabbir et al., 2009) and isotopically mixed Si nanowires (Mukherjee et al., 2015) was also reported. The phonon transport in the isotopically mixed $^{28}Si_x^{30}Si_{1-x}$ nanowires was compared to the isotopically pure ^{29}Si nanowires using Raman nanothermometry, as shown in Fig. 5B. A 25%–30% reduction in κ_L of the $^{28}Si_x^{30}Si_{1-x}$ nanowires relative to the ^{29}Si nanowires was demonstrated, close to room temperature (Mukherjee et al., 2015).

κ_L of alloyed $Si_{1-x}Ge_x$ nanowires was studied as a function of Ge content (Lee et al., 2012; Wang and Mingo, 2010). Fig. 5C shows the evolution of κ_L with temperature for Si nanowires of different diameters with different Ge contents (Lee et al., 2012). A comparison across different diameters and Ge contents becomes difficult at a first glance in Fig. 5C, since both of them can influence κ_L. The inset provides a good comparison where the empty blue circle represents $Si_{0.94}Ge_{0.06}$ nanowires of diameter 56 nm and the filled red circle represents $Si_{0.59}Ge_{0.41}$ nanowires of 65 nm diameter. The nanowires have comparable diameters and therefore should have comparable rates of phonon boundary scattering. Close to a 50% reduction in κ_L can be seen for the $Si_{0.59}Ge_{0.41}$ nanowires which can be attributed to alloy scattering. Obviously, the mass difference between Si and Ge is much more than between any two Si isotopes. So, the disorder-induced reduction in κ_L is much enhanced in SiGe nanowires than in isotopically disordered Si nanowires.

Homo- and heterointerfaces also induce phonon scattering and have been a subject of detailed theoretical investigation in bulk and nanowires (Chalopin et al., 2012; Chen et al., 2009; Dames and Chen, 2004;

Hu and Poulikakos, 2012; Schelling et al., 2002; Yang et al., 2008). In both $^{28}Si/^{29}Si$ isotopic superlattice (Yang et al., 2008) and Si/Ge superlattice nanowires (Chen et al., 2009), the simulations revealed that the presence of homo- and heterointerfaces reduces the κ_L compared to a nanowire without any interfaces. The result obtained for the Si/Ge superlattice nanowire is shown in Fig. 6A. For a fixed length of a nanowire, as the length of each period in the superlattice structure is reduced, κ_L reduces due to the increase in the number of interfaces. The decrease continues up to a certain critical length of each period which was 1.09 nm for the isotopic superlattice and 1.11 nm for Si/Ge superlattice. Below the critical thickness, κ_L increases sharply with a further decrease in the superlattice period length. The dominant phonon wavelength in most materials is about 1–2 nm, the same order as that of the critical superlattice period length. At the critical period length, there occurs a crossover from the incoherent particle-like phonons to the coherent Bloch wave-like phonons. Below the critical period length, transport is conducted by the coherent phonon Bloch waves which causes the sharp increase in κ_L. In order to experimentally observe this sharp increase in κ_L below the critical superlattice period length, the interfaces must be atomically abrupt. Interfacial mixing of atoms (isotopes) can lead to a crossover to incoherent phonon transport and the κ_L will keep decreasing monotonically with decreasing period length (Landry and McGaughey, 2009). However, it is still an open challenge even for the state-of-the-art nanowire synthesis technique, to achieve a period length down to ~1 nm and interfaces that are atomically abrupt. Experimentally, the evolution of κ_L of Si/SiGe superlattice nanowires of diameter 83 and 58 nm and period length ranging from 50 to 150 nm was measured and compared with Si/SiGe superlattice thin film and SiGe alloyed thin film (Li et al., 2003a). The results showed that the alloy scattering was the dominant scattering mode in the SiGe segment of the superlattice nanowires. Also, the enhanced boundary scattering in the nanowires contributed to additional reduction in κ_L relative to the thin films.

The effect of a core–shell nanowire structure on κ_L was also investigated theoretically (Hu et al., 2011a,b; Yang et al., 2005). For example, in Si/Ge core–shell nanowires, it was shown that only a couple of layers of Ge atoms at the shell can lower the κ_L by about 25% relative to Si nanowires of the same diameter but devoid of any shell, as shown in Fig. 6B (Hu et al., 2011a). The reason for the reduction in κ_L was discussed to be due to interface phonon scattering and the lower κ_L of Ge relative to Si. The localization of some long wavelength phonon modes at the Si/Ge interface could have

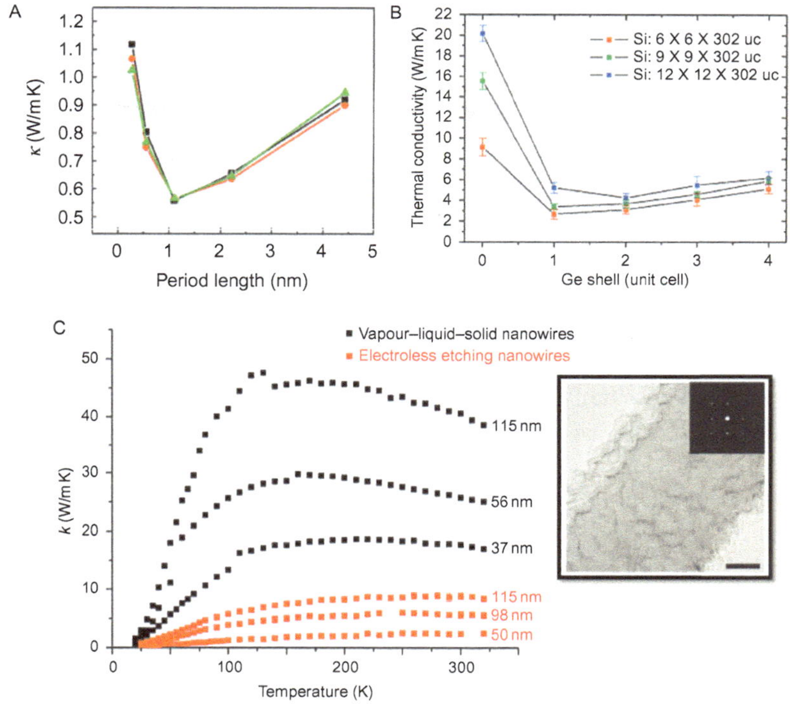

Fig. 6 (A) Calculated profile of κ_L at 300 K for Si/Ge superlattice nanowires as a function of superlattice period length. For the calculations, three different scenarios were considered: Si atoms touching both the heat baths (■), Ge atoms touching both the heat baths (●) and Si atoms touching one heat bath, and Ge atoms touching the other heat bath (▲). (B) Calculated overall thermal conductivity of the Si/Ge core–shell nanowire as a function of the number of layers in the Ge shell for different cross-sectional areas of the Si core. (C) Evolution of κ_L for Si nanowire as a function of temperature for smooth VLS-grown nanowires from the work of Li and coworkers (*black squares*) (Li et al., 2003b) and EE synthesized rough nanowires (*red squares*). *Inset*: Bright-field TEM image showing the rough surface of an EE synthesized Si nanowire. The scale bar denotes 20 nm. *Panel (A): Adapted from Chen, J., Zhang, G., Li, B., 2009. Tunable thermal conductivity of $Si_{1-x}Ge_x$ nanowires. Appl. Phys. Lett. 95, 73117. https://doi.org/10.1063/1.3212737. Panel (B): Adapted from Hu, M., Giapis, K.P., Goicochea, J.V., Zhang, X., Poulikakos, D., 2011. Significant reduction of thermal conductivity in Si/Ge core–shell nanowires. Nano Lett. 11, 618–623. https://doi.org/10.1021/nl103718a. Panel (C): Adapted from Hochbaum, A.I., Chen, R., Delgado, R.D., Liang, W., Garnett, E.C., Najarian, M., Majumdar, A., Yang, P., 2008. Enhanced thermoelectric performance of rough silicon nanowires. Nature 451, 163–167. https://doi.org/10.1038/nature06381.*

also played a role in the κ_L reduction. Having thicker Ge layers at the shell was found to increase the κ_L of the overall nanowire due to progressively larger contribution of the shell to the overall thermal conduction. Core–shell nanowires also allow the option to tune the κ_L by exploiting the phonon coherent resonance effect. In any given atomic plane within a core–shell nanowire, atoms at the core possess different vibrational frequencies and longitudinal sound velocity from that at the core. The difference is nonexistent in a single-component homogeneous nanowire, small if the nanowire is an isotopically modulated core–shell structure, and could be significant in a nanowire like Si/Ge where the core and the shell are made of atoms of two different elements. For single-component homogeneous nanowires and to a certain extent in isotopically modulated core–shell nanowires, the transverse vibration of atoms is decoupled from the longitudinal vibration. In a Si/Ge or Ge/Si core–shell nanowire, a resonance coupling of the localized transverse modes and the propagating longitudinal modes can occur when the frequency of the longitudinal mode is close to the quantized eigenfrequency of the transverse mode (Chen et al., 2011c). Since the transverse modes are localized, such a resonance coupling can perturb longitudinal thermal transport in the overall nanowire. Simulations showed close to a 40% reduction in the room temperature κ_L of Ge/Si core–shell nanowires of core to shell thickness ratio of 0.65, relative to Ge nanowires of the same overall diameter (Chen et al., 2011c). Other theoretical studies also indicated the reduction in κ_L of Ge/Si (Chen et al., 2012) and Si/Ge (Nika et al., 2013) core–shell nanowire as a consequence of the resonance-coupling effect.

In addition to the mass disorder addressed earlier, lattice defects such as twin boundaries can also influence heat transport in nanowires. A twin boundary is a single atomic plane that separates two neighboring crystal domains with very specific relative crystallographic orientations (Davidson et al., 2007). The twin planes in Si and Ge nanowires are often found to lie along the length of a nanowire that kink along the $\langle 121 \rangle$ direction at some point during the growth from the $\langle 111 \rangle$ direction (Davidson et al., 2007; Lopez et al., 2009, 2011). This is in contrast to the III–V nanowires where the twin planes are often found to lie perpendicular to the nanowire axis in the form of twinned axial homojunctions (Algra et al., 2008; Caroff et al., 2009). The effect of stacking faults and polytypic crystal phase on the κ_L of Si nanowires was studied theoretically (Zhan et al., 2014). Simulations showed that the presence of external and internal stacking faults as well as the rhombohedral (9R) polytypic crystal phase causes a 15% reduction in the room temperature κ_L, relative to a Si nanowire devoid of any faulted stacking layers.

Experimentally, the combined effect of the isotopic disorder and polytypic crystal phase on the κ_L in Si nanowires was studied using Raman nanothermometry (Mukherjee et al., under review). Isotopically pure and isotopically mixed Si nanowires of pure diamond cubic (3C) and polytypic (3C + 9R) crystal phases were grown by the VLS technique. The results showed that the reduction in κ_L of a polytypic nanowire depends on the fractional volume of the nanowire occupied by the polytypes. The reductions were observed to be about 6% and 14% relative to a 3C nanowire close to room temperature when the fractional volume of the 9R crystal phase within a polytypic nanowire is smaller than and equal to the volume of the 3C phase, respectively. Also, the κ_L of the isotopically mixed 9R crystal phase is reduced by about 40% close to room temperature relative to the isotopically pure 3C Si nanowire.

Surface engineering can also be an effective tool to optimize the κ_L by controlling the surface scattering rate of phonons. There are several theoretical studies in the literature aimed at surface engineering in Si nanowires (Martin et al., 2009a, 2010; Moore et al., 2008; Sansoz, 2011). The studies showed that the high-frequency phonons are more affected by the presence of a rough nanowire surface as compared to the low-frequency phonons (Martin et al., 2010). Also, the linear dependence of κ_L on the nanowire diameter (d) in Si nanowires with rough surface changes into a $(d/\Delta)^2$ dependence, where Δ is the root-mean-square height of the surface roughness (Martin et al., 2009a). Fig. 6C shows the evolution of κ_L as a function of temperature for VLS-grown Si nanowires from the work of Li et al. (2003b) (black squares) and rough Si nanowires synthesized by an electroless etch (EE) method (red squares) (Hochbaum et al., 2008). The bright-field transmission electron microscope (TEM) image of a segment of an EE Si nanowire, depicting the presence of a rough surface, is shown in the inset of Fig. 6C. The reduction in κ_L of the rough Si nanowires is quite evident. Room temperature κ_L of about 40 and 25 W/mK for the VLS-grown nanowires of diameter 115 and 56 nm reduces to below 10 and below 2 W/mK, respectively, in the EE rough Si nanowires of similar diameter. Recently, VLS-grown Si nanowires were imparted with controlled amount of surface roughness by the EE process and the effect on the root-mean square height (Δ) and covariance length (l_c) of the surface roughness on κ_L was studied (Lim et al., 2012). The covariance length is a statistical parameter that determines the mean distance between the roughness peaks at the surface of a nanowire. The study showed that in order to understand the dependence of κ_L on the surface roughness in Si nanowires, it is important to correlate κ_L not just to Δ but also to l_c and the power spectrum of the surface roughness in the wavelength range of $1-100$ nm. The power spectrum is the Fourier transform of the surface roughness autocovariance function (Goodnick, 1983; Lim et al., 2012; Martin et al., 2009a).

The effect of functionalization of the surface of a Si nanowire on its κ_L was also studied. It is well known that functionalization passivates the surface electronic states and can improve the electronic transport. Theoretical studies showed that functionalization can also lead to reduction in κ_L of Si nanowires (Li et al., 2011; Markussen et al., 2009). For example, simulations showed a room temperature reduction in κ_L by about 40% and by factors of 2.5 and 3.8 for Si nanowires which, respectively, had 14%, 29%, and 43% of the surface Si atoms functionalized with nitrogen, relative to Si nanowires without nitrogen functionalization (Li et al., 2011). The reduction in κ_L was attributed to the suppression of some phonon modes and surface defect-induced enhanced phonon scattering. The presence of an amorphous oxide shell around a Si nanowire can also affect its thermal behavior, as was found theoretically (Donadio and Galli, 2010). The study found quite contrasting behaviors of κ_L as a function of temperature between Si–SiO_2 core–shell-type nanowires and homogeneous Si nanowires of the same total diameter but devoid of any amorphous shell. While κ_L in the high-temperature regime (200–600 K) falls off as $\sim T^{-1}$ with increasing T for the homogeneous crystalline Si nanowires, κ_L was found to be temperature independent for the Si–SiO_2 core–shell-type nanowires. It was discussed that the presence of the crystalline–amorphous interface in the core–shell Si nanowires plays a major role in turning the propagating vibrational modes in the crystalline core into diffused modes of the amorphous shell. The combined action of these propagating and the diffused modes gave rise to the peculiar temperature-independent behavior of κ_L.

To sum up this section on thermal transport, it is important to mention that phonon engineering has shown to be a promising paradigm in tuning κ_L in Si nanowires. The general principles of phonon engineering apply to all types of nanowires. Theoretical and experimental studies found several parameters that can influence the phonon transport like the nanowire diameter, orientation, cross-sectional shape, presence of defects, isotopic modulation, alloying, and interfaces. Each of these scattering mechanisms dominates its influence on the phonon transport only in a certain frequency range. For example, the high-frequency phonons are sensitive to the surface roughness in nanowires, whereas the low-frequency phonons are not. The low-frequency phonons on the other hand are more sensitive to the coherent resonance in core–shell nanowires. The available experimental data, however, depart from theoretical predictions simply because nanowire dimensions which are often studied in theoretical simulations (like nanowires with a diameter of only few nm, atomically sharp interfaces with the period length of few nm in a superlattice nanowire) are still far out reach of what can currently be achieved in practice.

2.5 *ZT* Investigations in Semiconductor Nanowires

This section highlights some of the current theoretical and experimental investigations on the thermoelectric figure of merit in semiconductor nanowires. This includes the group IV nanowires and alloys as well as group III–V nanowires and others.

2.5.1 Group IV Nanowires (Elemental and Alloys)

ZT measurements were carried out on Si nanowires grown by the superlattice nanowire pattern transfer (SNAP) method from doped Si wafers (Boukai et al., 2008). In the SNAP process, the dopant type and concentration of the starting wafer are transferred on to the nanowires. Fig. 7A shows the evolution of ZT as a function of temperature of p-type Si nanowires of two different diameters of 10 and 20 nm and boron doping concentrations of 2×10^{20} and $7 \times 10^{19}\,\text{cm}^{-3}$, respectively (Boukai et al., 2008). Fig. 7A shows that the ZT at 200 K for the Si nanowires of 20 nm diameter reaches a value of unity. Fig. 7B shows the evolution of the power factor (S^2/ρ) and ZT as a function of temperature of rough Si nanowires synthesized by the EE technique (Hochbaum et al., 2008). Like the SNAP process, for the EE synthesized nanowires, the doping concentration of the starting wafer is transferred on to the nanowires. The rough Si nanowires of diameter close to 50 nm, root-mean-squared roughness of about 3 nm, and doping concentration of about $10^{19}\,\text{cm}^{-3}$ have a ZT of 0.6 at 300 K. Both the reported ZT values for the SNAP and EE synthesized Si nanowires at 200 and 300 K, respectively, are about two orders of magnitude higher than the ZT of bulk Si at the same temperatures. Theoretical studies conducted on the thermoelectric properties of Ge nanowires indicate that the power factor of Ge nanowires is only slightly smaller than that of Si nanowires of similar diameter and doping concentrations (Huang et al., 2010). The κ_L of Ge nanowires on the other hand can be much lower than Si nanowires of the same diameter and root-mean-squared surface roughness (Martin et al., 2010), simply because the κ_L of bulk Ge (56 W/m K at 300 K) is lower than that of bulk Si. Although the ZT of Ge nanowires can be expected to be similar to the Si nanowires, a detailed experimental study of the same is yet to be undertaken. SiGe alloyed nanowires reported earlier (Lee et al., 2012) were found to possess a κ_L that depends strongly on the Ge content and also weakly on the nanowire diameter.

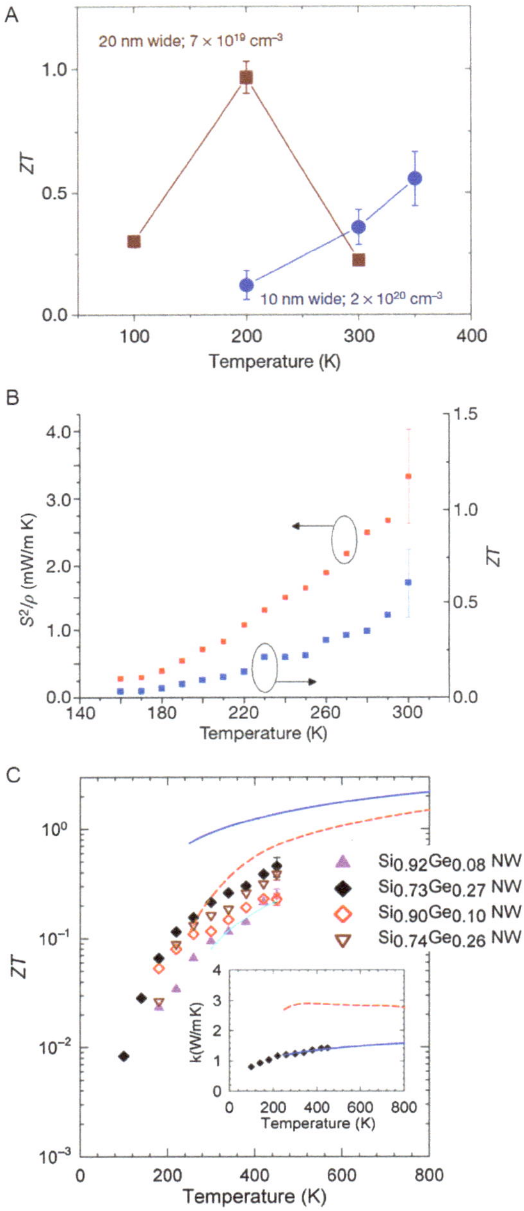

Fig. 7 (A) Evolution of *ZT* as a function of temperature of SNAP synthesized Si nanowires of two different diameters of 20 and 10 nm and of boron doping concentrations of 7×10^{19} and $2 \times 10^{20}\,cm^{-3}$, respectively. (B) Evolution of power factor (S^2/ρ) and *ZT* as a function of temperature of rough Si nanowires synthesized by the EE technique. The uncertainties in measuring ρ and S measurements propagate into the power factor

It was shown that a κ_L of SiGe nanowires can get as low as 1–2 W/m K at room temperature, which is much lower than the κ_L of 7–8 W/m K of bulk SiGe alloys. Nanostructured bulk SiGe alloys were found to have a ZT close to unity between 800°C and 900°C (Joshi et al., 2008). Fig. 7C shows a comparison between the measured ZT of individual SiGe nanowires as a function of temperature. The highest ZT of 0.46 at 450 K was reported for the $Si_{0.73}Ge_{0.27}$ nanowires (Lee et al., 2012). n-Type β-SiC nanowires of diameter between 70 and 90 nm were found to possess a κ_L of 5–12 W/m K between 190 and 370 K, an order of magnitude smaller than bulk SiC (Valentín et al., 2013). The highest S of about −70 μV/K at 370 K was found for the 90 nm diameter nanowires, which gave a ZT of about 3×10^{-3}.

2.5.2 III–V Nanowires

In the thermoelectric transport calculation from the work of Mingo, InSb nanowires were assigned with the highest room temperature ZT, GaAs, InAs, and InP being the other III–V nanowires in the list (Mingo, 2004). For InSb nanowires with diameter 30 nm, the room temperature ZT was shown to exceed unity, much higher than the value for bulk InSb. Thermoelectric measurements were conducted on VLS-grown InSb nanowires of diameters exceeding 100 nm and doping concentration in the order of 10^{17}–10^{18} cm^{-3} (Zhou et al., 2010). The measurements revealed that nanowires possess a lower κ_L and a comparable σ relative to Te-doped bulk InSb. But the S for the InSb nanowires was significantly lower making the room temperature ZT much smaller than what was reported in bulk InSb

and ZT which are denoted by error bars which are 21% for the power factor and 31% for ZT. (C) Evolution of ZT as a function of temperature of $Si_{0.92}Ge_{0.08}$, $Si_{0.90}Ge_{0.10}$, $Si_{0.74}Ge_{0.26}$, and $Si_{0.73}Ge_{0.27}$ nanowires and bulk SiGe alloys (*sky blue line*). The calculated ZT for the $Si_{0.74}Ge_{0.26}$ nanowires for two different doping concentrations of 9.0×10^{19} cm^{-3} (*deep blue line*) and 3.3×10^{20} cm^{-3} (*red dash line*). *Inset*: Evolution of κ_L as a function of temperature for the $Si_{0.74}Ge_{0.26}$ nanowires along with calculated κ_L with doping concentrations of 9.0×10^{19} cm^{-3} (*deep blue line*) and 3.3×10^{20} cm^{-3} (*red dash line*). *Panel (A): Adapted from Boukai, A.I., Bunimovich, Y., Tahir-Kheli, J., Yu, J.K., Goddard 3rd, W.A., Heath, J.R., 2008. Silicon nanowires as efficient thermoelectric materials. Nature 451, 168–171. https://doi.org/10.1038/nature06458. Panel (B): Adapted from Hochbaum, A.I., Chen, R., Delgado, R.D., Liang, W., Garnett, E.C., Najarian, M., Majumdar, A., Yang, P., 2008. Enhanced thermoelectric performance of rough silicon nanowires. Nature 451, 163–167. https://doi.org/10.1038/nature06381. Panel (C): Adapted from Lee, E.K., Yin, L., Lee, Y., Lee, J.W., Lee, S.J., Lee, J., Cha, S.N., Whang, D., Hwang, G.S., Hippalgaonkar, K., Majumdar, A., Yu, C., Choi, B.L., Kim, J.M., Kim, K., 2012. Large thermoelectric figure-of-merits from SiGe nanowires by simultaneously measuring electrical and thermal transport properties. Nano Lett. 12, 2918–2923. https://doi.org/10.1021/nl300587u.*

(Yamaguchi et al., 2005). The peak *ZT* of only 0.01 was observed for the InSb nanowires at 560 K. The authors attributed the low *ZT* value to the loss of In due to oxidation and the formation of Sb-doped nanowires. InAs nanowires with their high electron mobility (Dayeh et al., 2007) and lower κ_L relative to bulk InAs (Karg et al., 2013; Persson et al., 2009) are a promising material for thermoelectric applications. Although the σ and κ_L of single InAs nanowires of diameter between 20 and 80 nm were measured, the measurement of *S* and therefore the estimate of *ZT* were missing (Karg et al., 2013). In another study, the power factor in InAs nanowires of diameter between 50 and 70 nm was shown to be an improvement from the theoretical prediction (Wu et al., 2013), but measurement of κ_L and therefore the estimate of *ZT* were missing. A complete measurement was presented for suspended InAs nanowires, synthesized by the top-down approach on InAs thin film (Mavrokefalos et al., 2006). The nanowires were a few microns long and rectangular in cross-section with a width of about 150 nm and a thickness of about 40 nm. The reported *ZT* at room temperature was as low as 0.00032. The reason for the low *ZT* was attributed to the extremely low electronic mobility due to the strain-induced defects in the parent InAs films that were grown on lattice-mismatched GaAs substrate. Although there are a few theoretical studies on the thermoelectric properties of GaAs nanowires, no experimental verification of these studies exists. Bulk GaAs is well known to be an inefficient thermoelectric material, but calculation showed that thin GaAs nanowires of diameter 1.1 nm possess a room temperature *ZT* of about 1.34 which is two orders of magnitude higher than bulk GaAs (Zou et al., 2015). The theoretical study also shows that by surface roughness engineering the κ_L of GaAs nanowires of 56 nm diameter can get as low as 1 W/mK (Martin et al., 2009b). CVD-grown GaN nanowires of diameters between 97 and 181 nm showed a room temperature κ_L of $13-19$ W/mK, which is an order of magnitude reduction compared to bulk GaN (Guthy et al., 2008). Theoretical calculations also showed that GaN nanowires of diameter between 3 and 9 nm and doping concentrations between 10^{18} and $10^{20}\,cm^{-3}$ can attain a room temperature *ZT* of 0.2 for a 4 nm nanowire and a maximum *ZT* of 0.8 at 1000 K for a 3 nm nanowire (Davoody et al., 2014). No experimental study on the *ZT* in GaN nanowires exists.

2.5.3 Thermoelectric Properties of Other Nanowires

Table 1 lists a few noteworthy *ZT* measurements conducted on other types of nanowires like the telluride- and selenide-based nanowires.

Nanowire-based thermoelectric devices could be operated either as power generators or as refrigerators. A popular prototype for a semiconductor

Table 1 A Summary of *ZT* Measurements on Other Nanowires

Nanowire	References	*ZT*	Main Conclusions
Single n-type PbTe	Lee et al. (2011)	0.0054 (300 K)	Improvement from bulk PbTe which has *ZT* of 0.002 at 300 K
Single p-type Bi_2Te_3	Hsin et al. (2013)	0.2 (500 K)	*ZT* same as bulk Bi_2Te_3 due to poor electrical conductivity of the nanowires. Before phase transformation into Bi_2Te_3, BiTe nanowires were also measured. High-temperature *ZT* of Bi_2Te_3 nanowires is significantly higher than the BiTe nanowires
Single p-type PbSe	Liang et al. (2009)	0.12 (300 K)	For the as-grown PbSe nanowires, *ZT* was found to be 0.04 at 300 K. Presence of defects and surface states led to poor electronic mobility. Annealed PbSe nanowires showed improvement of defect density and had a room temperature *ZT* which was three times that of as-grown nanowires
Single p-type SnTe	Xu et al. (2015)	0.0379 (300 K)	σ did not show any strong diameter dependence but *S* did. Therefore, $S^2\sigma$ increased by a factor of 2 when the nanowire diameter was reduced from 913 nm to 218 nm. κ_L of the nanowires was significantly lower than bulk SnTe. Room temperature *ZT* of SnTe nanowires was almost three times that of bulk SnTe of comparable doping density
Single n-type ZnO	Liu et al. (2011)	0.10 (970 K)	κ_L of the ZnO nanowires was significantly lower than bulk ZnO of similar doping concentration. At the high-temperature regime, κ_L of the nanowires was found to fall faster with temperature than what is predicted by three-phonon Umklapp scattering, signifying the presence of four-phonon anharmonic scattering
Single p-type BiSbTe	Li et al. (2015)	1.14 (330 K)	The thermoelectric performance of BiSeTe nanowires grown by dc electrodeposition and pulsed electrodeposition was compared. The best performance was showed by pulsed electrodeposited $Bi_{0.5}Sb_{1.5}Te_3$ nanowires with *ZT* (280 K) of 0.48 and *ZT* (330 K) of 1.14. Comparatively, dc electrodeposited $Bi_{0.5}Sb_{1.5}Te_3$ nanowires had *ZT* (280 K) of 0.28 and *ZT* (330 K) of 0.74

nanowire-based thermoelectric device, as proposed by Abramson et al., is shown in Fig. 8 (Abramson et al., 2004). It consists of an assembly of vertical nanowires coated with a polymer, after the growth. Such a device based on an assembly of nanowires could support more heat and current fluxes than those based on a single nanowire. The vertical nanowires could be synthesized by any of the well-known bottom-up (like the VLS) or top-down (like EE) methods. The top of the polymer is then etched to expose the tips of the

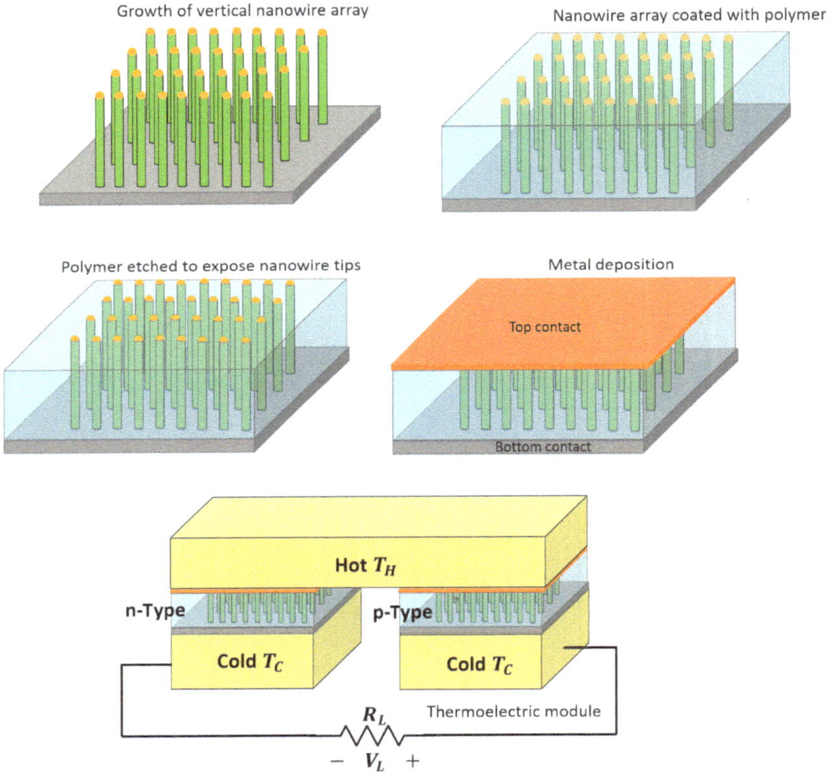

Fig. 8 Steps for fabricating a nanowire array-based thermoelectric device: growth of an array of vertical nanowires by either a bottom-up or a top-down method, coating the nanowire array with a polymer, etching the top of the polymer to expose the nanowire tips, laying metallic contacts on top, while the highly doped substrate form the bottom contact, connecting p-type and n-type legs to form a thermoelectric module. *Drawn from the concepts outlines in Abramson, A.R., Kim, W.C., Huxtable, S.T., Yan, H., Wu, Y., Majumdar, A., Tien, C.-L., Yang, P., 2004. Fabrication and characterization of a nanowire/ polymer-based nanocomposite for a prototype thermoelectric device. J. Microelectromech. Syst. 13, 505–513. https://doi.org/10.1109/JMEMS.2004.828742.*

nanowires onto which a metallic contact layer is deposited. The substrate for the nanowire growth needs to be conducting as well (like degenerately doped Si wafer) to form the bottom contact. Finally, p-doped and n-doped nanowire arrays can be coupled to form a thermoelectric module. Most of the nanowire-based devices reported so far are power generators (Curtin et al., 2012; Dávila et al., 2011; Li et al., 2012; Wang et al., 2005; Xu et al., 2012). However, as pointed out in the discussion by Majumdar (2009), the more demanding future application of nanowire-based thermoelectric devices could be for the purpose of refrigeration than power generators. These days more than one processor is being put on a single chip in dual- and multicore computers to attain more computing power at the same clock speed and input power. Although the average dissipated power over the whole chip stays relatively the same, there occur local hot spots with peak power dissipation reaching $100\,W/cm^2$, raising the local temperature at these hot spots beyond what can be cooled off by current heat management technologies. This is where nanowires could come in handy as efficient local thermoelectric refrigerators in multiprocessor chips.

3. SN-CONTAINING GROUP IV NANOWIRES AND THEIR POTENTIAL APPLICATIONS IN PHOTOVOLTAICS

As outlined in the previous section, group IV nanowires provide a rich platform to engineer phonon and heat transport. In the following, we unravel a fairly new approach to expand the capabilities of this family of nanowires and make them even more relevant to photovoltaic applications. The basic concept to be elaborated here is related to the growth of SiGeSn and GeSn nanostructure semiconductors. We first start by providing the broad context of these emerging semiconductors.

3.1 Monolithic SiGeSn 1.0 eV Subcell for Hybrid Multijunction III–V Solar Cells

Engineering light absorption and conversion in semiconductors has been a powerful paradigm to implement innovative and high-performance solar cells (Green and Bremner, 2017). III–V semiconductor multijunction technology is presently the industry standard with record efficiencies reaching 46% (Green, 2016; Green and Bremner, 2017; Green et al., 2017; Soitec et al., 2014; Takamoto et al., 2014). The interest in III–V multijunction solar cells is currently driven by space applications, but the largest potential market for these cells is certainly terrestrial. In this regard, terrestrial concentrator systems utilizing high-efficiency III–V multijunction solar cells have been increasingly considered for large-scale generation of electrical power (Green and Bremner, 2017). However, a breakthrough in this long-sought-for integration of III–V multijunction solar cells in large-scale consumer applications will depend primarily on the ability to establish new scalable technologies with higher efficiencies and significant material cost saving (Green, 2016; Green and Bremner, 2017; Green et al., 2017; Soitec et al., 2014; Takamoto et al., 2014). The current state-of-the-art multijunction solar cells are grown on bulk gallium arsenide (GaAs) or Ge substrates, which are prohibitively expensive. For instance, bulk Ge is currently used for the growth of both lattice-matched (Dimroth, 2006; King et al., 2007; Yamaguchi et al., 2006) and metamorphic (Guter et al., 2009) InGaP/InGaAs/Ge triple-junction solar cells. The competitiveness and scalability of these technologies are limited by the high cost of bulk Ge, which is expected to surge even higher in the near future due to an increasing demand that may outpace the world supply. This is mainly caused by the high military usage of Ge in IR missile guidance technologies and surveillance cameras, combined with the Ge production being mainly controlled by the Chinese export, which produces up to 80% of the world's demand. Therefore, the development of any bulk Ge-based technology needs to consider these factors in addition to the increasing demand on Ge in the multibillion-dollar microelectronics industry ("The International Road Map for Semiconductor Technologies," 2015).

A key strategy in developing a new family of multijunction solar cells is to introduce epitaxial silicon–germanium–tin (SiGeSn) in the fabrication of III–V solar cells. The fact that SiGeSn bandgap can be tuned around 1.0 eV while being lattice-matched to Ge and GaAs (Attiaoui and Moutanabbir, 2014; Fang et al., 2008) will enable an optimal absorption of solar spectrum, which is central to further enhance the conversion efficiency. Moreover, the fact that high crystalline quality SiGeSn and Ge layers can be grown on Si wafers (Fournier-Lupien et al., 2013; Wirths et al., 2016) will also eliminate the need for bulk Ge and enable the integration of multijunction solar cells onto mechanically robust and inexpensive Si substrates. This is key to achieve the cost effectiveness crucial for large-scale, terrestrial applications of III–V solar cells.

III–V triple-junction solar cells made of $In_{0.49}Ga_{0.51}P/In_{0.01}Ga_{0.99}As/Ge$ layers are commonly grown on Ge substrates as shown in Fig. 9 (King et al., 2007; Wang et al., 2017). Theoretical proposals suggest that the efficiency of this configuration can be further increased by incorporating a fourth junction between the Ge cell and the GaAs cell (Dimroth and Kurtz, 2007).

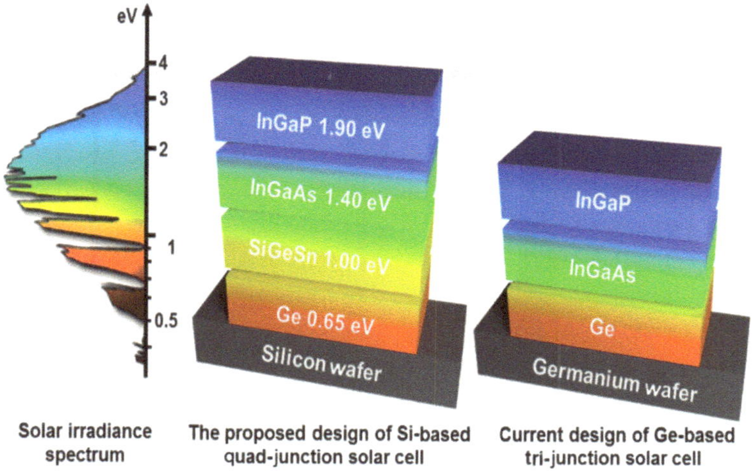

Fig. 9 Schematic illustration of the proposed design of lattice-matched 4*J* solar cell with an additional SiGeSn 1.0 eV subcell directly grown on a Si wafer.

The material in this fourth cell should have a direct bandgap close to 1.0 eV. This can be accomplished using three different strategies: inverted metamorphic growth (France et al., 2015, 2016; Patel et al., 2012; Stan et al., 2010); direct wafer bonding (Chiu et al., 2014; Dimroth et al., 2016); or monolithic growth (Jackrel et al., 2007; Ochoa et al., 2016; Tan et al., 2011; Thomas et al., 2014). The latter is highly attractive as it enhances device efficiency without the need for bonded wafers with poor structural and electrical properties at the interface (Essig et al., 2013; Moutanabbir and Gösele, 2010), or for metamorphic structures, which inherently contain many defects in the graded region and are generally thicker due to the buffer layer, thus making their fabrication more difficult (Essig et al., 2013; France et al., 2016; Jackrel et al., 2007; Moutanabbir and Gösele, 2010). Indeed, efficiencies of up to 44.0% at 942 suns were recently achieved using this kind of cells with a ~1.0 eV dilute nitride subcell (GaInNAsSb) grown on GaAs or Ge substrates (Solar Junction sets CPV cell efficiency record of 44%, 2012). Notwithstanding these significant achievements, the need for the costly GaAs and Ge substrates remains a huge burden that may limit the impact of this technology outside the current niche market.

By exploiting the lattice parameter and bandgap tunability of SiGeSn alloys, a $Si_{0.08}Ge_{0.90}Sn_{0.02}$ subcell is the ideal candidate for the proposed fourth junction (Fang et al., 2008). This alloy has a lattice parameter that is very close to both Ge and GaAs, reducing the strain in the system and the amount of defects across the multilayer stacking. The proposed design consists of a lattice-matched quad-junction (4-J) $In_{0.49}Ga_{0.51}P$/ $In_{0.01}Ga_{0.99}As$/$Si_{0.08}Ge_{0.90}Sn_{0.02}$/Ge directly grown on a Si wafer, as illustrated in Fig. 9. The success in implementing this monolithic 4-J will enable a new generation of solar cells compatible with Si processing, providing both cost effectiveness and high efficiency. These lattice-matched 4-J solar cells have the potential to reach very high efficiencies (+48%) (Wilson et al., 2016) that may enable their use in both terrestrial and space applications.

In addition to the control over the composition of the multijunction heterostructure, major research efforts have been focused on enhancing light trapping in the active layer using textured surfaces and coatings to improve the device performance (Nositschka et al., 2003; Schneider et al., 2014). The high aspect ratio and enhanced optical properties using nanowires are expected to boost the efficiency of nanowire-based solar cells. The nanowire geometry provides excellent antireflection properties over a large wavelength range (Diedenhofen et al., 2011; van Dam et al., 2016), with a reduced material consumption with respect to conventional planar schemes. The concentration of the incoming light due to the nanowire geometry leads to higher absorption, which strongly enhances the performances of nanowire-based solar cells (Diedenhofen et al., 2011; Kempa et al., 2012; Krogstrup et al., 2013; van Dam et al., 2016). In particular, Sn-rich SiGeSn alloys grown in a core/shell nanowire geometry are expected to enhance the near-IR (NIR) light absorption and boost the efficiency of Si-based nanostructures (Attiaoui et al., under review). The predicted NIR absorption enhancement in Si/SiGeSn core/shell nanowires is four orders of magnitude higher than in Si nanowires, which leads to an ~14-fold increase in photocurrent of a SiGeSn nanowire-based PV device (Attiaoui et al., under review). The possibility to independently tailor the diameter of the core and the thickness of the shell allows for spectral tuning of the absorption resonances, therefore providing an additional degree of freedom in the PV design. Furthermore, similar enhancement in the light absorption is also predicted when a Ge core is used, extending the operational wavelength range toward mid-IR (MIR) with Ge/SiGeSn core/shell nanowires (Attiaoui et al., under review).

The capabilities of SiGeSn nanowire-based PV devices can be further expanded by implementing innovative III–V/IV hybrid nanostructures based on the multijunction stacking discussed in Fig. 9. Based on this, a new type of hybrid nanowire solar cell can be designed using a Ge core and three different multishell materials: $Si_{0.08}Ge_{0.90}Sn_{0.02}$; $In_{0.01}Ga_{0.99}As$; and $In_{0.49}Ga_{0.51}P$, as sketched in Fig. 10. The enhanced strain relaxation

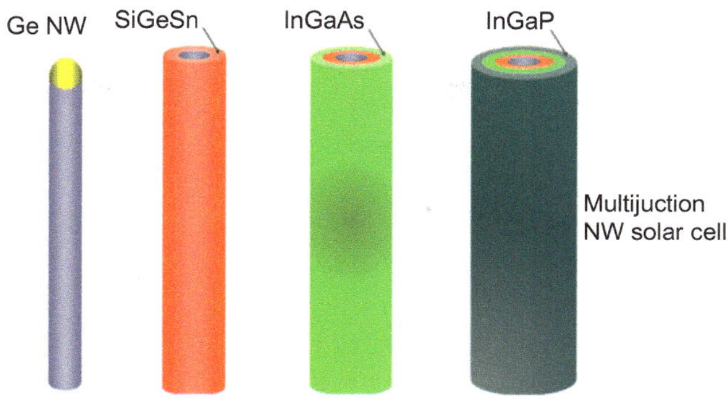

Fig. 10 Process steps for a radial nanowire four-junction solar cell. The axial modulation of the nanowire content can also be used to implement an alternative design of such nanowire solar cell.

in a nanowire geometry reduces the formation of defects at the interfaces between different materials (Assali et al., 2017), therefore minimizing losses due to carrier trapping. These coaxial multijunction nanowire PV elements provide a new nanoscale test bed to investigate energy and charge transport, possibly improving carrier collection and overall device efficiency with respect to single-crystal bulk systems (Tian et al., 2007).

3.2 Recent Progress in SiGeSn Semiconductors

The interest in developing SiGeSn semiconductors has been motivated mostly by the need for monolithic infrared (IR) light sources that are still missing to implement true Si photonics. One of the key elements in a Si photonic chip is the monolithically integrated light source, which converts the electrical signal into optical information. On-chip light sources are the key ingredients in achieving high integration density and scalability, and an enhanced energy efficiency for the development of multiple new functionalities. Integration of these light sources in a Si photonic chip will allow three major breakthroughs: (i) larger bandwidth via the smaller optical interconnects than in conventional electrical circuits; (ii) the elimination of losses due to the resistive heating will strongly reduce the energy consumption; and (iii) the quest for a *siliconized* photonics that will allow for CMOS-compatible and scalable sensing and imaging technologies. Despite the positive premises, the indirect bandgap in Si leads to poor light emission efficiency, which is a major challenge in the development of a scalable and cost-effective Si photonics (Iyer and Xie, 2012; Paniccia, 2010).

Multiple milestones in the components for Si photonics were recently achieved, such as low-loss waveguides (Dai et al., 2012), ultra-fast modulators (Thomson et al., 2012), and high bandwidth detectors (Vivien et al., 2012). In addition, a significant progress in the field of photonic integration was achieved, leading to the demonstration of high-density optical interconnects, operating at a bandwidth density above 30 Tbps cm^2 (Rickman, 2014; Urino et al., 2014) and designed with monolithic-integrated photonic interconnections (Chaisakul et al., 2014). The large light emission efficiency and good temperature stability demonstrated in *off*-chip light sources are key features for optical interconnections, but the large light-coupling losses with the Si chip and the high packaging costs limit their potential for integration (Heck and Bowers, 2014). To overcome these issues, a massive interest in developing light sources that are compatible with the Si-manufacturing process was generated. Multiple solutions were proposed to circumvent the Si indirect bandgap. Quantum confinement effects in porous and nanocrystalline Si (Koshida and Koyama, 1992; Pavesi et al., 2000), Si Raman lasing (Rong et al., 2005), and erbium-based light sources have been investigated (Franzò et al., 2003; Kenyon et al., 2003; Miritello et al., 2007; Yerci et al., 2010). Despite these efforts, these devices may not outperform the state-of-the-art III–V technology for optoelectronic devices in the telecommunication market (Fang et al., 2006; Park et al., 2005; Read et al., 2015; Sun et al., 2009; Tanabe et al., 2012). However, III–V/Si hybrid photonics still faces major difficulties in the scalability and fabrication (Moutanabbir and Gösele, 2010; Read et al., 2015). Thus, the development of group IV light emitters directly integrated on a Si chip is an attractive pathway for CMOS-compatible photonics (Pavesi and Lockwood, 2016; Soref, 2010; Soref et al., 2016; Zhou et al., 2015). The first attempts in this direction resulted in germanium (Ge) LEDs (Liu et al., 2007, 2010) and in the demonstration of an electrically pumped Ge-on-Si laser (Camacho-Aguilera et al., 2012). However, the high threshold and low emission efficiency due to the presence of an indirect bandgap in Ge strongly limit this application. In addition, the requirement for heavily n-type doping ($>10^{19}\,cm^{-3}$) and highly tensile-strained ($\sim$2%) Ge to achieve a direct bandgap emission further compromises the development of a Ge-based laser (Zhou et al., 2015).

To avoid the issues related to the presence of an indirect bandgap in Si, a new class of Si-compatible light sources operating at the nanoscale were recently developed using the emerging group IV semiconductors. Binary and ternary alloys made of Si, Ge, and tin (Sn) have a direct bandgap across a wide range of energies (Soref et al., 2016; Wirths et al., 2015, 2016), with the possibility to independently engineer the bandgap and lattice parameter, similar to the more developed III–V semiconductors. SiGeSn is directly grown on a Si wafer and this provides a powerful tool to achieve a scalable and cost-effective Si on-chip optoelectronics. By tuning composition and strain in SiGeSn-based heterostructures, MIR photonic devices operating in the 2–5 μm wavelength range (~0.25–0.62 eV) can be fabricated. Developing these MIR devices will create far-reaching new opportunities to interface electronics with biological and chemical sensing due to the presence of absorption bands associated with organic molecules in this wavelength range. This novel family of photonic devices will benefit from the compatibility with CMOS processing leading to a full exploitation of the current microelectronic and optoelectronic technologies.

3.3 Recent Progress in the Growth of GeSn Semiconductors

Metastable GeSn and SiGeSn alloys have been extensively studied in recent years as a promising candidate for direct bandgap emission in group IV semiconductors (He and Atwater, 1995, 1997; Kouvetakis et al., 2006; Ragan et al., 2003; Wegscheider et al., 1992; Wirths et al., 2016). Band structure calculations predict an indirect–to-direct transition in GeSn alloys for Sn incorporation higher than 9%, leading to decreasing bandgap values below 0.5 eV (i.e., at wavelengths longer than 2.5 μm) with increasing Sn content (Moontragoon et al., 2012; Wirths et al., 2016). Heteroepitaxy of GeSn and SiGeSn alloys is commonly performed on a Ge virtual substrate (VS) grown on a Si wafer using a chemical vapor deposition (CVD) reactor, using digermane (Ge_2H_6) and tin-tetrachloride ($SnCl_4$) precursors (Aubin et al., 2016; Wirths et al., 2013, 2015). However, multiple challenges in the growth of Sn-rich GeSn alloys are present:

- *Segregation and precipitation of Sn.* Ge–Sn forms an eutectic system above 231°C, where the mutual equilibrium solubility is only limited to ~1% (Olesinski and Abbaschian, 1984a). The difference in surface free energy combined with the extremely low Ge–Sn solubility leads to Sn segregation and precipitation at the surface during growth; hence GeSn alloys are metastable (He and Atwater, 1995; Ragan et al., 2003; Takeuchi et al., 2008; Wirths et al., 2013). The limited thermal stability of GeSn is a relevant issue when postgrowth annealing treatments are required, which can compromise the surface morphology and induce Sn clustering in the GeSn layer (Li et al., 2013; Wang et al., 2015a; Zhang et al., 2017).

- *Strain*. The large difference in lattice parameters between Sn (6.491 Å) and Ge (5.657 Å) can lead to strain up to ~15% (Moontragoon et al., 2012). This leads to the formation of a high density of defects at the GeSn–Ge interface (von den Driesch et al., 2015), which limit the performance of heterostructures based on GeSn. Furthermore, the presence of residual compressive strain in the GeSn layer leads to higher Sn contents required for the indirect-to-direct-bandgap transition (von den Driesch et al., 2015; Wirths et al., 2015).

The growth of Sn-rich GeSn alloys has been recently achieved at temperatures lower than 350°C to minimize segregation and precipitation of Sn (Assali et al., under review; Aubin et al., 2016, 2017a; Margetis et al., 2017a,b; Stange et al., 2015; von den Driesch et al., 2015; Wang et al., 2015b; Wirths et al., 2015). The direct bandgap emission in GeSn was experimentally demonstrated using bulk GeSn layers with Sn contents above 12%, as shown in Fig. 11, where a direct band optical emission is observed at energies lower than 0.55 eV (von den Driesch et al., 2015; Wirths et al., 2015). Furthermore, lasing emission in Fabry–Perot waveguide and microdisk cavities has been also observed up to temperatures lower than ~180 K (Reboud et al., 2017; Stange et al., 2016; Wirths et al., 2015). However, for increasing temperature the tunneling of electrons from the direct Γ-minimum into the indirect L-minimum strongly reduces the GeSn quantum efficiency (QE), suppressing the lasing emission and reducing the QE at room temperature (Stange et al., 2015; Wirths et al., 2015).

To circumvent the aforementioned difficulties, higher Sn content GeSn alloy with a Sn incorporation up to 15%–17% was recently developed using a GeSn-buffered heterostructure with different Sn contents (Assali et al., under review; Aubin et al., 2016, 2017b; Margetis et al., 2017b,c). When the GeSn layer is pseudomorphic with the Ge-VS substrate, the Sn composition is mainly tailored using the growth temperature (Aubin et al., 2016). However, at larger GeSn thickness the increase in strain energy in the system leads to a high amount of defects, and segregation of Sn at the surface occurs at Sn contents above 12% (Aubin et al., 2017a). The addition of step-graded GeSn buffer layers with increasing Sn content reduces the strain on the Sn-rich top layer, allowing higher Sn incorporation and reducing segregation (Aubin et al., 2017b; Margetis et al., 2017c). For many years the most common precursor for the growth of Sn-rich GeSn layers has been digermane (Ge_2H_6) (Aubin et al., 2016; Wirths et al., 2013, 2015). Ge_2H_6 is a highly expensive and low stability precursor when compared to the industrial standard monogermane (GeH_4) precursor. However, it provides a lower

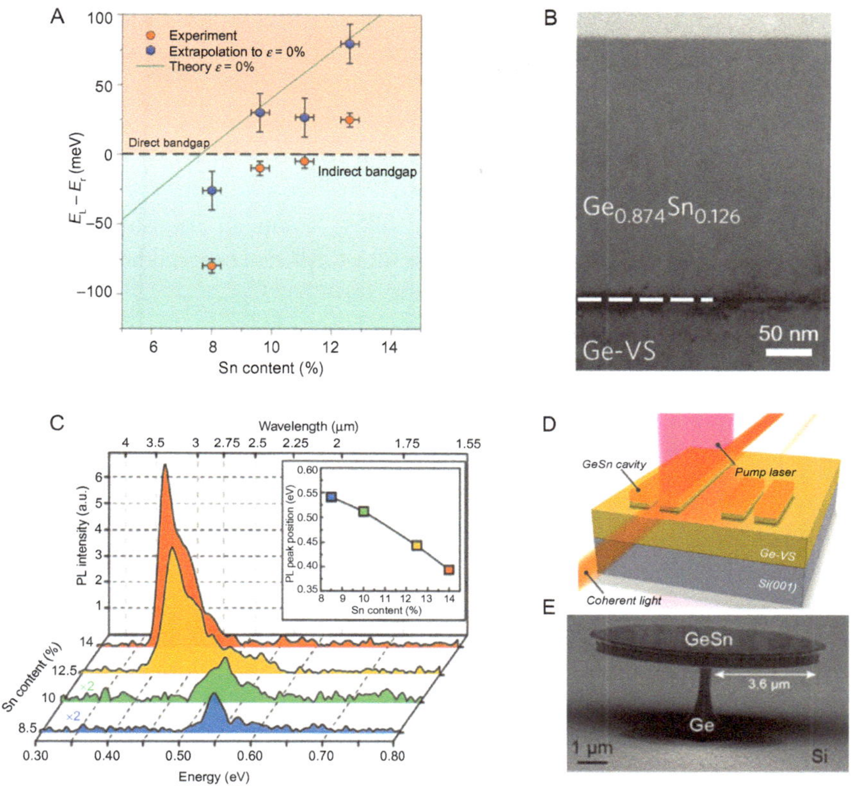

Fig. 11 (A) Conduction band offset ΔE between the minima of the Γ- and L-valleys as a function of Sn concentration. (B) Cross-sectional TEM image of $Ge_{0.874}Sn_{0.126}$/Ge-VS/Si layers. (C) Photoluminescence spectra for GeSn layers with Sn contents between 8.5% and 14%. (D) Sketch of the GeSn waveguide cavity excitation using a pump laser, while the lasing emission is collected from one of the etched facets. (E) SEM of 8 μm diameter $Ge_{0.875}Sn_{0.125}$ microdisk with the underlying Ge-VS undercut by 3.6 μm. *Panels (A), (B), and (D): Adapted from Wirths, S., Geiger, R., von den Driesch, N., Mussler, G., Stoica, T., Mantl, S., Ikonic, Z., Luysberg, M., Chiussi, S., Hartmann, J.M., Sigg, H., Faist, J., Buca, D., Grützmacher, D., 2015. Lasing in direct-bandgap GeSn alloy grown on Si. Nat. Photonics 9, 88–92. https://doi.org/10.1038/nphoton.2014.321. Panel (C): Adapted from von den Driesch, N., Stange, D., Wirths, S., Mussler, G., Holländer, B., Ikonic, Z., Hartmann, J.M., Stoica, T., Mantl, S., Grützmacher, D., Buca, D., 2015. Direct bandgap group IV epitaxy on Si for laser applications. Chem. Mater. 27, 4693–4702. https://doi.org/10.1021/acs.chemmater.5b01327. Panel (E): Adapted from Stange, D., Wirths, S., Geiger, R., Schulte-Braucks, C., Marzban, B., Driesch, N.V. Den, Mussler, G., Zabel, T., Stoica, T., Hartmann, J.-M., Mantl, S., Ikonic, Z., Grützmacher, D., Sigg, H., Witzens, J., Buca, D., 2016. Optically pumped GeSn microdisk lasers on Si. ACS Photonics 3, 1279–1285. https://doi.org/10.1021/acsphotonics.6b00258.*

decomposition temperature, allowing for higher growth rate and enhanced Sn incorporation (Aubin et al., 2017b; Vincent et al., 2011; von den Driesch et al., 2015). Recently, the growth of GeSn layers with a Sn content up to 17% using GeH_4 was demonstrated (Assali et al., under review; Margetis et al., 2017a,b,c), also using a growth rate that is at least 10 times lower than similar samples grown with Ge_2H_6 (Assali et al., under review). Thus, the possibility to grow Sn-rich GeSn alloys using an industrial standard GeH_4 precursor will allow a large-scale and CMOS-compatible integration of GeSn in the Si-manufacturing process.

3.4 Bandgap Tunability in SiGeSn Ternary Alloys

With the growth of SiGeSn alloys the direct bandgap can be further tuned to higher energies (up to ~0.9 eV) independent of the lattice parameter (Attiaoui and Moutanabbir, 2014; Du et al., 2017; Jiang et al., 2014; Margetis et al., 2017b; Moontragoon et al., 2012; von den Driesch et al., 2017; Wendav et al., 2016), with the advantage of decoupling strain and band structure effects, providing additional functionality in the design of materials (Alher et al., 2015; Fukuda et al., 2017; Kouvetakis et al., 2006; Wirths et al., 2016). However, due to the extremely low equilibrium solubility of Sn in Si (<0.1%) (Olesinski and Abbaschian, 1984b) the growth of Sn-rich SiGeSn ternary alloys is still very challenging. Si incorporation of 10%–12% was demonstrated, but at the expense of a reduced Sn incorporation up to 9%–10% (Du et al., 2017; Margetis et al., 2017b; von den Driesch et al., 2017), as shown in Fig. 12. The optical characterization of SiGeSn layers using photoluminescence (PL) measurements showed a tunable emission energy up to ~0.8 eV in $Si_{0.12}Ge_{0.84}Sn_{0.04}$, as shown in Fig. 12A (von den Driesch et al., 2017). The bandgap tunability allows for the growth of multiquantum well (MQW) structures made of SiGeSn barriers, which can lead to efficient carrier confinement in the optically active GeSn layers (Du et al., 2017; Gallagher et al., 2015; Stange et al., 2017). Multiple GeSn wells with 8.6% Sn content and SiGeSn barrier with 10.5% Si and 10.9% Sn contents can be grown in combination with B- and P-doped GeSn contact layers, resulting an MIR LED (Stange et al., 2017). The addition of multiple SiGeSn barrier layers enhances the electroluminescence (EL) emission compared to the bare GeSn, as shown in Fig. 12D, with a factor of ~2 increase in emission intensity at room temperature (Stange et al., 2017). Furthermore, the preliminary boost in performance of group IV optoelectronics has been shown in the first prototypes of photodetectors (Soref et al., 2016; Zheng et al., 2016), and FETs based on SiGeSn alloys (Lieten et al., 2013).

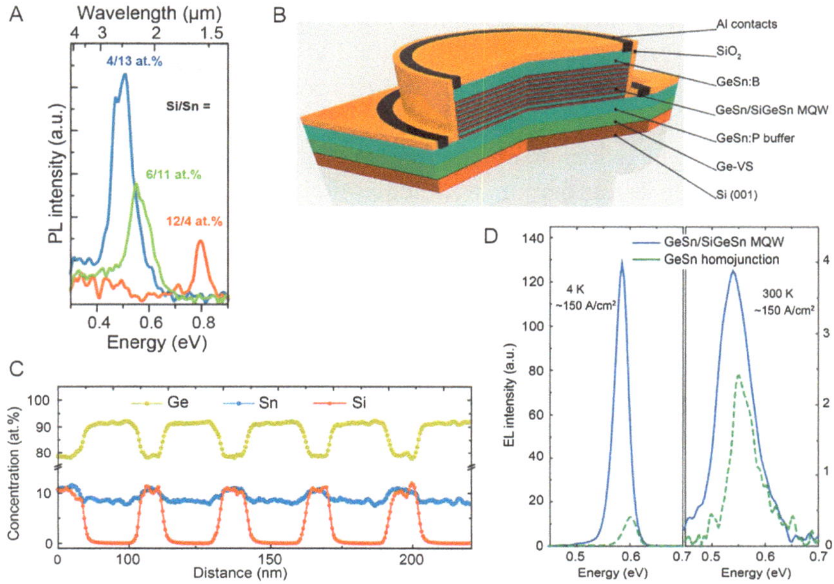

Fig. 12 (A) Room temperature tunability of the PL emission in SiGeSn alloys. (B) Schematic of a GeSn/SiGeSn MQW LED. (C) APT concentration profiles in the central part of the MQW LED. (D) EL spectra acquired at 4 and 300 K of a GeSn/SiGeSn MQW (*blue*) and GeSn homojunction (*dashed green*) LEDs. *Panel (A): Adapted from von den Driesch, N., Stange, D., Wirths, S., Rainko, D., Povstugar, I., Savenko, A., Breuer, U., Geiger, R., Sigg, H., Ikonic, Z., Hartmann, J.-M., Grützmacher, D., Mantl, S., Buca, D., 2017. SiGeSn ternaries for efficient group IV heterostructure light emitters. Small 13, 1603321. https://doi.org/10.1002/smll.201603321. Panels (B)–(D): Adapted from Stange, D., Von Den Driesch, N., Rainko, D., Roesgaard, S., Povstugar, I., Hartmann, J.-M., Stoica, T., Ikonić, Z., Mantl, S., Grützmacher, D., Buca, D., 2017. Short-wave infrared LEDs from GeSn/SiGeSn multiple quantum wells. Optica 4, 185–188. https://doi.org/doi.org/10.1364/OPTICA.4.000185.*

3.5 GeSn Nanowires

In the growth of GeSn alloys in a conventional planar geometry on a Ge-VS, the strain in the GeSn layer is progressively released by the formation of a high density of defects (up to $5\times10^{6}\,cm^{-2}$) (von den Driesch et al., 2015) at the Ge–GeSn interface. This network of dislocation acts as traps for charge carriers, reducing their mobility and the performance of optoelectronic devices. In addition, any residual compressive strain in the GeSn layer will further increase the Sn incorporation necessary for a direct bandgap emission (Attiaoui and Moutanabbir, 2014; von den Driesch et al., 2015; Wirths et al., 2015). Therefore, a minimum GeSn layer thickness of

400 nm is required for strain relaxation and to promote a direct bandgap at room temperature (von den Driesch et al., 2015). The one-dimensional nanowire geometry provides enhanced functionality in the growth of GeSn alloys, minimizing the impact of those limitations that are commonly observed in bulk GeSn. The reduction in dimensionality promotes strain relaxation and enhances the Sn incorporation well above the ~1% equilibrium solubility limit, without compromising the crystalline purity and optical properties of GeSn. In the following, recent progress in the growth of GeSn alloys using axial and radial nanowire heterostructures will be discussed.

3.5.1 Axial GeSn Nanowires

GeSn nanowires can be grown from colloidal AuAg alloy nanoparticles deposited onto Si (001) substrates using a liquid-injection CVD technique (Biswas et al., 2016). At a growth temperature of 440°C the AuAg–Sn–Ge eutectic liquid alloy is formed, but no precipitation of Sn is expected (Ciulik and Notis, 1993). However, by introducing a step cooling at 230°C a small region in the Sn-rich side of the Au–Sn and GeSn phase can be exploited to increase the Sn content up to $9.2 \pm 0.8\%$ (see Fig. 13), with no effect on the nanowire morphology. Nonequilibrium Sn incorporation from the supersaturated catalyst droplet is promoted at 230°C, leading to Sn contents much higher than the equilibrium solubility of Sn in Ge lower than 1% (Olesinski and Abbaschian, 1984a). The growth of a Sn-rich GeSn segment in a Ge nanowire provides a possible pathway for the fabrication of Ge–GeSn nanowire axial heterostructures, eventually leading to a nanowire-embedded quantum dot (QD) (Reimer et al., 2012) operating in the MIR.

The low-temperature PL measurements in Fig. 14 provide information on the optical bandgap of GeSn nanowires. At a low Sn content of 6% a broad emission peak at a wavelength of ~2200 nm is observed, while when the incorporation increases to 9.2%, the peak shits to 2233 nm and it reduces in linewidth. This indicates that an enhanced optical emission from the direct bandgap at the Γ-point with increasing Sn content is observed, as compared with similar Sn incorporations of 8%–10% in planar GeSn alloys (Ghetmiri et al., 2014; Gupta et al., 2013). GeSn nanowires with Sn contents well above the equilibrium solubility can also be grown directly in solution using Sn seeds in a microwave-supported solvent-based reactor (Barth et al., 2015; Seifner et al., 2015, 2017). At a growth temperature of 140°C β-Sn globular particles are formed, which then convert into the growth of GeSn nanowires with an α-Sn phase, as shown in Fig. 15A (Seifner et al., 2017). TEM images of the nanowire show a quasi-hemispherical segment and the

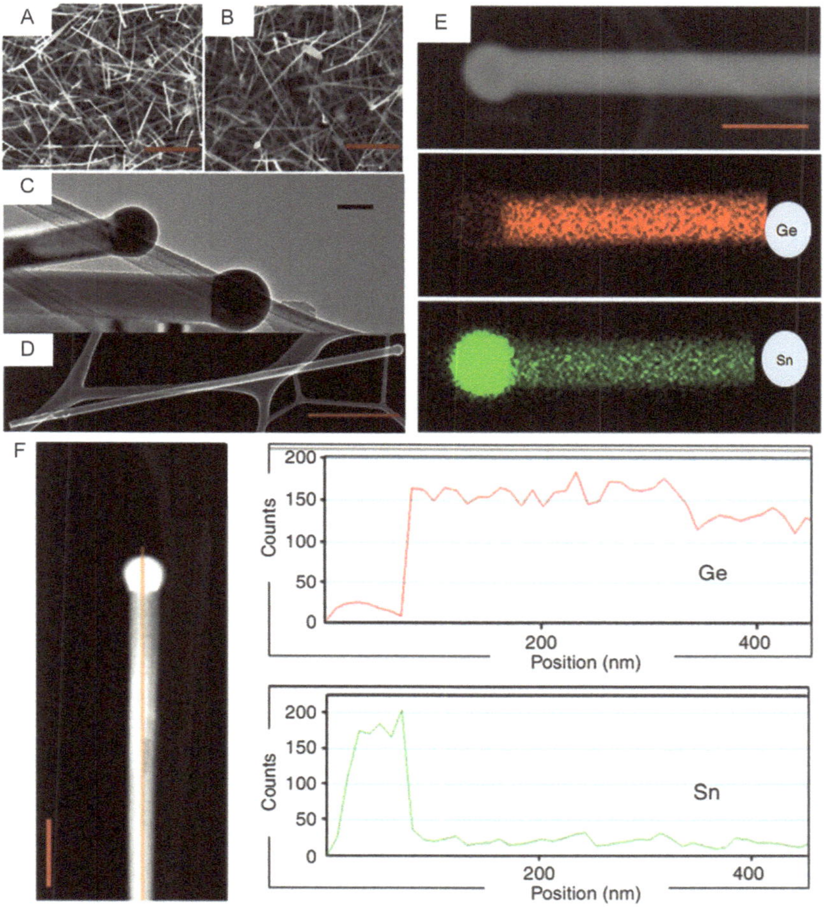

Fig. 13 (A, B) SEM images of GeSn nanowires grown from Au (A) and $Au_{0.90}Ag_{0.10}$ (B) catalysts (scale bar, 1 nm). (C) TEM image confirming the VLS growth mechanism with dark-contrast spherical seed at the tip of the nanowire with AuAg catalysts. Scale bar, 100 nm. (D) HAADF STEM image showing a negligible tapering in the nanowires. Scale bar, 1 mm. (E, F) Dark-field HAADF image and EDX mapping for Ge and Sn in a $Ge_{1-x}Sn_x$ nanowires with a uniform Sn content of 9.4%. The *red curve* denotes linescan for Ge, whereas the *green curve* for Sn. Scale bar, 50 nm (D) and 100 nm (E). *Adapted from Biswas, S., Doherty, J., Saladukha, D., Ramasse, Q., Majumdar, D., Upmanyu, M., Singha, A., Ochalski, T., Morris, M.A., Holmes, J.D., 2016. Non-equilibrium induction of tin in germanium: towards direct bandgap* $Ge_{1-x}Sn_x$ *nanowires. Nat. Commun. 7, 11405. https://doi.org/10.1038/ncomms11405.*

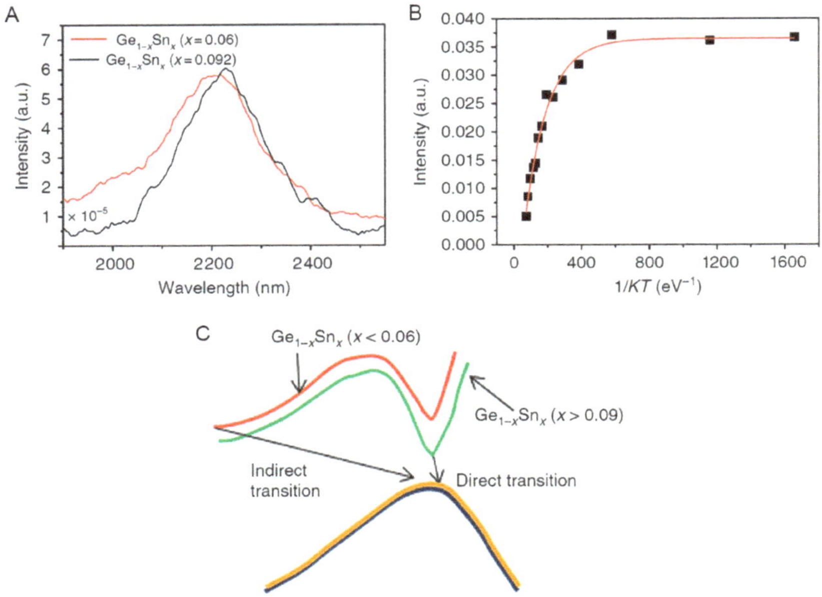

Fig. 14 (A). Photoluminescence spectra of GeSn nanowires with a Sn content of 6.0% and 9.2% measured at 7K. Broadened PL emission is observed in the 6% Sn sample, while a narrower peak is observed for 9.2% Sn nanowires. (B) Arrhenius plot from 9.2% Sn nanowires shows a single exponential decay of the PL emission with increasing temperature. (C) Schematics of possible paths for direct and indirect transitions for different Sn contents. *Adapted from Biswas, S., Doherty, J., Saladukha, D., Ramasse, Q., Majumdar, D., Upmanyu, M., Singha, A., Ochalski, T., Morris, M.A., Holmes, J.D., 2016. Non-equilibrium induction of tin in germanium: towards direct bandgap* $Ge_{1-x}Sn_x$ *nanowires. Nat. Commun. 7, 11405. https://doi.org/10.1038/ncomms11405.*

nanowire body with different diffraction contrasts in Fig. 15B, with a high crystallinity observed in the nanowire body when imaging in HRTEM. Two distinct sites of Sn enrichment located at both extremes on the nanowires are observed in the EDX map in Fig. 15C, with the bigger globular part located at the growth front of the nanowires (Seifner et al., 2015), while a Sn-rich segment is visible at the nucleation site. A 100–150 nm transition zone with slightly higher Sn content (~32%) is present in between the Sn-rich area and the constant $Ge_{0.72}Sn_{0.28}$ composition along the nanowire body. Therefore, the Sn-rich zone at the nucleation site can be associated with a remaining α-Sn segment acting as a template for the GeSn phase formation.

The thermal stability of GeSn alloy has been investigated using these $Ge_{0.72}Sn_{0.28}$ nanowires and recording the XRD patterns, while the

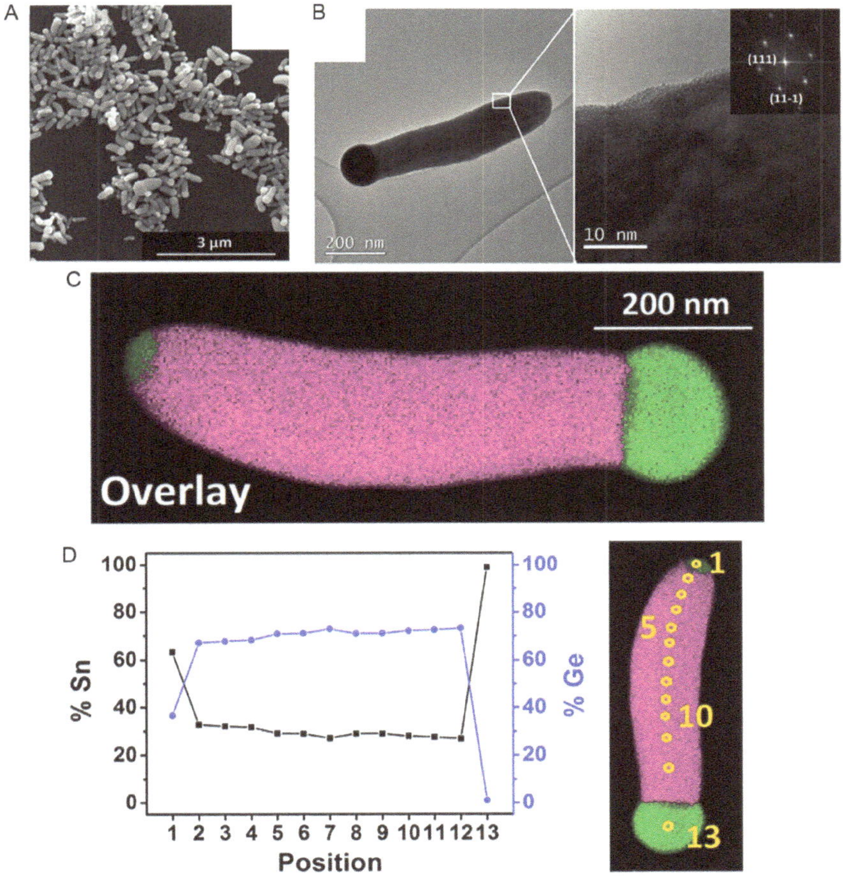

Fig. 15 (A) SEM images of GeSn nanowires grown at 140°C for 2 min. (B) TEM-HRTEM images of a GeSn nanowire. The corresponding fast-Fourier transform (FFT) image is shown in the *inset*. (C) STEM-EDX maps showing homogeneous Sn incorporation in the nanowire body and an accumulation of Sn at the nanowire extremes. (D) EDX linescan measurements showing a progressive transition from ~32% Sn content (point 13) down to a ~28% segment (points 5–12) after being ~150 nm away from the Sn-enriched nucleation site (point 1). *Adapted from Seifner, M.S., Hernandez, S., Bernardi, J., Romano-Rodriguez, A., Barth, S., 2017. Pushing the composition limit of anisotropic $Ge_{1-x}Sn_x$ nanostructures and determination of their thermal stability. Chem. Mater. 29, 9802–9813. https://doi.org/10.1021/acs.chemmater.7b03969.*

temperature is increased in 10°C steps (Seifner et al., 2017). In the XRD curves in Fig. 16A and B, no changes are visible up to 150°C, while above this temperature a gradual splitting of the α-Sn GeSn peak into two different peaks associated with $Ge_{0.90}Sn_{0.10}$ and the original $Ge_{0.72}Sn_{0.28}$ composition

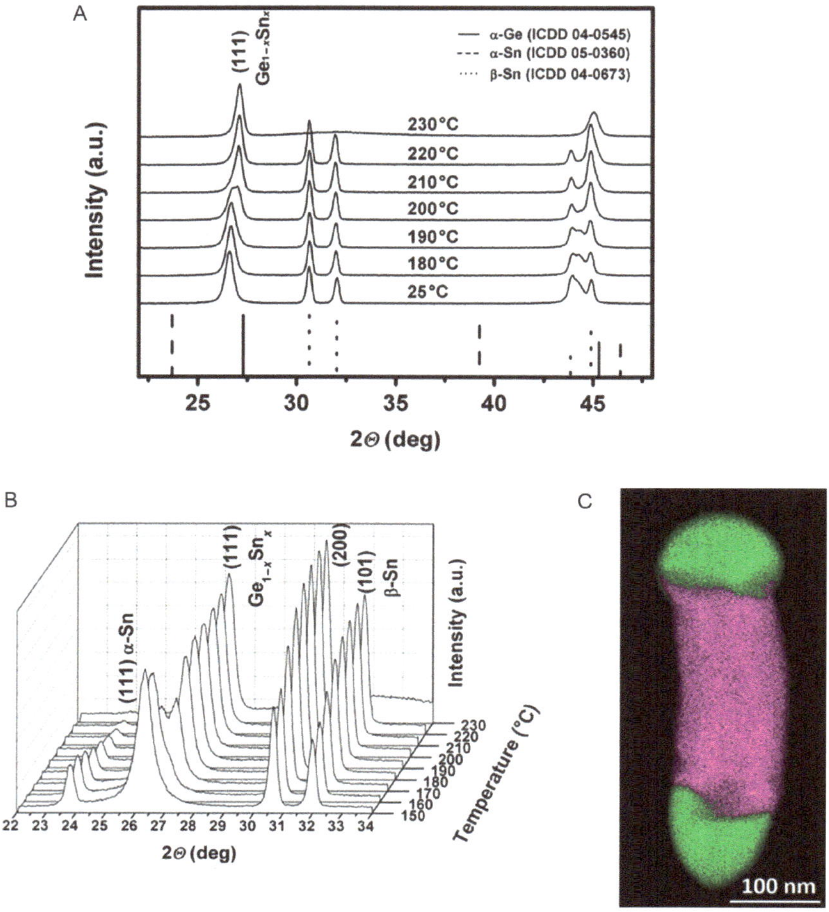

Fig. 16 (A) Temperature-dependent XRD measurements in the 25–230°C temperature range. The XRD curve for the sample annealed to 500°C and then cooled down to 25°C starting with $Ge_{0.72}Sn_{0.28}$ nanowires containing the Sn growth seeds is also shown. (B) 3D-XRD measurements acquired in the 150–230°C temperature range with 10°C increment. (C) STEM-EDX image of a $Ge_{0.72}Sn_{0.28}$ nanowire treated with 10 min annealing at 220°C. *Adapted from Seifner, M.S., Hernandez, S., Bernardi, J., Romano-Rodriguez, A., Barth, S., 2017. Pushing the composition limit of anisotropic $Ge_{1-x}Sn_x$ nanostructures and determination of their thermal stability. Chem. Mater. 29, 9802–9813. https://doi.org/10.1021/acs.chemmater.7b03969.*

is observed. When the temperature reaches 180–190°C, the higher Sn content peak is still visible with diminishing intensity, while only a single peak associated with ~6% Sn remains at 210°C. This behavior reflects the increase in β-Sn content in the alloy, as demonstrated in the 3D XRD representation of the 150–230°C temperature in Fig. 16B. In situ TEM measurements during heating cycles up to 220°C provide additional insight into the mass diffusion process by analyzing the change in contrast associated with the migrating atomic species. When the diffusing species from the larger Sn globular segment reach the initial α-Sn nucleation, a rapid change in diffraction contrast is observed, but the diffusion process does not stop, with further enrichment of Sn atoms diffusing from the bigger Sn particle. As a confirmation of this, the EDX map in Fig. 16C for $Ge_{0.72}Sn_{0.28}$ nanowires heated in the TEM to 220°C for ~10 min shows preferential Sn enrichment at the two ends of the nanowire. Additional temperature-dependent XRD measurements on the same nanowires after Sn seed removal show an increased thermal stability up to ~220°C, hence ~70°C higher than in what observed in the presence of Sn seeds. It is noteworthy that the Sn diffusion process takes place below the melting point of Sn (232°C), therefore indicating that GeSn alloys remain metastable also at the nanoscale, similar as what is observed for bulk GeSn (Aubin et al., 2017a; Takeuchi et al., 2008; Weisshaupt et al., 2017).

3.5.2 Radial GeSn Nanowires

Ge/GeSn core/shell nanowires can be grown using a Au catalyst with the conventional VLS growth in a CVD reactor (Assali et al., 2017; Meng et al., 2016). Arrays of 50-nm-wide untapered Ge nanowires are grown at 320°C using a germane (GeH_4) precursor as shown in Fig. 17A. The catalytic action of the Au nanoparticle (Verheijen et al., 2006) possibly enhances the GeH_4 decomposition at these low temperatures, therefore avoiding the use of the more expensive Ge_2H_6 precursor. Next, the temperature is reduced to 300°C, and tin chloride ($SnCl_4$) and hydrogen chloride (HCl) precursors are supplied for the GeSn shell growth, as shown in the SEM image in Fig. 17B (Assali et al., 2017). A tapered nanowire top section is visible, but the nanowires are not bent, indicating that a negligible strain is present. EDX compositional maps in combination with atom probe tomography (APT) measurements reveal a uniform Sn distribution in the GeSn shell along the nanowire growth axis, with no axial VLS GeSn growth observed (see Fig. 17C). By performing cross-sectional EDX-APT measurements in Fig. 17D and E an ~120-nm-thick GeSn shell is visible, with six wide {112} facets and six narrow {110} corner facets. Enhanced Sn incorporation along the ⟨112⟩ direction of the shell is present, reaching up to $13 \pm 1\%$ Sn content in the outer portion of the shell (Assali et al., 2017). Due to the large lattice mismatch between the GeSn shell and the Ge core a ~2% compressive

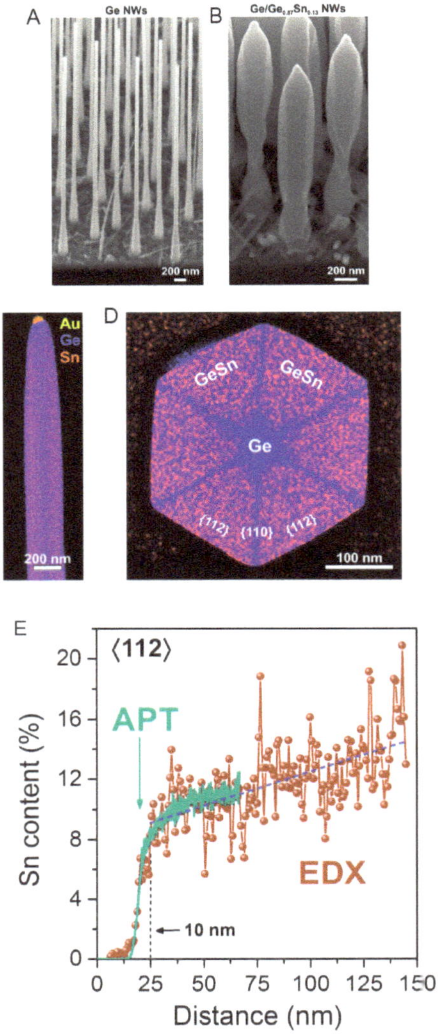

Fig. 17 (A) SEM picture of an array of Ge nanowires (tilting angle 30°). untapered middle-top section of the nanowires and a wider bottom segment are visible. (B) SEM image of $Ge/Ge_{0.87}Sn_{0.13}$ core/shell nanowire arrays (tilting angle 30°). (C) EDX compositional map of a single nanowire. A uniform Sn distribution in the GeSn shell along the nanowire growth axis is observed, and no axial growth of GeSn is visible. (D) Cross-sectional EDX compositional map of an ~120-nm-thick GeSn shell. Enhanced Sn incorporation along the {112} side-facets compared to {110} is present. An average Sn content of 13 ± 1% is observed in the outer part of the shell. (E) Sn content as a function of the distance acquired along the ⟨112⟩ radial direction using EDX and APT measurements. A 10 nm transition region from the Ge core into the GeSn shell is observed, followed by a gradual increase in Sn content toward the outer portion of the shell. The *blue dashed line* is the linear fit of the EDX profile. *Adapted from Assali, S., Dijkstra, A., Li, A., Koelling, S., Verheijen, M.A., Gagliano, L., von den Driesch, N., Buca, D., Koenraad, P.M., Haverkort, J.E.M., Bakkers, E.P.A.M., 2017. Growth and optical properties of direct band gap* $Ge/Ge_{0.87}Sn_{0.13}$ *core/shell nanowire arrays. Nano Lett. 17, 1538–1544. https://doi.org/10.1021/acs.nanolett.6b04627.*

strain in the shell is expected, which could lead to a large amount of defects during plastic relaxation (von den Driesch et al., 2015). However, in the high-resolution TEM images in Fig. 18 a pure diamond cubic crystal structure of the GeSn shell is observed, without the presence of protruding dislocations extending from the core–shell interface to the outer edge. From selective area electron diffraction patterns and finite element method calculations the strain

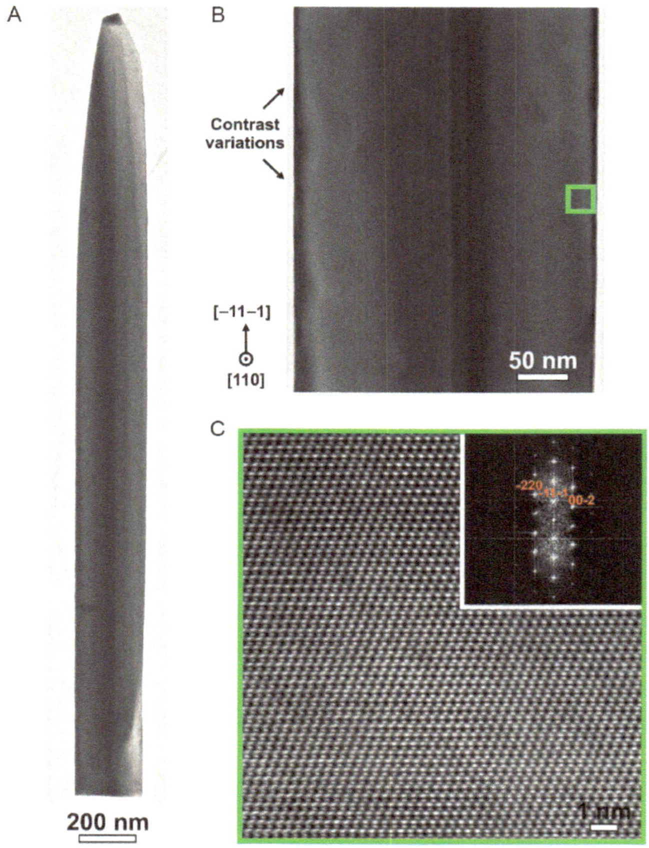

Fig. 18 (A, B) TEM image showing the entire (A) and central segment (B) of a Ge/$Ge_{0.87}Sn_{0.13}$ core/shell nanowire. The contrast variations in (B) are highlighted using a *dashed elliptical shape*. (C) Aberration-corrected HRTEM image of the outer portion of the GeSn shell (*green square* in B). *Inset*: Indexed FFT pattern calculated from (C). *Adapted from Assali, S., Dijkstra, A., Li, A., Koelling, S., Verheijen, M.A., Gagliano, L., von den Driesch, N., Buca, D., Koenraad, P.M., Haverkort, J.E.M., Bakkers, E.P.A.M., 2017. Growth and optical properties of direct band gap Ge/$Ge_{0.87}Sn_{0.13}$ core/shell nanowire arrays. Nano Lett. 17, 1538–1544. https://doi.org/10.1021/acs.nanolett.6b04627.*

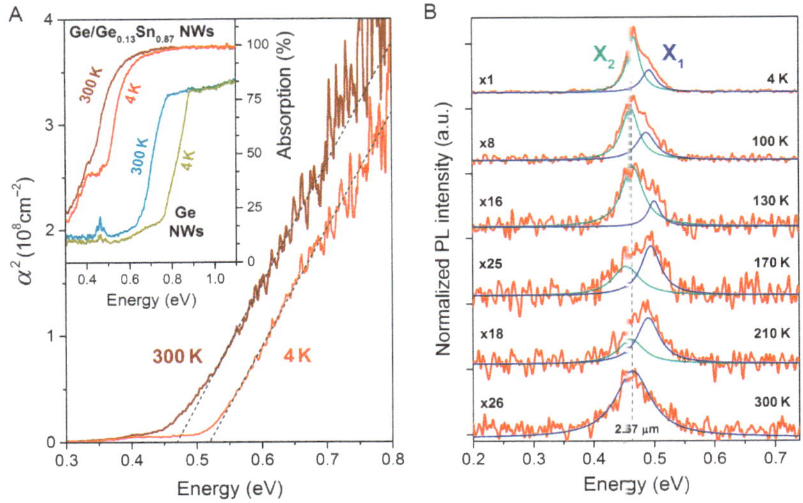

Fig. 19 (A) Absorption coefficient α^2 for Ge/$Ge_{0.87}Sn_{0.13}$ core/shell nanowires as a function of the energy, measured at 4 and 300 K. *Inset*: Absorption measurements for Ge nanowires and Ge/$Ge_{0.87}Sn_{0.13}$ core/shell nanowires acquired at 4 and 300 K. (B) Photoluminescence spectra of Ge/$Ge_{0.87}Sn_{0.13}$ core/shell nanowires in the 4–300 K temperature range. *Adapted from Assali, S., Dijkstra, A., Li, A., Koelling, S., Verheijen, M.A., Gagliano, L., von den Driesch, N., Buca, D., Koenraad, P.M., Haverkort, J.E.M., Bakkers, E.P.A.M., 2017. Growth and optical properties of direct band gap Ge/$Ge_{0.87}Sn_{0.13}$ core/shell nanowire arrays. Nano Lett. 17, 1538–1544. https://doi.org/10.1021/acs.nanolett.6b04627.*

level in the nanowires can be estimated being lower than −0.5% (Albani et al., under review; Assali et al., 2017). Therefore, the partial strain relaxation in the ∼120-nm-thick $Ge_{0.87}Sn_{0.13}$ shell results from the compositional Sn gradient, the one-dimensional nanowire geometry, and possibly from the plastic strain relaxation with the formation of misfit dislocations at the GeSn–Ge interface.

The direct bandgap in Ge/GeSn core/shell nanowires was investigated using temperature-dependent integrating-sphere Fourier transform infrared absorption measurements and PL, as shown in Fig. 19 (Assali et al., 2016, 2017; Leroux et al., 1999). A strong red shift of the Ge/$Ge_{0.87}Sn_{0.13}$ core/shell nanowire absorption with respect to the 50 nm diameter Ge core nanowires is observed in Fig. 19A. The enhanced absorption results from multiple factors, such as diameter increase, a tapered nanowire geometry, and possibly from the direct bandgap $Ge_{0.87}Sn_{0.13}$ shell (Attiaoui et al., 2018; Diedenhofen et al., 2011; Reimer et al., 2012). The linear increase of the square of the absorption coefficient (α^2) with energy in Ge/$Ge_{0.87}Sn_{0.13}$ nanowires clearly demonstrates the presence of a direct bandgap (Grundmann, 2006). In the PL emission spectrum of the Ge/$Ge_{0.87}Sn_{0.13}$ nanowires, two peaks at 0.493 eV (X_1)

and 0.468 eV (X_2) are visible at low temperature (4 K). The emission peaks X_1 and X_2 are 25 and 50 meV lower than the bandgap energy estimated from the 4 K absorption spectrum. These peaks are associated with bandgap-related localized states, such as alloy fluctuations (X_1) and acceptors (X_2), broadened by disorder, and could also be associated with, e.g., residual aluminum or oxygen in the nanowires or interstitial Sn atoms (Assali et al., 2016, 2017; Koelling et al., 2017; Leroux et al., 1999; Querales-Flores et al., 2016). When the temperature is increased above 170 K, the X_2 emission quenches, while the X_1 peak evolves toward the 0.465 eV GeSn direct bandgap emission at room temperature (Assali et al., 2017). Furthermore, when comparing the estimated PL QE of Ge/GeSn core/shell nanowires with a planar $Ge_{0.88}Sn_{0.12}$/Ge-VS/Si sample with similar composition (Stange et al., 2015; von den Driesch et al., 2015), comparable QE in the 10^{-3}–10^{-4} range is obtained in both samples (Assali et al., 2017). Since nonradiative channels reduce the GeSn radiative efficiency independent of the nanowires or planar geometry, this suggests that nonradiative channels are active in the bulk of the material rather than at the surface of the material. Thus, the reduced dimensionality of GeSn alloys when grown in a nanowire geometry results in high Sn incorporation and enhanced absorption, without compromising the radiative efficiency of the material.

3.6 GeSn and SiGeSn Nanowires for Optoelectronic Applications

The early-stage development in nanowire-based GeSn alloys shows that the reduction in dimensionality enhances the Sn incorporation well above the equilibrium solubility in Ge (~1%) (Assali et al., 2017; Barth et al., 2015; Biswas et al., 2016; Meng et al., 2016; Seifner et al., 2015, 2017). The tunable Sn content in the 7%–28% range, the high crystalline purity, and the direct bandgap emission (<0.5 eV) up to room temperature, combined with the enhanced light absorption in the nanowire geometry, show the tremendous potential on nanowire-based GeSn alloys when compared to conventional planar GeSn (Al-Kabi et al., 2016; Gallagher et al., 2015; Gassenq et al., 2017; Ghetmiri et al., 2014; Homewood and Lourenço, 2015; Kasper et al., 2012; Kouvetakis et al., 2006, 2010; Loo et al., 2012; Reboud et al., 2017; Soref, 2010; Soref et al., 2016; Stange et al., 2016; Su et al., 2011; Taoka et al., 2017; Wang et al., 2015b; Werner et al., 2011; Wirths et al., 2015, 2016; Yu et al., 2015; Zaima et al., 2015). GeSn nanowires provide the perfect tool for investigating new photonics effects and optoelectronic properties in the 2–5 μm MIR range, hence in a wavelength range that is commonly limited to the more expensive InAsSb alloyed nanowires (Farrell et al., 2015; Koblmüller et al., 2012; Wook Shin et al., 2015; Zhuang et al., 2017).

The nanowire geometry provides excellent antireflection properties over a large wavelength range (Diedenhofen et al., 2011; van Dam et al., 2016), with a reduced material consumption respect to conventional planar schemes. The concentration of the incoming light due to the nanowire geometry leads to higher absorption, which strongly enhances the performances of photodetectors and solar cells (Diedenhofen et al., 2011; Kempa et al., 2012; Krogstrup et al., 2013; van Dam et al., 2016). The possibility to independently control the size and composition in a core/shell nanowires geometry is strongly beneficial for enhancing the light outcoupling in nanowire-based LEDs (Dai et al., 2015; Qian et al., 2005; Ra et al., 2013). Furthermore, nanowires provide an additional degree of freedom in the control over the directionality and polarization of the emitted light, which is achieved without any external optics by tuning the nanowire diameter and tapering (Bulgarini et al., 2014; Grzela et al., 2012; Reimer et al., 2012; van Dam et al., 2015). Therefore, a large control over the in- and outcoupling of the light is possible using semiconductor nanowires, with a strong benefit in the design of optoelectronic devices like LEDs, lasers, photodetectors, and solar cells, as shown in Fig. 20. Lastly, future work is required for the development of SiGeSn-based nanowires, which can provide the required direct bandgap tunability (up to 0.9 eV, i.e., 1.4 μm wavelength) (Jiang et al., 2014; Moontragoon et al., 2012; Wendav et al., 2016) to enhance carrier confinement in MQWs. The boost in efficiency and the broader wavelength operational range when using SiGeSn alloys will increase the potential of SiGeSn optoelectronics.

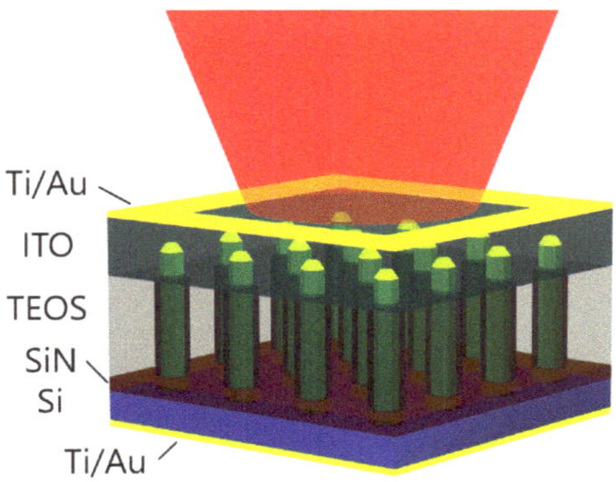

Fig. 20 Schematic of SiGeSn-based nanowire array mid-IR optoelectronic device. After growth, the nanowires are embedded in a transparent tetraethyl orthosilicate (TEOS) layer with indium–tin–oxide (ITO) front contact, and titanium/gold (Ti/Au) as top contact pad and bottom contact.

4. CONCLUSIONS

Si and Si-based technologies have been inarguably revolutionizing the way we communicate, collect, and transfer information, use and preserve natural resources, and interact with our local and global environments. As a matter of fact, the modern marvels of the digital world owe their very existence to the mastery of Si-enabled integrated circuits and computer chips. In the continuous strive to improve our existing technologies, one thing that must not be forgotten is the energy burden that modern-day devices are putting on our already fragile environment. Not surprising that engineers and material scientists have put considerable efforts in the last few decades in inventing ingenious ways to mitigate this predicament. This chapter strives to bring home to the readers some of these current ongoing endeavors related to the use of semiconductor nanowires in harnessing carbon-free clean energy. An overwhelming majority of these efforts are dedicated to synthesizing and studying the characteristics of novel families of Si-compatible nanowires, be it phonon engineering in Si nanowires or bandgap engineering in Sn-alloyed Si and SiGe nanowires.

While studies conducted on single nanowires provide key information regarding the electronic, thermal, and optical properties of the nanowires, a thermoelectric or photovoltaic device requires integrated arrays of nanowires to be practically relevant. This integration remains a formidable challenge holding the key to scalable technological applications. The studies conducted thus far on thermoelectric and photovoltaic properties of nanowires have all exploited the unique geometry of nanowires. Phonon engineering in Si nanowires has proven to be an important paradigm in optimizing the thermoelectric figure of merit by tuning the nanowire's κ, more so than any efforts thus far in optimizing the S or σ. In general, nanowires have a lower κ than their bulk counterparts due to surface scattering of phonons at the nanowire boundary, a phenomena which can be enhanced by surface engineering, alloying, isotopic modulation, or exploiting the resonance-coupling effects in core/shell nanowires. Si nanowires were found to possess a ZT of unity at 200 K and 0.6 at room temperature, close to the best ZT values obtained using scarce and toxic telluride- and antimonide-based nanowires. The advantage of a nanowire-based photovoltaic device is that the dimensions of nanowires are on the same scale as the wavelength of visible light and the minority carrier diffusion lengths, which can provide the device with a unique opportunity of improving the light absorption and charge collection efficiencies. Additionally, nanowire-based devices bear the potential for integration into flexible devices. The problems of monolithic integration and scalability of the light-emitting III–V semiconductors grown on Si substrates can be circumvented to a large extent by the progress in Sn-based binary and ternary group IV alloys. In parallel, the progress in the synthesis of Sn-alloyed GeSn and SiGeSn nanowires has opened a myriad of opportunities for nanowire-based photovoltaic devices with promises to outperform the Si-based photovoltaic devices.

Despite the tremendous prospect, on a device level, nanowires still fall far short of bulk, thin films, and bulk nanostructure semiconductors, which are the current industry standards. To outperform these *canonical* material systems, nanowires require a system-level engineering in parallel to advancing nanowire studies. The major focus in the near future should be toward improving the assembly of nanowires, engineering the nanobulk interfaces, and assembling them into complex 3D geometries hardly achievable in bulk structures. Techniques must be developed to engineer nanowire devices at the wafer scale, with high throughput, low cost, without compromising the material properties and performance. This certainly looks within the grasp of materials and device engineering with the current pace of advancement in nanowire synthesis. The use of earth-abundant materials on one hand and abstaining from high-temperature processing on the other are key advantages of group IV nanowire-based devices. Only the future will tell if this is enough to make the discussed potential technologies competitive enough in the marketplace.

REFERENCES

Abramson, A.R., Kim, W.C., Huxtable, S.T., Yan, H., Wu, Y., Majumdar, A., Tien, C.-L., Yang, P., 2004. Fabrication and characterization of a nanowire/polymer-based nanocomposite for a prototype thermoelectric device. J. Microelectromech. Syst. 13, 505–513. https://doi.org/10.1109/JMEMS.2004.828742.

Albani, M., Assali, S., Verheijen, M.A., Koelling, S., Bergamaschini, R., Pezzoli, F., Bakkers, E.P.A.M., Miglio, L., Critical strain for Sn incorporation in spontaneously graded Ge/GeSn core/shell nanowires (under review).

Algra, R.E., Verheijen, M.A., Borgström, M.T., Feiner, L.-F., Immink, G., van Enckevort, W.J.P., Vlieg, E., Bakkers, E.P.A.M., 2008. Twinning superlattices in indium phosphide nanowires. Nature 456, 369–372. https://doi.org/10.1038/nature07570.

Alher, M.A., Mosleh, A., Cousar, L., Dou, W., Grant, P., Ghetmiri, S.A., Al-Kabi, S., Du, W., Benamara, M., Li, B., Mortazavi, M., Yu, S.-Q., Naseem, H.A., 2015. CMOS compatible growth of high quality Ge, SiGe and SiGeSn for photonic device applications. ECS Trans. 69, 269–278. https://doi.org/10.1149/06905.0269ecst.

Ali, A., Chen, Y., Vasiraju, V., Vaddiraju, S., 2017. Nanowire-based thermoelectrics. Nanotechnology 28. 282001. https://doi.org/10.1088/1361-6528/aa75ae.

Al-Kabi, S., Ghetmiri, S.A., Margetis, J., Du, W., Mosleh, A., Dou, W., Sun, G., Soref, R.A., Tolle, J., Li, B., Mortazavi, M., Naseem, H.A., Yu, S.-Q., 2016. Study of high-quality GeSn alloys grown by chemical vapor deposition towards mid-infrared applications. J. Electron. Mater. 45, 6251–6257. https://doi.org/10.1007/s11664-016-5028-x.

Asen-Palmer, M., Bartkowski, K., Gmelin, E., Cardona, M., Zhernov, A.P., Inyushkin, A.V., Taldenkov, A., Ozhogin, V.I., Itoh, K.M., Haller, E.E., 1997. Thermal conductivity of germanium crystals with different isotopic compositions. Phys. Rev. B 56, 9431–9447. https://doi.org/10.1103/PhysRevB.56.9431.

Assali, S., Greil, J., Zardo, I., Belabbes, A., de Moor, M.W.A., Koelling, S., Koenraad, P.M., Bechstedt, F., Bakkers, E.P.A.M., Haverkort, J.E.M., 2016. Optical study of the band structure of wurtzite GaP nanowires. J. Appl. Phys. 120, 44304. https://doi.org/10.1063/1.4959147.

Assali, S., Dijkstra, A., Li, A., Koelling, S., Verheijen, M.A., Gagliano, L., von den Driesch, N., Buca, D., Koenraad, P.M., Haverkort, J.E.M., Bakkers, E.P.A.M., 2017. Growth and optical properties of direct band gap Ge/$Ge_{0.87}Sn_{0.13}$ core/shell nanowire arrays. Nano Lett. 17, 1538–1544. https://doi.org/10.1021/acs.nanolett.6b04627.

Assali, S., Nicolas, J., Mukherjee, S., Hamzaoui, A., Moutanabbir, O., Multi-step growth of Sn-rich Ge1-xSnx ($0.07 \leq x \leq 0.17$) semiconductors using monogermane as precursor (under review).

Attiaoui, A., Moutanabbir, O., 2014. Indirect-to-direct band gap transition in relaxed and strained $Ge_{1-x-y}Si_xSn_y$ ternary alloys. J. Appl. Phys. 116, 063712. https://doi.org/10.1063/1.4889926.

Attiaoui A., Wirths S., Blanchard-Dionne A.-P., Meunier M., Hartmann J.M., Buca D. and Moutanabbir O., Enhanced IR light absorption in group IV-SiGeSn core-shell nanowires (under review).

Aubin, J., Hartmann, J.M., Barnes, J.P., Pin, J.B., Bauer, M., 2016. Very low temperature epitaxy of heavily in-situ phosphorous doped Ge layers and high Sn content GeSn layers. ECS Trans. 75, 387–398. https://doi.org/10.1149/07508.0387ecst.

Aubin, J., Hartmann, J.M., Gassenq, A., Milord, L., Pauc, N., Reboud, V., Calvo, V., 2017a. Impact of thickness on the structural properties of high tin content GeSn layers. J. Cryst. Growth 473, 20–27. https://doi.org/10.1016/j.jcrysgro.2017.05.006.

Aubin, J., Hartmann, J.M., Gassenq, A., Rouviere, J.L., Robin, E., Delaye, V., Cooper, D., Mollard, N., Reboud, V., Calvo, V., 2017b. Growth and structural properties of step-graded, high Sn content GeSn layers on Ge. Semicond. Sci. Technol. 32, 94006. https://doi.org/10.1088/1361-6641/aa8084.

Balandin, A., Wang, K.L., 1998. Significant decrease of the lattice thermal conductivity due to phonon confinement in a free-standing semiconductor quantum well. Phys. Rev. B 58, 1544–1549. https://doi.org/10.1103/PhysRevB.58.1544.

Balandin, A.A., Ghosh, S., Bao, W., Calizo, I., Teweldebrhan, D., Miao, F., Lau, C.N., 2008. Superior thermal conductivity of single-layer graphene. Nano Lett. 8, 902–907. https://doi.org/10.1021/nl0731872.

Barth, S., Hernandez-Ramirez, F., Holmes, J.D., Romano-Rodriguez, A., 2010. Synthesis and applications of one-dimensional semiconductors. Prog. Mater. Sci. 55, 563–627. https://doi.org/10.1016/J.PMATSCI.2010.02.001.

Barth, S., Seifner, M.S., Bernardi, J., 2015. Microwave-assisted solution–liquid–solid growth of Ge_{1-x} Sn_x nanowires with high tin content. Chem. Commun. 51, 12282–12285. https://doi.org/10.1039/C5CC03639A.

Biswas, K.G., Sands, T.D., Cola, B.A., Xu, X., 2009. Thermal conductivity of bismuth telluride nanowire array-epoxy composite. Appl. Phys. Lett. 94. 223116. https://doi.org/10.1063/1.3143221.

Biswas, S., Doherty, J., Saladukha, D., Ramasse, Q., Majumdar, D., Upmanyu, M., Singha, A., Ochalski, T., Morris, M.A., Holmes, J.D., 2016. Non-equilibrium induction of tin in germanium: towards direct bandgap $Ge_{1-x}Sn_x$ nanowires. Nat. Commun. 7, 11405. https://doi.org/10.1038/ncomms11405.

Boukai, A.I., Bunimovich, Y., Tahir-Kheli, J., Yu, J.K., Goddard 3rd, W.A., Heath, J.R., 2008. Silicon nanowires as efficient thermoelectric materials. Nature 451, 168–171. https://doi.org/10.1038/nature06458.

Bracht, H., Eon, S., Frieling, R., Plech, A., Issenmann, D., Wolf, D., Lundsgaard Hansen, J., Nylandsted Larsen, A., Ager Iii, J.W., Haller, E.E., 2014. Thermal conductivity of isotopically controlled silicon nanostructures. New J. Phys. 16, 15021. https://doi.org/10.1088/1367-2630/16/1/015021.

Bulgarini, G., Reimer, M.E., Bouwes Bavinck, M., Jöns, K.D., Dalacu, D., Poole, P.J., Bakkers, E.P.A.M., Zwiller, V., 2014. Nanowire waveguides launching single photons in a Gaussian mode for ideal fiber coupling. Nano Lett. 14, 4102–4106. https://doi.org/10.1021/nl501648f.

Cahill, D.G., Braun, P.V., Chen, G., Clarke, D.R., Fan, S., Goodson, K.E., Keblinski, P., King, W.P., Mahan, G.D., Majumdar, A., Maris, H.J., Phillpot, S.R., Pop, E., Shi, L., 2014. Nanoscale thermal transport. II. 2003-2012. Appl. Phys. Rev. 1, 1–45. https://doi.org/10.1063/1.4832615.

Camacho-Aguilera, R.E., Cai, Y., Patel, N., Bessette, J.T., Romagnoli, M., Kimerling, L.C., Michel, J., 2012. An electrically pumped germanium laser. Opt. Express 20. 11316. https://doi.org/10.1364/OE.20.011316.

Cardona, M., Thewalt, M.L.W., 2005. Isotope effect on the optical spectra of semiconductors. Rev. Mod. Phys. 77, 1173–1224.

Caroff, P., Dick, K.A., Johansson, J., Messing, M.E., Deppert, K., Samuelson, L., 2009. Controlled polytypic and twin-plane superlattices in III-V nanowires. Nat. Nanotechnol. 4, 50–55. https://doi.org/10.1038/nnano.2008.359.

Carrete, J., Longo, R.C., Gallego, L.J., 2011. Prediction of phonon thermal transport in thin GaAs, InAs and InP nanowires by molecular dynamics simulations: influence of the interatomic potential. Nanotechnology 22. 185704. https://doi.org/10.1088/0957-4484/22/18/185704.

Chaisakul, P., Marris-Morini, D., Frigerio, J., Chrastina, D., Rouifed, M.-S., Cecchi, S., Crozat, P., Isella, G., Vivien, L., 2014. Integrated germanium optical interconnects on silicon substrates. Nat. Photonics 8, 482–488. https://doi.org/10.1038/nphoton.2014.73.

Chalopin, Y., Esfarjani, K., Henry, A., Volz, S., Chen, G., 2012. Thermal interface conductance in Si/Ge superlattices by equilibrium molecular dynamics. Phys. Rev. B 85, 195302. https://doi.org/10.1103/PhysRevB.85.195302.

Chen, G.P., 2005. Nanoscale Energy Transport and Conversion: A Parallel Treatment of Electrons, Molecules, Phonons, and Photons. Oxford University Press, New Jersey.

Chen, J., Zhang, G., Li, B., 2009. Tunable thermal conductivity of $Si_{1-x}Ge_x$ nanowires. Appl. Phys. Lett. 95, 73117. https://doi.org/10.1063/1.3212737.

Chen, K.-I., Li, B.-R., Chen, Y.-T., 2011a. Silicon nanowire field-effect transistor-based biosensors for biomedical diagnosis and cellular recording investigation. Nano Today 6, 131–154. https://doi.org/10.1016/J.NANTOD.2011.02.001.

Chen, J., Zhang, G., Li, B., 2011b. A universal gauge for thermal conductivity of silicon nanowires with different cross sectional geometries. J. Chem. Phys. 135. 204705. https://doi.org/10.1063/1.3663386.

Chen, J., Zhang, G., Li, B., 2011c. Phonon coherent resonance and its effect on thermal transport in core-shell nanowires. J. Chem. Phys. 135. 104508. https://doi.org/10.1063/1.3637044.

Chen, J., Zhang, G., Li, B., 2012. Impacts of atomistic coating on thermal conductivity of germanium nanowires. Nano Lett. 12, 2826–2832. https://doi.org/10.1021/nl300208c.

Chiu, P.T., Law, D.C., Woo, R.L., Singer, S.B., Bhusari, D., Hong, W.D., Zakaria, A., Boisvert, J., Mesropian, S., King, R.R., Karam, N.H., 2014. Direct semiconductor bonded 5j cell for space and terrestrial applications. IEEE J. Photovolt. 4, 493–497. https://doi.org/10.1109/JPHOTOV.2013.2279336.

Ciulik, J., Notis, M.R., 1993. The AuSn phase diagram. J. Alloys Compd. 191, 71–78. https://doi.org/10.1016/0925-8388(93)90273-P.

Colinge, J.-P., Lee, C.-W., Afzalian, A., Akhavan, N.D., Yan, R., Ferain, I., Razavi, P., O'Neill, B., Blake, A., White, M., Kelleher, A.-M., McCarthy, B., Murphy, R., 2010. Nanowire transistors without junctions. Nat. Nanotechnol. 5, 225–229. https://doi.org/10.1038/nnano.2010.15.

Curtin, B.M., Fang, E.W., Bowers, J.E., 2012. Highly ordered vertical silicon nanowire array composite thin films for thermoelectric devices. J. Electron. Mater. 41, 887–894. https://doi.org/10.1007/s11664-012-1904-1.

Dai, D., Bauters, J., Bowers, J.E., 2012. Passive technologies for future large-scale photonic integrated circuits on silicon: polarization handling, light non-reciprocity and loss reduction. Light Sci. Appl. 1. e1. https://doi.org/10.1038/lsa.2012.1.

Dai, X., Messanvi, A., Zhang, H., Durand, C., Eymery, J., Bougerol, C., Julien, F.H., Tchernycheva, M., 2015. Flexible light-emitting diodes based on vertical nitride nanowires. Nano Lett. 15, 6958–6964. https://doi.org/10.1021/acs.nanolett.5b02900.

Dames, C., Chen, G., 2004. Theoretical phonon thermal conductivity of Si/Ge superlattice nanowires. J. Appl. Phys. 95, 682–693. https://doi.org/10.1063/1.1631734.

Datta, S., 2012. Lessons From Nanoelectronics: A New Perspective on Transport. World Scientific Publishing Company, New Jersey.

Davidson, F.M., Lee, D.C., Fanfair, D.D., Korgel, B.A., 2007. Lamellar twinning in semiconductor nanowires. J. Phys. Chem. C 111, 2929–2935. https://doi.org/10.1021/JP0672205.

Dávila, D., Tarancón, A., Fernández-Regúlez, M., Calaza, C., Salleras, M., San Paulo, A., Fonseca, L., 2011. Silicon nanowire arrays as thermoelectric material for a power microgenerator. J. Micromech. Microeng. 21. 104007. https://doi.org/10.1088/0960-1317/21/10/104007.

Davoody, A.H., Ramayya, E.B., Maurer, L.N., Knezevic, I., 2014. Ultrathin GaN nanowires: electronic, thermal, and thermoelectric properties. Phys. Rev. B 89, 115313. https://doi.org/10.1103/PhysRevB.89.115313.

Dayeh, S.A., Aplin, D.P.R., Zhou, X., Yu, P.K.L., Yu, E.T., Wang, D., 2007. High electron mobility InAs nanowire field-effect transistors. Small 3, 326–332. https://doi.org/10.1002/smll.200600379.

Diedenhofen, S.L., Janssen, O.T.A., Grzela, G., Bakkers, E.P.A.M., Gómez Rivas, J., 2011. Strong geometrical dependence of the absorption of light in arrays of semiconductor nanowires. ACS Nano 5, 2316–2323. https://doi.org/10.1021/nn103596n.

Dimroth, F., 2006. High-efficiency solar cells from III-V compound semiconductors. Phys. stat. sol. 3, 373–379. https://doi.org/10.1002/pssc.200564172.

Dimroth, F., Kurtz, S., 2007. High-efficiency multijunction solar cells. MRS Bull. 32, 230–235. https://doi.org/10.1557/mrs2007.27.

Dimroth, F., Tibbits, T.N.D., Niemeyer, M., Predan, F., Beutel, P., Karcher, C., Oliva, E., Siefer, G., Lackner, D., Fus-Kailuweit, P., Bett, A.W., Krause, R., Drazek, C., Guiot, E., Wasselin, J., Tauzin, A., Signamarcheix, T., 2016. Four-junction wafer-bonded concentrator solar cells. IEEE J. Photovolt. 6, 343–349. https://doi.org/10.1109/JPHOTOV.2015.2501729.

Dismukes, J.P., Ekstrom, L., Steigmeier, E.F., Kudman, I., Beers, D.S., 1964. Thermal and electrical properties of heavily doped Ge-Si alloys up to 1300°K. J. Appl. Phys. 35, 2899–2907. https://doi.org/10.1063/1.1713126.

Donadio, D., Galli, G., 2010. Temperature dependence of the thermal conductivity of thin silicon nanowires. Nano Lett. 10, 847–851. https://doi.org/10.1021/nl903268y.

Du, W., Ghetmiri, S.A., Margetis, J., Al-Kabi, S., Zhou, Y., Liu, J., Sun, G., Soref, R.A., Tolle, J., Li, B., Mortazavi, M., Yu, S.-Q., 2017. Investigation of optical transitions in a SiGeSn/GeSn/SiGeSn single quantum well structure. J. Appl. Phys. 122. 123102. https://doi.org/10.1063/1.4986341.

Dughaish, Z.H., 2002. Lead telluride as a thermoelectric material for thermoelectric power generation. Phys. B Condens. Matter 322, 205–223. https://doi.org/10.1016/S0921-4526(02)01187-0.

Eilertsen, J., Li, J., Rouvimov, S., Subramanian, M.A., 2011. Thermoelectric properties of indium-filled InxRh4Sb12 skutterudites. J. Alloys Compd. 509, 6289–6295. https://doi.org/10.1016/J.JALLCOM.2011.03.057.

Essig, S., Moutanabbir, O., Wekkeli, A., Nahme, H., Oliva, E., Bett, A.W., Dimroth, F., 2013. Fast atom beam-activated n-Si/n-GaAs wafer bonding with high interfacial transparency and electrical conductivity. J. Appl. Phys. 113. 203512. https://doi.org/10.1063/1.4807905.

Fang, A.W., Hyundai Park, Jones, R., Cohen, O., Paniccia, M.J. Bowers, J.E., 2006. A continuous-wave hybrid AlGaInAs-silicon evanescent laser. IEEE Photon. Technol. Lett. 18, 1143–1145. https://doi.org/10.1109/LPT.2006.874690.

Fang, Y.-Y., Xie, J., Tolle, J., Roucka, R., D'Costa, V.R., Chizmeshya, A.V.G., Menendez, J., Kouvetakis, J., 2008. Molecular-based synthetic approach to new group IV materials for high-efficiency, low-cost solar cells and Si-based optoelectronics. J. Am. Chem. Soc. 130, 16095–16102. https://doi.org/10.1021/ja806636c.

Farrell, A.C., Lee, W.-J., Senanayake, P., Haddad, M.A., Prikhodko, S.V., Huffaker, D.L., 2015. High-quality InAsSb nanowires grown by catalyst-free selective-area metal–organic chemical vapor deposition. Nano Lett. 15, 6614–6619. https://doi.org/10.1021/acs.nanolett.5b02389.

Ferry, D.K., 2008. Materials science. Nanowires in nanoelectronics. Science 319, 579–580. https://doi.org/10.1126/science.1154446.

Fournier-Lupien, J.-H., Mukherjee, S., Wirths, S., Pippel, E., Hayazawa, N., Mussler, G., Hartmann, J.M., Desjardins, P., Buca, D., Moutanabbir, O., 2013. Strain and composition effects on Raman vibrational modes of silicon-germanium-tin ternary alloys. Appl. Phys. Lett. 103. 263103. https://doi.org/10.1063/1.4855436.

France, R.M., Geisz, J.F., Garcia, I., Steiner, M.A., McMahon, W.E., Friedman, D.J., Moriarty, T.E., Osterwald, C., Scott Ward, J., Duda, A., Young, M., Olavarria, W.J., 2015. Quadruple-junction inverted metamorphic concentrator devices. IEEE J. Photovolt. 5, 432–437. https://doi.org/10.1109/JPHOTOV.2014.2364132.

France, R.M., Geisz, J.F., Garcia, I., Steiner, M.A., McMahon, W.E., Friedman, D.J., Moriarty, T.E., Osterwald, C., Ward, J.S., Duda, A., Young, M., Olavarria, W.J., 2016. Design flexibility of ultrahigh efficiency four-junction inverted metamorphic solar cells. IEEE J. Photovolt. 6, 578–583. https://doi.org/10.1109/JPHOTOV.2015.2505182.

Franzò, G., Boninelli, S., Pacifici, D., Priolo, F., Iacona, F., Bongiorno, C., 2003. Sensitizing properties of amorphous Si clusters on the 1.54-μm luminescence of Er in Si-rich SiO2. Appl. Phys. Lett. 82, 3871–3873. https://doi.org/10.1063/1.1579555.

Fukuda, M., Watanabe, K., Sakashita, M., Kurosawa, M., Nakatsuka, O., Zaima, S., 2017. Control of $Ge_{1-x-y}Si_xSn_y$ layer lattice constant for energy band alignment in $Ge_{1-x}Sn_x$/ $Ge_{1-x-y}Si_xSn_y$ heterostructures. Semicond. Sci. Technol. 32, 104008. https://doi.org/10.1088/1361-6641/aa80ce.

Gallagher, J.D., Senaratne, C.L., Wallace, P.M., Menendez, J., Kouvetakis, J., 2015. Electroluminescence from $Ge_{1-y}Sn_y$ diodes with degenerate pn junctions. Appl. Phys. Lett. 107, 1–4. https://doi.org/10.1063/1.4931707.

Garnett, E.C., Brongersma, M.L., Cui, Y., McGehee, M.D., 2011. Nanowire solar cells. Annu. Rev. Mater. Res. 41, 269–295. https://doi.org/10.1146/annurev-matsci-062910-100434.

Gassenq, A., Milord, L., Aubin, J., Pauc, N., Guilloy, K., Rothman, J., Rouchon, D., Chelnokov, A., Hartmann, J.M., Reboud, V., Calvo, V., 2017. Raman spectral shift versus strain and composition in GeSn layers with 6%–15% Sn content. Appl. Phys. Lett. 110. 112101. https://doi.org/10.1063/1.4978512.

Gelbstein, Y., Dashevsky, Z., Dariel, M.P., 2005. High performance n-type PbTe-based materials for thermoelectric applications. Phys. B Condens. Matter 363, 196–205. https://doi.org/10.1016/J.PHYSB.2005.03.022.

Ghetmiri, S.A., Du, W., Margetis, J., Mosleh, A., Cousar, L., Conley, B.R., Domulevicz, L., Nazzal, A., Sun, G., Soref, R.A., Tolle, J., Li, B., Naseem, H.A., Yu, S.-Q., 2014. Direct-bandgap GeSn grown on silicon with 2230 nm photoluminescence. Appl. Phys. Lett. 105. 151109. https://doi.org/10.1063/1.4898597.

Goldsmid, H.J., Douglas, R.W., 1954. The use of semiconductors in thermoelectric refrigeration. Br. J. Appl. Phys. 5, 386–390. https://doi.org/10.1088/0508-3443/5/11/303.

Goldsmid, H.J., Sheard, A.R., Wright, D.A., 1958. The performance of bismuth telluride thermojunctions. Br. J. Appl. Phys. 9, 365–370. https://doi.org/10.1088/0508-3443/9/9/306.

Goodnick, S.M., 1983. Surface roughness scattering at the Si–SiO_2 interface. J. Vac. Sci. Technol. B Nanotechnol. Microelectron. 1, 803. https://doi.org/10.1116/1.582696.

Green, M.A., 2016. Commercial progress and challenges for photovoltaics. Nat. Energy 1, 1–4. https://doi.org/10.1038/nenergy.2015.15.

Green, M.A., Bremner, S.P., 2017. Energy conversion approaches and materials for high-efficiency photovoltaics. Nat. Mater. 16, 23–34. https://doi.org/10.1038/nmat4676.

Green, M.A., Hishikawa, Y., Warta, W., Dunlop, E.D., Levi, D.H., Hohl-Ebinger, J., Ho-Baillie, A.W.H., 2017. Solar cell efficiency tables (version 50). Prog. Photovolt. Res. Appl. 25, 668–676. https://doi.org/10.1002/pip.2909.

Grundmann, M., 2006. The Physics of Semiconductors. Springer, New York.

Grzela, G., Paniagua-Domínguez, R., Barten, T., Fontana, Y., Sánchez-Gil, J.A., Rivas, J.G., 2012. Nanowire antenna emission. Nano Lett. 12, 5481–5486.

Gupta, S., Chen, R., Huang, Y.-C., Kim, Y., Sanchez, E., Harris, J.S., Saraswat, K.C., 2013. Highly selective dry etching of germanium over germanium–tin ($Ge_{1-x}Sn_x$): a novel route for $Ge_{1-x}Sn_x$ nanostructure fabrication. Nano Lett. 13, 3783–3790. https://doi.org/10.1021/nl4017286.

Guter, W., Schöne, J., Philipps, S.P., Steiner, M., Siefer, G., Wekkeli, A., Welser, E., Oliva, E., Bett, A.W., Dimroth, F., 2009. Current-matched triple-junction solar cell reaching 41.1% conversion efficiency under concentrated sunlight. Appl. Phys. Lett. 94. 223504. https://doi.org/10.1063/1.3148341.

Guthy, C., Nam, C.-Y., Fischer, J.E., 2008. Unusually low thermal conductivity of gallium nitride nanowires. J. Appl. Phys. 103, 64319. https://doi.org/10.1063/1.2894907.

Harman, T.C., Taylor, P.J., Walsh, M.P., LaForge, B.E., 2002. Quantum dot superlattice thermoelectric materials and devices. Science 297, 2229–2232. https://doi.org/10.1126/science.1072886.

He, G., Atwater, H.A., 1995. Synthesis of epitaxial Sn_xGe_{1-x} alloy films by ion-assisted molecular beam epitaxy. Appl. Phys. Lett. 68, 664–666.

He, G., Atwater, H.A., 1997. Interband transitions in Sn_xGe_{1-x} alloys. Phys. Rev. Lett. 79, 1937–1940. https://doi.org/10.1063/1.4829697.

Heck, M.J.R., Bowers, J.E., 2014. Energy efficient and energy proportional optical interconnects for multi-core processors: driving the need for on-chip sources. IEEE J. Sel. Top. Quantum Electron. 20, 332–343. https://doi.org/10.1109/JSTQE.2013.2293271.
Hicks, L., Dresselhaus, M., 1993a. Thermoelectric figure of merit of a one-dimensional conductor. Phys. Rev. B 47, 16631–16634. https://doi.org/10.1103/PhysRevB.47.16631.
Hicks, L., Dresselhaus, M., 1993b. Effect of quantum-well structures on the thermoelectric figure of merit. Phys. Rev. B 47, 12727–12731. https://doi.org/10.1103/PhysRevB.47.12727.
Hochbaum, A.I., Yang, P., 2010. Semiconductor nanowires for energy conversion. Chem. Rev. 110, 527–546. https://doi.org/10.1021/cr900075v.
Hochbaum, A.I., Chen, R., Delgado, R.D., Liang, W., Garnett, E.C., Najarian, M., Majumdar, A., Yang, P., 2008. Enhanced thermoelectric performance of rough silicon nanowires. Nature 451, 163–167. https://doi.org/10.1038/nature06381.
Homewood, K.P., Lourenço, M.A., 2015. Optoelectronics: the rise of the GeSn laser. Nat. Photonics 9, 78–79. https://doi.org/10.1038/nphoton.2015.1.
Hsin, C.-L., Wingert, M., Huang, C.-W., Guo, H., Shih, T.-J., Suh, J., Wang, K., Wu, J., Wu, W.-W., Chen, R., 2013. Phase transformation and thermoelectric properties of bismuth-telluride nanowires. Nanoscale 5, 4669. https://doi.org/10.1039/c3nr00876b.
Hu, M., Poulikakos, D., 2012. Si/Ge superlattice nanowires with ultralow thermal conductivity. Nano Lett. 12, 5487–5494. https://doi.org/10.1021/nl301971k.
Hu, Y., Churchill, H.O., Reilly, D.J., Xiang, J., Lieber, C.M., Marcus, C.M., 2007. A Ge/Si heterostructure nanowire-based double quantum dot with integrated charge sensor. Nat. Nanotechnol. 2, 622–625. https://doi.org/10.1038/nnano.2007.302.
Hu, M., Giapis, K.P., Goicochea, J.V., Zhang, X., Poulikakos, D., 2011a. Significant reduction of thermal conductivity in Si/Ge core–shell nanowires. Nano Lett. 11, 618–623. https://doi.org/10.1021/nl103718a.
Hu, M., Zhang, X., Giapis, K.P., Poulikakos, D., 2011b. Thermal conductivity reduction in core-shell nanowires. Phys. Rev. B 84, 85442. https://doi.org/10.1103/PhysRevB.84.085442.
Huang, W., Koong, C.S., Liang, G., 2010. Theoretical study on thermoelectric properties of Ge nanowires based on electronic band structures. IEEE Electron Device Lett. 31, 1026–1028. https://doi.org/10.1109/LED.2010.2053190.
Iyer, S.S., Xie, Y.-H., 2012. Light emission from silicon. Science 260, 40–46.
Jackrel, D.B., Bank, S.R., Yuen, H.B., Wistey, M.A., Harris, J.S., Ptak, A.J., Johnston, S.W., Friedman, D.J., Kurtz, S.R., 2007. Dilute nitride GaInNAs and GaInNAsSb solar cells by molecular beam epitaxy. J. Appl. Phys. 101, 1–8. https://doi.org/10.1063/1.2744490.
Jeong, C., Kim, R., Luisier, M., Datta, S., Lundstrom, M., 2010. On Landauer versus Boltzmann and full band versus effective mass evaluation of thermoelectric transport coefficients. J. Appl. Phys. 107, 23707. https://doi.org/10.1063/1.3291120.
Jiang, L., Xu, C., Gallagher, J.D., Favaro, R., Aoki, T., Mennedez, J., Kouvetakis, J., 2014. Development of light emitting group IV ternary alloys on Si platforms for long wavelength optoelectronic applications. Chem. Mater. 26, 2522–2531. https://doi.org/10.1021/cm403801b.
Joshi, G., Lee, H., Lan, Y., Wang, X., Zhu, G., Wang, D., Gould, R.W., Cuff, D.C., Tang, M.Y., Dresselhaus, M.S., Chen, G., Ren, Z., 2008. Enhanced thermoelectric figure-of-merit in nanostructured p-type silicon germanium bulk alloys. Nano Lett. 8, 4670–4674. https://doi.org/10.1021/nl8026795.
Karg, S., Mensch, P., Gotsmann, B., Schmid, H., Kanungo, P.D., Ghoneim, H., Schmidt, V., Björk, M.T., Troncale, V., Riel, H., 2013. Measurement of thermoelectric properties of single semiconductor nanowires. J. Electron. Mater. 42, 2409–2414. https://doi.org/10.1007/s11664-012-2409-7.

Kasper, E., Werner, J., Oehme, M., Escoubas, S., Burle, N., Schulze, J., 2012. Growth of silicon based germanium tin alloys. Thin Solid Films 520, 3195–3200. https://doi.org/10.1016/j.tsf.2011.10.114.

Kelzenberg, M.D., Boettcher, S.W., Petykiewicz, J.A., Turner-Evans, D.B., Putnam, M.C., Warren, E.L., Spurgeon, J.M., Briggs, R.M., Lewis, N.S., Atwater, H.A., 2010. Enhanced absorption and carrier collection in Si wire arrays for photovoltaic applications. Nat. Mater. 9, 239–244. https://doi.org/10.1038/nmat2635.

Kempa, T.J., Cahoon, J.F., Kim, S.-K., Day, R.W., Bell, D.C., Park, H.-G., Lieber, C.M., 2012. Coaxial multishell nanowires with high-quality electronic interfaces and tunable optical cavities for ultrathin photovoltaics. Proc. Natl. Acad. Sci. U.S.A. 109, 1407–1412. https://doi.org/10.1073/pnas.1120415109.

Kenyon, A.J., Steinman, E.A., Pitt, C.W., Hole, D.E., Vdovin, V.I., 2003. The origin of the 0.78 eV luminescence band in dislocated silicon. J. Phys. Condens. Matter 15, S2843–S2850. https://doi.org/10.1088/0953-8984/15/39/009.

Kim, W., Zide, J., Gossard, A., Klenov, D., Stemmer, S., Shakouri, A., Majumdar, A., 2006. Thermal conductivity reduction and thermoelectric figure of merit increase by embedding nanoparticles in crystalline semiconductors. Phys. Rev. Lett. 96, 45901. https://doi.org/10.1103/PhysRevLett.96.045901.

Kim, R., Datta, S., Lundstrom, M.S., 2009. Influence of dimensionality on thermoelectric device performance. J. Appl. Phys. 105, 34506. https://doi.org/10.1063/1.3074347.

King, R.R., Law, D.C., Edmondson, K.M., Fetzer, C.M., Kinsey, G.S., Yoon, H., Sherif, R.A., Karam, N.H., 2007. 40% efficient metamorphic GaInP / GaInAs / Ge multijunction solar cells. Appl. Phys. Lett. 90. 183516. https://doi.org/10.1063/1.2734507.

Kittel, C., 2005. Introduction to Solid State Physics. Wiley, New York.

Klemens, P.G., 1955. The scattering of low-frequency lattice waves by static imperfections. Proc. Phys. Soc. Sect. A Lond. 68, 1113–1128. https://doi.org/10.1088/0370-1298/68/12/303.

Koblmüller, G., Vizbaras, K., Hertenberger, S., Bolte, S., Rudolph, D., Becker, J., Döblinger, M., Amann, M.-C., Finley, J.J., Abstreiter, G., 2012. Diameter dependent optical emission properties of InAs nanowires grown on Si. Appl. Phys. Lett. 101, 53103. https://doi.org/10.1063/1.4739001.

Koelling, S., Li, A., Cavalli, A., Assali, S., Car, D., Gazibegovic, S., Bakkers, E.P.A.M., Koenraad, P.M., 2017. Atom-by-atom analysis of semiconductor nanowires with parts per million sensitivity. Nano Lett. 17, 599–605. https://doi.org/10.1021/acs.nanolett.6b03109.

Koshida, N., Koyama, H., 1992. Visible electroluminescence from porous silicon. Appl. Phys. Lett. 60, 347–349. https://doi.org/10.1063/1.106652.

Kouvetakis, J., Menendez, J., Chizmeshya, A.V.G., 2006. Tin-based group IV semiconductors: new platforms for opto- and microelectronics on silicon. Annu. Rev. Mater. Res. 36, 497–554. https://doi.org/10.1146/annurev.matsci.36.090804.095159.

Kouvetakis, J., Mathews, J., Roucka, R., Chizmeshya, A.V.G., Tolle, J., Menendez, J., 2010. Practical materials chemistry approaches for tuning optical and structural properties of group IV semiconductors and prototype photonic devices. IEEE Photonics J. 2, 924–941. https://doi.org/10.1109/JPHOT.2010.2081357.

Krogstrup, P., Jørgensen, H.I., Heiss, M., Demichel, O., Holm, J.V., Aagesen, M., Nygard, J., Fontcuberta i Morral, A., 2013. Single-nanowire solar cells beyond the Shockley-Queisser limit. Nat. Photonics 7, 1–5. https://doi.org/10.1038/nphoton.2013.32.

Landauer, R., 1957. Spatial variation of currents and fields due to localized scatterers in metallic conduction. IBM J. Res. Dev. 1, 223–231. https://doi.org/10.1147/rd.13.0223.

Landauer, R., 1989. Conductance determined by transmission: probes and quantised constriction resistance. J. Phys. Condens. Matter 1, 8099–8110. https://doi.org/10.1088/0953-8984/1/43/011.

Landry, E.S., McGaughey, A.J.H., 2009. Effect of interfacial species mixing on phonon transport in semiconductor superlattices. Phys. Rev. B 79, 75316. https://doi.org/10.1103/PhysRevB.79.075316.

Lee, S.H., Shim, W., Jang, S.Y., Roh, J.W., Kim, P., Park, J., Lee, W., 2011. Thermoelectric properties of individual single-crystalline PbTe nanowires grown by a vapor transport method. Nanotechnology 22. 295707. https://doi.org/10.1088/0957-4484/22/29/295707.

Lee, E.K., Yin, L., Lee, Y., Lee, J.W., Lee, S.J., Lee, J., Cha, S.N., Whang, D., Hwang, G.S., Hippalgaonkar, K., Majumdar, A., Yu, C., Choi, B.L., Kim, J.M., Kim, K., 2012. Large thermoelectric figure-of-merits from SiGe nanowires by simultaneously measuring electrical and thermal transport properties. Nano Lett. 12, 2918–2923. https://doi.org/10.1021/nl300587u.

Leroux, M., Grandjean, N., Beaumont, B., Nataf, G., Semond, F., Massies, J., Gibart, P., 1999. Temperature quenching of photoluminescence intensities in undoped and doped GaN. J. Appl. Phys. 86, 3721–3728.

Li, D., Wu, Y., Fan, R., Yang, P., Majumdar, A., 2003a. Thermal conductivity of Si/SiGe superlattice nanowires. Appl. Phys. Lett. 83, 3186–3188. https://doi.org/10.1063/1.1619221.

Li, D., Wu, Y., Kim, P., Shi, L., Yang, P., Majumdar, A., 2003b. Thermal conductivity of individual silicon nanowires. Appl. Phys. Lett. 83, 2934. https://doi.org/10.1063/1.1616981.

Li, H.P., De Sarkar, A., Zhang, R.Q., 2011. Surface-nitrogenation–induced thermal conductivity attenuation in silicon nanowires. EPL 96, 56007. https://doi.org/10.1209/0295-5075/96/56007.

Li, Y., Buddharaju, K., Singh, N., Lee, S.J., 2012. Top-down silicon nanowire-based thermoelectric generator: design and characterization. J. Electron. Mater. 41, 989–992. https://doi.org/10.1007/s11664-012-1901-4.

Li, H., Cui, Y.X., Wu, K.Y., Tseng, W.K., Cheng, H.H., Chen, H., 2013. Strain relaxation and Sn segregation in GeSn epilayers under thermal treatment. Appl. Phys. Lett. 102. 251907. https://doi.org/10.1063/1.4812490.

Li, L., Xu, S., Li, G., 2015. Enhancement of thermoelectric properties in Bi-Sb-Te alloy nanowires by pulsed electrodeposition. Energ. Technol. 3, 825–829. https://doi.org/10.1002/ente.201500071.

Liang, W., Rabin, O., Hochbaum, A.I., Fardy, M., Zhang, M., Yang, P., 2009. Thermoelectric properties of p-type PbSe nanowires. Nano Res. 2, 394–399. https://doi.org/10.1007/s12274-009-9039-2.

Liang, G., Huang, W., Koong, C.S., Wang, J.-S., Lan, J., 2010. Geometry effects on thermoelectric properties of silicon nanowires based on electronic band structures. J. Appl. Phys. 107, 14317. https://doi.org/10.1063/1.3273485.

Lieten, R.R., Maeda, T., Jevasuwan, W., Hattori, H., Uchida, N., Miura, S., Tanaka, M., Locquet, J.P., 2013. Tensile-strained GeSn metal-oxide-semiconductor field-effect transistor devices on Si(111) using solid phase epitaxy. Appl. Phys. Express 6, 4–7. https://doi.org/10.7567/APEX.6.101301.

Lim, J., Hippalgaonkar, K., Andrews, S.C., Majumdar, A., Yang, P., 2012. Quantifying surface roughness effects on phonon transport in silicon nanowires. Nano Lett. 12, 2475–2482. https://doi.org/10.1021/nl3005868.

Lin, H., Chen, R., Lu, W., Huo, Y., Kamins, T.I., Harris, J.S., 2012. Structural and optical characterization of $Si_xGe_{1-x-y}Sn_y$ alloys grown by molecular beam epitaxy. Appl. Phys. Lett. 100. 141908. https://doi.org/10.1063/1.3701732.

Liu, J., Sun, X., Pan, D., Wang, X., Kimerling, L.C., Koch, T.L., Michel, J., 2007. Tensile-strained, n-type Ge as a gain medium for monolithic laser integration on Si. Opt. Express 15. 11272. https://doi.org/10.1364/OE.15.011272.

Liu, J., Sun, X., Camacho-Aguilera, R., Kimerling, L.C., Michel, J., 2010. Ge-on-Si laser operating at room temperature. Opt. Lett. 35, 679. https://doi.org/10.1364/OL.35.000679.

Liu, Y., Zhang, Z., Wei, X., Li, Q., Peng, L.-M., 2011. Simultaneous electrical and thermoelectric parameter retrieval via two terminal current-voltage measurements on individual ZnO nanowires. Adv. Funct. Mater. 21, 3900–3906. https://doi.org/10.1002/adfm.201100701.

Liu, W., Jie, Q., Kim, H.S., Ren, Z., 2015. Current progress and future challenges in thermoelectric power generation: from materials to devices. Acta Mater. 87, 357–376. https://doi.org/10.1016/J.ACTAMAT.2014.12.042.

Loo, R., Vincent, B., Gencarelli, F., Merckling, C., Kumar, A., Eneman, G., Witters, L., Vandervorst, W., Caymax, M., Heyns, M., Thean, A., 2012. $Ge_{1-x}Sn_x$ materials: challenges and applications. ECS J. Solid State Sci. Technol. 2, N35–N40. https://doi.org/10.1149/2.039301jss.

Lopez, F.J., Hemesath, E.R., Lauhon, L.J., 2009. Ordered stacking fault arrays in silicon nanowires. Nano Lett. 9, 2774–2779. https://doi.org/10.1021/nl901315s.

Lopez, F.J., Givan, U., Connell, J.G., Lauhon, L.J., 2011. Silicon nanowire polytypes: identification by Raman spectroscopy, generation mechanism, and misfit strain in homostructures. ACS Nano 5, 8958–8966.

Lundstrom, M., Jeong, C., 2013. Near-Equilibrium Transport: Fundamentals and Applications. World Scientific Publishing Company, New Jersey.

Ma, L., Mei, R., Zhao, X., Sun, H., 2017. Monte Carlo simulation of single-crystalline PbSe nanowire thermal conductivity using first-principle phonon properties. Semicond. Sci. Technol. 32, 95008. https://doi.org/10.1088/1361-6641/aa7c15.

Majumdar, A., 2004. Materials science. Thermoelectricity in semiconductor nanostructures. Science 303, 777–778. https://doi.org/10.1126/science.1093164.

Majumdar, A., 2009. Helping chips to keep their cool. Nat. Nanotechnol. 4, 214–215. https://doi.org/10.1038/nnano.2009.65.

Margetis, J., Mosleh, A., Al-Kabi, S., Ghetmiri, S.A., Du, W., Dou, W., Benamara, M., Li, B., Mortazavi, M., Naseem, H.A., Yu, S.-Q., Tolle, J., 2017a. Study of low-defect and strain-relaxed GeSn growth via reduced pressure CVD in H2 and N2 carrier gas. J. Cryst. Growth 463, 128–133. https://doi.org/10.1016/j.jcrysgro.2017.01.041.

Margetis, J., Mosleh, A., Ghetmiri, S.A., Al-Kabi, S., Dou, W., Du, W., Bhargava, N., Yu, S.-Q., Profijt, H., Kohen, D., Loo, R., Vohra, A., Tolle, J., 2017b. Fundamentals of $Ge_{1-x}Sn_x$ and $Si_yGe_{1-x-y}Sn_x$ RPCVD epitaxy. Mater. Sci. Semicond. Process. 70, 38–43. https://doi.org/10.1016/j.mssp.2016.12.024.

Margetis, J., Yu, S.-Q., Bhargava, N., Li, B., Du, W., Tolle, J., 2017c. Strain engineering in epitaxial $Ge_{1-x}Sn_x$: a path towards low-defect and high Sn-content layers. Semicond. Sci. Technol. 32. 124006. https://doi.org/10.1088/1361-6641/aa7fc7.

Markussen, T., Jauho, A.-P., Brandbyge, M., 2008. Heat conductance is strongly anisotropic for pristine silicon nanowires. Nano Lett. 8, 3771–3775. https://doi.org/10.1021/nl8020889.

Markussen, T., Jauho, A.-P., Brandbyge, M., 2009. Surface-decorated silicon nanowires: a route to high-ZT thermoelectrics. Phys. Rev. Lett. 103, 55502. https://doi.org/10.1103/PhysRevLett.103.055502.

Martin, P., Aksamija, Z., Pop, E., Ravaioli, U., 2009a. Impact of phonon-surface roughness scattering on thermal conductivity of thin Si nanowires. Phys. Rev. Lett. 102, 125503. https://doi.org/10.1103/PhysRevLett.102.125503.

Martin, P.N., Aksamija, Z., Pop, E., Ravaioli, U., 2009b. Prediction of reduced thermal conductivity in nano-engineered rough semiconductor nanowires. J. Phys. Conf. Ser. 193, 12010. https://doi.org/10.1088/1742-6596/193/1/012010.

Martin, P.N., Aksamija, Z., Pop, E., Ravaioli, U., 2010. Reduced thermal conductivity in nanoengineered rough Ge and GaAs nanowires. Nano Lett. 10, 1120–1124. https://doi.org/10.1021/nl902720v.

Mavrokefalos, A., Pettes, M.T., Saha, S., Zhou, F., Shi, L., 2006. In: Combined thermoelectric and structure characterizations of patterned nanowires. 25th International Conference on Thermoelectrics. IEEE, pp. 234–237. https://doi.org/10.1109/ICT.2006.331358.

Meng, A.C., Fenrich, C.S., Braun, M.R., McVittie, J.P., Marshall, A.F., Harris, J.S., McIntyre, P.C., 2016. Core/shell germanium/germanium-tin nanowires exhibiting room temperature direct- and indirect-gap photoluminescence. Nano Lett. 16, 7521–7529. https://doi.org/10.1021/acs.nanolett.6b03316.

Mingo, N., 2004. Thermoelectric figure of merit and maximum power factor in III–V semiconductor nanowires. Appl. Phys. Lett. 84, 2652–2654. https://doi.org/10.1063/1.1695629.

Mingo, N., Broido, D.A., 2004. Lattice thermal conductivity crossovers in semiconductor nanowires. Phys. Rev. Lett. 93, 246106. https://doi.org/10.1103/PhysRevLett.93.246106.

Miritello, M., Lo Savio, R., Iacona, F., Franzò, G., Irrera, A., Piro, A.M., Bongiorno, C., Priolo, F., 2007. Efficient luminescence and energy transfer in erbium silicate thin films. Adv. Mater. 19, 1582–1588. https://doi.org/10.1002/adma.200601692.

Moontragoon, P., Soref, R.A., Ikonic, Z., 2012. The direct and indirect bandgaps of unstrained $Si_xGe_{1-x-y}Sn_y$ and their photonic device applications. J. Appl. Phys. 112, 73106. https://doi.org/10.1063/1.4757414.

Moore, A.L., Saha, S.K., Prasher, R.S., Shi, L., 2008. Phonon backscattering and thermal conductivity suppression in sawtooth nanowires. Appl. Phys. Lett. 93, 83112. https://doi.org/10.1063/1.2970044.

Moutanabbir, O., Gösele, U., 2010. Heterogeneous integration of compound semiconductors. Annu. Rev. Mater. Res. 40, 469–500. https://doi.org/10.1146/annurev-matsci-070909-104448.

Moutanabbir, O., Senz, S., Zhang, Z., Gösele, U., 2009. Synthesis of isotopically controlled metal-catalyzed silicon nanowires. Nano Today 4, 393–398. https://doi.org/10.1016/j.nantod.2009.08.009.

Mukherjee, S., Givan, U., Senz, S., Bergeron, A., Francoeur, S., de la Mata, M., Arbiol, J., Sekiguchi, T., Itoh, K.M., Isheim, D., Seidman, D.N., Moutanabbir, O., 2015. Phonon engineering in isotopically disordered silicon nanowires. Nano Lett. 15, 3885–3893. https://doi.org/10.1021/acs.nanolett.5b00708.

Mukherjee, S., Givan, U., Senz, S., de la Mata, M., Arbiol, J., Moutanabbir, O., Reduction in thermal conductivity in silicon nanowires by combined engineering of crystal phase and isotope disorder (under review).

Nika, D.L., Cocemasov, A.I., Crismari, D.V., Balandin, A.A., 2013. Thermal conductivity inhibition in phonon engineered core-shell cross-section modulated Si/Ge nanowires. Appl. Phys. Lett. 102. 213109. https://doi.org/10.1063/1.4807389.

Nositschka, W.A., Beneking, C., Voigt, O., Kurz, H., 2003. Texturisation of multicrystalline silicon wafers for solar cells by reactive ion etching through colloidal masks. Sol. Energy Mater. Sol. Cells 76, 155–166. https://doi.org/10.1016/S0927-0248(02)00214-3.

Ochoa, M., Garcia, I., Lombardero, I., Ayllon, L., Cifuentes, L., Rey-Stolle, I., Algora, C., Johnson, A.D., Davies, J.I., Tan, K.H., Loke, W.K., Wicaksono, S., Yoon, S.F., Ochoa, E., Gabas, M., Thomas, T., Ekins-Daukes, N.J., Garcia, H., Castan, H., Duenas, S., Estrade, S., Peiro, F., 2016. In: Advances towards 4J lattice-matched including dilute nitride subcell for terrestrial and space applications. IEEE 43rd Photovoltaic Specialists Conference (PVSC). IEEE, pp. 0052–0057. https://doi.org/10.1109/PVSC.2016.7749407.

Olesinski, R.W., Abbaschian, G.J., 1984a. The Ge-Sn (germanium-tin) system. Bull. All. Pha. Diag. 5, 265–271. https://doi.org/10.1007/BF02868550.

Olesinski, R.W., Abbaschian, G.J., 1984b. The Si-Sn (silicon-tin) system. Bull. All. Pha. Diag. 5, 273–276. https://doi.org/10.1007/BF02868552.

Palacios, T., 2012. Nanowire electronics comes of age. Nature 481, 152–153. https://doi.org/10.1038/481152a.

Paniccia, M., 2010. Integrating silicon photonics. Nat. Photonics 4, 498–499. https://doi.org/10.1038/nphoton.2010.189.

Park, H., Fang, A.W., Kodama, S., Bowers, J.E., 2005. Hybrid silicon evanescent laser fabricated with a silicon waveguide and III-V offset quantum wells. Opt. Express 13, 9460. https://doi.org/10.1364/OPEX.13.009460.

Park, D., Park, S., Jeong, K., Jeong, H.-S., Song, J.Y., Cho, M., 2016. Thermal and electrical conduction of single-crystal Bi_2Te_3 nanostructures grown using a one step process. Sci. Rep. 6, 19132. https://doi.org/10.1038/srep19132.

Patel, P., Aiken, D., Boca, A., Cho, B., Chumney, D., Clevenger, M.B., Cornfeld, A., Fatemi, N., Lin, Y., McCarty, J., Newman, F., Sharps, P., Spann, J., Stan, M., Steinfeldt, J., Strautin, C., Varghese, T., 2012. Experimental results from performance improvement and radiation hardening of inverted metamorphic multijunction solar cells. IEEE J. Photovolt. 2, 377–381. https://doi.org/10.1109/JPHOTOV.2012.2198048.

Silicon photonics III: systems and applications. Pavesi, L., Lockwood, D.J. (Eds.), 2016. In: Topics in Applied Physics. Springer Berlin Heidelberg, Berlin, Heidelberg. https://doi.org/10.1007/978-3-642-10503-6.

Pavesi, L., Dal Negro, L., Mazzoleni, C., Franzo, G., Pirlo, F., 2000. Optical gain in silicon nanocrystals. Nature 408, 440–444. https://doi.org/10.1038/35044012.

Pennelli, G., Macucci, M., 2013. Optimization of the thermoelectric properties of nanostructured silicon. J. Appl. Phys. 114. 214507. https://doi.org/10.1063/1.4842835.

Persson, A.I., Koh, Y.K., Cahill, D.G., Samuelson, L., Linke, H., 2009. Thermal conductance of InAs nanowire composites. Nano Lett. 9, 4484–4488. https://doi.org/10.1021/nl902809j.

Pomeranchuk, I.Y., 1942. Thermal conductivity of dielectrics below Debye temperatures. Zh. Eksp. Teor. Fiz. 12, 245. https://doi.org/10.1103/PhysRev.60.820.

Poudel, B., Hao, Q., Ma, Y., Lan, Y., Minnich, A., Yu, B., Yan, X., Wang, D., Muto, A., Vashaee, D., Chen, X., Liu, J., Dresselhaus, M.S., Chen, G., Ren, Z., 2008. High-thermoelectric performance of nanostructured bismuth antimony telluride bulk alloys. Science 320, 634–638. https://doi.org/10.1126/science.1156446.

Qian, F., Gradečak, S., Li, Y., Wen, C.Y., Lieber, C.M., 2005. Core/multishell nanowire heterostructures as multicolor, high-efficiency light-emitting diodes. Nano Lett. 5, 2287–2291. https://doi.org/10.1021/nl051689e.

Querales-Flores, J.D., Ventura, C.I., Fuhr, J.D., Barrio, R.A., 2016. The two gap transitions in $Ge_{1-x}Sn_x$: effect of non-substitutional complex defects. J. Appl. Phys. 120. 105705. https://doi.org/10.1063/1.4962381.

Ra, Y., Navamathavan, R., Park, J., Lee, C., 2013. Coaxial $InxGa_{1-x}N$/GaN multiple quantum well nanowire arrays on Si(111) substrate for high-performance light-emitting diodes. Nano Lett. 13, 3506–3516.

Ragan, R., Ahn, C.C., Atwater, H.A., 2003. Nonlithographic epitaxial Sn_xGe_{1-x} dense nanowire arrays grown on Ge(001). Appl. Phys. Lett. 82, 3439–3441. https://doi.org/10.1063/1.1563834.

Read, G.W., Marko, I.P., Hossain, N., Sweeney, S.J., 2015. Physical properties and characteristics of III-V lasers on silicon. IEEE J. Sel. Top. Quantum Electron. 21, 377–384. https://doi.org/10.1109/JSTQE.2015.2424923.

Reboud, V., Gassenq, A., Pauc, N., Aubin, J., Milord, L., Thai, Q.M., Bertrand, M., Guilloy, K., Rouchon, D., Rothman, J., Zabel, T., Armand Pilon, F., Sigg, H., Chelnokov, A., Hartmann, J.M., Calvo, V., 2017. Optically pumped GeSn micro-disks with 16% Sn lasing at 3.1 μm up to 180 K. Appl. Phys. Lett. 111, 92101. https://doi.org/10.1063/1.5000353.

Reimer, M.E., Bulgarini, G., Akopian, N., Hocevar, M., Bavinck, M.B., Verheijen, M.A., Bakkers, E.P.A.M., Kouwenhoven, L.P., Zwiller, V., 2012. Bright single-photon sources in bottom-up tailored nanowires. Nat. Commun. 3, 1–5. https://doi.org/10.1038/ncomms1746.

Rickman, A., 2014. The commercialization of silicon photonics. Nat. Photonics 8, 579–582. https://doi.org/10.1038/nphoton.2014.175.

Rong, H., Jones, R., Liu, A., Cohen, O., Hak, D., Fang, A., Paniccia, M., 2005. A continuous-wave Raman silicon laser. Nature 433, 725–728. https://doi.org/10.1038/nature03346.

Sansoz, F., 2011. Surface faceting dependence of thermal transport in silicon nanowires. Nano Lett. 11, 5378–5382. https://doi.org/10.1021/nl2029688.

Satterthwaite, C.B., Ure, R.W., 1957. Electrical and thermal properties of Bi_2Te_3. Phys. Rev. 108, 1164–1170. https://doi.org/10.1103/PhysRev.108.1164.

Schelling, P.K., Phillpot, S.R., Keblinski, P., 2002. Phonon wave-packet dynamics at semiconductor interfaces by molecular-dynamics simulation. Appl. Phys. Lett. 80, 2484–2486. https://doi.org/10.1063/1.1465106.

Schneider, B.W., Lal, N.N., Baker-Finch, S., White, T.P., 2014. Pyramidal surface textures for light trapping and antireflection in perovskite-on-silicon tandem solar cells. Opt. Express 22, A1422–A1430. https://doi.org/10.1364/oe.22.0a1422.

Schwab, K., Henriksen, E.A., Worlock, J.M., Roukes, M.L., 2000. Measurement of the quantum of thermal conductance. Nature 404, 974–977. https://doi.org/10.1038/35010065.

Seifner, M.S., Biegger, F., Lugstein, A., Bernardi, J., Barth, S., 2015. Microwave-assisted $Ge_{1-x}Sn_x$ nanowire synthesis: precursor species and growth regimes. Chem. Mater. 27, 6125–6130. https://doi.org/10.1021/acs.chemmater.5b02757.

Seifner, M.S., Hernandez, S., Bernardi, J., Romano-Rodriguez, A., Barth, S., 2017. Pushing the composition limit of anisotropic $Ge_{1-x}Sn_x$ nanostructures and determination of their thermal stability. Chem. Mater. 29, 9802–9813. https://doi.org/10.1021/acs.chemmater.7b03969.

Snyder, G.J., Christensen, M., Nishibori, E., Caillat, T., Iversen, B.B., 2004. Disordered zinc in Zn_4Sb_3 with phonon-glass and electron-crystal thermoelectric properties. Nat. Mater. 3, 458–463. https://doi.org/10.1038/nmat1154.

Soini, M., Zardo, I., Uccelli, E., Funk, S., Koblmüller, G., Fontcuberta i Morral, A., Abstreiter, G., 2010. Thermal conductivity of GaAs nanowires studied by micro-Raman spectroscopy combined with laser heating. Appl. Phys. Lett. 97, 263107. https://doi.org/10.1063/1.3532848.

Soitec, CEA-Leti, ISE, F.I. for S.E.S, 2014. New World Record for Solar Cell Efficiency at 46% French-German Cooperation Confirms Competitive Advantage of European Photovoltaic Industry. https://www.ise.fraunhofer.de/en/press-media/press-releases/2014/new-world-record-for-solar-cell-efficiency-at-46-percent.html.

Solar Junction sets CPV cell efficiency record of 44%, 2012. https://cleantechnica.com/2012/12/30/solar-pv-efficiency-record-broken-now-44-thanks-to-nrel-solar-junction/.

Soref, R., 2010. Mid-infrared photonics in silicon and germanium. Nat. Photonics 4, 495–497. https://doi.org/10.1038/nphoton.2010.171.

Soref, R.A., Perry, C.H., 1991. Predicted band gap of the new semiconductor SiGeSn. J. Appl. Phys. 69, 539–541. https://doi.org/10.1063/1.347704.

Soref, R., Buca, D., Yu, S.-Q., 2016. Group IV photonics: driving integrated optoelectronics. Opt. Photonics News 27, 32–39. https://doi.org/10.1364/OPN.27.1.000032.
Stan, M., Aiken, D., Cho, B., Cornfeld, A., Ley, V., Patel, P., Sharps, P., Varghese, T., 2010. High-efficiency quadruple junction solar cells using OMVPE with inverted metamorphic device structures. J. Cryst. Growth 312, 1370–1374. https://doi.org/10.1016/j.jcrysgro.2009.10.059.
Stange, D., Wirths, S., Von Den Driesch, N., Mussler, G., Stoica, T., Ikonic, Z., Hartmann, J.M., Mantl, S., Grützmacher, D., Buca, D., 2015. Optical transitions in direct-bandgap $Ge_{1-x}Sn_x$ alloys. ACS Photonics 2, 1539–1545. https://doi.org/10.1021/acsphotonics.5b00372.
Stange, D., Wirths, S., Geiger, R., Schulte-Braucks, C., Marzban, B., Den Driesch, N.V., Mussler, G., Zabel, T., Stoica, T., Hartmann, J.-M., Mantl, S., Ikonic, Z., Grützmacher, D., Sigg, H., Witzens, J., Buca, D., 2016. Optically pumped GeSn microdisk lasers on Si. ACS Photonics 3, 1279–1285. https://doi.org/10.1021/acsphotonics.6b00258.
Stange, D., Von Den Driesch, N., Rainko, D., Roesgaard, S., Povstugar, I., Hartmann, J.-M., Stoica, T., Ikonić, Z., Mantl, S., Grützmacher, D., Buca, D., 2017. Short-wave infrared LEDs from GeSn/SiGeSn multiple quantum wells. Optica 4, 185–188. https://doi.org/doi.org/10.1364/OPTICA.4.000185.
Su, S., Cheng, B., Xue, C., Wang, W., Cao, Q., Xue, H., Hu, W., Zhang, G., Zuo, Y., Wang, Q., 2011. GeSn p-i-n photodetector for all telecommunication bands detection. Opt. Express 19, 6400. https://doi.org/10.1364/OE.19.006400.
Sun, X., Liu, J., Kimerling, L.C., Michel, J., 2009. Direct gap photoluminescence of n-type tensile-strained Ge-on-Si. Appl. Phys. Lett. 95, 11911. https://doi.org/10.1063/1.3170870.
Swinkels, M.Y., van Delft, M.R., Oliveira, D.S., Cavalli, A., Zardo, I., van der Heijden, R.W., Bakkers, E.P.A.M., 2015. Diameter dependence of the thermal conductivity of InAs nanowires. Nanotechnology 26. 385401. https://doi.org/10.1088/0957-4484/26/38/385401.
Takamoto, T., Washio, H., Juso, H., 2014. In: Application of InGaP/GaAs/InGaAs triple junction solar cells to space use and concentrator photovoltaic. IEEE 40th Photovoltaic Specialist Conference (PVSC). IEEE, pp. 0001–0005. https://doi.org/10.1109/PVSC.2014.6924936.
Takeuchi, S., Sakai, A., Nakatsuka, O., Ogawa, M., Zaima, S., 2008. Tensile strained Ge layers on strain-relaxed $Ge_{1-x}Sn_x$/virtual Ge substrates. Thin Solid Films 517, 159–162. https://doi.org/10.1016/j.tsf.2008.08.068.
Tan, K.H., Wicaksono, S., Loke, W.K., Li, D., Yoon, S.F., Fitzgerald, E.A., Ringel, S.A., Harris Jr., J.S., 2011. Molecular beam epitaxy grown GaNAsSb 1 eV photovoltaic cell. J. Cryst. Growth 335, 66–69. https://doi.org/10.1016/j.jcrysgro.2011.09.023.
Tanabe, K., Watanabe, K., Arakawa, Y., 2012. III-V/Si hybrid photonic devices by direct fusion bonding. Sci. Rep. 2, 349. https://doi.org/10.1038/srep00349.
Taoka, N., Capellini, G., Schlykow, V., Montanari, M., Zaumseil, P., Nakatsuka, O., Zaima, S., Schroeder, T., 2017. Electrical and optical properties improvement of GeSn layers formed at high temperature under well-controlled Sn migration. Mater. Sci. Semicond. Process. 70, 139–144. https://doi.org/10.1016/j.mssp.2017.07.013.
Tarun, A., Hayazawa, N., Ishitobi, H., Kawata, S., Reiche, M., Moutanabbir, O., 2011. Mapping the "forbidden" transverse-optical phonon in single strained silicon (100) nanowire. Nano Lett. 11, 4780–4788. https://doi.org/10.1021/nl202599q.
The International Road Map for Semiconductor Technologies, 2015. http://www.itrs2.net/.
Thomas, T., Fuhrer, M., Alvarez, D.A., Ekins-Daukes, N., Tan, K.H., Wicaksono, S., Loke, W.K., Yoon, S.F., Johnson, A., 2014. In: GaNAsSb 1-eV solar cells for use in lattice-matched multi-junction architectures. IEEE 40th Photovoltaic Specialist Conference (PVSC). IEEE, pp. 0550–0553. https://doi.org/10.1109/PVSC.2014.6924980.

Thomson, D.J., Gardes, F.Y., Fedeli, J.-M., Zlatanovic, S., Hu, Y., Kuo, B.P.P., Myslivets, E., Alic, N., Radic, S., Mashanovich, G.Z., Reed, G.T., 2012. 50-Gb/s silicon optical modulator. IEEE Photon. Technol. Lett. 24, 234–236. https://doi.org/10.1109/LPT.2011.2177081.

Tian, B., Zheng, X., Kempa, T.J., Fang, Y., Yu, N., Yu, G., Huang, J., Lieber, C.M., 2007. Coaxial silicon nanowires as solar cells and nanoelectronic power sources. Nature 449, 885–889. https://doi.org/10.1038/nature06181.

Urino, Y., Usuki, T., Fujikata, J., Ishizaka, M., Yamada, K., Horikawa, T., Nakamura, T., Arakawa, Y., 2014. High-density optical interconnects by using silicon photonics. In: Proc. SPIE 9010, Next-Generation Optical Networks for Data Centers and Short-Reach Links. 901006. https://doi.org/10.1117/12.2041418.

Valentín, L.A., Betancourt, J., Fonseca, L.F., Pettes, M.T., Shi, L., Soszyński, M., Huczko, A., 2013. A comprehensive study of thermoelectric and transport properties of β-silicon carbide nanowires. J. Appl. Phys. 114. 184301. https://doi.org/10.1063/1.4829924.

van Dam, D., Abujetas, D.R., Paniagua-Domínguez, R., Sánchez-Gil, J.A., Bakkers, E.P.A.M., Haverkort, J.E.M., Gómez Rivas, J., 2015. Directional and polarized emission from nanowire arrays. Nano Lett. 15, 4557–4563. https://doi.org/10.1021/acs.nanolett.5b01135.

van Dam, D., van Hoof, N.J.J., Cui, Y., van Veldhoven, P.J., Bakkers, E.P.A.M., Gómez Rivas, J., Haverkort, J.E.M., 2016. High-efficiency nanowire solar cells with omnidirectionally enhanced absorption due to self-aligned indium–tin–oxide Mie scatterers. ACS Nano 10, 11414–11419. https://doi.org/10.1021/acsnano.6b06874.

Venkatasubramanian, R., Siivola, E., Colpitts, T., O'Quinn, B., 2001. Thin-film thermoelectric devices with high room-temperature figures of merit. Nature 413, 597–602. https://doi.org/10.1038/35098012.

Verheijen, M.A., Immink, G., De Smet, T., Borgstrom, M.T., Bakkers, E.P.A.M., 2006. Growth kinetics of heterostructured GaP-GaAs nanowires. J. Am. Chem. Soc. 128, 1353–1359.

Vincent, B., Gencarelli, F., Bender, H., Merckling, C., Douhard, B., Petersen, D.H., Hansen, O., Henrichsen, H.H., Meersschaut, J., Vandervorst, W., Heyns, M., Loo, R., Caymax, M., 2011. Undoped and in-situ B doped GeSn epitaxial growth on Ge by atmospheric pressure-chemical vapor deposition. Appl. Phys. Lett. 99, 15–17. https://doi.org/10.1063/1.3645620.

Vivien, L., Polzer, A., Marris-Morini, D., Osmond, J., Hartmann, J.M., Crozat, P., Cassan, E., Kopp, C., Zimmermann, H., Fédéli, J.M., 2012. Zero-bias 40Gbit/s germanium waveguide photodetector on silicon. Opt. Express 20, 1096. https://doi.org/10.1364/OE.20.001096.

Volz, S.G., Chen, G., 1999. Molecular dynamics simulation of thermal conductivity of silicon nanowires. Appl. Phys. Lett. 75, 2056–2058. https://doi.org/10.1063/1.124914.

von den Driesch, N., Stange, D., Wirths, S., Mussler, G., Holländer, B., Ikonic, Z., Hartmann, J.M., Stoica, T., Mantl, S., Grützmacher, D., Buca, D., 2015. Direct bandgap group IV epitaxy on Si for laser applications. Chem. Mater. 27, 4693–4702. https://doi.org/10.1021/acs.chemmater.5b01327.

von den Driesch, N., Stange, D., Wirths, S., Rainko, D., Povstugar, I., Savenko, A., Breuer, U., Geiger, R., Sigg, H., Ikonic, Z., Hartmann, J.-M., Grützmacher, D., Mantl, S., Buca, D., 2017. SiGeSn ternaries for efficient group IV heterostructure light emitters. Small 13. 1603321. https://doi.org/10.1002/smll.201603321.

Wallentin, J., Anttu, N., Asoli, D., Huffman, M., Aberg, I., Magnusson, M.H., Siefer, G., Fuss-Kailuweit, P., Dimroth, F., Witzigmann, B., Xu, H.Q., Samuelson, L., Deppert, K., Borgström, M.T., 2013. InP nanowire array solar cells achieving 13.8% efficiency by exceeding the ray optics limit. Science 339, 1057–1060. https://doi.org/10.1126/science.1230969.

Wang, Z., Mingo, N., 2010. Diameter dependence of SiGe nanowire thermal conductivity. Appl. Phys. Lett. 97. 101903. https://doi.org/10.1063/1.3486171.
Wang, W., Jia, F., Huang, Q., Zhang, J., 2005. A new type of low power thermoelectric micro-generator fabricated by nanowire array thermoelectric material. Microelectron. Eng. 77, 223–229. https://doi.org/10.1016/J.MEE.2004.11.005.
Wang, W., Li, L., Tok, E.S., Yeo, Y.-C., 2015a. Self-assembly of tin wires via phase transformation of heteroepitaxial germanium-tin on germanium substrate. J. Appl. Phys. 117. 225304. https://doi.org/10.1063/1.4922423.
Wang, W., Zhou, Q., Dong, Y., Tok, E.S., Yeo, Y.-C., 2015b. Critical thickness for strain relaxation of GeSn grown by molecular beam epitaxy on Ge(001). Appl. Phys. Lett. 106. 232106. https://doi.org/10.1063/1.4922529.
Wang, Y., Ren, Z., Thway, M., Lee, K., Yoon, S.F., Peters, I.M., Buonassisi, T., Fizgerald, E.A., Tan, C.S., Lee, K.H., 2017. Fabrication and characterization of single junction GaAs solar cells on Si with As-doped Ge buffer. Sol. Energy Mater. Sol. Cells 172, 140–144. https://doi.org/10.1016/j.solmat.2017.07.028.
Wegscheider, W., Olajos, J., Menczigar, U., Dondl, W., Abstreiter, G., 1992. Fabrication and properties of epitaxially stabilized Ge / α-Sn heterostructures on Ge(001). J. Cryst. Growth 123, 75–94. https://doi.org/10.1016/0022-0248(92)90012-8.
Weisshaupt, D., Jahandar, P., Colston, G., Allred, P., Schulze, J., Myronov, M., 2017. In: Impact of Sn segregation on Ge1 − xSnx epi-layers growth by RP-CVD. 40th International Convention on Information and Communication Technology, Electronics and Microelectronics (MIPRO). IEEE, pp. 43–47. https://doi.org/10.23919/MIPRO.2017.7973388.
Wendav, T., Fischer, I.A., Montanari, M., Zoellner, M.H., Klesse, W., Capellini, G., von den Driesch, N., Oehme, M., Buca, D., Busch, K., Schulze, J., 2016. Compositional dependence of the band-gap of $Ge_{1-x-y}Si_xSn_y$ alloys. Appl. Phys. Lett. 108. 242104. https://doi.org/10.1063/1.4953784.
Werner, J., Oehme, M., Schmid, M., Kaschel, M., Schirmer, A., Kasper, E., Schulze, J., 2011. Germanium-tin p-i-n photodetectors integrated on silicon grown by molecular beam epitaxy. Appl. Phys. Lett. 98, 61108. https://doi.org/10.1063/1.3555439.
Wilson, T., Thomas, T., Führer, M., Ekins-Daukes, N.J., Roucka, R., Clark, A., Johnson, A., Hoffman, R., Begarney, D., 2016. Single and multi-junction solar cells utilizing a 1.0 eV SiGeSn junction. AIP Conference Proceedings, p. 60006. https://doi.org/10.1063/1.4962096.
Wirths, S., Buca, D., Mussler, G., Tiedemann, A.T., Hollander, B., Bernardy, P., Stoica, T., Grutzmacher, D., Mantl, S., 2013. Reduced pressure CVD growth of Ge and $Ge_{1-x}Sn_x$ alloys. ECS J. Solid State Sci. Technol. 2, N99–N102. https://doi.org/10.1149/2.006305jss.
Wirths, S., Geiger, R., von den Driesch, N., Mussler, G., Stoica, T., Mantl, S., Ikonic, Z., Luysberg, M., Chiussi, S., Hartmann, J.M., Sigg, H., Faist, J., Buca, D., Grützmacher, D., 2015. Lasing in direct-bandgap GeSn alloy grown on Si. Nat. Photonics 9, 88–92. https://doi.org/10.1038/nphoton.2014.321.
Wirths, S., Buca, D., Mantl, S., 2016. Si–Ge–Sn alloys: from growth to applications. Prog. Cryst. Growth Charact. Mater. 62, 1–39. https://doi.org/10.1016/j.pcrysgrow.2015.11.001.
Witas, P., Kalinowski, L., Goraus, J., Fijalkowski, M., Slebarski, A., 2015. Thermoelectric properties of heavy fermion compound Ce3Co4Sn13. ACTA Phys. Pol. A 127, 309–311. https://doi.org/10.12693/APhysPolA.127.309.
Wook Shin, H., Jun Lee, S., Gun Kim, D., Bae, M.-H., Heo, J., Jin Choi, K., Jun Choi, W., Choe, J., Cheol Shin, J., 2015. Short-wavelength infrared photodetector on Si employing strain-induced growth of very tall InAs nanowire arrays. Sci. Rep. 5, 10764. https://doi.org/10.1038/srep10764.

Wu, P.M., Gooth, J., Zianni, X., Svensson, S.F., Gluschke, J.G., Dick K.A., Thelander, C., Nielsch, K., Linke, H., 2013. Large thermoelectric power factor enhancement observed in InAs nanowires. Nano Lett. 13, 4080–4086. https://doi.org/10.1021/nl401501j.

Xiao, N., Dong, X., Song, L., Liu, D., Tay, Y., Wu, S., Li, L.-J., Zhao, Y., Yu, T., Zhang, H., Huang, W., Hng, H.H., Ajayan, P.M., Yan, Q., 2011. Enhanced thermopower of graphene films with oxygen plasma treatment. ACS Nano 5, 2749–2755. https://doi.org/10.1021/nn2001849.

Xie, P., Xiong, Q., Fang, Y., Qing, Q., Lieber, C.M., 2012. Local electrical potential detection of DNA by nanowire–nanopore sensors. Nat. Nanotechnol. 7, 119–125. https://doi.org/10.1038/nnano.2011.217.

Xu, B., Li, C., Thielemans, K., Myronov, M., Fobelets, K., 2012. Thermoelectric performance of $Si_{0.8}Ge_{0.2}$ nanowire arrays. IEEE Trans. Electron Devices 59, 3193–3198. https://doi.org/10.1109/TED.2012.2215610.

Xu, E.Z., Li, Z., Martinez, J.A., Sinitsyn, N., Htoon, H., Li, N., Swartzentruber, B., Hollingsworth, J.A., Wang, J., Zhang, S.X., 2015. Diameter dependent thermoelectric properties of individual SnTe nanowires. Nanoscale 7, 2869–2876. https://doi.org/10.1039/C4NR05870D.

Yamaguchi, S., Matsumoto, T., Yamazaki, J., Kaiwa, N., Yamamoto, A., 2005. Thermoelectric properties and figure of merit of a Te-doped InSb bulk single crystal. Appl. Phys. Lett. 87. 201902. https://doi.org/10.1063/1.2130390.

Yamaguchi, M., Takamoto, T., Araki, K., 2006. Super high-efficiency multi-junction and concentrator solar cells. Sol. Energy Mater. Sol. Cells 90, 3068–3077. https://doi.org/10.1016/j.solmat.2006.06.028.

Yamashita, O., Tomiyoshi, S., 2004. High performance *n*-type bismuth telluride with highly stable thermoelectric figure of merit. J. Appl. Phys. 95, 6277–6283. https://doi.org/10.1063/1.1712013.

Yamashita, O., Tomiyoshi, S., Makita, K., 2003. Bismuth telluride compounds with high thermoelectric figures of merit. J. Appl. Phys. 93, 368–374. https://doi.org/10.1063/1.1525400.

Yan, H., Choe, H.S., Nam, S., Hu, Y., Das, S., Klemic, J.F., Ellenbogen, J.C., Lieber, C.M., 2011. Programmable nanowire circuits for nanoprocessors. Nature 470, 240–244. https://doi.org/10.1038/nature09749.

Yang, P., 2012. Semiconductor nanowire building blocks: from flux line pinning to artificial photosynthesis. MRS Bull. 37, 806–813. https://doi.org/10.1557/mrs.2012.200.

Yang, R., Chen, G., Dresselhaus, M.S., 2005. Thermal conductivity modeling of core–shell and tubular nanowires. Nano Lett. 5, 1111–1115. https://doi.org/10.1021/NL0506498.

Yang, N., Zhang, G., Li, B., 2008. Ultralow thermal conductivity of isotope-doped silicon nanowires. Nano Lett. 8, 276–280. https://doi.org/10.1021/nl0725998.

Yerci, S., Li, R., Dal Negro, L., 2010. Electroluminescence from Er-doped Si-rich silicon nitride light emitting diodes. Appl. Phys. Lett. 97, 81109. https://doi.org/10.1063/1.3483771.

Yu, C., Shi, L., Yao, Z., Deyu Li, A., Majumdar, A., 2005. Thermal conductance and thermopower of an individual single-wall carbon nanotube. Nano Lett. 5, 1842–1846. https://doi.org/10.1021/NL051044E.

Yu, S.-Q., Soref, R.A., Du, W., Li, B., Sun, G., Pham, T.N., Tolle, J., Conley, B.R., Margetis, J., 2015. Si-based Ge0.9Sn0.1 photodetector with peak responsivity of 2.85 A/W and longwave cutoff at 2.4 μm. Electron. Lett. 51, 854–856. https://doi.org/10.1049/el.2015.0331.

Zaima, S., Nakatsuka, O., Taoka, N., Kurosawa, M., Takeuchi, W., Sakashita, M., 2015. Growth and applications of GeSn-related group-IV semiconductor materials. Sci. Technol. Adv. Mater. 16, 43502. https://doi.org/10.1088/1468-6996/16/4/043502.

Zhan, H.F., Zhang, Y.Y., Bell, J.M., Gu, Y.T., 2014. Thermal conductivity of Si nanowires with faulted stacking layers. J. Phys. D Appl. Phys. 47, 15303. https://doi.org/10.1088/0022-3727/47/1/015303.

Zhang, Z.P., Song, Y.X., Li, Y.Y., Wu, X.Y., Zhu, Z.Y.S., Han, Y., Zhang, L.Y., Huang, H., Wang, S.M., 2017. Effect of thermal annealing on structural properties of GeSn thin films grown by molecular beam epitaxy. AIP Adv. 7. 105020. https://doi.org/10.1063/1.5005970.

Zheng, J., Wang, S., Liu, Z., Cong, H., Xue, C., Li, C., Zuo, Y., Cheng, B., Wang, Q., 2016. GeSn p-i-n photodetectors with GeSn layer grown by magnetron sputtering epitaxy. Appl. Phys. Lett. 108, 33503. https://doi.org/10.1063/1.4940194.

Zhou, F., Moore, A.L., Pettes, M.T., Lee, Y., Seol, J.H., Ye, Q.L., Rabenberg, L., Shi, L., 2010. Effect of growth base pressure on the thermoelectric properties of indium antimonide nanowires. J. Phys. D Appl. Phys. 43, 25406. https://doi.org/10.1088/0022-3727/43/2/025406.

Zhou, Z., Yin, B., Michael, J., 2015. On-chip light sources for silicon photonics. Light Sci. Appl. 4. e358.

Zhuang, Q.D., Alradhi, H., Jin, Z.M., Chen, X.R., Shao, J., Chen, X., Sanchez, A.M., Cao, Y.C., Liu, J.Y., Yates, P., Durose, K., Jin, C.J., 2017. Optically efficient InAsSb nanowires for silicon-based mid-wavelength infrared optoelectronics. Nanotechnology 28. 105710. https://doi.org/10.1088/1361-6528/aa59c5.

Zou, J., Balandin, A., 2001. Phonon heat conduction in a semiconductor nanowire. J. Appl. Phys. 89, 2932. https://doi.org/10.1063/1.1345515.

Zou, X., Chen, X., Huang, H., Xu, Y., Duan, W., 2015. Enhanced thermoelectric figure of merit in thin GaAs nanowires. Nanoscale 7, 8776–8781. https://doi.org/10.1039/C5NR01892G.

Zuev, Y.M., Chang, W., Kim, P., 2009. Thermoelectric and magnetothermoelectric transport measurements of graphene. Phys. Rev. Lett. 102, 96807. https://doi.org/10.1103/PhysRevLett.102.096807.

CHAPTER FOUR

Nanoscale Analyses Applied to Nanowire Devices

Valerio Piazza, Lorenzo Mancini, Hung-Ling Chen, Stéphane Collin, Maria Tchernycheva[1]

Centre de Nanosciences et de Nanotechnologies, UMR 9001 CNRS, Université Paris Sud, Université Paris-Saclay, Avenue de la Vauve, 91120 Palaiseau, France

[1]Corresponding author: e-mail address: maria.tchernycheva@u-psud.fr

Contents

1. INTRODUCTION

Nanoscale materials are building blocks for many device applications aiming at energy efficiency in various domains (energy harvesting, photonics, electronics, communication, environment, health, etc.). Indeed,

Semiconductors and Semimetals, Volume 98
ISSN 0080-8784
https://doi.org/10.1016/bs.semsem.2018.03.001

nanostructuration not only can solve the existing problems of the bulk and thin film devices but also can add new functionalities. For example, different nanostructures such as porous materials, dye-sensitized nanocrystalline TiO_2 layers, Si, CdSe, or CdS quantum dots are today investigated as active materials for solar cells. Among all these systems, one-dimensional semiconductor nano-objects—nanowires (NWs), also called nanocolumns or nanorods—attract a particular interest. Semiconductor NWs with a controlled chemical composition, doping incorporation, size, and morphology can be synthesized using a bottom-up approach. In the following, we will illustrate the potential advantages of these nanomaterials compared to the bulk or thin film technologies on the examples of solar cells and of light-emitting diodes (LEDs). Although the working principles of these devices are different (implying the light absorption and extraction of photogenerated carriers in solar cells and injection of carriers, radiative recombination and light extraction in LEDs), both types of devices rely on optical–electrical conversion using a p–n junction; therefore, they face similar material challenges.

1.1 Nanowires for Photovoltaics

The existing photovoltaic (PV) technologies can be roughly classified into three categories depending on the material and the targeted application. The largest market share belongs today to crystalline Si solar cells, which typically have efficiencies from 20% to 26% and are characterized by a relatively high cost. Si solar cells are used in flat-plate applications and under light concentration. Despite the present domination of Si, its efficiency is limited by the fundamental Shockley–Queisser limit and Auger effects to $\sim$29% (Kowalczewski et al., 2016; Richter et al., 2013). The second category aiming at very high efficiencies are tandem solar cells, which use a stack of light-absorbing materials with different bandgaps to efficiently convert different spectral ranges of the solar spectrum. Tandem III–V cells hold today the conversion efficiency record of 46% demonstrated for a four-junction device under light concentration (Dimroth, 2014). However, the high cost of multijunction solar cells limits their widespread application, so they are only used in niche markets (e.g., for satellites and rarely for high light concentration). Finally, the third category englobes different "cheap" technologies based on thin films, e.g., copper indium gallium diselenide, cadmium telluride, perovskite, and organic devices. The conversion efficiencies typically reach 22% for laboratory cells (Green et al., 2017). These cells, some of which can be mechanically bendable, are used in applications where a large surface is available, so that the produced electrical power per square meter is less important (e.g., on large roof tops).

NWs can potentially address the two main PV challenges, namely, the efficiency enhancement and the reduction of production costs. First, NW arrays have very attractive optical properties such as a low optical reflectance and light trapping leading to an increased absorption in comparison to thin films (Tian et al., 2007). This allows to achieve a high absorption with a reduced amount of active material and therefore a decrease of the associated material costs. This advantage is particularly relevant in applications targeting high efficiency, which make use of expensive crystalline materials such as III–V semiconductors. Second, NWs with radial (core–shell) p–n junction enable a separate optimization of the design requirements for the light absorption and for the carrier extraction, which are the two key conditions for efficient energy conversion in a PV device (Kuang et al., 2013). This decoupling of the absorption and carrier collection directions is particularly beneficial for cheap materials (e.g., amorphous Si), which suffer from short minority carrier diffusion lengths. Third, thanks to the strain accommodation by the free lateral side, defect-free NWs can be grown on lattice-mismatched substrates (Lewis et al., 2017; Singh et al., 2011). On one hand, this property enables synthesis of high-quality NWs on inexpensive substrates (e.g., glass or metal foils), which is of interest for low cost, but also mechanically flexible solar cells. On the other hand, it also makes possible integration of materials with different bandgaps to overcome the Shockley–Queisser efficiency limit. Indeed, planar multijunction devices are typically constrained by the lattice-matching requirement necessary for high-quality growth of the materials. Therefore, it is difficult to achieve an optimal bandgap sequence for solar spectrum conversion using a monolithic growth. Strain relaxation in NWs allows to integrate III–V nanomaterials on a planar Si cell or to combine highly mismatched materials in an axial tandem structure.

1.2 Nanowires for LEDs

LEDs represent a mature technology invented back in 1960s (Holonyak and Bevacqua, 1962). A standard LED consists of a planar p–i–n junction with an insertion of a lower bandgap semiconductor layer in the intrinsic region. First, LEDs with emission in the red, infrared, and orange spectral domains were demonstrated using ternary and quaternary alloys of III–V semiconductors. In 1993, the first blue LED using nitride semiconductors was realized. This major achievement enabled the creation of white LEDs using phosphor downconversion of the blue light and blue-yellow color mixing. Today white LEDs for solid-state lighting represent the main commercial application of semiconductor LEDs.

Despite the relative maturity of the LED technology, certain issues for thin film LEDs remain unsolved. The most important one is the efficiency droop at high injection. This is a key problem in high power LEDs, which operate beyond the optimal efficiency point. The principal physical phenomenon responsible for the efficiency droop is believed to be Auger recombinations (Iveland et al., 2013). These are three-body processes (mainly, interactions implying two electrons and one hole). The losses due to Auger recombination can only be reduced by reducing the carrier density. The second important problem is the so-called green gap, which designates the decrease of efficiency in LEDs with emission wavelengths in the yellow-green spectral range. For the short-wavelength side, this is due to the increased defect density and increased internal field in In-rich InGaN/GaN quantum wells. On the long wavelength side, the loss of efficiency comes from the carrier leakage through the barriers, which turn to be indirect bandgap when the composition is tuned toward higher bandgaps. For blue and white LEDs, there are additional problems related to nitride semiconductors. These are the high defect density and the presence of internal electric field in nitride layers. Indeed, nitride thin films suffer from threading dislocations originating mainly from heteroepitaxy on foreign substrates with a high lattice mismatch. In addition, the polar nature of the nitride wurtzite crystal leads to a high internal field in InGaN/GaN quantum wells, which spatially separates electrons and holes and decreases the radiative efficiency via quantum confined Stark effect.

The above-mentioned open issues of thin film LEDs have motivated a strong research effort on nanostructured LEDs changing the device architecture from two-dimensional films to three-dimensional NWs. NWs present a large number of potential advantages for high efficiency LEDs. As mentioned before, thanks to their small footprint, NWs can accommodate the strain (Glas, 2006; Lewis et al., 2017; Singh et al., 2011) arising from the lattice mismatch with the substrate or in the active region (which is especially beneficial for axial In-rich InGaN heterostructures) and thus to reduce the defect density. The reduction of dislocations in NWs is predicted to lead to an enhancement of the quantum efficiency of NW devices compared to thin films. More generally, this property of NWs enables integration of III–V emitters on silicon and on SOI (silicon-on-insulator) platforms (Kim et al., 2016; Stettner et al., 2017).

Additional advantages for LEDs are provided by the core–shell geometry. This geometry can significantly increase the active area keeping the same device footprint as for a thin film LED (Li and Waag, 2012; Waag et al., 2011). This leads to a smaller carrier density in the quantum wells for a given injected current and thus to a droop reduction via lowering Auger effects. These properties have motivated a strong research on core/shell NWs for high power high efficiency LEDs. In addition, for core/shell InGaN LEDs the active region is usually grown on nonpolar m-plane sidewalls, which allows to suppress the quantum confined Stark effect in InGaN/GaN quantum wells.

In addition to improving the existing device performance, NWs open the way for new functionalities. One of them is the integration of light emitters on unconventional substrates such as metal (May et al., 2016) or amorphous (Chung et al., 2014) substrates. The small NW footprint allows for NW transfer and fabrication of mechanically flexible LEDs (Dai et al., 2015; Guan et al., 2016). A monolithic integration of red, green, and blue emitters grown simultaneously on the same substrate using different nanopatterns opens new prospects for displays (Kishino et al., 2015). Finally, single NW LEDs can act as bright nanoscale sources for communication or sensing applications (Tchernycheva et al., 2014a).

1.3 NW Device Architecture

The three-dimensional nature of the NW devices opens the way for new architectures. NW solar cells and LEDs have typically one of the two following geometries with the NWs containing either an axial or a radial p–(i)–n junction.[a]

The axial geometry is similar to thin film devices. For solar cells, it contains base and emitter NW segments and eventually an intrinsic segment along the wire axis as illustrated in Fig. 1A. It may also contain a top window

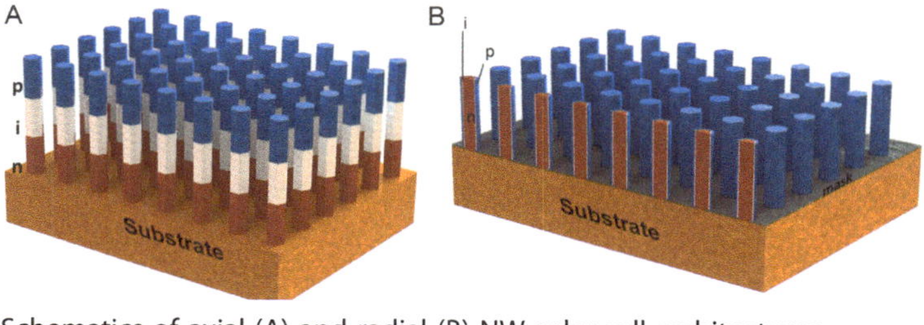

Fig. 1 Schematics of axial (A) and radial (B) NW solar cell architectures.

[a] We do not consider here the cases where the junction is between the NWs and the substrate.

layer and a heavily doped back surface field region close to the substrate. For LEDs, the active region in the form of quantum discs is inserted between the p- and n-NW segments to localize the injected carriers and increase the radiative recombination. For both LEDs and solar cells, a surface passivation and an electrical insulation of the junction are required. For device fabrication, the NWs are typically encapsulated into a transparent spin-on glass, which serves as an electrical insulating layer and as a mechanical support for the top transparent contact.

The radial architecture illustrated in Fig. 1B is strongly different from thin films. Indeed, the p–n junction is realized all around the NW so that the carrier collection/injection direction becomes perpendicular to the light propagation direction. As previously mentioned, for core/shell solar cells, the light absorption can be optimized independently from the carrier collection. This is particularly important for indirect bandgap semiconductors like Si. In standard Si solar cells, the weak absorption requires a large thickness of the absorbing base layer. The minority carrier diffusion length should be as large as the base thickness, which is strongly demanding for the material quality. In a core/shell NW solar cell, an efficient carrier collection in the radial direction can be achieved even with a small minority carrier diffusion length (Kayes et al., 2005). The same is true for low quality low cost materials like amorphous Si (Togonal et al., 2016; Yu et al., 2013; Zhu et al., 2009). For LEDs, as previously mentioned, the core/shell geometry allows to increase the active area, reduce the carrier density, and thus lessen the non-radiative Auger losses (Iveland et al., 2013; Waag et al., 2011). For both types of devices, the NW core is electrically connected to the substrate, whereas the shell is insulated from the substrate. A conformal transparent contact is deposited on the shell.

1.4 Present Status of NW Solar Cells and LEDs

The exploration of NW solar cells started with an intense research on Si NWs using both bottom-up and top-down fabrication methods and various architectures (Ghosh and Giri, 2017; Yu et al., 2016). Today the conversion efficiency of NW Si solar cells is around 10%. The record of 22% is reached for devices using Si NWs as a passive antireflection coating (Savin et al., 2015).

Si research was followed by numerous demonstrations of III–V NW solar cells. For single bandgap devices, GaAs and InP NWs are mostly explored. Indeed, according to the detailed balance limit, GaAs and InP bandgaps are close to the optimal value giving the maximum light to electric power conversion efficiency (Rühle, 2016). The record efficiencies for GaAs NW solar cells using core/shell architecture are 7.43% (Mariani et al., 2013), and 15.3% for axial junction devices (Aberg et al., 2016). For InP NWs with a radial junction, the efficiency of 6.3% was reported (Yoshimura et al., 2013). For axial InP junctions, 13.8% efficiency is demonstrated for bottom-up NWs (Wallentin et al., 2013). Top-down fabrication yields today a better result. Conical InP NWs with an axial p–n junction formed by dry etching hold the efficiency record for NW solar cells of 17.8% (van Dam et al., 2016).

For comparison, the record conversion efficiency under a nonconcentrated AM1.5G spectrum for thin film solar cells is ~28% for GaAs (Kayes et al., 2011, p. 6) and 22.1% for InP (Green et al., 2017). We note that the performance of the best NW solar cells is today still below the results for 2D devices.

Solar cells formed with organized NWs systematically yield a better conversion efficiency compared to devices built with self-assembled NWs. This is due to an improved homogeneity of the electrical and optical properties, which is important since individual NWs in a solar cell are connected in parallel.

Regarding the research on NW LEDs, it mainly focuses on two objectives. The first one is the integration of light sources on Si (Deshpande et al., 2017; Dimakis et al., 2014; Svensson et al., 2008; Tomioka et al., 2010), in particular, using CMOS-compatible procedures to achieve emission at telecom wavelengths (Deshpande et al., 2017). Core/shell NW LEDs using AlGaAs/GaAs, GaAs/InGaP,(In,Ga)As/GaAs, or InGaAs/InP material systems were demonstrated (Deshpande et al., 2017; Dimakis et al., 2014; Svensson et al., 2008; Tomioka et al., 2010). However, these LED prototypes remain exploratory. The publications do not always cite the relevant performance indicators. It is therefore difficult to benchmark the NW LED performance or to compare it with the parameters of conventional III–V sources wafer-bonded to Si.

The second objective of the NW LED research is the development of novel light sources for lighting and display applications (Guan et al., 2018; Monemar et al., 2016). In particular, NW LEDs are expected to improve the efficiency of conventional devices used in solid-state lighting. The research is mainly focused on nitride materials. Both core/shell (Schimpke et al., 2016) and axial (Kishino et al., 2015) LED architectures are explored.

The comparison of the internal quantum efficiency (IQE) and the external quantum efficiency (EQE) of nitride NW LEDs with the corresponding values for thin film devices shows that the NW technology today is still behind the conventional one. For example, for blue LEDs, the IQE of 44% was reported in core/shell NW devices (Mounir et al., 2016) to be compared with 80% in m-plane quantum wells on bulk 2D GaN substrates. For the green spectral range (520 nm), the performance of NW LEDs approaches the one of thin films but is not yet there: the EQE of 25% was measured in core/shell NW LEDs (Monemar et al., 2016), while the values around 30% are typically reported for 2D devices (Liu et al., 2015).

We note that despite the lower efficiency, light sources with new functionalities can be fabricated using NWs. For example, selective area growth of InGaN/GaN NWs allows for a monolithic integration of LEDs with different colors on the same substrate (Kishino et al., 2015). Moreover, the small NW diameter can be exploited to fabricate mechanically flexible LEDs either by NW lift-off and transfer (Guan et al., 2016) or by direct growth on metal foils (May et al., 2016).

Although the NW devices have shown an impressive progress in the last 10 years, their performance still remains below theoretical predictions. The envisioned improvement of performance with respect to two-dimensional devices has not been demonstrated either. This can be due to many factors related to design, NW inhomogeneities, surface recombination, and processing issues. It is clear that the three-dimensional geometry of the active region increases the complexity of the device design and fabrication. Moreover, the device optimization requires as a prerequisite the fundamental understanding of the physical phenomena taking place in NWs that have not been fully investigated yet. In this perspective, a reliable measurement of material, electrical, and optical properties of nano-objects to be used as input parameters for advanced modeling is a key challenge.

1.5 Nanoscale Characterization Techniques

Standard macroscopic characterization averaged over millions of nano-objects does not provide all the information necessary to understand the device physics and to optimize the performance. It is important to push the analyses down to the nanometric scale and to probe individual NWs in order to evaluate the material quality, to assess the wire-to-wire homogeneity, and to detect eventual electrical or optical failures. Combination of structural and optoelectronic characterization down to the nanoscale should guide the device design and fabrication and show the ways for performance optimization.

In the following, we focus on several techniques probing the homogeneity, quantum efficiency, non-radiative recombination, doping, minority carrier diffusion lengths, etc., in functional NWs.

1.5.1 Luminescence Techniques

Photoluminescence (PL) and cathodoluminescence (CL) are complementary nondestructive characterization methods. They are based on the spectroscopic analysis of radiative recombination of carriers. PL makes use of laser excitation with photon energy above the material bandgap. It can provide resonant excitation conditions required for accurate analyses of IQE of LEDs. The spatial resolution is limited by the Abbe diffraction limit. However, the use of a confocal PL setup with a pinhole can improve the resolution and allows probing PL signal from different positions in a NW (Mounir et al., 2016).

In CL, an electron beam excites charge carriers. It can be applied to materials with very different bandgap energies (including wide bandgap semiconductors). CL provides much higher spatial resolution than PL thanks to the small beam diameter (<10 nm). It enables independent probing of the optical emission of individual NWs in a dense array, but also mapping intrawire variations of luminescence. The resolution can be further enhanced down to 5 nm in a scanning transmission electron microscope with high accelerating voltage (Kociak and Zagonel, 2017).

Besides steady-state excitation used in PL and CL, many optoelectronic properties of NWs are determined by their charge dynamics and by an interplay between radiative and non-radiative relaxation channels. Non-radiative phenomena related to eventual defects or surface recombination can be probed by pulsed excitation and measurement of the luminescence decay as a function of the temperature. Time-resolved photoluminescence (TRPL) uses a pulsed laser and a streak camera to achieve picosecond resolution. It can be applied to NW arrays (Dagytė et al., 2017) or to individual NWs (Smith et al., 2010). Intrawire lifetime variations can be measured by time-resolved cathodoluminescence (TRCL). Short electron pulses are obtained either by photoemission controlled by a pulsed laser (Merano et al., 2005a) or by a subnanosecond beam blanking (Meuret et al., 2016). TRCL combines a picosecond time resolution and spatial resolution of a few tens of nanometers. It is well suited to study carrier dynamics in NWs at the submicrometer scale.

1.5.2 Charge Collection Microscopy Techniques

Current mapping allows to study spatial distribution of electrical properties of a nanoscale device through the generation and collection of an electrical current. In charge collection microscopies, the current map is constructed by scanning the surface of an electrically contacted sample with an excitation source, which can be an optical beam (*laser beam-induced current* (LBIC)) or an electron beam (*electron beam-induced current* (EBIC)). The excitation locally generates excess carriers, which can be collected if electric fields are present within the device. Both LBIC and EBIC were applied to analyze NW device, namely, solar cells (Yu et al., 2013) and LED's (Tchernycheva et al., 2015; Zhang et al., 2016). In LBIC, carrier generation by photon absorption is similar to solar excitation; therefore, LBIC experiments can be compared with EQE measurements but performed on a micron scale. Carrier generation in EBIC implies electron beam/matter interaction, so the results cannot be directly related with the EQE measurements. However, EBIC offers excellent spatial resolution, especially for STEM-based setup with a probe size down to 0.5 nm (Progl et al., 2008). Therefore, EBIC allows to address intrawire electrical inhomogeneities and local failures. EBIC maps can be complemented with voltage contrast analyses, which provide information on the variation of doping (Tsurumi and Hamada, 2013). Both LBIC and EBIC are nondestructive characterization methods, which can provide information on the properties of individual nanostructures composing a macroscopic device.

1.5.3 Structural Characterization

A large number of structural characterization techniques was applied to functional NWs to probe their crystal structure and composition. These are, namely, transmission electron microscopy in different forms, X-ray diffraction, energy dispersive X-ray spectroscopy, etc. In this chapter, we have chosen to highlight one less-known technique—the *atom probe tomography* (APT). This is a destructive technique consisting in field evaporation of atoms (assisted with a femtosecond laser heating) from the surface of a NW. Using position-resolved time-of-flight (TOF) detection, the atomic structure of the specimen is reconstructed. APT offers both three-dimensional image of the NW structure and the chemical composition with atomic resolution. In particular, it can evidence alloy inhomogeneities, quantum dot formation, and assess the concentration of doping impurities.

1.6 Summary of the Chapter

In this chapter, we review the application of the above-listed characterization techniques to functional NW devices. The techniques will be discussed together with the present challenges of NW solar cells and LEDs. We will show how the ability to assess the structural, electrical, and optical properties on a single NW level gives access to the key nanomaterial parameters, allows to identify origins of failures, and can help solving the existing issues of these devices. Fundamental analyses of individual NWs as well as the characterization of fully operational devices will be discussed.

In Section 2, the operation principle of the characterization techniques will be presented. In Section 3, the application of these techniques to the study of single NWs and to fully operation NW array devices will be discussed addressing the following points: evaluation of compositional homogeneity, visualization of heterostructures, analyses of defects, estimation of doping, measurement of minority carrier diffusion length, assessment of non-radiative recombination, and surface recombination velocity.

2. NANOSCALE CHARACTERIZATION TECHNIQUES

2.1 Scanning Electron Beam Techniques for Nanoscale Analysis

The use of a scanning electron beam enables investigation of the electronic properties in semiconductor materials with a nanoscale resolution. High-energy electrons (3–30 kV) undergo numerous interactions after entering the material. Their trajectory is deviated by elastic and inelastic scattering. Scattering phenomena result in the emission of secondary electrons (SE), backscattered electrons, Auger electrons, and X-rays. They arise from different regions of the material and can be recorded for morphological analysis (SE images) or compositional analysis by energy-dispersive X-ray spectroscopy (XEDS).

In Monte-Carlo simulations, the energy losses and trajectories are calculated by considering random scattering events in a stepwise manner. The shape and size of the resulting *interaction volume* depend on the primary electron energy and on the material (elements, density, scattering cross section). Pear, spherical, and hemispherical shapes are obtained for increasing atomic numbers (Reimer, 2013; Yacobi and Holt, 2013). Fig. 2 shows the interaction volume in bulk GaAs for normal incidence primary electrons with an

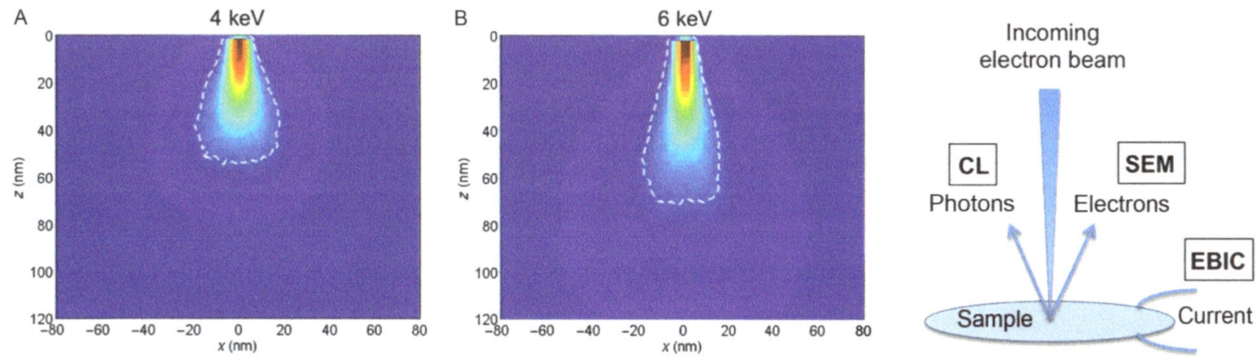

Fig. 2 (A and B) Maps of the density of energy dissipated by an electron beam in GaAs, with an incident energy of (A) 4 keV and (B) 6 keV (*white dashed line*: 80% of the maximum) calculated by Monte-Carlo simulation (Casino software). (C) Schematic principle of a CL setup with simultaneous CL, SEM and EBIC measurements. *Panels (A) and (B) adapted with permission from Chen, H.-L., Himwas, C., Scaccabarozzi, A., Rale, P., Oehler, F., Lemaître, A., Lombez, L., Guillemoles, J.-F., Tchernycheva, M., Harmand, J.-C., Cattoni, A., Collin, S., 2017. Determination of n-type doping level in single GaAs nanowires by cathodoluminescence. Nano Lett. 17, 6667–6675. https://doi.org/10.1021/acs.nanolett.7b02620. Copyright 2017 American Chemical Society.*

energy of (A) 4 keV and (B) 6 keV. With an electron beam diameter of 10 nm, the lateral extension of the interaction volume is about 35 nm (cut-off at 80% of the maximum energy) in both cases. The depth extension is 50 nm for 4 keV and 70 nm for 6 keV. These characteristics are perfectly suitable to probe NWs with diameters down to 100 nm. For thinner NWs, the exact shape of the material should be taken into account in the simulation. For thicker NWs, higher electron energies may be required to excite charge carriers at the very core of the NW.

In semiconductor materials, e^-/h^+ pairs are excited in the interaction volume, enabling the possibility to probe electronic properties with a spatial resolution as low as a few tens of nanometers. Therefore, it is possible to have a nanoscale injection of excess carrier, offering precious tools for the design, characterization, and understanding of physical phenomena in NWs. However, the actual spatial resolution of CL depends on carrier diffusion which can be boosted or hindered by bandgap gradients, charge barriers, heterostructures, or passivated surfaces.

In this section, we will describe three scanning probe techniques: EBIC, secondary electrons voltage contrast (SEVC) and CL.

2.1.1 EBIC and SEVC Technique

EBIC is a probe scanning technique which enables investigation of internal electric fields. The drift of excess carriers induced by the internal field offers the possibility to extract a so-called induced current in an external circuit (Fig. 3A). By scanning the electron beam on the sample surface, the EBIC map allows to visualize the internal field at the nanoscale.

EBIC working principle is similar to the photovoltaic effect, except that light excitation is substituted by an electron beam. The main advantage of using the electron beam is the possibility to reach a spatial resolution of a few tens of nm depending on the structures and the material properties. On the other hand, electron excitation can deliver to the sample a higher energy density compared to light excitation. Different models were proposed to describe carrier generation in semiconductor materials under electron beam irradiation. For detailed analysis, refer to works by Leamy (1982), Donolato (1981)), and Pasemann (1981).

Excess carrier density can affect transport phenomena in the materials, as pointed out by Cavalcoli and Cavallini (1994). If the excess carrier density is comparable or higher than the equilibrium carrier concentration (either the intrinsic concentration (n_0) or the intentional doping level in doped semiconductors), current extracted during EBIC measurements is governed by

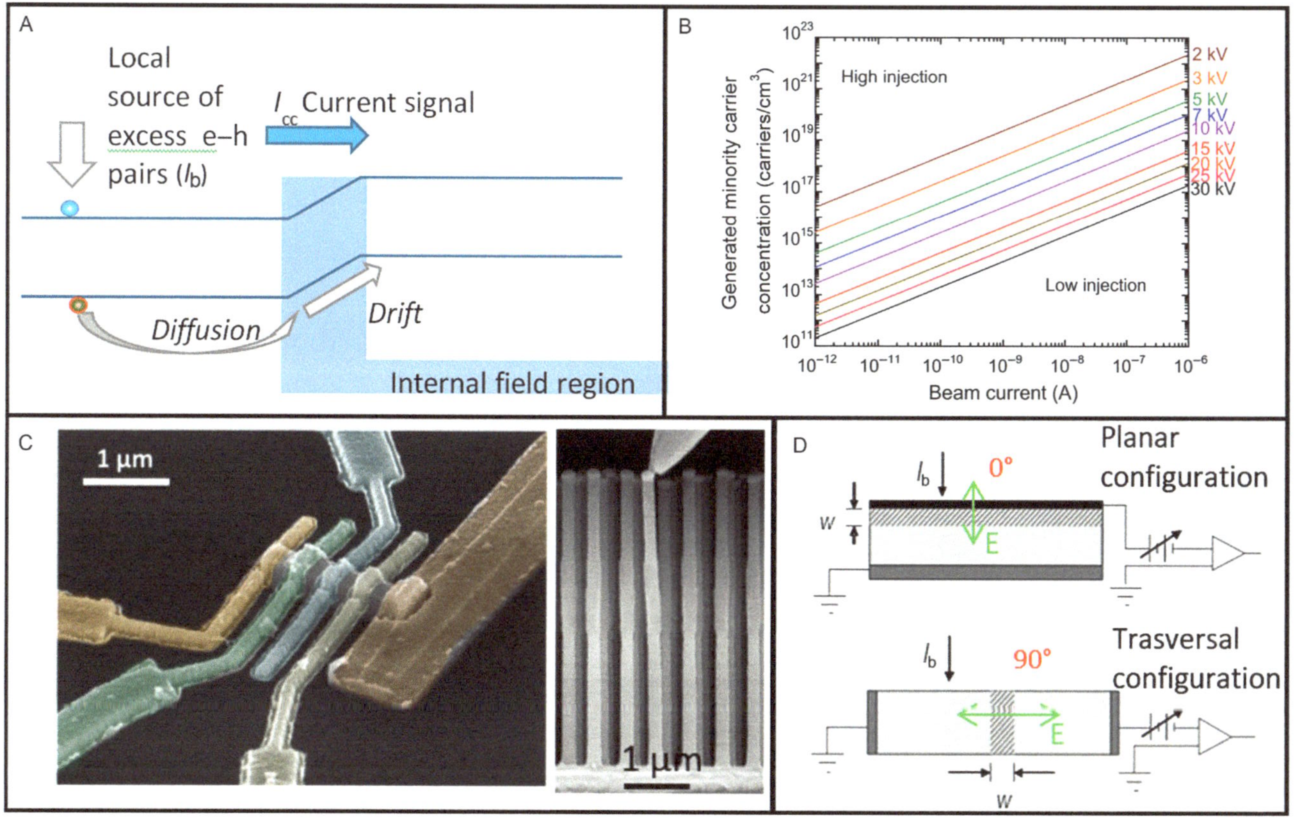

Fig. 3 (A) Schematic representation of EBIC mapping working principle. (B) Examples of generated carriers as a function of electron beam current and acceleration voltage. (C) Examples of electrical contacts on single NWs by metal deposition and by in situ nanomanipulators. (D) Schematic of planar and transversal modes of operation: the *black arrows* indicate the direction of the electron beam flux and the *green arrows* indicate the electric field direction (*W* indicates the extension of the space charge region). *Adapted from Bunker, K.L., 2004. Development and application of electron beam induced current and cathodoluminescence analytical techniques for characterization of gallium*

ambipolar diffusion. This situation is referred to as *high injection regime* (*HIR*). On the contrary, if the generated carrier density is much lower than the equilibrium concentration, measurements are performed in *low injection regime* (*LIR*), and collected current is given by minority carrier diffusion. For a given material, it is possible to calculate the excess carrier density as a function of the injected current and the acceleration voltage and to define HIR and LIR. Fig. 3B shows an example of the calculated minority carrier density in a nitride LED when diffusion is neglected (Bunker, 2004). We note that the local carrier density is increasing with a decrease of the acceleration voltages because the interaction volume scales faster than the generation rate.

In EBIC microscopy two configurations are typically used depending on the angle between the electron beam and the direction of the internal electric field: a planar (0 degrees) and a transversal (90 degrees) mode of operation. Planar configuration (Fig. 3D) is mainly used for defect analysis in thin films. Indeed, EBIC map of a defect-free sample shows a constant induced current independently on the position of the beam. 1D defect appears in this configuration as a local decrease of the current (Blumtritt et al., 1979; Donolato, 1981; Leamy, 1982; Pasemann, 1981), indicating the defect position and the defect surface density can be estimated. (Depth extension of the defects can be evaluated by varying the acceleration voltage to increase the e^- beam penetration depth and thus the analyzed volume.)

Transversal mode of operation (Fig. 3D) enables qualitative and quantitative analysis of the electric field: in this case by analyzing the induced current variation in the direction perpendicular to the field information about the space charge region extension and on the carrier diffusion length can be obtained. Intermediate configurations ($0° < \alpha < 90°$) can be evaluated considering geometrical factors; however, the extraction of material parameters is difficult.

In order to study single NWs by EBIC, the sample geometry has to be taken into account. Scanning along the NW axis, transversal configuration is obtained for NWs containing axial junction(s). In the case of core–shell NWs (containing a radial junction) it is necessary to cleave the wire either along its axis or perpendicular to the axis to expose the junction for EBIC analyses (for details, refer to Section 3.3.3).

Concerning the setup, besides the scanning electron beam, electrical contacts are required to collect the induced current. Since the top contacts are influencing the shape of the interaction volume and the energy of electrons penetrating through the contact into the semiconductor, contact material and thickness should be adjusted to find a compromise between a low resistance and low attenuation of the electron beam. In NW array devices, top contact is usually made of transparent conductive oxides (typically ITO), which has a good optical transparency over a large spectral range; however, it strongly attenuates the electron beam (Bright, 2007; Coutts et al., 1999).

To analyze single NWs, the contacts can be fabricated by electron beam lithography on dispersed NWs with a few tens of nm resolution (Fig. 3C), but the process requires NW localization, development of a dedicated design and a slow e-beam writing, which is strongly time-consuming (Hultin et al., 2016; Lavenus et al., 2014). Another possible approach is to implement micro- or nanomanipulators into the SEM chamber to contact single NWs both as dispersed and standing on a substrate (Fig. 3C). The second approach is faster and is more flexible, since the contact points can be easily changed. However, it is difficult to control the contact area and the contact resistance, which depends on the mechanical pressure applied, on the chemistry of the surface and on the workfunctions of the semiconductor and of the metallic tip. This approach typically results in a formation of a Schottky barrier, which can perturb the measurement.

Together with the induced electrical current or optical emission, EBIC or CL setups record secondary electrons using the SE detector of the microscope. Morphological image is important in the interpretation of EBIC and CL maps. Moreover, SEM images can be used to study the variation of SE emission with applied bias, which allows to analyze the doping contrast.

There are many parameters influencing the SE emission in solid materials, such as density, roughness, electrical resistivity, or electrostatic charging. Concerning semiconductor materials, it has been found experimentally that p-doped semiconductors have an enhanced SE emission (resulting in a bright contrast) with respect to n-doped materials (resulting in a dark contrast). Theoretical explanation for this experimental evidence is controversial in the literature. The model proposed by Perovic et al. (1995) attributes the effect to the band bending at the surface induced by surface states: charges accumulated on the surface increase (in n-doped semiconductors) or decrease (in p-doped semiconductors) the energy required for electron emission. This results in a higher SE emission in p-doped semiconductors. On the contrary Sealy et al. (2000) considered the ionization energy of SEs as primarily cause of the doping-related contrast. The authors stated that the SE energy lies in between the vacuum energy level for p-doped and n-doped semiconductors; therefore, the ionization energy difference (ΔE^{SEI}) for the two types of doping can be estimated to be $\Delta E^{SEI} = 1/2E_{bi}$, where E_{bi} is the built-in potential arising from the hypothetical p–n junction among the two materials. In this theory, the effect of the surface states is to reduce the ionization energy difference and thus reduce the SE contrast. Similarly, Volotsenko et al. (2010) proposed a theory based on SE emission yield and surface band bending extracted from KPFM measurements: in this work they associate the SE contrast to the SE energy distribution and effective electron affinity as a function of the doping level in good agreement with experimental data. More details can be found in the literature (El-Gomati et al., 2005).

Investigation of the SE doping-related contrast and enhancement of the contrast by application of an external bias led to the development of SEVC technique to localize the p–n interface. Concerning the effect of the voltage on the doping contrast, the mechanism is not fully understood, but following the path of the theory by Sealy et al. (2000) and Volotsenko et al. (2010), Narchi et al. found a good experimental correlation between KPFM measurements and SE contrast profiles at different applied biases (Narchi et al., 2017).

The combination of EBIC maps and SEVC analyses is a powerful tool to investigate the field and the doping level, making these techniques suitable for nanoscale analysis of homojunction and heterojunctions, as well as of Schottky diodes and electrical inhomogeneities.

2.1.2 CL Techniques

A scanning electron microscope is the backbone of a CL setup. A sketch of a complete CL tool equipped with EBIC and time-resolved capabilities is shown in Fig. 4. Emitted light is collected by a parabolic mirror, directed out of the microscope, and analyzed by a monochromator and a detector array (a silicon charge-coupled device (CCD) is used for visible and near-infrared wavelengths up to 1.1 μm, and InGaAs detector arrays can be used for wavelengths up to 1.7 μm). The electron beam current is usually in the nA range. However, it can be tuned from a few tens of pA up to a few tens of nA, depending on the radiative efficiency and the HIR/LIR targeted. A He cryostat allows cooling down the sample to a temperature of 5–10 K. The stability of the sample stage is a key factor for high-resolution CL maps of single NWs with low radiative efficiencies. The drift can reach values as low as 200 nm/h, allowing high-resolution mapping of single NWs with total integration times up to several tens of minutes.

Synchronization of the photon detection with the electron beam scanning results in hyperspectral CL maps in which a full spectrum is recorded

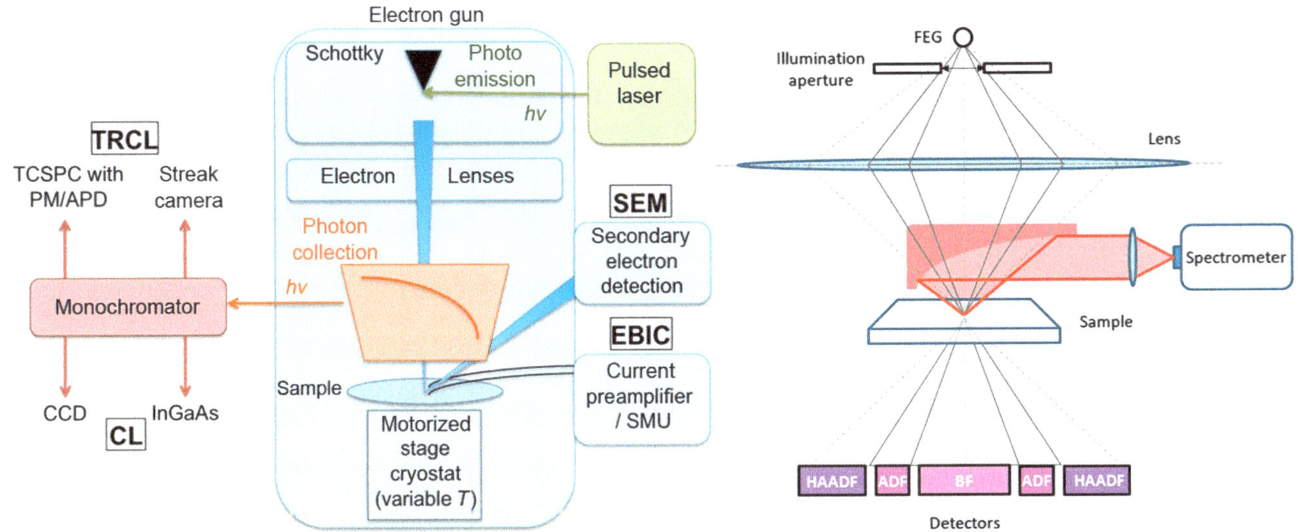

Fig. 4 Simplified layout of a SEM–CL/TRCL/EBIC setup (*left*) and STEM–CL setup (*right*). The electron probe is generated by the electron source (Schottky/field-emission gun—FEG) and focused on the sample by means of a diaphragm characterized by the illumination aperture and by the magnetic lenses. The scattered electrons are then collected by one of the available detectors. The CL emission generated in the far field by the probe/sample interaction (represented by the *red lines*) is collected by dedicated optics and sent to a spectrometer.

for each pixel. SEM and EBIC images can be recorded simultaneously. The information obtained by CL is typically extracted in the form of energy filtered images. They are obtained associating to each pixel of the image the CL intensity integrated in a specific spectral range.

The energy of the emitted photons reflects the local band structure which may depend on the local composition and strain state. The presence of quantum confining structures and shallow defects within the analyzed volume can contribute to shift the emission energy with respect to the local band gap. The main advantages of CL with respect to PL are the capability of probing simultaneously the structure and the morphology of the system, and the possibility of resolving variations of the optical properties at the nanoscale. As discussed previously, the spatial resolution depends on the interaction volume, the geometry of the sample, and the diffusion lengths. For instance, carrier diffusion toward a low bandgap region will result in a red-shift and a time lag of photon emission (Merano et al., 2005b). Both spectral and temporal filtering can help to improve the spatial resolution of CL images.

TRCL is achieved either by beam blanking or by pulsed photoemission. In the latter case, a pulsed laser is focused on the tip of the electron gun to enable extraction of electrons by photoemission (Merano et al., 2005a). As an example, electron pulses with temporal width of less than 10 ps can be emitted with a repetition rate of 12 ns and a beam current divided by three orders of magnitude as compared to steady-state excitation. Time-resolved detection should be synchronized with the pulsed laser. It can be done by time-correlated single photon counting using a single fast detector (photomultiplier or avalanche photodiode) or with a streak camera to record the full spectrum as a function of time.

To further improve resolution, many variations of the SEM–CL technique have been developed, as STEM–CL and TEM–CL. Using nanometric-thick lamellae of samples (obtained by FIB cutting) or NWs dispersed on membranes, interaction volume is strongly reduced, leading to even higher resolutions (in the order of Å) and lower heating effects and sample damage with respect to SEM–CL. The high energy of the electrons and the thinness of the sample needed for STEM (or TEM) structural characterization lead to a strong reduction of the interaction volume and of the amount of energy transferred to the material. The spatial resolution of the technique is limited by the diffusion length of the carriers generated within the volume excited by the angstrom-sized electron probe. Due to the weaker electrons/sample interaction, heating and nonlinear effects are also reduced, along with the energy shift and spectral broadening of the CL emission.

CL techniques allow for investigating different properties of semiconductor nanostructures. The main application is the mapping of the optical transition energies and intensities (Kociak and Zagonel, 2017). Variations of these quantities can in fact be related to the presence of quantum confining heterostructures (Griffiths et al., 2015; Lim et al., 2009; Tizei et al., 2014; Tourbot et al., 2012; Zagonel et al., 2011, 2012) or to inhomogeneous composition and alloy fluctuations (Pantzas et al., 2015; Pierret et al., 2013a,b). It is worth noting that in such cases the spatial resolution of the system is determined by the size of the confining structure: the volume that the charge carriers generated by the probe can explore before recombining is limited by their diffusion length within the local band structure, which in turn is determined by the composition and morphology of the confining structure. Another important application is the assessment of defects. Stacking faults, dislocations, and point defects create levels within the band gap which capture the charge carriers by allowing them to recombine at lower energies. This leads either to a quenching of the CL emission if the recombination process is non-radiative or to a shift in the emission energy in case of radiative recombination. STEM–CL was demonstrated to be able to characterize the optical properties of 2D and 1D radiative and non-radiative defects with unmatched spatial resolution (Müller et al., 2016; Urban et al., 2015). Non-radiative 0D defects are more difficult to address due to the necessity of using particularly thin samples and thus to the corresponding low CL signal intensities. Other applications include the study of nonlinear effects such as the efficiency drop of nanoemitters due to the high density of carriers and the corresponding Auger effect (Lefebvre and Gayral, 2008; Zagonel et al., 2016), and quantum optical effects such as the emission of single photon emitters by measuring time anticorrelation functions (Tizei and Kociak, 2013). The lifetime of quantum confining emitters with nanometric spatial resolution has been addressed by TRCL measurements (Fu et al., 2014; Grundmann et al., 1989; Merano et al., 2005b).

2.2 PL/TRPL Techniques

The physical phenomenon known as PL defines the emission of light following the absorption of photons. In a PL experiment, sufficiently energetic photons interact with a semiconductor sample by promoting electrons from the valence bands into the conduction band. The PL emission stems from the incoherent, spontaneous interband recombination of electrons and holes. In a typical PL spectroscopy setup, excitation is provided by a laser source in continuous or pulsed mode having a controlled wavelength and incident power, which is focused on the analyzed sample. The resulting emitted PL signal is collected, dispersed with a diffraction grid of a spectrometer and sent to a detector (e.g., a CCD camera). This allows for analyzing the signal intensity as a function of the wavelength (obtaining a time-integrated, energetically resolved spectrum). Control over additional experimental conditions and acquisition parameters (e.g., the excitation wavelength and power, sample temperature, polarization of the laser or of the PL signal) provides further understanding of the physical processes in the NWs as will be later discussed.

A spatially resolved analysis of the properties of the system can be performed by localizing the excitation on a desired position of the sample surface: high magnification objectives can be used in order to focus the laser in a spot whose lowest dimensions are defined by the diffraction limit of the system (which is in turn determined by the wavelength of the excitation and by the numerical aperture of the objective). Moving the sample with respect to the focal point of the focusing objective allows for scanning the surface of the analyzed object and selectively exciting regions of the size of less than 1 μm. The experimental technique implementing this configuration is commonly referred to as microphotoluminescence (μPL). Higher spatial resolution can be obtained by means of confocal μPL: the employment of a high numerical aperture objective and a pin-hole within the collection optics allows for excluding higher orders of the diffraction pattern (better in-plane resolution) and for excluding out of focus component of the light signal (submicrometric resolution along the optical axis). High spatial resolution confocal PL mapping can be obtained by scanning the sample by means of a piezoelectric stage with nanometric step resolution.

The previously discussed setups allow for resolving PL signals spatially and energetically. In order to have access to the dynamics of the radiative recombination processes, the possibility of resolving PL signals both energetically and temporally is needed. In TRPL the excitation of the sample is provided by a pulsed source, and the collection of the emitted signal is synchronized, so that the detection system can address the time evolution of the PL spectrum. Typical TRPL detection systems consist of a streak camera which collects the PL signal, disperses it on a diffraction grating, and converts it to photogenerated electrons by means of a photocathode.

The so generated electrons are deflected within a time-dependent process and detected by a two-dimensional CCD. The time-dependent deflection of photogenerated electrons is obtained by modulating the electric potential between two plates accordingly to a trigger signal sent to the streak camera, characterized by the same pulse frequency of the excitation laser. The obtained spectra consist of 2D datasets in which each data point corresponds to a detected photon (or to a packet of photons), the abscissa reporting their energy dispersion, and the ordinate the excitation/emission delay. Variations of the emitted photons energy in time can be directly visualized on the 2D maps, while the integration of PL signal along selected energy ranges (decay curves) gives insight into the decay dynamics of selected recombination processes.

2.3 APT Technique

Atom probe tomography is a destructive microscopy technique which allows for analyzing the 3D structure of materials with 3D near-atomic spatial resolution and maximal chemical sensitivity close to one atomic part per million. The working principle of the technique is schematically represented in Fig. 5. Single surface atoms of a needle-shaped specimen are field evaporated one-by-one under the effect of a high electric surface field (typically tens of volts per nanometer, induced by the potential applied to the specimen and by its nanometric apex radius) within a triggered process and eventually detected on a 2D position sensitive detector which records the position and time of arrival with respect to the trigger signal. The atoms impact position and order of evaporation allow reconstructing their 3D position within the analyzed volume, while their chemical nature can be

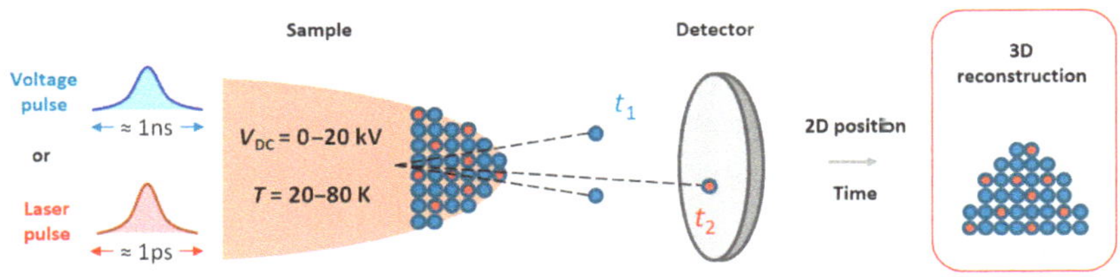

Fig. 5 Schematic representation of the working principle of APT. Single atoms (*red* and *blue dots*) are field evaporated as ions from a field emission tip under the effect of voltage or laser pulses and fly to the detector under the effect of the applied DC field. Atoms of different species reach the detector with different delays with respect to the trigger pulse. The 2D spatial information of the position of the impacts and the recorded times of flight are used for obtaining a 3D reconstruction of the sample.

deduced by TOF spectroscopy of the ions accelerated by the constant electric field. Samples for APT analysis are commonly prepared by focused ion beam milling (Miller et al., 2007) which allows for obtaining field emission tips with an apex radius smaller than 100 nm. Systems whose geometry naturally allows for high electric surface fields under bias such as NWs do not need to be processed by FIB and have to be only fixed on a suitable support (Isheim et al., 2012; Qu et al., 2015).

APT analyses can be performed within two operation modes which differ in the physical nature of the signal which triggers the field evaporation of adatoms. The so-called voltage pulse mode relies on the superposition of potential pulses with a temporal width of the order of few ns to a constant potential applied to the sample. In such operation mode the additional pulsed field considerably increases the probability of the most protruding surface atoms to field-evaporate during the short duration of the pulse by modifying the local potential landscape and lowering the adatoms confining energy barrier. For a detailed description of the physics underlying the field-evaporation process, refer to Müller and Bahadur (1956), McMullen and Perdew (1987), McMullen et al. (1982), and Kreuzer and Nath (1987). The voltage pulse mode is the operation mode of choice for the analysis of metallic materials but can perform poorly when applied for the study of semiconductors and insulators as, due to the low conductivity of the system, the pulses are strongly distorted (lowered maximal amplitude and longer duration) so that they are no longer able to properly trigger the field evaporation. The introduction of the "laser pulse operating mode," and of laser-assisted atom probe tomography (LA-APT), opened the possibility of analyzing high resistivity materials such as semiconductors (Lauhon et al., 2009), oxides (Oberdorfer et al., 2007), and polymers (Prosa et al., 2006). In LA-APT the interaction of sub-ps long laser pulses with the specimen locally increases the surface temperature in a phonon-mediated absorption process along with the thermal energy of surface ions, consequently increasing their probability of tunneling through the confining energy barrier or passing over it (Silaeva et al., 2014, 2015). For a thorough description of the laser-matter interaction and the complementary mechanisms involved in the field evaporation process, refer to Ding et al. (2005), Tsong and Kinkus (1984), Tsong (1984), Shen (2003), and Vella et al. (2006).

Under the effect of electric or laser pulses (trigger), single adatoms leave the surface of the specimen. The field evaporated ions are then subjected to the strong electric field induced by the potential difference existing between the specimen surface and the detector. The TOF of the ion, corresponding to the time needed for traveling from the surface to the detector, allows associating to each one of them the corresponding mass-over-charge ratio and thus their chemical species. The histogram of all measured ratios defines what is referred to as mass spectrum. The interpretation of mass spectra is one of the critical points of APT, as peaks indexing, spectrum statistical corrections and noise treatment determine the mass resolution and chemical sensitivity of the technique. For more detailed information on these topics, refer to Miller (2000) and Lefebvre et al. (2016).

APT data allow for reconstructing the 3D position of each detected atom within the original probed volume by transforming the data triplet associated to each event (two space coordinates and evaporation ordering) into its coordinates within the Cartesian reference system of the 3D reconstructed space. This transformation can be performed by means of back-projection algorithms characterized by different degrees of complexity. Commonly used transformation procedures require some basic knowledge of the geometry of the specimen such as apex radius and shank angle. For heterostructured systems additional information on the structure of the specimen (e.g., thickness of different layers) is needed in order to cope with reconstruction artifacts which stem from the coexistence of different phases within the probed volume. Such information is usually provided by complementary STEM or XRD analysis of the system. For a properly calibrated analysis, atomic position reconstruction resolutions of the order of 1 nm and few angstroms are reported perpendicularly and along the reconstruction axis, respectively. It follows that the correct crystallography of the system and structural defects (e.g., stacking faults) cannot be reconstructed. For a thorough discussion of the physical phenomena and parameters which play a role in the determination of the reconstruction resolution, refer to Chen and Seidman (1971), Gault et al. (2010), Moody et al. (2014), Waugh et al. (1976), Vurpillot et al. (2004), Gault et al. (2008)). Spatial resolution (in reconstructing the position of atoms) and chemical resolution (possibility of correctly indexing peaks in the mass spectrum) are not the only two factors which determine the total resolution of the technique. In fact, APT acquisitions are characterized by chemical species-dependent detection efficiencies: due to the stochastic nature of the field evaporation process and to the different behaviors of adatoms of different elements in different phases, not all atoms evaporated are actually detected and relative detection efficiencies may depend on experimental parameters and sample geometry. Thus, careful calibration of the experimental conditions and interpretation of APT data is needed when analyzing semiconductor alloys and heterostructured systems. In-depth discussion of APT composition quantification artifacts in semiconductors can be found in Mancini et al. (2014a,b), Agrawal et al. (2011), and Riley et al. (2014).

The APT reconstruction of a typical sample consists of millions of data quartets made of three spatial coordinates (extracted by the position sensitive detector) and an ion species index (extracted by ToF analysis), each one representing a detected atomic or molecular ion. Due to the considerable size of the data ensemble and to the limited spatial resolution and detection efficiency with which single ions are detected, a statistical treatment of the data is usually preferred in order to extract the information provided by APT analysis. When analyzing semiconductor alloys and heterostructured systems, the assessment of the relative abundances of different atomic species constituting the sample is one of the most important results achieved by APT. Given an ensemble of detected atoms, the atomic fractions corresponding to each detected element can be determined with an incertitude which is limited by fundamental counting statistics. The field evaporation and consecutive reconstruction of large homogeneous volumes allow for a precise estimation of the composition of alloys and for a high detection sensitivity (fundamental for assessing the concentration of impurities as discussed thoroughly in Section 3.3.4). The APT capability of resolving spatially the original position of evaporated atoms can be exploited by calculating relative abundancies within specific subensembles of data. For example, linear composition profiles can be computed for characterizing heterostructures or systems showing compositional gradients along specific directions. Moreover, 2D and 3D chemical mappings of the analyzed system can be obtained by defining a mesh in the reconstructed space and computing atomic fractions within each grid point. In this case, a good compromise should be found between the spatial resolution and the incertitude of the measured atomic fractions: a finer binning of the reconstructed volume allows for chemical maps of higher spatial resolution, but the number of detected events within each bin will be smaller implying a less accurate estimation of the local composition. Note that such approach is particularly powerful when analyzing III–V semiconductor systems, as it allows addressing local variations of composition at the typical length scale of alloy fluctuations in ternary alloys. Additional approaches for APT data analysis include the definition of concentration isosurfaces (Lorensen and Cline, 1987) and proximity histograms (Hellman et al., 2000), allowing, respectively, for addressing the morphology of heterostructures volumes and interfaces. The average composition and the amplitude of composition fluctuations can be assessed by studying the frequency of occurrence of subensembles of events with fixed compositions. The found distribution can eventually be compared to the one representing a randomly distributed alloy.

Recently APT has proven to be an extremely powerful tool for the characterization of III–V NWs. The number of systems studied is still limited, with most of the work addressing either InGaN/GaN (Mancini et al., 2016; Rigutti et al., 2014; Riley et al., 2013) or AlGaAs/GaAs (Mancini et al., 2014a,b; Morkötter et al., 2015; Loitsch et al., 2016b) NWs (with some exceptions, e.g., InGaAs and InAs; Perea et al., 2006; Koelling et al., 2017; Devaraj et al., 2018). Nevertheless, APT was shown to be able to address most of the main features related to the optical properties of these systems, namely, heterostructures geometry and interfaces definition, composition and alloy fluctuations, dopants distribution, and clustering. The ensemble of these works shows also the complementarity of APT with respect to more traditional techniques: APT synergies well with structural characterization techniques such as quantitative STEM or XEDS. Indeed, its 3D nature allows addressing features which cannot be assessed by 2D techniques, while relies on complementary structural and crystallographic information for a comprehensive description of the analyzed system. Moreover, the consistency between the optical properties addressed by PL or CL and the structural characterization is often used as a posteriori confirmation of the correctness of the APT analysis.

3. NANOSCALE ANALYSES ADDRESSING OPEN ISSUES OF NANOWIRE DEVICES

3.1 Alloy Composition, Inhomogeneity, Heterostructures, and Defects in III–V Nanowires

The ability to tune the band gap by adjusting the composition of a ternary alloy is a key point to control the LED emission wavelength. It is also important in multijunction solar cells to optimally design the absorption edge of individual subcells. Because of their 3D geometry, the control over the composition during the elaboration of NWs is more challenging than in thin films. Alloy inhomogeneity within a single nano-object can lead to localization of carriers and induce optical and electrical irregularities detrimental for the conversion efficiency (Chichibu et al., 2006; Himwas et al., 2014; Pierret et al., 2013a).

Concerning NW-based solar cells, the actual record conversion efficiencies for III–V semiconductors were achieved with binary NWs, for which the alloy is homogeneous by nature of the binary compound

(Aberg et al., 2016; van Dam et al., 2016; Wallentin et al., 2013). However, the development of multijunction solar cells similar to thin film devices (Dimroth et al., 2014, 2016) requires ternary alloy NWs. III–V NWs can be coupled with a Si-based 2D cell or can form a multijunction cell using different III–V alloys within the same NW. Theoretical optimization of a III–V/Si two-junction device indicates the optimum band gap for the nanostructured top cell around 1.7 eV (LaPierre, 2011), achievable with different ternary compounds (e.g., AlGaAs, GaAsP, InGaP, InGaN). The alloy composition defines the voltage and current generated by the cell. Intrawire or wire-to-wire alloy variations would lead to current redistribution and efficiency loss. In this context, a control over the alloy composition is a key step to obtain high conversion efficiency devices.

3.1.1 PL and CL for Probing the Composition and Alloy Homogeneity of III–V Nanowires

The most widespread way to assess the composition in ternary alloy NWs for PV application is the PL spectroscopy, which allows to extract the composition from the luminescence energy. PL can give an average information on an as-grown NW array, on a dispersed NW ensemble (as in Metaferia et al., 2016) or on dispersed single NWs; however, it is difficult to selectively probe different regions within a single NW. Commonly PL is coupled with morphological, structural, and/or compositional analysis (i.e., TEM, STEM, EDX, XEDS) to assess intrawire inhomogeneity (Shin et al., 2011; Zhang et al., 2014).

A comprehensive example is given by Metaferia et al. (2016), who performed a systematic investigation on GaAsP NWs grown by Aerotaxy. Aerotaxy is a continuous gas phase mass production process, in which NWs are grown from the aerosol catalyst particles (Heurlin et al., 2012; Yang et al., 2015). In order to investigate the effect of the precursor gas ratio ($X_g = [PH_3]/([AsH_3] + [PH_3])$) on the composition in a specific growth temperature range (between 530°C and 570°C), low temperature PL spectra (4 K) were acquired on NWs dispersed on a substrate (Fig. 6A). Using as a reference low-T PL spectra of pure GaAs NWs grown in the same conditions, P incorporation as a function of the concentration of PH_3 precursor was analyzed. Almost no influence of the growth temperature on the composition was observed for unbalanced gas mixture ($X_g = 0.25$ and $X_g = 0.75$), while for $X_g = 0.5$ alloy composition varied from $x = 0.21$ to $x = 0.27$ depending on the growth temperature, corresponding to a band gap variation from 1.71 to 1.77 eV at RT. Since the involved energy gap

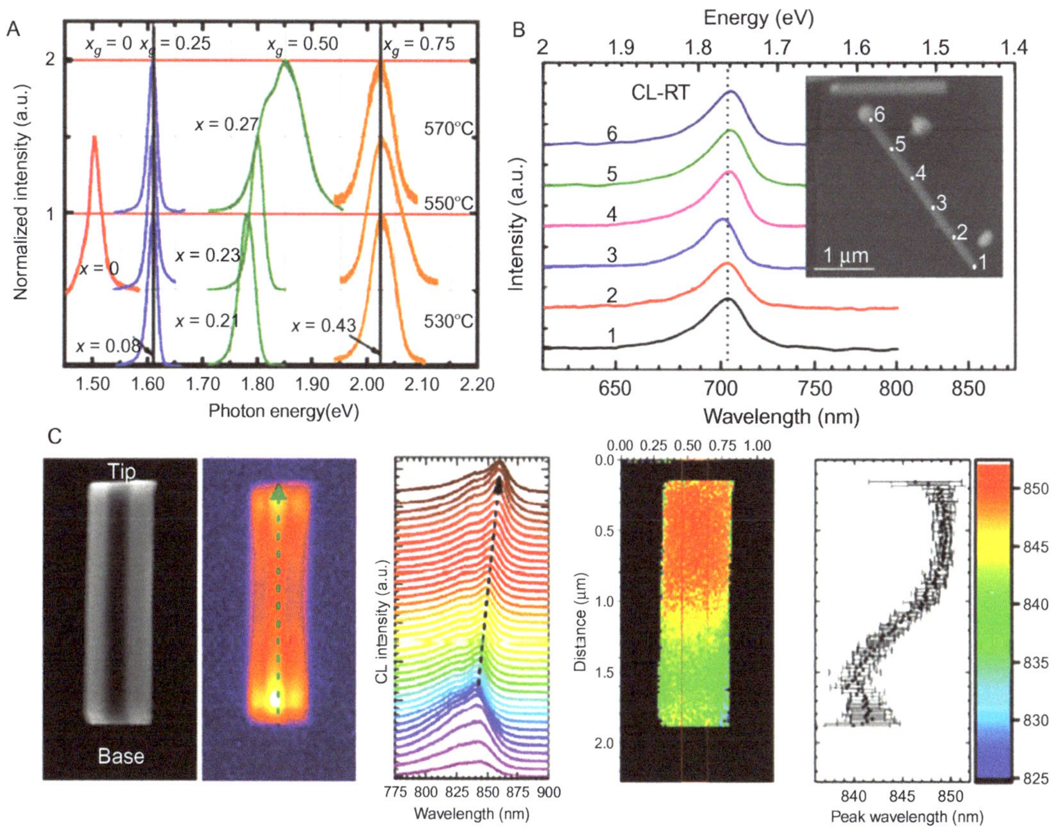

Fig. 6 See figure legend on oposite page.

range is critical for dual-junction solar cells, further investigations on intrawire homogeneity by means of XEDS were performed to determine the exact alloy composition. XEDS showed that the composition was homogeneous along the wire axis.

To gain a better resolution, the approach consists in using CL (SEM–CL or STEM–CL) mapping to acquire luminescence spectra at different positions. Himwas et al. (2017) used SEM–CL analysis to demonstrate intrawire composition homogeneity of GaAsP NWs grown by Ga-catalyzed MBE. Single NWs with a bandgap energy adjusted to ~1.7 eV were dispersed on a SiO_2/Si substrate, and the CL spectra were acquired for different excitation points along the wire axis (Fig. 6B). No significant peak shift could be observed proving a good alloy homogeneity.

Berg et al. (2017) studied InGaP NWs grown by selective area MOVPE in different conditions (changing the temperature, fluxes, and selective area pattern) and used CL to determine the average composition and the compositional gradient. The emission wavelength was tuned in the range from 870 to 800 nm with shorter wavelengths favored at higher growth temperatures. For InGaP NWs with a relatively low Ga composition a possible contamination from the Ga beam during FIB preparation may lead to falsified composition determination by TEM–XEDS. Therefore, CL was used to evaluate the composition homogeneity of the NWs as shown in Fig. 6C).

Fig. 6 (A) Low temperature PL spectra from GaAsP NWs grown with different gas phase ratios and at three different temperatures. The spectrum from intrinsic GaAs ($X_g = 0$) NWs grown at 550°C is included as a reference. (B) Low temperature CL spectra for a single $GaAs_{0.7}P_{0.3}$ NW passivated with a GaP shell obtained for different excitation positions from the bottom to the top. The *inset* shows SEM image of the wire collected during CL acquisition showing the excitation positions. (C) RT CL analysis on a single InGaP NW. From *left* to *right*, SEM image, integrated CL image, CL spectra taken along the center of the NW, CL map showing the emission peak wavelength of the entire NW and average peak wavelength along the length of the NW with error bars indicating standard deviation of the wavelength variation in the lateral direction. *Panel (A) reproduced from Metaferia, W., Persson, A.R., Mergenthaler, K., Yang, F., Zhang, W., Yartsev, A., Wallenberg, R., Pistol, M.-E., Deppert, K., Samuelson, L., Magnusson, M.H., 2016. GaAsP nanowires grown by aerotaxy. Nano Lett. 16, 5701–5707. http://pubs.acs.org/doi/abs/10.1021%2Facs.nanolett.6b02367. Further permissions related to the material excerpted should be directed to the ACS. Panel (C) reprinted by permission from Springer Nature: Springer Nature, Nano Research. Berg, A., Caroff, P., Shahid, N., Lockrey, M.N., Yuan, X., Borgström, M.T., Tan, H.H., Jagadish, C., 2017. Growth and optical properties of InxGa1−xP nanowires synthesized by selective-area epitaxy. Nano Res. 10, 672–682. https://doi.org/10.1007/s12274-016-1325-1. Copyright 2017.*

The wavelength shift from 849 nm for the top part to 840 nm for the lower part was observed and attributed to a composition gradient along the NW of ~2%.

3.1.2 APT and Optical Analyses of AlGaAs/GaAs NWs

3.1.2.1 Assessment of Al Segregation and Confinement

AlGaAs is a common material to provide an epitaxial passivation of the GaAs NWs (Ghahfarokhi et al., 2016; Lee et al., 2012); however, AlGaAs overgrowth on GaAs NWs can yield an inhomogeneous Al composition (Mancini et al., 2014a,b; Jeon et al., 2015). APT was effectively used for the study of localization effects stemming from alloy fluctuations within AlGaAs/GaAs multiquantum well core–shell NWs. Mancini et al. (2014a, b) demonstrated by APT that the segregation of Al within the AlGaAs barriers leads to the formation of Al-depleted regions with typical sizes around 5 nm inducing zero-dimensional quantum confinement of the charge carriers. This result was obtained following a procedure analogous to the one described in Mancini et al. (2016). A mesh with submicron resolution was defined within the APT reconstructed space, and an Al fraction composition map was defined. The APT reconstructed QDs highlighted by AlAs alloy fraction isosurfaces, the 2D Al elemental concentration map, and the 3D AlAs fraction map are reported in Fig. 7A–C, respectively. The effective mass calculation of the emission energy corresponding to a reconstructed Al-depleted region was demonstrated to be in good agreement with typical emission energies observed by μPL analysis. Further information about the composition inhomogeneities extended along larger volumes and about the definition of the core–shell structure interfaces was provided by considering Al fraction isosurfaces and proximity histogram analyses.

Similar NW structures were analyzed by CL in Heiss et al. (2013). Monochromatic CL maps of single NWs revealed localized emissions along the length of the wire (Fig. 7D). The authors ascribed the emission to QD-like structures self-assembling at the interface between the alternating GaAs and AlGaAs layers observed by cross-sectional STEM imaging (Fig. 7E). The APT analysis did not allow identifying these QDs. On the other hand, it suggested the existence of a different QD formation mechanism consistent with the results of CL characterization

Jeon et al. (2015) reached similar conclusions through the APT analysis of an analogous AlGaAs/GaAs multi-QW NW system. A more in-depth study of the quantum confining properties of the QD-like structures was performed addressing their composition, size distribution, and interface

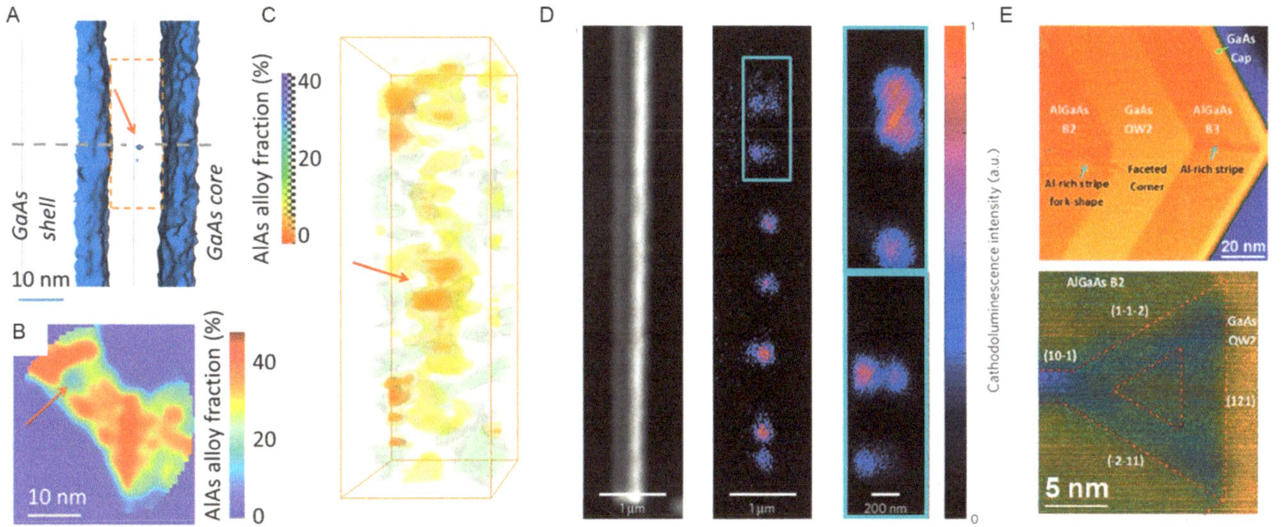

Fig. 7 (A) APT reconstruction of a QD in the AlGaAs shell, showing the isosurfaces at 14% AlAs alloy fraction. The GaAs core and the external GaAs shell appear on the *right*- and on the *left-hand side*, respectively. (B) 2D map of the Al elemental concentration, extracted from the 1 nm thick slice corresponding to the *gray dashed line* in (A). The *red arrows* point to the QD position. (C) 3D visualization of the AlAs alloy fraction distribution in the region of interest highlighted in (A). (D) SEM image of a single NW and corresponding CL panchromatic maps. (E) Cross-sectional STEM image of the alternating AlGaAs/GaAs layers (*top*) and zoom-in of a self-assembled QD (*bottom*). *Panels (A)–(C) are reprinted from Mancini, L., Fontana, Y., Conesa-Boj, S., Blum, I., Vurpillot, F., Francaviglia, L., Russo-Averchi, E., Heiss, M., Arbiol, J., Morral, A.F. i, Rigutti, L., 2014. Three-dimensional nanoscale study of Al segregation and quantum dot formation in GaAs/AlGaAs core-shell nanowires. Appl. Phys. Lett. 105, 243106. https://doi.org/10.1063/1.4904952, with the permission of AIP Publishing (License Number: 4275280815666). Panels (D) and (E) are reprinted by permission from Springer Nature: Springer Nature, Nature Materials, Heiss, M., Fontana, Y., Gustafsson, A., Wüst, G., Magen, C., O'Regan, D.D., Luo, J.W., Ketterer, B., Conesa-Boj, S., Kuhlmann, A.V., Houel, J., Russo-Averchi, E., Morante, J.R., Cantoni, M., Marzari, N., Arbiol, J., Zunger, A., Warburton, R.J., Fontcuberta i Morral, A., 2013. Self-assembled quantum dots in a nanowire system for quantum photonics. Nat. Mater. 12, 439–444, Copyright 2013.*

characteristics. The optical signatures of the system were correlated with the morphology of the QDs and with their spatial distribution within the sample, ascribing the absence of PL emission from QDs with maximal diameter lower than 4 nm and the broadening of the spectral lines corresponding to large clusters to the tunneling of charge carriers between different clusters. This interpretation was supported by effective mass calculations performed on interacting ellipsoidal QDs whose typical morphology and composition were inferred by APT characterization.

3.1.2.2 Analyses of GaAs/AlGaAs Interface in NWs

APT can be used to address the composition and interface definition. Chen et al. (2013) demonstrated that the combined use of APT, PL, and photocurrent spectroscopy allowed for the assessment of the band diagram of individual GaAs/AlGaAs core–shell NWs and corresponding band offsets. APT analysis was employed for characterizing the core–shell structure interfaces. In particular, the dimensions and the compositional gradient along the interface were determined. PL spectroscopy allowed for observing, along with band edge emissions from GaAs and AlGaAs region, the emission corresponding to interface recombination. Their relative spectral broadenings were correlated with the structural properties assessed by APT showing that transitions with narrower distributions of energy correspond to thinner junctions. The interface transition nature of the observed signal was confirmed by temperature-dependent PL assuming a constant band offset with temperature. Photocurrent spectroscopy was finally used for a precise determination of the heterostructure band offsets, overcoming the limitation of PL imposed by wire-to-wire inhomogeneities and multiple energy level impurity luminescence. The linear fitting of normalized photocurrent slopes allowed for the determination of transition energies and corresponding conduction band offsets.

3.1.3 CL Analyses of Intentional and Unintentional Heterostructures in III–V NWs

Embedded quantum dots in NWs have attracted a strong attention to fabricated quantum emitter devices (Borgström et al., 2005) or explore new concepts for third-generation photovoltaics. CL mapping provides a powerful tool to visualize localized emissions. CL intensity increases and the wavelength is shifted in the NW regions with quantum dot or quantum well-like structures, where carriers are confined by wide bandgap barriers. For example, CL spectroscopy and mapping was applied to analyze

InP/InAs NWs (Zhang et al., 2015b, p. 6). The InP NWs were synthesized using In-catalyzed MOVPE approach, and three InAs insertions of different length were embedded in the wire. By using SEM–CL, the authors could probe the emission wavelength of the individual segments (in the 1.1–1.6 μm range), visualize their location, and correlate with the growth time and size. Similarly, CL was also used to investigate the alloy composition in InGaAs NWs and GaAs NWs with InGaAs axial insertions elaborated by self-catalyst MBE growth with InGa catalyst (Heiß et al., 2009).

Quantum-confined structures can be implemented unintentionally due to alloy inhomogeneity (Moratis et al., 2016) or to crystal defects (Loitsch et al., 2016a,b). In Moratis et al. (2016), GaAs/InGaAs core–shell NWs were grown by MBE. The purpose of the highly mismatched InGaAs shell was to apply a strain on the NW core and to induce a piezoelectric field. This field was proposed to be used for carrier separation in NW solar cells (Boxberg et al., 2010). CL mapping of core/shell GaAs/InGaAs NWs shown in Fig. 8 revealed a strain-induced shift of the core emission and a localized emission from InGaAs quantum dot-like structures with up to 30% In content at the NW tips. The In content also exhibited a gradient along the NW axis, with a higher In content at the tip compared with the lateral facets.

3.1.4 CL Analyses of Defects in Nanowires

CL intensity is sensitive to the density of non-radiative recombination centers such as dislocations, grain boundaries, and unpassivated surface. Spatially and spectrally resolved CL mapping at the nanoscale reveals rich information on the material properties. Shallow defects can be identified through the unique photon emission energy, and the location of deep levels is usually detected by the quenching of the CL intensity compared to defect-free region.

3.1.4.1 Impact of Doping Impurities

Doping impurities may introduce non-radiative recombination centers in a heavily doped material. In Yao et al. (2014), a solar cell using GaAs axial n–i–p NW array was reported with a conversion efficiency of 7.58%. The authors grew GaAs NWs by MOCVD on patterned n + GaAs(111B) substrate, employing disilane for n-type dopant and diethylzinc for p-type dopant. The length of the top p-GaAs segment was varied between 100 and 400 nm. The EQE was compared for NWs with different junction depth defined by the length of the p-GaAs segment. Lower EQE is observed in the short-wavelength part because most of the carriers are generated close

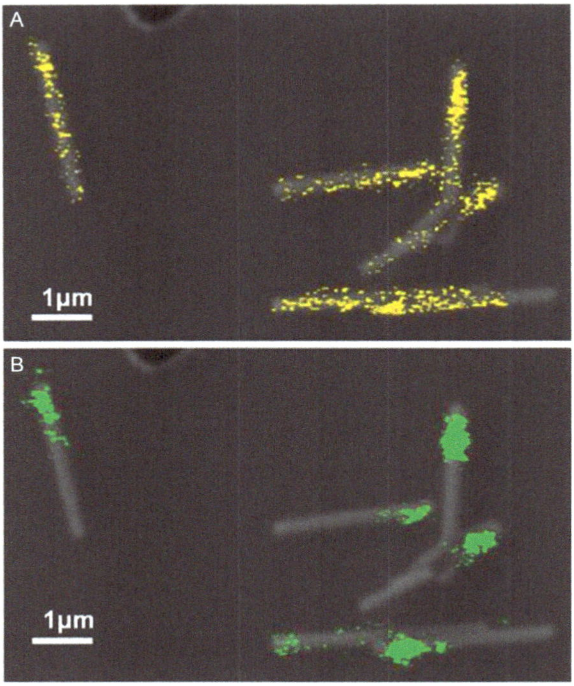

Fig. 8 CL mappings at $T = 5\,K$ from dispersed GaAs/InGaAs core/shell NWs spectrally filtered at (A) $1.508 \pm 0.005\,eV$, where a relatively uniform emission attributed to the GaAs core is observed and (B) $1.475 \pm 0.005\,eV$, where strong emission from InGaAs quantum dot-like structures is obtained. *Panels (A) and (B) are reproduced by permission from Moratis, K., Tan, S.L., Germanis, S., Katsidis, C., Androulidaki, M., Tsagaraki, K., Hatzopoulos, Z., Donatini, F., Cibert, J., Niquet, Y.-M., Mariette, H., Pelekanos, N.T., 2016. Strained GaAs/InGaAs core-shell nanowires for photovoltaic applications. Nanoscale Res. Lett. 11, 176. https://doi.org/10.1186/s11671-016-1384-y, https://creativecommons.org/licenses/by/4.0/legalcode.*

to the top of the NWs, and non-radiative recombination processes are dominant in the strongly p-doped top regions. This conclusion was confirmed from the CL intensity maps on single NWs: the intensity decreased continuously from the junction position to the top region (see Fig. 14C and D from Section 3.2.1). In this case, a shallow junction design had a better short-circuit current than a deep junction because of a better collection efficiency of photogenerated carriers.

CL was also applied to investigate the intrawire variation of the optical emission in GaInP/AlGaInP red-orange LEDs (Hussain et al., 2017). A core/shell GaInP/AlGaInP quantum well produced the electroluminescence at 1.8 eV as expected from the design, but a defect-related emission at about 1 eV was also detected. The authors used CL spectroscopy to investigate this phenomenon. Spatially inhomogeneous GaInP bandgap emission was observed and attributed to the variation of the quantum well thickness or composition. The deep level emission and CL quenching at 1.8 eV were ascribed to several causes: growth-induced defects related to disordered AlGaInP barriers, oxygen-related defects, phosphorous-vacancy, Zn dopant-induced defects, or extended dislocations.

3.1.4.2 Impact of Stacking Faults and Wurtzite/Zinc-Blende Switching

For most of the III-arsenide and III-phosphide semiconductors, the bulk crystals adopt the zinc-blende structure. When growing NWs with high surface-to-volume ratio, the crystal structure may switch between zinc-blende (cubic) and wurtzite (hexagonal) phases because wurtzite has a lower surface energy compared to zinc-blende. The occurrence of both crystal phases and eventually formation of twins and stacking faults have an impact on the optoelectronic properties of the NWs.

GaAs/AlGaAs NWs were studied by TEM and CL spectroscopy to correlate optical properties with the atomic structures (Bolinsson et al., 2014). High density of stacking faults lower considerably the emission efficiency and decrease the diffusion length. Combining CL and TEM, near-band edge (NBE) emission of GaAs at 1.52 eV was observed only at the region of sufficiently long pure zinc-blende segment without twins. The minimum length to observe clear 1.52 eV emission was about 25–35 nm, comparable with the exciton diameter in GaAs.

The electronic band structure of the wurtzite III–V crystals is not totally understood yet. The fundamental bandgap of wurtzite GaAs is still debated. Here we illustrate several studies of the wurtzite GaAs through different characterization techniques. In Spirkoska et al. (2009), polytypic GaAs NWs were studied by means of TEM, PL, CL, and Raman spectroscopy. The zinc-blende/wurtzite GaAs is believed to form a type II band alignment where electrons are confined in the zinc-blende segment and holes in the wurtzite segment (see Fig. 9A). Depending on the thickness of the confined segment, the ground states of the electron and hole shift in energy. In Fig. 9B), SEM and monochromatic CL maps for polytypic (Fig. 9A–E) and zinc-blende (Fig. 9F and G) GaAs NWs are shown. Luminescence spectra of polytypic NWs usually exhibit red-shifted peaks compared to the bandgap of both pure zinc-blende and wurtzite material due to spatially indirect recombination of electrons and holes. CL maps reveal the localized nature of lower energy emission due to confinement of electron and holes.

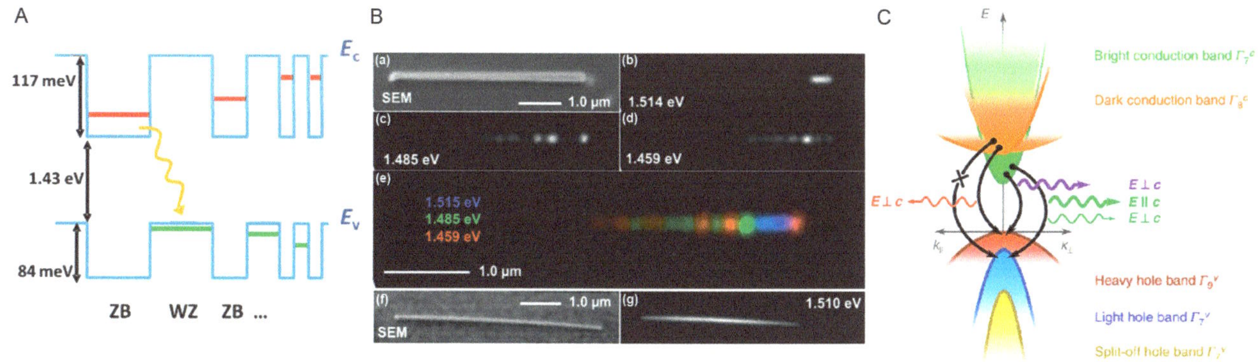

Fig. 9 (A) Schematic for GaAs type II band alignment in zinc-blende/wurtzite NW heterostructures. (B) SEM images and CL maps of polytypic GaAs NWs (a–e) and zinc-blende GaAs NWs (f and g). Selected emitted photon energies are indicated in the monochromatic CL maps, and the panel (e) constitutes a combination of panels (b–d), showing the spatial confinement of electron and hole in polytypic NW segments. (C) Illustration of wurtzite GaAs band structure and the symmetry of the states. *Panels (A) and (B) are reprinted with permission from Spirkoska, D., Arbiol, J., Gustafsson, A., Conesa-Boj, S., Glas, F., Zardo, I., Heigoldt, M., Gass, M.H., Bleloch, A.L., Estrade, S., Kaniber, M., Rossler, J., Peiro, F., Morante, J.R., Abstreiter, G., Samuelson, L., Fontcuberta i Morral, A., 2009. Structural and optical properties of high quality zinc-blende/wurtzite GaAs nanowire heterostructures. Phys. Rev. B, 80, 245325. https://doi.org/10.1103/PhysRevB.80.245325. Copyright (2009) by the American Physical Society. Panel (C) reprinted by permission from Springer Nature: Springer Nature, Nature Communications, Signorello, G., Lörtscher, E., Khomyakov, P.A., Karg, S., Dheeraj, D.L., Gotsmann, B., Weman, H., Riel, H., 2014. Inducing a direct-to-pseudodirect bandgap transition in wurtzite GaAs nanowires with uniaxial stress. Nat. Commun. 5, 3655. https://doi.org/10.1038/ncomms4655. Copyright (2009).*

This explanation is also in agreement with TRPL measurement on polytypic GaAs NWs: luminescence intensity is lower but has a longer lifetime due to the spatial separation of electron and hole.

Later on, luminescence measurements were analyzed to extract somewhat different wurtzite GaAs bandgap and band offset values for zinc-blende/wurtzite heterostructure (Heiss et al., 2011). Higher-energy luminescence emission than zinc-blende bandgap was also observed in wurtzite-rich GaAs NWs, so it was attempted to attribute a higher bandgap to wurtzite GaAs (Ahtapodov et al., 2012; Jahn et al., 2012). On the other hand, resonant Raman spectroscopy was used to identify the critical points of the band structure, and similar GaAs bandgap of wurtzite and zinc-blende was concluded (Ketterer et al., 2011). In Fig. 9C, a band structure for wurtzite GaAs was proposed (Signorello et al., 2014): the heavy-hole and light-hole band split, and the zone folding of hexagonal symmetry leads to the existence of a supplementary conduction band. Selection rules also dictate that the transition involved in the wurtzite bandgap should be polarized perpendicular to the *c*-axis.

3.1.5 Alloy Composition and Homogeneity in InGaN/GaN Nanowire LEDs

The control over the alloy composition of the InGaN QW in NWs is important since it determines the operating wavelength of the LED. Composition fluctuation in the QW is detrimental for the LED performance. Indeed, In-rich regions locally change the parameters of the p–n junction and favor current injection in these locations (Tchernycheva et al., 2014b). Because of the current redistribution, the emission wavelength shifts with increasing injection (Schimpke et al., 2016; Tchernycheva et al., 2014b). Therefore, the usual assumption of a homogeneous injection made in core/shell NW LED modeling is not fulfilled in the experiments. An inhomogeneous current density with a high carrier concentration in the In-rich regions results in an early onset of Auger recombinations and thus in efficiency droop. In addition, the emission wavelength of these regions is red-shifted with respect to the wavelength for which the LED was designed.

3.1.5.1 APT Structural Characterization of InGaN/GaN QWs in NWs for LED Application

In segregation in planar quantum wells leading to the formation of quantum confining centers is an important issue affecting the properties of thin-film nitride LEDs. It has been extensively addressed by APT analyses of the structure and composition of InGaN/GaN planar quantum wells (Bennett et al., 2012; Griffiths et al., 2016; Ren et al., 2015; Schulz et al., 2015; Watson-Parris et al., 2011). Recently several studies were published addressing the analogous issue in core–shell NWs. Riley et al. (2013) employed APT for the analysis of GaN/InGaN core–shell NW LED arrays. The considered NWs consisted of a GaN core surrounded by an alternation of MOCVD grown nonpolar InGaN QWs and GaN barriers along the radial direction, and by semipolar InGaN QWs and GaN barriers on the top planes. The top part of the NW array was covered with a p-doped InGaN capping layer as illustrated in Fig. 10A. APT was demonstrated to allow addressing composition inhomogeneities between the different regions of single NWs and within the same QW (Fig. 10B and C). Frequency distribution analysis of reconstructed volumes extracted from nonpolar and semipolar QWs was performed in order to estimate the average In composition highlighting the different In incorporation between the two types of QWs. Nonpolar QWs were observed to incorporate more In than the semipolar ones (around 13.5% and 8% In mole fraction, respectively). Moreover, the statistical analysis of the frequency distribution allowed for excluding In clustering. Even if locally the In composition is consistent with a random alloy, a gradient of the In mole fraction along the growth axis of the NWs and a high In concentration region at the apex of the NWs (crossing of the semipolar facets) were observed. APT allowed also excluding the existence of quantum confining mechanisms related to QW interface fluctuations: the definition of In composition isosurfaces delimiting each single well on both sides indicated width fluctuations equal or lower than 0.5 nm, thus not strong enough for inducing localization of the charge carriers or scattering effects within the NW QWs. Comparable width localization was previously observed in planar geometries and demonstrated not to affect the optical properties of the system (Detchprohm et al., 2010). The composition information extracted by APT analysis was compared to the composition analysis performed by EDX and to the results of spatially resolved optical characterization performed by CL. This comparison showed that the In concentrations measured by the two former techniques are comparable and that the expected wavelength emissions on the basis of measured composition are in good agreement with experimentally assessed optical properties. The observation of the In gradient and an In-rich layer at the polar/semipolar junction is a general problem in NW LEDs, also reported by other groups using XEDS and CL characterization techniques

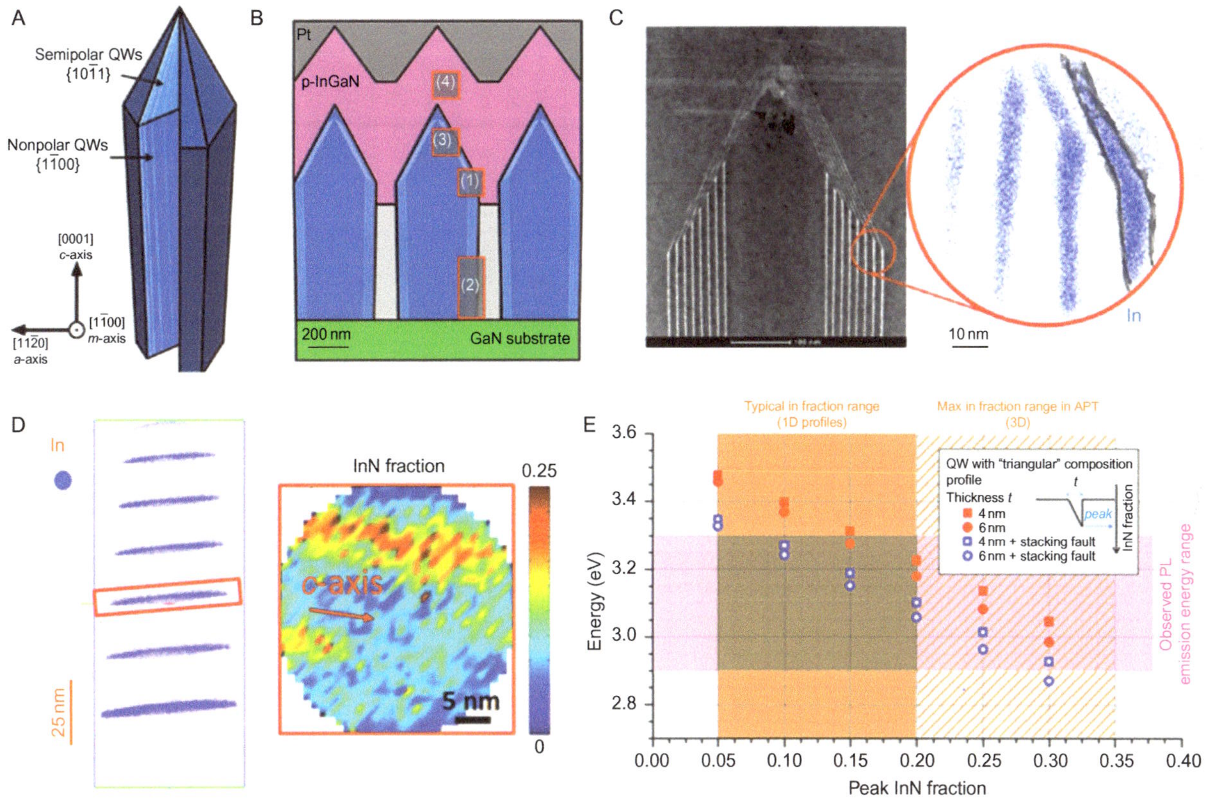

Fig. 10 See figure legend on next page.

(Krause et al., 2016; Schimpke et al., 2016; Tchernycheva et al., 2015). The characterization of the NW LED was completed by the APT assessment of the dopants spatial distribution within the Mg-doped p-InGaN canopy layer. A more detailed discussion of the results of this characterization can be found in Section 3.3.

Further insight into the properties of InGaN/GaN NW LEDs was more recently provided in the work of Rigutti et al. (2014) and Mancini et al. (2016). Both studies addressed the optical and structural properties of the NWs by correlating APT, STEM, and μPL analyses. The authors characterized nitrogen-polar 20–30 μm long wires spontaneously formed on a sapphire substrate by MOVPE. Only nonpolar InGaN/GaN radial QWs grown along the m-plane of the template GaN NW were considered for analysis. The approach used in Rigutti et al. (2014) for interpreting APT reconstructions differed from the one used in Riley et al. (2013) in order to account for the dissimilar properties of the wells. In the case of Riley et al. frequency distribution analysis and In composition isosurfaces depicted homogeneous QWs with flat and smooth surfaces, whereas the system considered by Rigutti et al. is characterized by strongly inhomogeneous QWs. 1D In composition profiles defined along the *m*-axis showed an In

Fig. 10 (A) Schematic illustrating a single InGaN/GaN core–shell NW with nonpolar side facet and semipolar apex facet QWs and (B) the full LED device layout. The approximate regions from which samples for APT analysis were extracted are also indicated. (C) Cross-sectional HAADF-STEM image of a representative NW and associated APT reconstruction depicting both nonpolar and semipolar QWs. The QW surface is rendered in *gray*. (D) APT reconstructed position of In atoms of a field emission tip extracted from one of the side facets of a NW. The 2D InN alloy fraction map extracted from the forth QW from the top of the sample is also shown. (E) Transmission energies as a function of the peak InN fraction in the QW obtained from simulations considering QWs with composition profile varying along the m-crystal direction only. The wells are either free from defects (*red symbols*) or intersecting a SF (*blue symbols*). The ranges of measured In fraction and observed PL emission energies are indicated. *Panels (A)–(C) are reprinted (adapted) with permission from Riley, J.R., Padalkar, S., Li, Q., Lu, P., Koleske, D.D., Wierer, J.J., Wang, G.T., Lauhon, L.J., 2013. Three-dimensional mapping of quantum wells in a GaN/InGaN core-shell nanowire light-emitting diode array. Nano Lett. 13, 4317–4325. https://doi.org/10.1021/nl4021045. Copyright (2013) American Chemical Society. Panels (D) and (E) are reprinted from Mancini, L., Hernández-Maldonado, D., Lefebvre, W., Houard, J., Blum, I., Vurpillot, F., Eymery, J., Durand, C., Tchernycheva, M., Rigutti, L., 2016. Multi-microscopy study of the influence of stacking faults and three-dimensional In distribution on the optical properties of m-plane InGaN quantum wells grown on microwire sidewalls. Appl. Phys. Lett. 108, 042102. https://doi.org/10.1063/1.4940748, with the permission of AIP Publishing.*

incorporation gradient within each well along the *m*-axis (decreasing In content going from the core to the surface). Moreover, 2D In composition maps of single QWs demonstrated that In segregates within stretched regions within the QW plane (Fig. 10D). The complementary information obtained by STEM allowed for determining the crystallographic direction along which the high In concentration regions elongate and thus for showing how these structures intercept perpendicularly the SF planes. The structural information was correlated with the optical signature of the wires assessed by μPL by simulating QWs on the basis of the APT composition profiles. Reasonable agreement between the APT analysis and the optical spectroscopy was demonstrated. The work by Mancini et al. (2016) explored further the potential of the APT in characterizing multi-QW system by fully exploiting its 3D chemical mapping capabilities. The In composition map of single QWs was used as a direct input for 3D effective mass calculations aiming to assess the energy of radiative recombination processes. The correlated APT (composition) and STEM (crystallography) information allowed for obtaining an even better agreement between calculated and experimentally assessed emission energies and thus for addressing the effect of the presence of defects on the optical properties of the QWs (Fig. 10E).

3.1.5.2 PL for Addressing Local Optical Properties and IQE Within Single InGaN/GaN NW LEDs

PL has been proven to be an extremely interesting technique for characterizing heterostructured NW systems for light-emitting applications. While not having sufficient spatial resolution for addressing local variations of the optical properties at the typical scale of alloy fluctuations, it allows characterizing signals emitted from different regions of single NWs. Moreover, PL is polarization sensitive and different regions of the system can be probed selectively by means of resonant excitation allowing studying the response of the system as a function of the excitation density. The ensemble of these features was exploited by Mounir et al. (2016, 2017) for addressing the optical properties of nonpolar and semipolar facets of dispersed InGaN/GaN core–shell NWs and for quantifying the IQE of the system. In Mounir et al. (2016), the authors performed confocal μPL on single NWs grown under different conditions. RT PL highlighted that the emission from the NW tip semipolar QWs is less intense than the one from the lateral m-plane QWs and that the PL signal blue-shifts from the tip toward the base of the wires. Moreover, the average spectra of single NWs (obtained averaging PL spectra acquired at different positions along the side of the wire) were shown to be almost identical to the macro-PL spectrum of NWs on their native growth substrate, confirming that the emission broadening of the macroscale sample follows from intrawire inhomogeneities. Single NWs issued from growth batches obtained under different conditions were compared showing that low growth temperatures with high In/III flux ratios lead to a blue-shift at the base and low emission intensity at the tip, while high T and low fluxes ensure a better homogeneity but worse average intensity. The ensemble of these results indicated that for obtaining a good device, the NWs should be grown within a narrow window of temperatures and fluxes in order to compromise between the homogeneity and the total emission intensity, and that their tips should be either removed or passivated (Zhang et al., 2015a). The authors performed excitation density-dependent μPL on the NWs showing the best optical properties for quantifying the m-polar QWs IQE and for identifying the main factors limiting the efficiency. The absence of blue-shift with increasing excitation density proved the absence of strong quantum confined Stark effect which could limit the emission efficiency. RT IQE dependence on excitation density was estimated by normalizing RT PL intensity with respect to the one obtained at 10K for the same excitation density (thus assuming LT IQE equal to one). Efficiency dependence over position along the wire was assessed, showing lower IQE toward the tip (red-shifted emission) and increasing efficiency toward the base of the wire. A maximal IQE of $44 \pm 5\%$ at 430nm was measured. Independently of the position, saturation of the IQE was observed within the low excitation density regime and ascribed to activated donors within the n-doped GaN core in the case of RT emission. The IQE was finally expressed as a function of the excess carrier density within the QW and fitted within the framework of a model taking into account Schokley-Read-Hall radiative and non-radiative recombination rates. Simulations allowed for excluding the background carrier density of the doped core as the efficiency limiting factor and for identifying the radiative recombination coefficient as the main factor limiting the IQE. On the basis of this result the authors concluded that substituting single QW with multiple QWs could improve the performance of the LED NWs.

Following the insight provided by Mounir et al. (2016), Mounir et al. analyzed by confocal excitation-dependent μPL single MQW NWs (Mounir et al., 2017). An even better insight into the different contributions limiting the IQE was obtained by adding the information issued by resolving the degree of linear polarization of the emission. High-resolution PL mappings showed that growing NW with thinner multiple QWs instead of a single broader QW allows for a spectrally narrower emission with a better homogeneity along the m-plane facets. Still, the In incorporation gradient was observed, leading to a red-shift from the bottom to the tip of the NWs and causing the broad emission observable by macro-PL analysis of the array of NWs. Information about the diameter dependence of single NWs on the position along the growth axis was extracted by studying, for two perpendicular polarization directions, the spectral position of the Fabry-Perot-like interference peak induced by the presence of two parallel m-plane facets. Excitation-dependent measurement highlighted an IQE dependence over excess carrier density similar to the one observed in Mounir et al. (2016), namely, saturating at low excitation density at both room and low temperatures. Moreover, IQE was demonstrated to be the highest close to the center of the m-plane facets and decreasing toward the base and tip of the wire (ascribed to lower material quality). A maximal IQE of $73 \pm 7\%$ was observed. In order to assess the relative contributions of the different recombination mechanisms, the authors considered both the IQE and DLP dependencies over the excitation density. The two datasets were simultaneously fitted estimating the excess carrier density modifying the model used in Mounir et al. (2016) for taking into account the effect of phase space filling and simulating the DLP evolution within the framework of k·p band theory. The simultaneous fitting of the IQE and DLP curves ensured a robust estimation of minority carrier lifetimes and radiative recombination coefficients. The demonstrated large IQEs, narrow spectral emissions and large radiative recombination coefficients indicated that multiple-QW NW LEDs are a promising system for competing with thin film devices.

3.1.5.3 SEM–CL Analyses of Nitride Nanowires LEDs

CL can be used to analyze compositional gradients and In-accumulation regions in NW LEDs. Tchernycheva et al. (2014a) investigated the emission of the InGaN/GaN radial QWs in NW LEDs. These devices presented undesirable low-energy emission peak and a spectral shift of the electroluminescence with injected current. To understand the origin of these phenomena, CL spectroscopy and mapping of NWs were performed in a top view and cross-sectional configurations. CL spectra at different acceleration voltage (Fig. 11A) revealed the presence of a parasitic low energy peak, which was found to originate from a region close to the top NW surface. From top view CL maps the authors concluded that the parasitic low energy emission was present in all NWs close to their perimeter (Fig. 11B and C).

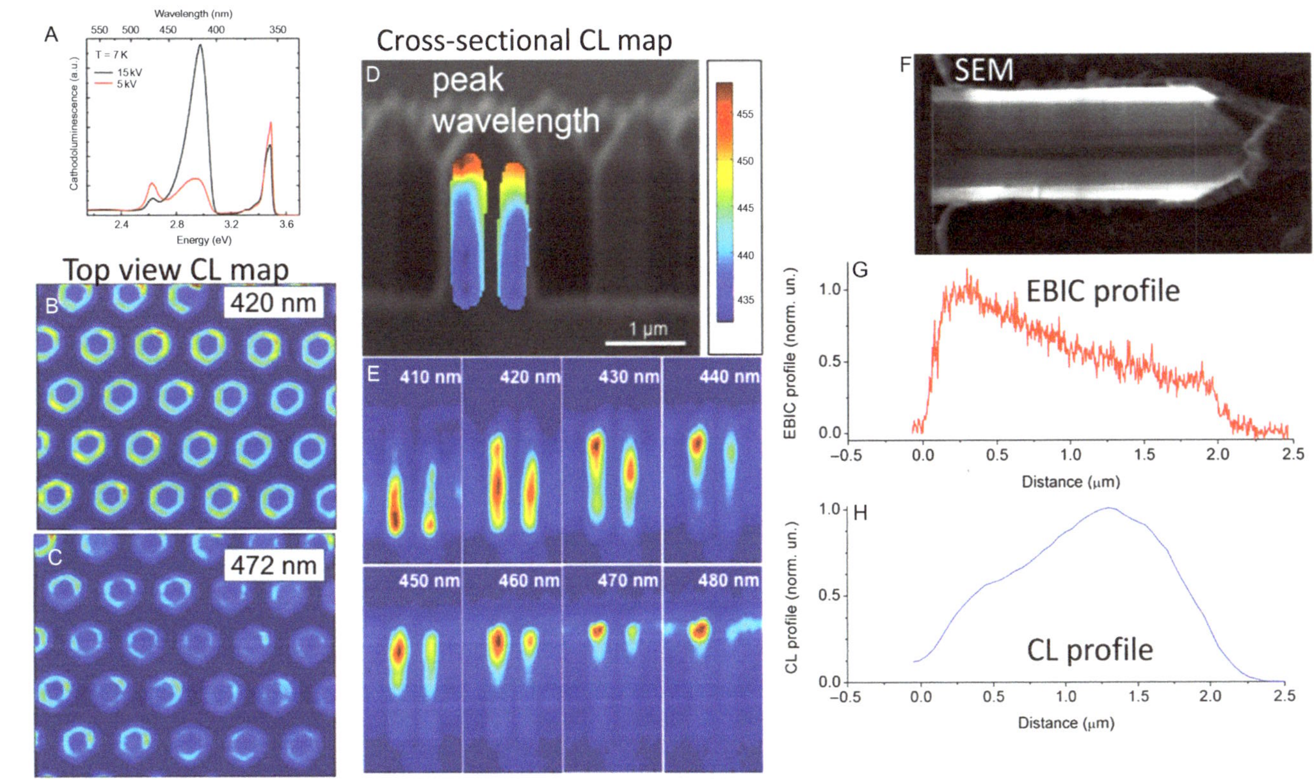

Fig. 11 See figure legend on oposite page.

The NW LED was cleaved and the QW emission was analyzed on the cross section. A shift of the CL peak wavelength (Fig. 11D and E) was observed and attributed to the In compositional gradient in agreement with the XEDS results. In addition, In-rich region was identified in the QW at the m-plane to semipolar junction, which yielded a localized emission spectrally corresponding to the low-energy parasitic peak (Fig. 11E). The CL maps were further compared with EBIC results (Tchernycheva et al., 2015). It was found that the CL intensity increased toward the top of the NW, whereas the EBIC current decreased (Fig. 11F–H). This opposite trend was related with the compositional gradient in the InGaN QW and in the AlGaN electron blocking layer, making the carrier extraction easier in the lower NW part, where the QW was shallow and the barrier height was reduced.

3.1.5.4 STEM–CL Analyses of Nitride Nanowires Heterostructures

Recently, STEM–CL has been proven to provide extremely valuable insight into the existing relationship between structural properties and optical signature of NW structures for light emission applications. Thanks to the STEM capability of resolving local variations of the periodicity of the crystallographic structure, the CL emission energy and intensity can be directly correlated to the presence of defects such as stacking faults and threading

Fig. 11 (A) Low temperature CL spectrum collected at 5 kV (shallow penetration) and 15 kV (deep penetration) acceleration voltage. (B) and (C) CL maps recorded for a detection wavelength of 420 and 472 nm, respectively. (D) Cross-sectional SEM image of the NW array containing a cleaved NW. The superimposed map shows the average CL wavelength. (E) CL maps collected at 7 K with 5 kV acceleration voltage for fixed detection wavelength from 410 to 480 nm. (F) SEM image of a cleaved NW (*vertical lines* indicate the positions corresponding to the NW base and to the end of the m-plane, respectively). (G) Normalized EBIC profile parallel to the wire axis. (H) Normalized CL profile spectrally filtered for quantum well emission wavelengths from 390 to 550 nm parallel to the wire axis. *Panels (A)–(E) are reprinted (adapted) with permission from Tchernycheva, M., Lavenus, P., Zhang, H., Babichev, A.V., Jacopin, G., Shahmohammadi, M., Julien, F.H., Ciechonski, R., Vescovi, G., Kryliouk, O., 2014a. InGaN/GaN core–shell single nanowire light emitting diodes with graphene-based P-contact. Nano Lett. 14, 2456–2465. https://doi.org/10.1021/nl5001295. Copyright 2014 American Chemical Society. Panels (F)–(H) are published by The Royal Society of Chemistry. Tchernycheva, M., Neplokh, V., Zhang, H., Lavenus, P., Rigutti, L., Bayle, F., Julien, F.H., Babichev, A., Jacopin, G., Largeau, L., Ciechonski, R., Vescovi, G., Kryliouk, O., 2015. Core–shell InGaN/GaN nanowire light emitting diodes analyzed by electron beam induced current microscopy and cathodoluminescence mapping. Nanoscale 7, 11692–11701. https://doi.org/10.1039/C5NR00623F.*

dislocations within the optically active region of single nanostructures. Notable examples in the literature can be found in the work of Urban et al. (2015) and Mohajerani et al. (2016b). In Urban et al. (2015), the authors ascribed variations of the CL emission energy along the growth direction of single Ga-polar GaN NWs to the presence of basal plane stacking faults (BSFs) acting as crystal phase heterostructures. The comparison between defect-free NWs and NWs showing BSFs allowed demonstrating that, while the former show CL emissions corresponding solely to the GaN NBE emission along the whole length of the nanostructure, the CL signal red-shifts when the excited charge carriers recombine in proximity of the planar defects (see the STEM image of a single NW, the corresponding panchromatic CL intensity image, and the CL linescan in Fig. 12A–C, respectively). The STEM–CL analysis allowed moreover attributing the emission to the recombination of excitons bound to BSF of type I_1. In Mohajerani et al. (2016a,b)) the authors studied the correlations between structural and optical properties of InGaN/GaN single quantum well core–shell NW LEDs. In particular, the coupled HAADF observation of single NWs and the electron beam excitation localized within the QW allowed ascribing a quenching of the CL signal intensity to the presence of extended defects intersecting perpendicularly the QW. Defects intersecting the r-plane QW at the tip of the NW were on the other hand showed to red-shift the CL emission while preserving the signal intensity (this behavior was attributed to In clustering). A general red-shift of the emission from the bottom to the top of the wire was also observed and ascribed to higher In content and thicker QWs (the STEM image of a NW and the contour plot of the CL–STEM linescan along the InGaN QW are reported in Fig. 12D and E).

Another interesting application of STEM–CL for the characterization of structural and optical properties of NW LEDs was demonstrated by Müller et al. (2016). The STEM–CL analysis of single GaN core–shell NWs allowed, along with studying the effect of defects on radiative recombination processes, for assessing the incorporation of impurities during the HVPE overgrowth (shell) of the MOVPE template (core). Panchromatic CL maps of the cross section of single NWs showed that threading dislocations propagating along the *c*-axis through the MOVPE core and HVPE shell partially or completely quench the CL emission. BSFs propagating into the lateral growing shell were also shown to affect negatively the CL intensity, while still allowing for a good CL emission in low defect density regions (Fig. 12F). CL linescans performed along the [0001] direction (Fig. 12G, top)

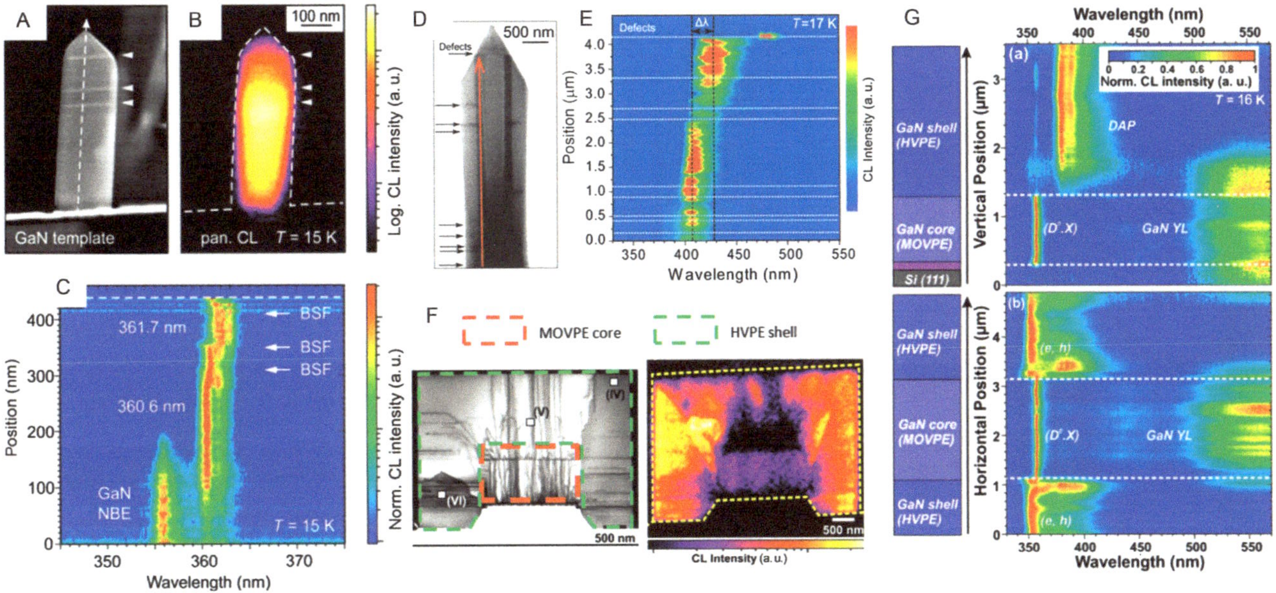

Fig. 12 See figure legend on next page.

demonstrated a CL emission dominated by the donor-bound exciton recombination in the GaN core (strain-free MOVPE template) and by the donor acceptor pair recombination in the upper GaN shell. CL linescans along the radial direction (Fig. 12G, bottom) on the contrary showed an emission dominated by free-excitons recombination processes. The prevalence of DAP and electron–hole emissions, respectively, in the upper and lateral HVPE shells were ascribed to a lower incorporation of impurities in the former with respect to the latter.

STEM–CL was also demonstrated to be able to address nonlinear effects in NW systems. Recently, Zagonel et al. (2016) studied the droop of the emission efficiency of single GaN/AlN quantum discs (QDiscs) structures embedded in GaN NWs. Similar GaN/AlN QDiscs in individual NWs were previously studied in Zagonel et al. (2012) by STEM–CL. The optical properties of the individual quantum emitters as small as 1 nm and separated from each other by only a few nanometers were assessed and correlated to the morphology of the single QDiscs, proving the ultimate resolution of the technique. In Zagonel et al. (2011), further insight into the physics of radiative recombination processes was obtained by studying the dependence of

Fig. 12 STEM image (A) of a single NW and (B) the corresponding panchromatic CL intensity image. The *white arrows* highlight three SFs. The *dashed arrow* indicates the direction of the CL linescan showed in (C). (D) STEM image and (E) the contour plot of CL–STEM linescan along the InGaN QW. Defects and the direction of the CL linescan are indicated in (D) by *black* and *red arrows*, respectively. (F) Cross-sectional bright-field STEM image of the core–shell NW and the corresponding panchromatic CL intensity image taken at 16 K. The MOVPE core and HVPE shell are highlighted by *red* and *green dashed boxes*, respectively. (G) Vertical (*top*) and horizontal (*bottom*) CL linescans across the core–shell NW with the corresponding schematic drawing indicating the scan direction (Müller et al., 2016). *Panels (A)–(C) are reprinted (adapted) with permission from Urban, A., Müller, M., Karbaum, C., Schmidt, G., Veit, P., Malindretos, J., Bertram, F., Christen, J., Rizzi, A., 2015. Optical emission of individual GaN nanocolumns analyzed with high spatial resolution. Nano Lett. 15, 5105–5109. https://doi.org/10.1021/acs.nanolett.5b01278. Copyright (2015) American Chemical Society. Panels (D) and (E) are under Copyright 2016 The Japan Society of Applied Physics. Mohajerani, M.S., Müller, M., Hartmann, J., Zhou, H., Wehmann, H.-H., Veit, P., Bertram, F., Christen, J., Waag, A., 2016. Direct correlations of structural and optical properties of three-dimensional GaN/InGaN core/shell micro-light emitting diodes. Jpn. J. Appl. Phys. 55, 05FJ09. https://doi.org/10.7567/JJAP.55.05FJ09. Panels (F) and (G) are under Copyright 2016 The Japan Society of Applied Physics. Müller, M., Schmidt, G., Metzner, S., Veit, P., Bertram, F., Krylyuk, S., Debnath, R., Ha, J.-Y., Wen, B., Blanchard, P., Motayed, A., King, M.R., Davydov, A.V., Christen, J., 2016. Structural and optical nanoscale analysis of GaN core–shell microrod arrays fabricated by combined top-down and bottom-up process on Si(111). Jpn. J. Appl. Phys. 55, 05FF02. https://doi.org/10.7567/JJAP.55.05FF02.*

the CL emission intensity as a function of the excitation power. The emission efficiency of single QDiscs was shown to increase significantly at moderate excitation consistently with the saturation of trap-related non-radiative channels and field screening effects. For higher electron beam currents the efficiency was observed to decrease abruptly. In analogy to the well-known droop effect in nitride LEDs, such behavior was ascribed to Auger recombination becoming dominant at high charge carrier densities.

3.2 Probing the Electrical Activity at the Nanoscale

Probing the localization and extension of internal electric field and measuring the minority carrier diffusion length in NW solar cells is important for device design and optimization. Indeed, the collection volume is given by the internal field region enlarged by the volumes from which the minority carriers can diffuse to the field region. If the light is absorbed outside the collection volume, it will not produce any output power. Visualization of electric fields can also be used to analyze doping at a nanometric level as described in Section 3.3.3.

Concerning LEDs, the localization of the active region with respect to the built-in field is important to design the band structure. Assessment of the electrical field distribution related to the doping homogeneity is a key point to optimize carrier injection in the active region. Presence of electrical inhomogeneities can lead to carrier crowding and loss of efficiency within individual NWs and on a macroscopic level.

3.2.1 Visualization of the p–n Junction Electrical Activity in NW LEDs

Being sensitive to geometrical factors, EBIC maps can give different information on core/shell and axial structures. In core/shell NWs, electrical junction is not directly exposed to the electron beam and can be probed in projection. Nevertheless, in this configuration EBIC mapping can be used to gather quantitative information on the junction homogeneity along the NW axis and on the resistivity of the core and of the shell. This topic will be addressed in Section 3.3.2.

A way to visualize electrical activity of core/shell junction is to cut single NWs in order to open the inner part and to expose it to the electron beam. Cleavage of single NWs is possible if the diameter is sufficiently large and the NWs can maintain their connection with the substrate. Otherwise, FIB cutting can be used.

An example is given by Zhang et al. (2015b), who performed EBIC mapping on core/shell InGaN/GaN NW LEDs. These devices were designed to emit at 460 nm but presented an undesired green emission peak in EL spectra originating from the In-rich region in the top part of the m-plane InGaN QW as illustrated in Fig. 13A. The parasitic peak could be suppressed by inhibiting the electrical activity of the top pyramidal part of the NWs by removing the ITO contact and treating the top with a CHF_3 plasma. EBIC microscopy on cleaved NWs was used to visualize the "electrical deactivation" of the top part of the NW LEDs. Fig. 13B and C show a standard LED SEM and EBIC map, while Fig. 13D and E demonstrate a strong reduction of the EBIC signal in the top part after ITO etching and plasma treatment of the p-GaN top part.

The same group used EBIC mapping to analyze the homogeneity of the electric field along the NWs axis (Tchernycheva et al., 2015). In addition, observation and analysis of a transitory EBIC signal showed the presence of charged traps at the interface between the axially grown n-GaN NW part and the radially grown n-GaN underlayer, which lead to an inhomogeneous injection in the NW array (Tchernycheva et al., 2015).

Another application of EBIC mapping on cleaved GaN core/shell p–n junctions was performed by Tchoulfian et al. (2014). Fig. 13F and G show SEM and EBIC images of the NW array demonstrating the activity of the core/shell p–n junction. In particular, the authors evidenced the electron beam exposure effect on the activation of Mg dopants in the GaN shell depending on beam parameters. Fig. 13I shows a close-up view of a single NW, for which no annealing step to activate Mg dopants was performed. The top part was exposed to the electron beam, and the bottom part was not activated by the electron beam. The increase of the p-doping concentration after Mg activation is clearly seen from the reduction of the EBIC signal extension in the top part of the wire. In addition, in this work SEVIC technique was used to localize the position of the p–n junction with a resolution below 10 nm. Combination of EBIC and SEVIC for doping estimation across the p–n junction is discussed in Section 3.3.2.

3.2.2 Visualization of the p–n Junction Electrical Activity in NW Solar Cells

EBIC analysis of axial p–n junction NWs was performed by Aberg et al. (2016) who fabricated a GaAs NWs-based solar cell reaching 15.3% efficiency. In addition to the I–V characteristics and EQE measurements, carrier separation in dispersed single NWs is visualized by EBIC. Fig. 14A shows the SEM and EBIC images, and an EBIC axial profile for a NW similar to the wires studied in Aberg et al. (2016). Analysis of the profile (Fig. 14B) allowed to localize the junction position. The doping sequence

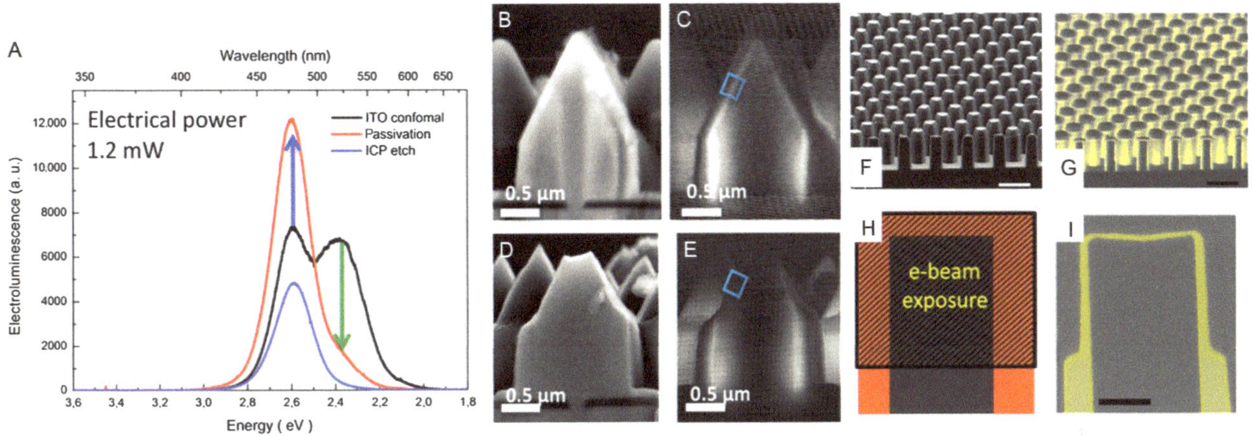

Fig. 13 (A) Electroluminescence spectra for conformally contacted NW LED, for ITO contact removed from the top and for ICP-etched top part. (B)–(E) SEM images of a cleaved InGaN/GaN NW LED and the corresponding EBIC map. (B) and (C) A single NW with conformal ITO with two EL emission peaks; (D) and (E) A single NW with ITO etched from the semipolar planes showing a single EL emission peak. The *blue rectangles* in (C) and (E) indicate the different activity in semipolar planes. (F) SEM image of cleaved InGaN/GaN NWs array and (G) corresponding EBIC map. (H) Schematic of the area exposed by electron beam corresponding to (I) EBIC map where Mg dopants activation is observed by the difference in the depletion region width. *Panels (A)–(E) are reproduced with permission of IOP Publishing in the format Book via Copyright Clearance Center. Zhang, H., Jacopin, G., Neplokh, V., Largeau, L., Julien, F.H., Kryliouk, O., Maria Tchernycheva, 2015. Color control of nanowire InGaN/GaN light emitting diodes by post-growth treatment. Nanotechnology 26, 465203. https://doi.org/10.1088/0957-4484/26/46/465203. Panels (F)–(I) are reprinted (adapted) with permission from Tchoulfian, P., Donatini, F., Levy, F., Dussaigne, A., Ferret, P., Pernot, J., 2014. Direct imaging of p–n junction in core–shell GaN wires. Nano Lett. 14, 3491–3498. https://doi.org/10.1021/nl5010493. Copyright (2014) American Chemical Society.*

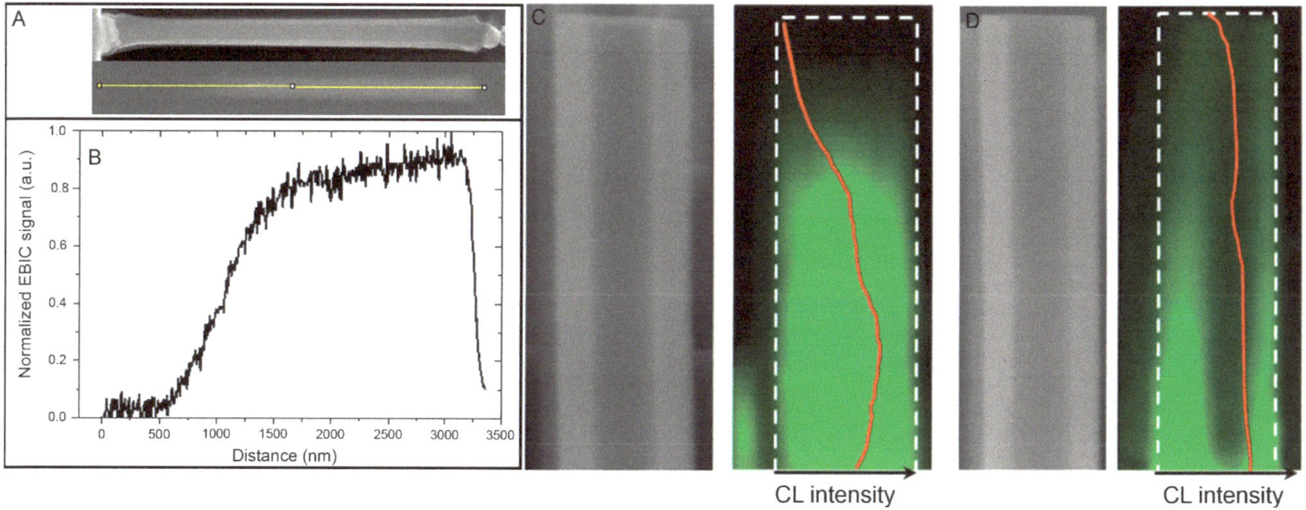

Fig. 14 (A) SEM images of a single GaAs NW and the corresponding EBIC map. *Line* indicates the position corresponding to the EBIC profile in (B). (B) Normalized EBIC profile as a function of the distance from the bottom to the top of the NW. SEM images and normalized CL mapping at wavelength of 865 nm for nanowires with 400 nm deep junction (C) and 100 nm deep junction (D). *Red curves* are relative CL intensity profiles along the NW axis. *Panels (A) and (B) unpublished figures, courtsey of I. Aberg. Panels (C) and (D) are reprinted (adapted) with permission from Yao, M., Huang, N., Cong, S., Chi, C.-Y., Seyedi, M.A., Lin, Y.-T., Cao, Y., Povinelli, M.L., Dapkus, P.D., Zhou, C., 2014. GaAs nanowire array solar cells with axial p-i-n junctions. Nano Lett. 14, 3293–3303. https://doi.org/10.1021/nl500704r. Copyright (2014) American Chemical Society.*

was identified to be $p/p^-/n^+$, and the maximum of the EBIC signal is peaked at the p^-/n^+ interface close to the NW top. This design with a very short n^+ segment is favorable to reduce the losses in the top heavily doped segment.

CL mapping can also be used to analyze the p–n junction in single NWs. Due to carrier separation within the depletion region, radiative recombination rate decreases in the space charge region, which can be evidenced by spatially resolved CL maps. However, CL intensity may be reduced not only in the field region, but in some cases in highly doped parts due to unintentional defects.

In Yao et al. (2014)), CL maps of axial n–i–p GaAs NWs were analyzed. CL intensity maps and profiles (displayed in Fig. 14) are shown to continuously decrease from the junction position to the top region. Signal reduction is due to carrier separation in the junction and to a strong non-radiative recombination in heavily doped p top region. In this case, shallow junction design (Fig. 14D) had smaller losses and thus a better short-circuit current than a deep junction (Fig. 14C).

3.2.3 Electrical Homogeneity in NW Array Solar Cells

Electric fields can also be visualized in a fully processed operational device. For instance, Nakai et al. (2015) used cross-sectional EBIC mapping of an InGaAs NW-based solar cell to assess the photovoltaic activity at the nano-scale. Fig. 15 shows the cross-sectional view of a processed nanostructured device overlapped with the EBIC map (a) and EBIC profile along the axis (b). In this way, electrical activity of single NWs was analyzed, indicating that current is collected from the whole structure. Moreover, comparing the activity among different nanostructures it was possible to gather information on NW-to-NW homogeneity. The analysis allowed to improve conversion efficiency from 4.66% to 7.14%.

In a similar way, Togonal et al. (2016) investigated a Si NW array on top of a heterojunction with intrinsic thin layer solar cell with 12.4% conversion efficiency. Fig. 15C shows EBIC maps for the whole array in cross section and top view configurations. Extension of the region where the current was collected was measured from the cross-sectional map, showing that the NWs generated signal all over their length. However, a certain number of passive NWs (referred to as "dead wires") was observed. By means of top view EBIC maps a statistical analysis of the percentage and distribution of active and inactive NWs was performed and the device failure was correlated with the processing steps.

An example of EBIC mapping to detect failure in a solar cell was reported by Tchernycheva et al. (2012). The authors analyzed the current generation in an array of Au-catalyzed MOVPE grown GaAsP NWs using LBIC

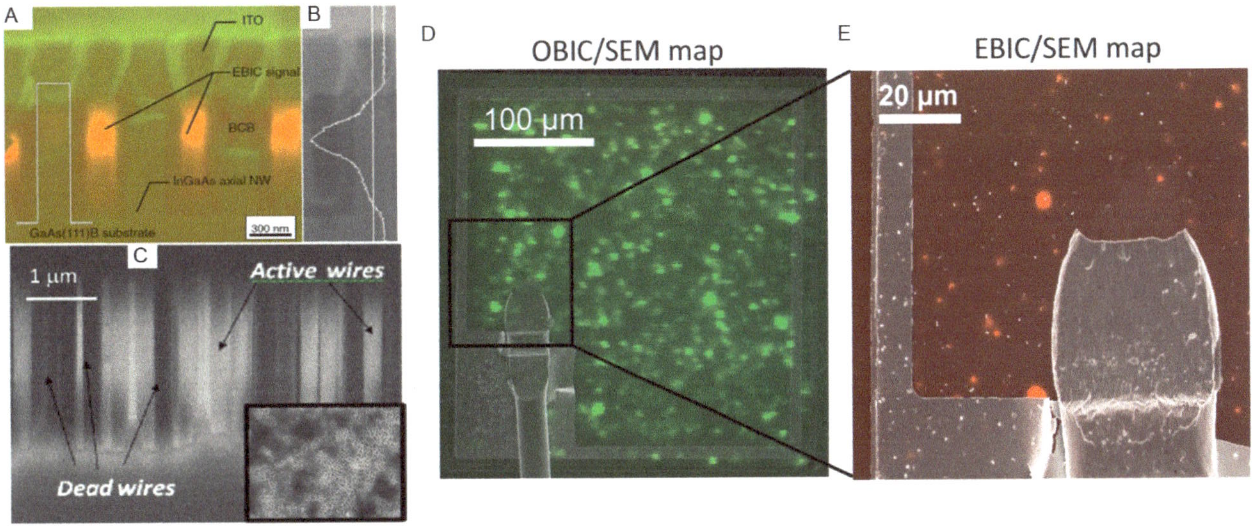

Fig. 15 (A) Overlapped SEM image/EBIC map of InGaAs NWs and (B) the corresponding EBIC profile along the axis. (C) EBIC map acquired on Si NWs indicating activity of the junction and the presence of passive NWs in the array. *Inset* shows the EBIC map of the same NW array in top view configuration. (D) Optical beam-induced current map (*green*) superimposed with an SEM image of a GaAsP NW solar cell. (E) EBIC map (*red*) superimposed with an SEM image of the corresponding region (corresponding to the *black rectangle* in (D)). *Panels (A) and (B) are under Copyright 2016 The Japan Society of Applied Physics. Nakai, E., Chen, M., Yoshimura, M., Tomioka, K., Fukui, T., 2015. InGaAs axial-junction nanowire-array solar cells. Jpn. J. Appl. Phys. 54, 015201. https://doi.org/10.7567/JJAP.54.015201. Panel (C) reprinted with permission from Togonal, A.S., Foldyna, M., Chen, W., Wang, J.X., Neplokh, V., Tchernycheva, M., Nassar, J., Roca i Cabarrocas, P., Rusli, 2016. Core–shell heterojunction solar cells based on disordered silicon nanowire arrays. J. Phys. Chem. C 120, 2962–2972. https://doi.org/10.1021/acs.jpcc.5b09618. Copyright (2016) American Chemical Society. Panel (E) unpublished figures by M. Tchernycheva.*

(Fig. 15D) and EBIC (Fig. 15E) mapping. They found that induced current strongly varied from wire to wire and that only 1 over 6 contacted NWs generated current. No morphological singularities were observed in passive NWs which could explain the absence of generation. The authors attributed the detected electrical failure to the interface between GaAsP NWs and the Si substrate. Indeed, a low-resistance connection between the substrate and the NW is crucial for an efficient charge collection.

3.2.4 Electrical Homogeneity in NW Array LEDs

Fluctuations of the In composition and of the electrical parameters from wire to wire lead to a non-uniform current distribution in the NW array. This effect was evidenced by combining EBIC, CL, and electroluminescence mapping by Limbach et al. (2012). The authors analyzed axial NW LEDs based on self-assembled MBE grown NWs. By imaging individual NWs, a strong shift of the wavelength was observed. It was found that only 1% of NWs produced electroluminescence, whereas a much larger amount of NWs yielded a CL signal. CL and EBIC maps were recorded on a cleaved edge of the LED. Higher CL homogeneity compared to electroluminescence showed that the limiting factor was not the fluctuation of the IQE of individual wires, but the electrical injection. EBIC maps revealed that all NWs were correctly contacted. Therefore, the absence of electroluminescence in 99% of NWs was attributed to the differences in the individual currents in the NWs connected in parallel, which are due to fluctuations of individual series resistances and turn-on voltages. This result clearly evidences that the electrical injection issue is the main factor of losses in NW LEDs: despite a good IQE most of the wires remain inactive under electrical excitation. The wire-to-wire injection homogeneity improves in organized NW arrays; however, electroluminescence intensity fluctuations can still be observed (Schimpke et al., 2016).

3.3 Evaluation of Doping in Nanowires

Doping control is essential for LEDs and PV devices. However, characterization of doping at the nanoscale is a challenging task. Standard TOF secondary ion mass spectrometry (ToF-SIMS) method cannot be directly applied to nanostructures because of their small size. ToF-SIMS has been used to evaluate Te and Be dopant concentration in GaAs/GaP NW ensembles embedded in a Cyclotene mask (Chia et al., 2013). However, the encapsulation material may introduce artifacts, so a careful calibration using top-down etched NWs from thin films with a known doping profile is required (Chia et al., 2015). ToF-SIMS can be performed on single microwires as it was demonstrated in Koester et al. (2011) to analyze the InGaN/GaN core/shell quantum wells; however, a planar reference was still necessary for calibrations. Because of these difficulties, many alternative tools including electrical methods, optical methods, and atomic scale techniques are developed to access the NW doping and the homogeneity.

3.3.1 Doping Measurement by Electrical Methods

Determination of doping levels in thin films is traditionally performed with electrical methods such as Hall measurement. It usually requires four contacts, and the layer thickness is known in advance. The charge carrier type (electrons or holes), carrier concentration, and mobility can be determined from the measurement of the Hall voltage, which results from the externally applied magnetic field to an electrical current. In a NW configuration, it is a huge technical challenge to contact a horizontally lying NW on both longitudinal direction and the sidewall facets. An experimental realization is shown in Fig. 16. The Hall measurement on single NWs was reported

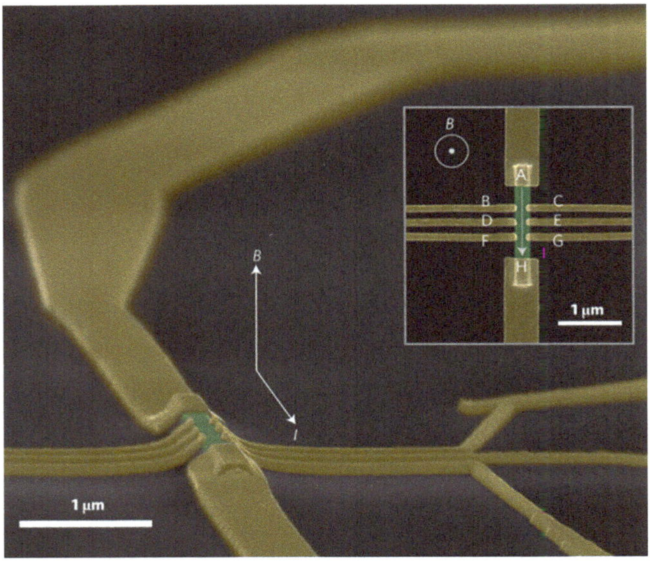

Fig. 16 *Tilted* and *top view* (*inset*) SEM images of Hall effect measurement setup on a single NW. Electrical current circulates along the NW (from contact A–H), and the magnetic field is applied perpendicular to the substrate. Hall voltages are measured across the NW (contacts BC, DE, and FG). *Reprinted by permission from Springer Nature: Springer Nature, Nature Nanotechnology. Storm, K., Halvardsson, F., Heurlin, M., Lindgren, D., Gustafsson, A., Wu, P.M., Monemar, B., Samuelson, L., 2012. Spatially resolved Hall effect measurement in a single semiconductor nanowire. Nat. Nanotechnol. 7, 718. https://doi.org/10.1038/nnano.2012.190. Copyright (2012).*

for core–shell InP NWs, and spatial variations of carrier concentration were analyzed by placing multiple contacts along the wire and by combining Hall measurement with CL spectra (Storm et al., 2012). The precise determination of carrier concentration requires the knowledge of the electrically active area depending on the exact NW geometry and eventually the depletion zone in a complex three-dimensional structure.

The conductivity and field effect measurement are alternative ways to determine the doping concentration. The contact scheme is simpler: source and drain are contacted at the two ends of the NW, and the gate is typically taken through a dielectric layer and a conductive substrate. The conductivity of NWs is obtained by sweeping the current–voltage through the source and drain, which reveals a combined information of carrier mobility and concentration. Field effect measurements further provide the carrier type and mobility by monitoring the source–drain response to the gate voltage, hence determining the doping concentration. Comparison of Hall effect and field effect measurements was reported for InAs (Blömers et al., 2012) and InP (Hultin et al., 2016) NWs. The accuracy of field effect measurement is more questionable because it relies on a good ohmic contact with low contact resistance, and the estimation of the gate capacitance remains approximate and needs a complete three-dimensional modeling (Khanal and Wu, 2007; Wunnicke, 2006). Moreover, high surface state density in many III-arsenide materials results in the Fermi level pinning at the surface, making electrical measurements and interpretations even more difficult.

3.3.2 Doping Measurement by Optical Methods

Contactless optical methods are desirable to provide rapid feedback to NW growth without additional processing steps. Many techniques are proposed to investigate the NW doping such as Raman spectroscopy, terahertz spectroscopy, PL, and CL. Raman scattering usually provides insight on the crystalline quality (broadening of Raman peaks) and lattice strain (shift in the characteristic phonon energy). Moreover, the coupling of longitudinal phonon with plasmon in polar semiconductors gives rise to characteristic Raman scattering modes that have been used to evaluate the doping concentration in n-type GaN NWs (Jeganathan et al., 2009; Mohajerani et al., 2016a,b). Raman spectroscopy was also used to evidence local vibrational modes of Si atoms in GaAs lattice to investigate whether Si dopants occupy As sites (p-type doping) (Ketterer et al., 2010) or Ga sites (n-type doping) (Dimakis et al., 2012). Terahertz spectroscopy was also used to probe the doping concentration of NW ensemble (Boland et al., 2016). Terahertz radiation excites the collective oscillation of free charge carriers, and the complex conductivity spectra can be extracted to estimate the free carrier density. Optical pump terahertz probe spectroscopy was further developed to reveal the dynamics of photogenerated carriers and extract important parameters such as carrier lifetime and surface recombination velocity in semiconductor NWs (Joyce et al., 2012, 2016). However, the experimental setup is quite complicated, and the calculation is not straightforward and depends on several assumptions such as the area density of NW ensemble.

Luminescence-based techniques like PL and CL are another family of characterization tools. PL or CL spectra can be modeled by the generalized Planck's law (Wurfel, 1982). The important parameters are the absorption near bandgap and the electron and hole quasi-Fermi levels. The spectra usually broaden with increasing doping concentration due to enhanced carrier scattering and overlap of the impurity wavefunction. The full-width at half maximum of the emission spectra can be used to monitor the doping level at low concentration. At high doping concentration, the increased carrier–carrier interaction and disordered potential fluctuations of ionized dopants cause the bandgap shrinkage or narrowing. As a consequence, the emission spectra are continuously red-shifted with increasing doping concentration. Fig. 17A shows simulated spectra of p-type GaAs with increased bandgap narrowing at high doping levels. Empirical bandgap narrowing values for many III–V semiconductors are tabulated in the review paper (Jain et al., 1990). For n-type III–V semiconductors (GaAs, InP, etc.), this bandgap narrowing is in competition with the electron occupation in the conduction band. For example, the emission spectra broaden and blue-shift (Burstein–Moss shift) for GaAs with electron concentration above the degenerative threshold of about $5 \times 10^{17}\,cm^{-3}$. Fig. 17B shows simulated spectra of n-type GaAs with different electron concentrations, or equivalently, different electron Fermi levels E_{fc} relative to the conduction band minimum (Chen et al., 2017). Blue-shift of luminescence spectra has already been observed in GaAs (Arab et al., 2016) and InP (Lindgren et al., 2015; Storm et al., 2012) NWs, and quantitative results were obtained by careful analysis of the emission spectra. CL mapping and full fit of the CL spectra were applied to precisely determine the electron concentration in single GaAs NWs (Chen et al., 2017). CL also offers superior spatial resolution compared to μPL, suitable for studying doping homogeneity in NWs. For example, local Mg dopant activity in core–shell GaN nanorods was studied by CL mapping to reveal doping mechanism in a complex nanostructure system (Hortelano et al., 2016).

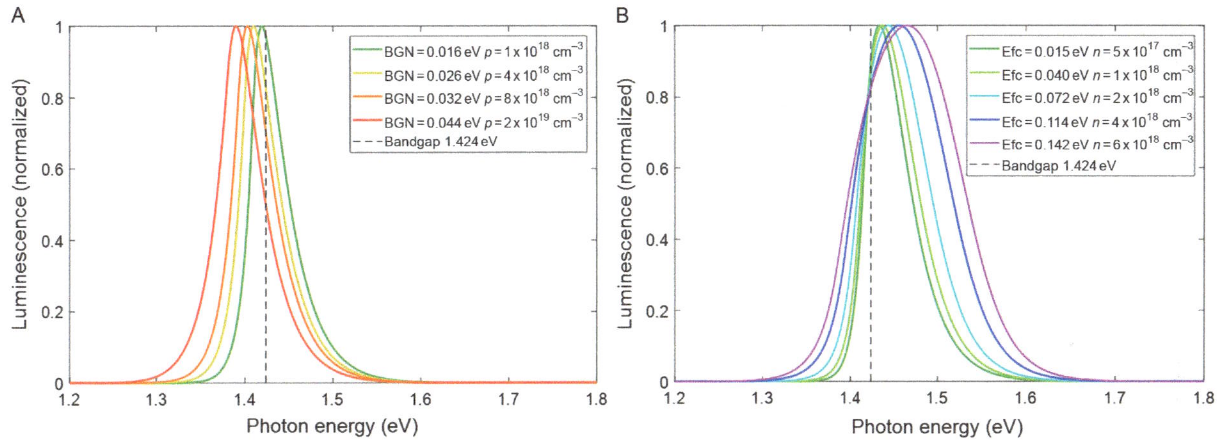

Fig. 17 Calculated luminescence spectra for (A) p-type GaAs (bandgap narrowing effect) and (B) n-type GaAs (bandgap narrowing and degenerate electron filling in the conduction band). *Reprinted (adapted) with permission from Chen, H.-L., Himwas, C., Scaccabarozzi, A., Rale, P., Oehler, F., Lemaître, A., Lombez, L., Guillemoles, J.-F., Tchernycheva, M., Harmand, J.-C., Cattoni, A., Collin, S., 2017. Determination of n-type doping level in single GaAs nanowires by cathodoluminescence. Nano Lett. 17, 6667–6675. https://doi.org/10.1021/acs.nanolett.7b02620. Copyright (2017) American Chemical Society.*

3.3.3 Doping Measurement by EBIC Microscopy

EBIC microscopy can provide information about the doping type and the carrier concentration. For that EBIC profiles in the vicinity of a p–n junction or of a Schottky contact are analyzed. In particular, the depletion region width (W) can be estimated with a nanometric precision. The doping level can be evaluated from the measurement of the space charge region at different external biases using a model of the junction.

In this paragraph, the simplest case is described: a bulk p–n homojunction, with a pure drift regime within the space charge region and a pure diffusive transport elsewhere. This model can be also applied to single NWs (Tchoulfian et al., 2014). The depletion region width in a planar p–n junction writes:

$$W = \left[\frac{2\varepsilon(V_{bi} - U)}{q} \times \left(\frac{1}{N_{eff}}\right)\right]^{(1/2)} \tag{1}$$

where W is the extension of the depletion region, ε is the material permittivity, V_{bi} is the built-in field across the junction, U is the applied external bias, q is the electron charge and N_{eff} is the effective doping, defined as $N_{eff} = N_a N_d/(N_a + N_d)$, where N_d and N_a are the acceptor and donor doping level across the junction. The built-in potential depends on the doping level as.

$$V_{bi} = \frac{kT}{q} \ln \frac{N_d N_a}{n_i^2} \tag{2}$$

where k indicates Boltzmann's constant, T is temperature, and n_i is the intrinsic carrier concentration. It is easy to verify that combination of Eqs. (1) and (2) can provide a set of $[N_a, N_d]$ values but not a unique solution (for instance, when inverting N_a and N_d values, equations do not change). To unambiguously extract the N_d and N_a values, additional relation is required.

To this end, SEVC can be used because it offers the possibility to localize the position of the junction interface. SE contrast in p-doped and n-doped segments is characterized by an opposite dependence on applied bias, thus SE contrast profile across the junction vs external voltage allows to localize the position of the junction. With the information on the position of the p–n interface, the extension of the space charge region in the n-segment (W_N) and in the p-segment (W_P) can be measured. Considering that $W = W_N + W_P$ and the relation:

$$N_a W_P = N_d W_N \tag{3}$$

acceptor and donor doping levels can be calculated from EBIC and SEVC profiles.

Accuracy of the evaluation depends mainly on the precision of the experimental measurements and the applicability of the model for a given system. The error on the space charge region width estimation depends on the lateral extension of the electron beam/sample interaction volume, which is limiting the spatial resolution of EBIC maps, while localization of the p–n interface by SEVC is limited by either the Debye length (related to the spatial influence of electrostatic potential) or the SE microscope resolution.

A comprehensive example is given by Tchoulfian et al. (2014) who applied successfully the above-described model to n-core/p-shell GaN NWs. In this work EBIC profiles across the p–n junction were evaluated at different applied bias: through numerical fitting, depletion region width W was evaluated together with the minority carrier diffusion length in p- and n-doped regions (L_n and L_p). By plotting the extracted parameters as a function of the external voltage, W was found to be proportional to ($U^{1/2}$), while L_n and L_p were independent on the voltage (Fig. 18A). The agreement with the model indicates that the system can well be described by a pure drift regime in the space charge region and a pure diffusive regime outside the junction.

This work has been enriched by a detailed analysis of SE contrast variation, showing that the contrast is maximal at a strong reverse bias coherently with the increase of the built-in field across the junction. Comparison between quantitative EBIC and SEVC profiles allows to localize the p–n interface and estimate the doping levels $N_a = 3{*}10^{18}\,\text{cm}^{-3}$; $N_d = 3.5{*}10^{18}\,\text{cm}^{-3}$ (Fig. 18B).

The described model can be adapted to any kind of region containing internal field, including a Schottky contact and a heterojunction. In all cases, the reliability is affected by a correct modeling of the field region. For metal/semiconductor interfaces, it is evident to localize the maximum of the internal field, which is located at the surface. However, the height of the Schottky barrier is a critical parameter, which is not easy to estimate

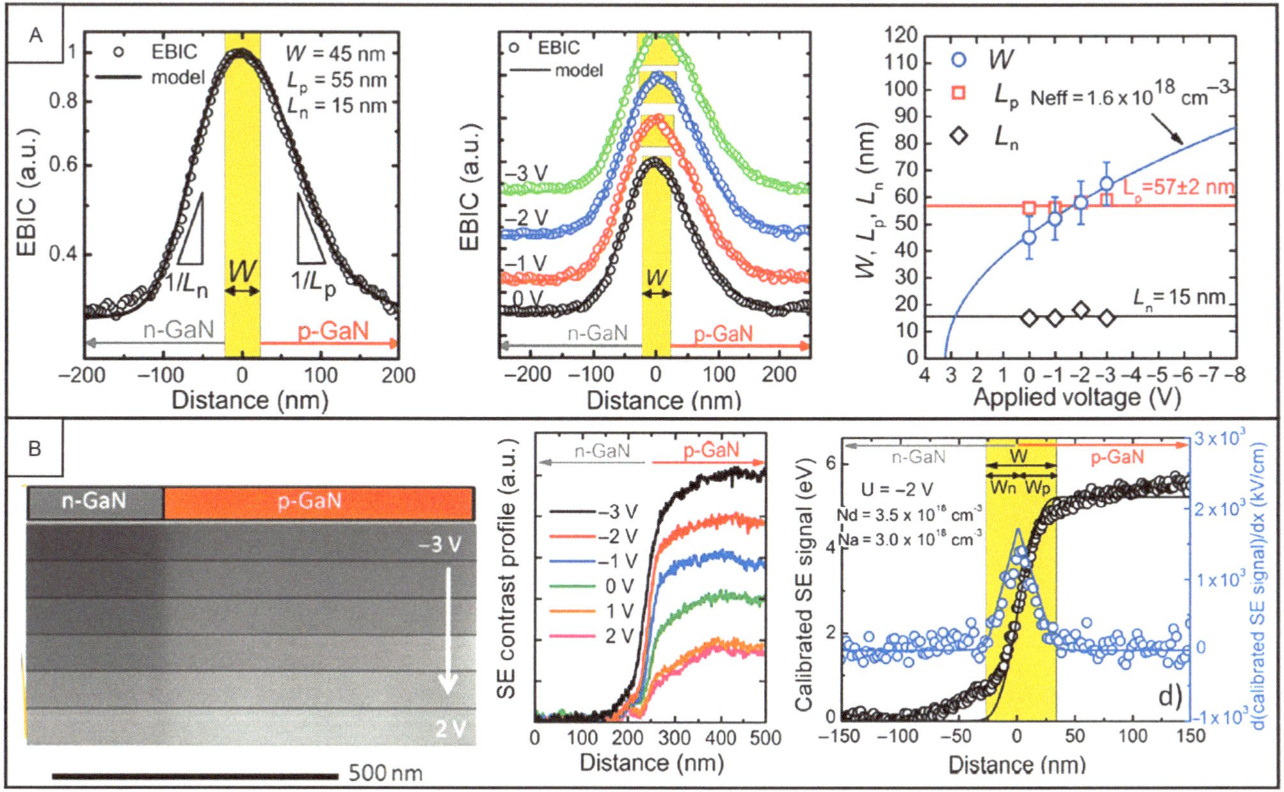

Fig. 18 (A) Evaluation of diffusion length and space charge region width as a function of the voltage by EBIC mapping: the plot on the *right* shows that on both sides of the junction L_{diff} is independent on the bias and W is proportional to $U^{1/2}$ as expected from the model. (B) Results on contrast variation across the junction at different applied bias: the plot on the *right* shows the application of both techniques to estimate p- and n-doping level. *Reprinted (adapted) with permission from Tchoulfian, P., Donatini, F., Levy, F., Dussaigne, A., Ferret, P., Pernot, J., 2014. Direct imaging of p–n junction in core–shell GaN wires. Nano Lett. 14, 3491–3498. https://doi.org/10.1021/nl5010493. Copyright (2014) American Chemical Society.*

since it is influenced by surface states and can be altered by a surface contamination. Nevertheless, EBIC signal arising from Schottky contacts can offer information on the doping type since the band bending on the surface has the opposite sign for a p-doped or an n-doped region. In the case of a heterojunction between two different materials (e.g., between a NW and its substrate), the additional parameter that should be considered in the Eq. (1) in addition to V_{bi} is the band offset.

3.3.4 Measurement of the Impurity Concentration by APT

The above-described techniques probe the carrier concentration resulting from intentional doping. On the contrary, the APT was demonstrated to be a valuable tool for assessing the incorporation of impurities and distribution of doping atoms; however, it cannot provide direct information about the resulting concentration of free carriers.

Recently Koelling et al. (2017) explored the impurities detection capabilities of APT considering several different technologically relevant semiconductor NW systems, namely, P and Au impurities in Ge NWs, Al in GaP, P in InAs, and Si in InP, demonstrating that the detection and mapping of impurities in these systems can be obtained by APT analysis with ppm sensitivity, corresponding to a concentration as low as 1×10^{17} atoms/cm^3. The geometry of the analyzed systems allowed for achieving such a high sensitivity. In fact, NWs with diameter of the order of 150 nm or lower can be field-evaporated for virtually unlimited depths without changing experimental conditions (the tapering angle of typical NWs is equal to zero; Mancini et al., 2014a,b) and volumes containing tens of millions of atoms can be reconstructed. The sensitivity with which APT is able to detect impurities is limited by fundamental counting statistics, implying that ppm sensitivity can be obtained only when analyzing homogeneous large reconstructed volumes. For higher doping concentrations, the distribution of the impurities can be spatially resolved over smaller reconstructed volumes. Examples in the literature can be found in the work of Morkötter et al. (2015), where the authors demonstrated the existence of a Si dopants layer within the AlGaAs shell with concentration of around 6×10^{19} cm^{-3} extending along a few nanometers wide region around 5 nm from the core–shell interface (this spatial resolution was achieved by statistical treatment of the APT data by isosurfaces and proximity histograms). Another example is given in the work by Du et al. (2013), where 1D relative concentration profiles of impurities along the radial direction of GaAs NWs were considered for addressing the incorporation of pollutants (H, O, and C atoms) and Si doping atoms at the surface of the wires.

We note that several other techniques can be applied to measure the doping in nanostructures. For example, off-axis electron holography employs electron interference to measure the electrical potential in a thin specimen, and recently demonstrated the doping inhomogeneity in core–shell GaAs NW (Dastjerdi et al., 2017). A review on NW doping characterizations can be found in Wallentin and Borgström (2011).

3.4 Analyses of Carrier Dynamics at the Nanoscale

Diffusion length is one of the critical parameters in optimizing the performance of solar cells. It is particularly important in axial junction NW solar cells because the carrier collection takes place in the direction of light propagation. Contrary to the radial junction architecture, the minority carriers need to travel over a considerable distance before being collected. The diffusion length is related to the carrier recombination rate. Diffusion length in NWs is largely reduced compared to bulk values due to their high surface-to-volume ratio and thus to surface recombination. High surface recombination leads to a short minority carrier diffusion length and thus to carrier loss. We note that the use of a p–i–n axial junction instead of a p–n junction increases the extension of the internal field region, where carrier transport is governed by drift and not by diffusion. This design was shown to yield a higher conversion efficiency in NW solar cells.

In this section, we describe the analyses of carrier diffusion by EBIC and CL and the estimation of surface recombination by time-resolved techniques.

3.4.1 Measurement of the Diffusion Length in NWs

EBIC and CL maps can give a similar information about the carrier diffusion. In both CL and EBIC, carriers are generated locally within the interaction volume of the electron beam excitation. Carriers then move due to concentration gradient (diffusion) or potential gradient (drift) before they recombine radiatively or non-radiatively. In a region with a built-in field, the maximum of the EBIC signal corresponds to a minimum of the CL intensity if the non-radiative recombination processes are constant.

Depending on the material, the temperature, and the doping level, different types of diffusion can take place. For doped semiconductors at low injection the key parameter is the diffusion of the minority carriers (unipolar process), in intrinsic materials the diffusion is ambipolar implying that the diffusive transport involves electron-hole pairs or, in certain condition, excitons (e.g. at low temperature).

EBIC microscopy can be used to estimate minority carrier diffusion length L_{diff} by fitting EBIC profiles outside the built-in filed region. Donatini et al. (2016) proposed a comparison between the EBIC and CL analyses applied to a single ZnO NW. In this work, the presence of a Schottky barrier arising at the metal contact interface was investigated by CL and EBIC mapping. Comparison between the maps (Fig. 19A) showed clearly the complementarity of the two techniques: the region with a strong built-in field produced a high EBIC signal accompanied by a reduced CL intensity. Moreover, axial profiles at different applied bias were used to extract the minority carrier diffusion length (Fig. 19B), showing an excellent agreement between the results obtained by the two techniques.

For core/shell structures, measurement of the L_{diff} requires cleavage of the nanostructure in order to expose the junction to the electron beam, as discussed in the previous sections. Without cleaving, qualitative information on the homogeneity of the junction can be obtained but the built-in field region extension and the L_{diff} can only be estimated by changing the interaction volume, which is not sufficiently precise (Lavenus et al., 2014). EBIC mapping of cleaved InGaN/GaN NW LEDs was reported by Tchernycheva et al. (2015). The SEM contrast allowed to locate the junction position, and the EBIC map was used to visualize the depletion region and the carrier diffusion. The authors showed that the EBIC profiles at different positions along the wire axis (Fig. 19C) had the same shape (i.e., the depletion region width and the L_{diff} values were the same at the bottom and the top of the NW). This demonstrated that the material parameters and in particular the doping did not vary along the wire axis, which is important for homogeneous injection in a core/shell LED.

Wallentin et al. (2011) applied EBIC to analyze the electrical properties of single n–p InP NW tunnel diodes. The negative differential resistance typical of a tunnel junction was found in the I—V characteristics. EBIC profile analysis showed that depletion region width is small (below 100 nm) in agreement with the high doping of the p and n segments. The electron diffusion length in the p-side (~160 nm) was systematically higher than the L_{diff} for holes in the n-side (~65 nm), even though a large spread of the values was found among different NWs due to difference in the surface recombination.

Minority carrier diffusion length strongly depends on the surface recombination velocity. Chang et al. (2012) investigated the effect of surface

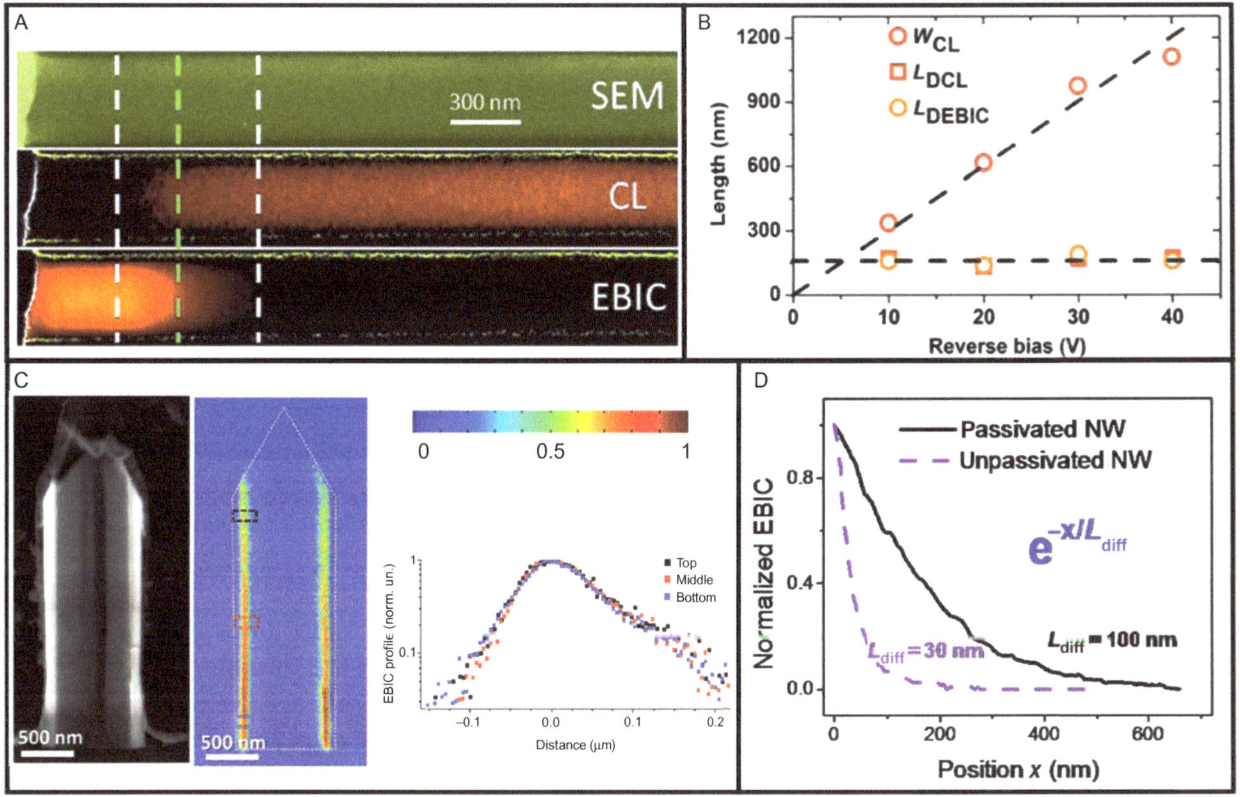

Fig. 19 See figure legend on opposite page.

passivation on the L_{diff} and on the emission properties. GaAs NWs with and without an AlGaAs passivating shell were grown by selective area epitaxy by MOCVD. Single NWs were contacted by electron beam lithography to create a Schottky contact. The comparison of EBIC profiles for passivated and non-passivated NWs (Fig. 19D) demonstrated a sixfold increase of the L_{diff} with passivation (from 30 to 180 nm). TRPL measurements of the photogenerated carrier lifetime confirmed the passivating effect of the shell.

Gutsche et al. (2012) used the L_{diff} obtained from EBIC maps to quantitatively estimate the surface recombination rate in unpassivated GaAs NWs. In bulk semiconductors, diffusion length is related to carrier recombination rate (and thus to carrier lifetime) as.

$$L_i = \sqrt{D_i \tau_i} \tag{4}$$

where L is the diffusion length, D is the diffusion coefficient, τ is the carrier lifetime, and the index i indicates the diffusing carriers. In order to take into account surface recombination, which has a strong influence on carrier lifetime in high aspect ratio structure, the authors modify this model considering an effective minority carrier diffusion length.

Fig. 19 (A) SEM, CL, and EBIC maps for a ZnO NW: on the *left end*, a metal contact creates a Schottky barrier producing a strong EBIC current and reducing the CL intensity. (B) Space charge region width and minority carrier diffusion length vs applied voltage extracted from CL and EBIC profiles fitting. (C) From the *left* to the *right*: SEM image of a cleaved InGaN/GaN NW LED, the corresponding EBIC map (normalized current scale aside) and the EBIC profiles at different positions. (D) Normalized EBIC profile along the axis of a GaAs NW with and without AlGaAs passivation shell. *Panels (A) and (B) are reprinted (adapted) with permission from Donatini, F., de Luna Bugallo, A., Tchoulfian, P., Chicot, G., Sartel, C., Sallet, V., Pernot, J., 2016. Comparison of Three E-beam techniques for electric field imaging and carrier diffusion length measurement on the same nanowires. Nano Lett. 16, 2938–2944. https://doi.org/10.1021/acs.nanolett.5b04710. Copyright (2016) American Chemical Society. Panel (C) published by The Royal Society of Chemistry. Tchernycheva, M., Neplokh, V., Zhang, H., Lavenus, P., Rigutti, L., Bayle, F., Julien, F.H., Babichev, A., Jacopin, G., Largeau, L., Ciechonski, R., Vescovi, G., Kryliouk, O., 2015. Core–shell InGaN/GaN nanowire light emitting diodes analyzed by electron beam induced current microscopy and cathodoluminescence mapping. Nanoscale 7, 11692–11701. https://doi.org/10.1039/C5NR00623F. Panel (D) reprinted (adapted) with permission from Chang, C.-C., Chi, C.-Y., Yao, M., Huang, N., Chen, C.-C., Theiss, J., Bushmaker, A.W., LaLumondiere, S., Yeh, T.-W., Povinelli, M.L., Zhou, C., Dapkus, P.D., Cronin, S.B., 2012. Electrical and optical characterization of surface passivation in GaAs nanowires. Nano Lett. 12, 4484–4489. https://doi.org/10.1021/nl301391h. Copyright (2012) American Chemical Society.*

$$L_i^* = \sqrt{D_i \tau_i^*} \tag{5}$$

where τ^* is the effective carrier lifetime, which can be expressed as.

$$\frac{1}{\tau_i^*} = \frac{1}{\tau_i} + \frac{4S}{d} \tag{6}$$

where S is the surface recombination rate and d is the diameter of the NW.

The authors then performed a systematic study on unpassivated GaAs NWs with different diameters, containing an axial p–n junction and effective minority carrier diffusion length was measured by EBIC microscopy. Plotting L_i^* vs d, they estimated the surface recombination rate around $S = 4^*10^6$ cm/s for p-doped segment and around $S = 3^*10^5$ cm/s for the n-doped semijunction. In addition, they analyzed the effect of surface passivation using ammonium sulfide and showed that the passivation increased the minority carrier diffusion length. To probe the time stability of the passivation, EBIC experiments were performed 10 h after the passivation treatment and repeated within 4 days. The L_{diff} values were reduced showing that the ammonium sulfide passivation is not stable without encapsulation.

CL was also applied to analyze the diffusion length of electron–hole pairs in NWs (Gustafsson et al., 2010). GaAs NWs with a low-bandgap InGaAs insertion and a GaAs/AlGaAs axial junction were investigated. The material with a lower bandgap, which has a CL signature well separated from other emissions of the wire, is used as a "marker": the emission intensity of the marker is traced as a function of the excitation position and the obtained profile is fitted to extract the carrier diffusion length. For example, diffusion in GaAs NWs was analyzed by inspecting the CL signal from the InGaAs segment or from the carbon acceptors in the substrate. The authors found the low-temperature diffusion length in unpassivated GaAs NWs to be in the 100–200 nm range. The diffusion length in AlGaAs segment of the NW surrounded with a higher bandgap AlGaAs shell was larger (~700–750 nm). Finally, the diffusion length in GaAs capped with AlGaAs shell was found to increase to ~900 nm evidencing the passivating effect of the shell. These analyses were further developed in Bolinsson et al. (2011).

3.4.2 Surface Passivation Analyses by TRPL

Because of a high surface-to-volume ratio in NWs, surface passivation is critical for device operation, especially in materials like arsenides with a high surface recombination velocity. The existing passivation strategies consist of either in situ overgrowth of the NWs with a high bandgap shell material (e.g., AlGaAs on GaAs NWs (Chang et al., 2012; Dagytė et al., 2017; Jiang et al., 2013), InGaP on GaAs NWs (Yao et al., 2014), AlInP on InGaAs NWs (Nakai et al., 2015)), or with an ex situ surface treatment. This can be a wet chemical treatment (e.g., passivation with ammonium sulfide $(NH_4)_2S$ or sodium sulfide (Na_2S) solutions (Skromme et al., 1987), or nitridation with a hydrazine–sulfide solution (Alekseev et al., 2015)) or an ALD deposition of a dielectric layer (e.g., AlN, TiN, Al_2O_3, GaN, and TiO_2 shells on GaAs and InP (Dhaka et al., 2016) or POx/Al_2O_3 shell on InP (Black et al., 2017)).

The effect of passivation manifests as an increase of luminescence yield and as a longer carrier lifetime, which can be directly probed by time-resolved techniques. An example is given by Chang et al. (2012) who used TRPL measurements to evaluate the effect of AlGaAs passivation on GaAs NWs, showing a minority carrier lifetime (τ) between 0.2 and 1.3 ns (for unpassivated NWs, $\tau < 0.06$ ns). Moreover, measurements showed that high doping levels lead to a decrease of the carrier lifetime ($\tau = 0.1$–0.9 ns) due to increase of impurity scattering (Fig. 20A). Similar values (τ between 0.2 and 1.2 ns) were found by Jiang et al. (2013), who investigated the effect of the shell thickness on the carrier lifetime. Surprisingly, minority carrier lifetime trend at different shell thickness was found to be nonmonotonic, with a maximum value $\tau = 1.2$ ns for 16 nm thick AlGaAs shell (Fig. 20B). In Fig. 20B the experimental values are compared with the probability of carrier tunneling to the surface (in green, P_e) and the τ variation with shell thickness, simulated taking into account P_e.

The high saturation τ value and the sharper increase suggested the presence of defects in the passivation shell which can "leak" carriers to the surface. To investigate the τ decrease for thickness higher than 16 nm, interdiffusion phenomena at the AlGaAs/GaAs interface were taken into account: comparing TRPL measurements performed at different AlGaAs shell growth rate, it was found that τ decreased as a function of shell growth duration (Fig. 20C). The authors suggested to use a high growth rate to obtain highly performing passivation shell: increasing the growth rate by a factor of 5, a value $\tau = 1.9$ ns was found for 16 nm thick shell (corresponding to an increase in lifetime around 60%).

We note that for all the existing passivation strategies the carrier lifetime in GaAs NWs is still far from the state-of-the-art planar GaAs with high-quality surface passivation by AlGaAs or GaInP. Improving the lifetime by optimizing the growth and passivation is mandatory to compete with and outperform thin film devices.

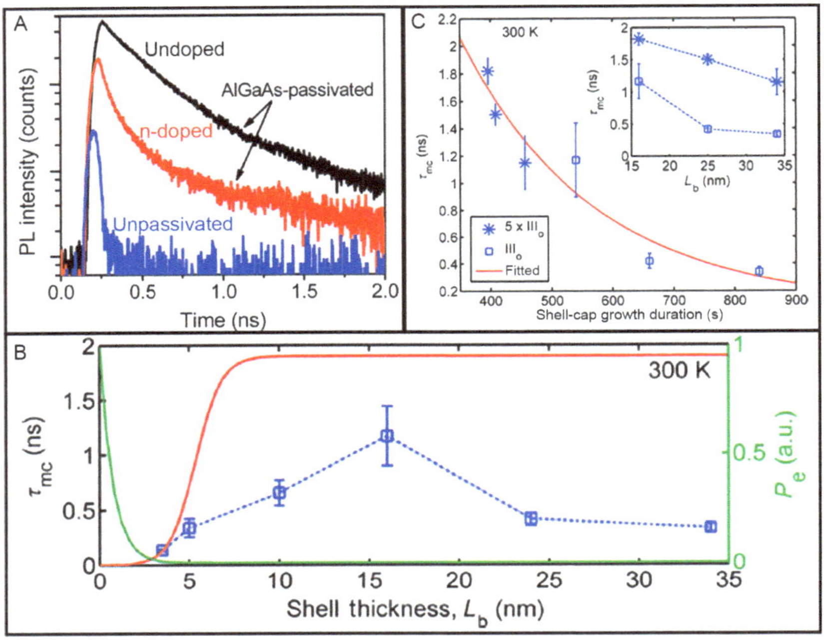

Fig. 20 (A) TRPL measurements on GaAs NWs at different doping and passivation condition. (B) Experimental carrier lifetime at different passivation shell thickness (*blue squares*), compared to simulations of carrier tunneling probability (in *green*) and of the resulting carrier lifetime. (C) Carrier lifetime at different shell growth duration, indicating the influence of interdiffusive phenomena on recombination rate. *Panel (A) reprinted (adapted) with permission from Chang, C.-C., Chi, C.-Y., Yao, M., Huang, N., Chen, C.-C., Theiss, J., Bushmaker, A.W., LaLumondiere, S., Yeh, T.-W., Povinelli, M.L., Zhou, C., Dapkus, P.D., Cronin, S.B., 2012. Electrical and optical characterization of surface passivation in GaAs nanowires. Nano Lett. 12, 4484–4489. https://doi.org/10.1021/nl301391h. Copyright (2012) American Chemical Society. Panels (B) and (C) are reprinted (adapted) with permission from Jiang, N., Gao, Q., Parkinson, P., Wong-Leung, J., Mokkapati, S., Breuer, S., Tan, H.H., Zheng, C.L., Etheridge, J., Jagadish, C., 2013. Enhanced minority carrier lifetimes in GaAs/AlGaAs core–shell nanowires through shell growth optimization. Nano Lett. 13, 5135–5140. https://doi.org/10.1021/nl4023385. Copyright (2013) American Chemical Society.*

3.4.3 TRCL to Assess Exciton Diffusion in Nanowire LEDs

TRCL is a unique tool to probe carrier recombination dynamics with nanoscale spatial resolution. Due to its capability of addressing the carrier recombination dynamics at the typical spatial scales of excitons diffusion lengths in III–V alloys, the technique is well suited for getting an insight into the contributions of different recombination processes and to identify the features which can limit the efficiency of devices. The first TRCL with picosecond temporal resolution was demonstrated in Merano et al. (2005a,b). InGaAs/AlGaAs pyramids were used as an example of nano-objects for the TRCL measurement. The pyramids presented complex structures (quantum dot at the top, quantum wire at the edges, and quantum well in the facets) where carriers can travel and recombine. For example, monochromatic CL map of quantum dot emission shows not only the location of the quantum dot but also any point from which carriers can diffuse to the quantum dot and then recombine. TRCL technique unambiguously reveals the pathway of carrier diffusion by measuring the delay time when exciting different places of the pyramid.

Recently Shahmohammadi et al. (2016) employed TRCL at temperatures between 4 and 240 K for addressing the excitonic diffusion in InGaN/GaN core–shell NW LEDs (a schematic representation of the structure is reported in Fig. 21A). Core/shell NWs have been proposed as promising building blocks for LED devices; however, the practical realization of efficient devices is questioned following the observation of inhomogeneous incorporation of In in the active region (Shahmohammadi et al., 2016). The alloy variations may lead to inhomogeneous carrier injection and consequently to the loss of efficiency. Spatially and spectrally resolved CL mappings of single NWs in Shahmohammadi et al. (2016) indeed confirmed the indium gradient within the m-plane QWs, yielding a CL blue-shift from the tip to the base of the NW (see CL spectra corresponding to different position along the NW in Fig. 21B). T-dependent TRCL allowed for addressing the thermalization effects on the motion of charge carriers. At low temperature ($T < 40$ K), the excitons were frozen and trapped at the first shallow localization center. In particular, they were unable to move following the energy gradient induced by the varying In content resulting in no spectral shift after the electron pulse excitation. A red-shift of the CL was observed at higher T and ascribed to a thermally activated carrier motion process (see the time-dependent CL intensity of emissions at different energies collected from a single NW at 200 K and the streak images obtained at different temperatures in Fig. 21C and D, respectively). The drift along the energy gradient was excluded as a possible cause of the red-shift. On the contrary, a strong shift at early delays followed by a smaller shift at longer delays was understood as a signature of a hopping process between randomly distributed localization centers. The hopping motion process was modeled taking into account the time delay-dependent volume explored by excitons before recombination, the temperature dependence of their tunneling distance, and the decrease of the carrier density with the time delay. The good

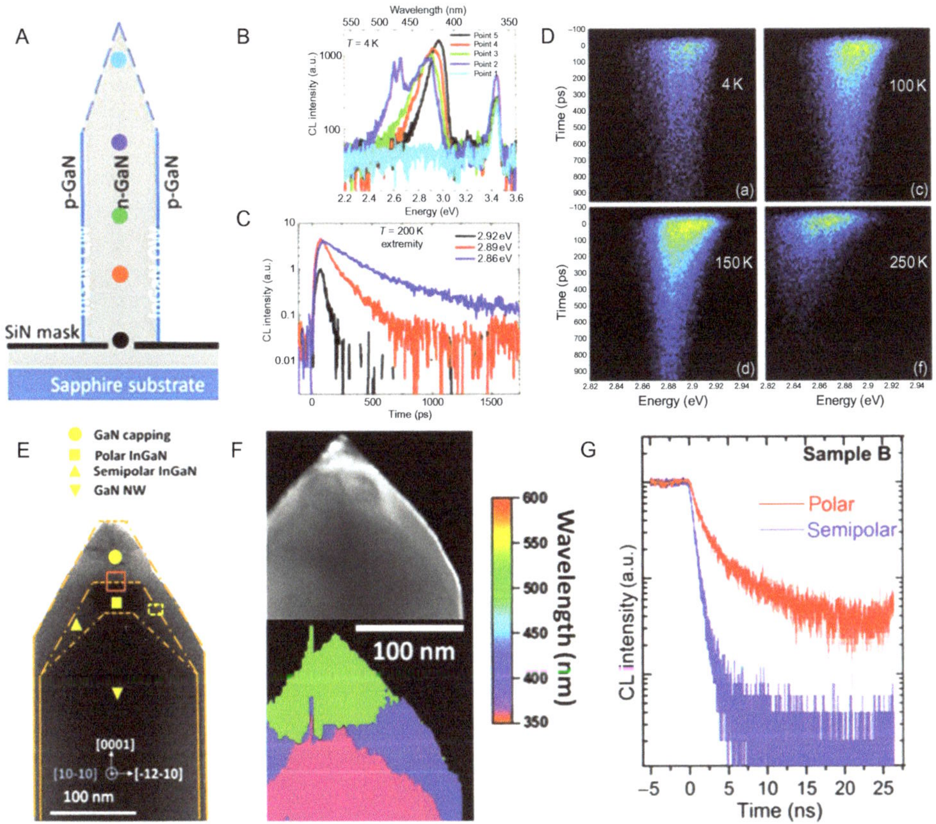

Fig. 21 See figure legend on oposite page.

agreement between TRCL experimental results and simulations confirmed that the excitons dynamic was dominated by a hopping process between localization centers, which in turn prevented any directional drift along the In gradient. This result shows that the In gradient does not necessarily lead to carrier accumulation in the In-rich region, which is important for LED performance. Moreover, a tentative quantification of the excitons temperature-dependent diffusion length in the system was provided by observing the variation of the CL decay times with respect to the distance of the probe region from the cleaved base of dispersed NWs (the free surface acted as a non-radiative defect).

Another example showing the potential of TRCL as a tool for separately addressing the optical properties and carrier dynamics of different regions of single nanostructures is provided in the recent work of Gačević et al. (2016). In the study, axial InGaN/GaN heterostructures used as the active region in LEDs are analyzed. The authors employed TRCL for disentangling the effects of strain, spontaneous polarization, and concentration gradients on the radiative recombination of carriers within semipolar and polar InGaN QDiscs embedded in GaN NWs. STEM and XEDS allowed for addressing the geometry and composition of the different optically active regions (the STEM image of a single NW is reported in Fig. 21E). STEM–CL was used for assessing the excitonic recombination energy of the InGaN semipolar and polar QDiscs, and of the GaN core (see the CL color coded map

Fig. 21 (A) InGaN/GaN core–shell NW LED structure studied by Shahmohammadi et al. (B) CL spectra acquired from different positions along a single NW. The on-wire position corresponding to each spectrum is indicated by the *circles* in (A). (C) Time-dependent CL intensity of emissions at different energies collected from a single NW at 200 K. (D) Streak images of the TRCL analysis obtained at different temperatures. (E) STEM image of semipolar and polar InGaN QDiscs embedded in GaN NWs studied by Gačević et al. (F) CL color-coded map obtained at 15 K from a single NW. (G) Time-dependent CL intensity of emissions corresponding to polar (*red*) and semipolar (*blue*) QDiscs. *Panels (A)–(D) are reprinted (adapted) with permission from Shahmohammadi, M., Ganière, J.-D., Zhang, H., Ciechonski, R., Vescovi, G., Kryliouk, O., Tchernycheva, M., Jacopin, G., 2016. Excitonic diffusion in InGaN/GaN core–shell nanowires. Nano Lett. 16, 243–249. https://doi.org/10.1021/acs.nanolett.5b03611. Copyright (2016) American Chemical Society. Panels (E)–(G) are reprinted with permission from Gačević, Ž., Vukmirović, N., García-Lepetit, N., Torres-Pardo, A., Müller, M., Metzner, S., Albert, S., Bengoechea-Encabo, A., Bertram, F., Veit, P., Christen, J., González-Calbet, J.M., Calleja, E., 2016. Influence of composition, strain, and electric field anisotropy on different emission colors and recombination dynamics from InGaN nanodisks in pencil-like GaN nanowires. Phys. Rev. B 93, 125436. https://doi.org/10.1103/PhysRevB.93.125436. Copyright (2016) by the American Physical Society, https://doi.org/10.1103/PhysRevB.93.125436.*

obtained at 15 K from a single NW in Fig. 21F). These were shown to emit within the green, blue, and UV range, respectively. Excitation-dependent PL analyses of whole NWs allowed for determining the dependence of the relative intensities of semipolar and polar QDisc emission on the carrier density. The emission energy of the semipolar structure was shown to be unaffected by the variation of carrier density, while the polar QDisc emission was shown to blue-shift with increasing excitation power. TRCL allowed for further studying the radiative recombination dynamics, highlighting radiative decay times a half order of magnitude longer for the polar QDiscs with respect to the semipolar ones (see the time-dependent CL intensity of emissions corresponding to polar (red) and semipolar (blue) QDiscs in Fig. 21). The ensemble of these results was interpreted as the signature of screening and descreening effects of strong electric fields affecting recombination processes within the polar QDiscs. Additional simulations within the framework of the virtual crystal approximation allowed for ascribing the slower recombination to the following contributions reported in order of decreasing importance: composition inhomogeneities along NW growth direction, the spontaneous polarization electric field in the proximity of polar intracolumnar interfaces and the highly anisotropic strain distribution across the nanodisc volume.

4. CONCLUSIONS

In summary, NWs are intensively explored for energy harvesting and lighting applications. NW devices have made a tremendous progress during the last 10 years; however, they have not yet entered the market. Indeed, the complexity of their three-dimensional architecture brings new challenges to the material growth and device processing. Understanding of the processes taking place in individual nano-objects is necessary to boost the device efficiency. Nanoscale characterization techniques give access to the key physical parameters, which are required to improve the design and fabrication. EBIC microscopy and photo- and cathodeluminescence are used to detect failures at different scales starting from individual NWs and up to a macroscopic device level. These techniques provide information on the electrical and optical inhomogeneities in NW solar cells and LEDs. Together with structural characterizations, they give a crucial feedback to guide the device optimization.

ACKNOWLEDGMENTS

The authors acknowledge technical support from Nan Guan and fruitful discussions with Gwénolé Jacopin. This work has been partially financially supported by ANR programmes 'GaNeX' (ANR-11-LABX-2014), 'NanoSaclay' (ANR-10-LABX-0035), HETONAN (ANR-15-CE05-0009), and NANOCELL (ANR-15-CE05-0026), by EU ERC project 'NanoHarvest' (Grant no. 639052) and by H2020 ITN Marie Curie project 'INDEED' (Grant no. 722176).

REFERENCES

Aberg, I., Vescovi, G., Asoli, D., Naseem, U., Gilboy, J.P., Sundvall, C., Dahlgren, A., Svensson, K.E., Anttu, N., Bjork, M.T., Samuelson, L., 2016. A GaAs nanowire array solar cell with 15.3% efficiency at 1 sun. IEEE J. Photovoltaics. 6, 185–190. https://doi.org/10.1109/JPHOTOV.2015.2484967.

Agrawal, R., Bernal, R.A., Isheim, D., Espinosa, H.D., 2011. Characterizing atomic composition and dopant distribution in wide band gap semiconductor nanowires using laser-assisted atom probe tomography. J. Phys. Chem. C 115, 17688–17694. https://doi.org/10.1021/jp2047823.

Ahtapodov, L., Todorovic, J., Olk, P., Mjåland, T., Slåttnes, P., Dheeraj, D.L., van Helvoort, A.T.J., Fimland, B.-O., Weman, H., 2012. A story told by a single nanowire: optical properties of wurtzite GaAs. Nano Lett. 12, 6090–6095. https://doi.org/10.1021/nl3025714.

Alekseev, P.A., Dunaevskiy, M.S., Ulin, V.P., Lvova, T.V., Filatov, D.O., Nezhdanov, A.V., Mashin, A.I., Berkovits, V.L., 2015. Nitride surface passivation of GaAs nanowires: impact on surface state density. Nano Lett. 15, 63–68. https://doi.org/10.1021/nl502909k.

Arab, S., Yao, M., Zhou, C., Daniel Dapkus, P., Cronin, S.B., 2016. Doping concentration dependence of the photoluminescence spectra of n-type GaAs nanowires. Appl. Phys. Lett. 108, 182106. https://doi.org/10.1063/1.4947504.

Bennett, S.E., Smeeton, T.M., Saxey, D.W., Smith, G.D.W., Hooper, S.E., Heffernan, J., Humphreys, C.J., Oliver, R.A., 2012. Atom probe tomography characterisation of a laser diode structure grown by molecular beam epitaxy. J. Appl. Phys. 111, 053508. https://doi.org/10.1063/1.3692569.

Berg, A., Caroff, P., Shahid, N., Lockrey, M.N., Yuan, X., Borgström, M.T., Tan, H.H., Jagadish, C., 2017. Growth and optical properties of InxGa1 − xP nanowires synthesized by selective-area epitaxy. Nano Res. 10, 672–682. https://doi.org/10.1007/s12274-016-1325-1.

Black, L.E., Cavalli, A., Verheijen, M.A., Haverkort, J.E.M., Bakkers, E.P.A.M., Kessels, W.M.M., 2017. Effective surface passivation of InP nanowires by atomic-layer-deposited Al2O3 with POx interlayer. Nano Lett. 17, 6287–6294. https://doi.org/10.1021/acs.nanolett.7b02972.

Blömers, C., Grap, T., Lepsa, M.I., Moers, J., Trellenkamp, S., Grützmacher, D., Lüth, H., Schäpers, T., 2012. Hall effect measurements on InAs nanowires. Appl. Phys. Lett. 101, 152106. https://doi.org/10.1063/1.4759124.

Blumtritt, H., Gleichmann, R., Heydenreich, J., Johansen, H., 1979. Combined scanning (EBIC) and transmission electron microscopic investigations of dislocations in semiconductors. Phys. Status Solidi A 55, 611–620.

Boland, J.L., Casadei, A., Tütüncüoglu, G., Matteini, F., Davies, C.L., Jabeen, F., Joyce, H.J., Herz, L.M., Fontcuberta i Morral, A., Johnston, M.B., 2016. Increased photoconductivity lifetime in GaAs nanowires by controlled n-type and p-type doping. ACS Nano 10, 4219–4227. https://doi.org/10.1021/acsnano.5b07579.

Bolinsson, J., Ek, M., Trägårdh, J., Mergenthaler, K., Jacobsson, D., Pistol, M.-E., Samuelson, L., Gustafsson, A., 2014. GaAs/AlGaAs heterostructure nanowires studied by cathodoluminescence. Nano Res. 7, 473–490. https://doi.org/10.1007/s12274-014-0414-2.

Bolinsson, J., Mergenthaler, K., Samuelson, L., Gustafsson, A., 2011. Diffusion length measurements in axial and radial heterostructured nanowires using cathodoluminescence. In: 15th International Conference on Metalorganic Vapor Phase Epitaxy (ICMOVPE-XV)Journal of Crystal Growth, Vol. 315, pp. 138–142. https://doi.org/10.1016/j.jcrysgro.2010.08.054.

Borgström, M.T., Zwiller, V., Müller, E., Imamoglu, A., 2005. Optically bright quantum dots in single nanowires. Nano Lett. 5, 1439–1443. https://doi.org/10.1021/nl050802y.

Boxberg, F., Søndergaard, N., Xu, H.Q., 2010. Photovoltaics with Piezoelectric core – shell nanowires. Nano Lett. 10, 1108–1112. https://doi.org/10.1021/nl9040934.

Bright, C.I., 2007. Review of Transparent Conductive Oxides (TCO). In: Mattox, D.M., Mattox, V.H. (Eds.), Chapter 7: 50 Years of Vacuum Coating Technology and the Growth of the Society of Vacuum Coaters. Society of Vacuum Coaters.

Bunker, K.L., 2004. Development and Application of Electron Beam Induced Current and Cathodoluminescence Analytical Techniques for Characterization of Gallium Nitride-Based Devices. PhD Thesis, N. C. State Univ.

Cavalcoli, D., Cavallini, A., 1994. Evaluation of diffusion length at different excess carrier concentrations. Mater. Sci. Eng. A 98–100.

Chang, C.-C., Chi, C.-Y., Yao, M., Huang, N., Chen, C.-C., Theiss J., Bushmaker, A.W., LaLumondiere, S., Yeh, T.-W., Povinelli, M.L., Zhou, C., Dapkus, P.D., Cronin, S.B., 2012. Electrical and optical characterization of surface passivation in GaAs nanowires. Nano Lett. 12, 4484–4489. https://doi.org/10.1021/nl301391h.

Chen, G., Sun, G., Ding, Y.J., Prete, P., Miccoli, I., Lovergine, N., Shtrikman, H., Kung, P., Livneh, T., Spanier, J.E., 2013. Direct measurement of band edge discontinuity in individual core-shell nanowires by photocurrent spectroscopy. Nano Lett. 13, 4152–4157. https://doi.org/10.1021/nl401737u.

Chen, H.-L., Himwas, C., Scaccabarozzi, A., Rale, P., Oehler, F., Lemaître, A., Lombez, L., Guillemoles, J.-F., Tchernycheva, M., Harmand, J.-C., Cattoni A., Collin, S., 2017. Determination of n-type doping level in single GaAs nanowires by cathodoluminescence. Nano Lett. 17, 6667–6675. https://doi.org/10.1021/acs.nanolett.7b02620.

Chen, Y.C., Seidman, D.N., 1971. On the atomic resolution of a field ion microscope. Surf. Sci. 26, 61–84. https://doi.org/10.1016/0039-6028(71)90114-2.

Chia, A.C.E., Boulanger, J.P., LaPierre, R.R., 2013. Unlocking doping and compositional profiles of nanowire ensembles using SIMS. Nanotechnology 24, 045701. https://doi.org/10.1088/0957-4484/24/4/045701.

Chia, A.C.E., Dhindsa, N., Boulanger, J.P., Wood, B.A., Saini, S.S., LaPierre, R.R., 2015. Nanowire dopant measurement using secondary ion mass spectrometry. J. Appl. Phys. 118, 114306. https://doi.org/10.1063/1.4931148.

Chichibu, S.F., Uedono, A., Onuma, T., Haskell, B.A., Chakraborty, A., Koyama, T., Fini, P.T., Keller, S., DenBaars, S.P., Speck, J.S., Mishra, U.K., Nakamura, S., Yamaguchi, S., Kamiyama, S., Amano, H., Akasaki, I., Han, J., Sota, T., 2006. Origin of defect-insensitive emission probability in In-containing (Al,In,Ga)N alloy semiconductors. Nat. Mater. 5, 810–816. https://doi.org/10.1038/nmat1726.

Chung, K., Beak, H., Tchoe, Y., Oh, H., Yoo, H., Kim, M., Yi, G.-C., 2014. Growth and characterizations of GaN micro-rods on graphene films for flexible light emitting diodes. APL Mater. 2, 092512. https://doi.org/10.1063/1.4894780.

Coutts, T.J., Mason, T.O., Perkins, J.D., Ginley, D.S., 1999. Transparent conducting oxides: status and opportunities in basic research. Proc. Electrochem Soc 99, 274–288.

Dagytė, V., Barrigón, E., Zhang, W., Zeng, X., Heurlin, M., Otnes, G., Anttu, N., Borgström, M.T., 2017. Time-resolved photoluminescence characterization of GaAs nanowire arrays on native substrate. Nanotechnology 28, 505706. https://doi.org/10.1088/1361-6528/aa974b.

Dai, X., Messanvi, A., Zhang, H., Durand, C., Eymery, J., Bougerol, C., Julien, F.H., Tchernycheva, M., 2015. Flexible light-emitting diodes based on vertical nitride nanowires. Nano Lett. 15, 6958–6964. https://doi.org/10.1021/acs.nanolett.5b02900.

Dastjerdi, M.H.T., Fiordaliso, E.M., Leshchenko, E.D., Akhtari-Zavareh, A., Kasama, T., Aagesen, M., Dubrovskii, V.G., LaPierre, R.R., 2017. Three-fold symmetric doping mechanism in GaAs nanowires. Nano Lett. 17, 5875–5882. https://doi.org/10.1021/acs.nanolett.7b00794.

Deshpande, S., Bhattacharya, I., Malheiros-Silveira, G., Wei Ng, K., Schuster, F., Cook, K., Chang-Hasnian, C., 2017. Ultra-compact position-controlled InP nanopillar LEDs on silicon with brigth electroluminescence at communication wavelength. ACS Photonics 4, 695–702.

Detchprohm, T., Zhu, M., Li, Y., Zhao, L., You, S., Wetzel, C., Preble, E.A., Paskova, T., Hanser, D., 2010. Wavelength-stable cyan and green light emitting diodes on nonpolar m-plane GaN bulk substrates. Appl. Phys. Lett. 96, 051101. https://doi.org/10.1063/1.3299257.

Devaraj, A., Perea, D.E., Liu, J., Gordon, L.M., Prosa, T.J., Parikh, P., Diercks, D.R., Meher, S., Kolli, R.P., Meng, Y.S., Thevuthasan, S., 2018. Three-dimensional nanoscale characterisation of materials by atom probe tomography. Int. Mater. Rev. 63, 68–101. https://doi.org/10.1080/09506608.2016.1270728.

Dhaka, V., Perros, A., Naureen, S., Shahid, N., Jiang, H., Kakko, J.-P., Haggren, T., Kauppinen, E., Srinivasan, A., Lipsanen, H., 2016. Protective capping and surface passivation of III-V nanowires by atomic layer deposition. AIP Adv. 6, 015016. https://doi.org/10.1063/1.4941063.

Dimakis, E., Jahn, U., Ramsteiner, M., Tahraoui, A., Grandal, J., Kong, X., Marquardt, O., Trampert, A., Riechert, H., Geelhaar, L., 2014. Coaxial multishell (In,Ga)As/GaAs nanowires for near-infrared emission on Si substrates. Nano Lett. 14, 2604–2609. https://doi.org/10.1021/nl500428v.

Dimakis, E., Ramsteiner, M., Tahraoui, A., Riechert, H., Geelhaar, L., 2012. Shell-doping of GaAs nanowires with Si for n-type conductivity. Nano Res. 5, 796–804. https://doi.org/10.1007/s12274-012-0263-9.

Dimroth, F., 2014. New World Record for Solar Cell Efficiency at 46%—Fraunhofer ISE. WWW Document, Fraunhofer Institute for Solar Energy Systems ISE. URL, https://www.ise.fraunhofer.de/en/press-media/press-releases/2014/new-world-record-for-solar-cell-efficiency-at-46-percent.html. Accessed 15 December 2017.

Dimroth, F., Grave, M., Beutel, P., Fiedeler, U., Karcher, C., Tibbits, T.N.D., Oliva, E., Siefer, G., Schachtner, M., Wekkeli, A., Bett, A., Krause, R., Piccin, M., Blanc, N., Drazek, C., Guiot, E., Ghyselen, B., Salvetat, T., Tauzin, A., Schwarzburg, K., 2014. Wafer bonded four-junction GaInP/GaAs//GaInAsP/GaInAs concentrator solar cells with 44.7% efficiency. Prog. Photovolt. Res. Appl. 22, 277–282. https://doi.org/10.1002/pip.2475.

Dimroth, F., Tibbits, T.N.D., Niemeyer, M., Predan, F., Beutel, P., Karcher, C., Oliva, E., Siefer, G., Lackner, D., Fuß-Kailuweit, P., Bett, A.W., Krause, R., Drazek, C., Guiot, E., Wasselin, J., Tauzin, A., Signamarcheix, T., 2016. Four-junction wafer-bonded concentrator solar cells. IEEE J. Photovoltaics. 6, 343–349. https://doi.org/10.1109/JPHOTOV.2015.2501729.

Ding, G.-H., Chan, C.T., Zhang, Z.Q., Sheng, P., 2005. Resonance-enhanced optical annealing of silicon nanowires. Phys. Rev. B 71, 205302. https://doi.org/10.1103/PhysRevB.71.205302.

Donatini, F., de Luna Bugallo, A., Tchoulfian, P., Chicot, G., Sartel, C., Sallet, V., Pernot, J., 2016. Comparison of three E-beam techniques for electric field imaging and carrier diffusion length measurement on the same nanowires. Nano Lett. 16, 2938–2944. https://doi.org/10.1021/acs.nanolett.5b04710.

Donolato, C., 1981. An analytical model of SEM and STEM charge collection images of dislocations in thin semiconductor layers: I. Minority carrier generation, diffusion, and collection. Phys. Status Solidi A 65, 649–658.

Du, S., Burgess, T., Gault, B., Gao, Q., Bao, P., Li, L., Cui, X., Kong Yeoh, W., Liu, H., Yao, L., Ceguerra, A.V., Hoe Tan, H., Jagadish, C., Ringer, S.P., Zheng, R., 2013. Quantitative dopant distributions in GaAs nanowires using atom probe tomography. Ultramicroscopy 132, 186–192. https://doi.org/10.1016/j.ultramic.2013.02.012.

El-Gomati, M., Zaggout, F., Jayacody, H., Tear, S., Wilson, K., 2005. Why is it possible to detect doped regions of semiconductors in low voltage SEM: a review and update. Surf. Interface Anal. 37, 901–911. https://doi.org/10.1002/sia.2108.

Fu, X., Jacopin, G., Shahmohammadi, M., Liu, R., Benameur, M., Ganière, J.-D., Feng, J., Guo, W., Liao, Z.-M., Deveaud, B., Yu, D., 2014. Exciton drift in semiconductors under uniform strain gradients: application to bent ZnO microwires. ACS Nano 8, 3412–3420. https://doi.org/10.1021/nn4062353.

Gačević, Ž., Vukmirović, N., García-Lepetit, N., Torres-Pardo, A., Müller, M., Metzner, S., Albert, S., Bengoechea-Encabo, A., Bertram, F., Veit, P., Christen, J., González-Calbet, J.M., Calleja, E., 2016. Influence of composition, strain, and electric field anisotropy on different emission colors and recombination dynamics from InGaN nanodisks in pencil-like GaN nanowires. Phys. Rev. B 93, 125436. https://doi.org/10.1103/PhysRevB.93.125436.

Gault, B., de Geuser, F., Stephenson, L.T., Moody, M.P., Muddle, B.C., Ringer, S.P., 2008. Estimation of the reconstruction parameters for atom probe tomography. Microsc. Microanal. 14, 296–305. https://doi.org/10.1017/S1431927608080690.

Gault, B., Müller, M., Fontaine, A.L., Moody, M.P., Shariq, A., Cerezo, A., Ringer, S.P., Smith, G.D.W., 2010. Influence of surface migration on the spatial resolution of pulsed laser atom probe tomography. J. Appl. Phys. 108, 044904. https://doi.org/10.1063/1.3462399.

Ghahfarokhi, O.M., Anttu, N., Samuelson, L., Åberg, I., 2016. Performance of GaAs nanowire array solar cells for varying incidence angles. IEEE J. Photovoltaics. 6, 1502–1508. https://doi.org/10.1109/JPHOTOV.2016.2604564.

Ghosh, R., Giri, P.K., 2017. Silicon nanowire heterostructures for advanced energy and environmental applications: a review. Nanotechnology 28, 012001. https://doi.org/10.1088/0957-4484/28/1/012001.

Glas, F., 2006. Critical dimensions for the plastic relaxation of strained axial heterostructures in free-standing nanowires. Phys. Rev. B 74, 121302. https://doi.org/10.1103/PhysRevB.74.121302.

Green, M.A., Emery, K., Hishikawa, Y., Warta, W., Dunlop, E.D., Levi, D.H., Ho-Baillie, A.W.Y., 2017. Solar cell efficiency tables (version 49): Solar cell efficiency tables (version 49). Prog. Photovolt. Res. Appl. 25, 3–13. https://doi.org/10.1002/pip.2855.

Griffiths, J.T., Oehler, F., Tang, F., Zhang, S., Fu, W.Y., Zhu, T., Findlay, S.D., Zheng, C., Etheridge, J., Martin, T.L., Bagot, P.A.J., Moody, M.P., Sutherland, D., Dawson, P., Kappers, M.J., Humphreys, C.J., Oliver, R.A., 2016. The microstructure of non-polar a-plane (1120) InGaN quantum wells. J. Appl. Phys. 119, 175703. https://doi.org/10.1063/1.4948299.

Griffiths, J.T., Zhang, S., Rouet-Leduc, B., Fu, W.Y., Bao, A., Zhu, D., Wallis, D.J., Howkins, A., Boyd, I., Stowe, D., Kappers, M.J., Humphreys, C.J., Oliver, R.A., 2015. Nanocathodoluminescence reveals mitigation of the stark shift in InGaN quantum wells by Si doping. Nano Lett. 15, 7639–7643. https://doi.org/10.1021/acs.nanolett.5b03531.

Grundmann, M., Christen, J., Bimberg, D., Fischer-Colbrie, A., Hull, R., 1989. Misfit dislocations in pseudomorphic In0.23Ga0.77As/GaAs quantum wells: influence on lifetime

and diffusion of excess excitons. J. Appl. Phys. 66, 2214–2216. https://doi.org/10.1063/1.344288.
Guan, N., Dai, X., Julien, F.H., Eymery, J., Durand, C., Tchernycheva, M., 2018. Nitride nanowires for light emitting diodes. In: Light-Emitting Diodes—Materials, Processes, Devices and Applications, Solid State Lighting Technologies and Applications, Springer (to be published).
Guan, N., Dai, X., Messanvi, A., Zhang, H., Yan, J., Gautier, E., Bougerol, C., Julien, F.H., Durand, C., Eymery, J., Tchernycheva, M., 2016. Flexible white light emitting diodes based on nitride nanowires and nanophosphors. ACS Photonics 3, 597–603. https://doi.org/10.1021/acsphotonics.5b00696.
Gustafsson, A., Bolinsson, J., Sköld, N., Samuelson, L., 2010. Determination of diffusion lengths in nanowires using cathodoluminescence. Appl. Phys. Lett. 97, 072114. https://doi.org/10.1063/1.3473829.
Gutsche, C., Niepelt, R., Gnauck, M., Lysov, A., Prost, W., Ronning, C., Tegude, F.-J., 2012. Direct determination of minority carrier diffusion lengths at axial GaAs nanowire p–n junctions. Nano Lett. 12, 1453–1458. https://doi.org/10.1021/nl204126n.
Heiss, M., Conesa-Boj, S., Ren, J., Tseng, H.-H., Gali, A., Rudolph, A., Uccelli, E., Peiró, F., Morante, J.R., Schuh, D., Reiger, E., Kaxiras, E., Arbiol, J., Fontcuberta i Morral, A., 2011. Direct correlation of crystal structure and optical properties in wurtzite/zinc-blende GaAs nanowire heterostructures. Phys. Rev. B 83, 045303. https://doi.org/10.1103/PhysRevB.83.045303.
Heiss, M., Fontana, Y., Gustafsson, A., Wüst, G., Magen, C., O'Regan, D.D., Luo, J.W., Ketterer, B., Conesa-Boj, S., Kuhlmann, A.V., Houel, J., Russo-Averchi, E., Morante, J.R., Cantoni, M., Marzari, N., Arbiol, J., Zunger, A., Warburton, R.J., Fontcuberta i Morral, A., 2013. Self-assembled quantum dots in a nanowire system for quantum photonics. Nat. Mater. 12, 439–444. https://doi.org/10.1038/nmat3557.
Heiß, M., Gustafsson, A., Conesa-Boj, S., Peiró, F., Morante, J.R., Abstreiter, G., Arbiol, J., Samuelson, L., Morral, A.F.i., 2009. Catalyst-free nanowires with axial In x Ga 1 − x As/ GaAs heterostructures. Nanotechnology 20, 075603. https://doi.org/10.1088/0957-4484/20/7/075603.
Hellman, n., Vandenbroucke, n., Rüsing, n., Isheim, n., Seidman, n., 2000. Analysis of three-dimensional atom-probe data by the proximity histogram. Microsc. Microanal. 6, 437–444. https://doi.org/10.1017.S1431927600000635.
Heurlin, M., Magnusson, M., Lindgren, D., Ek, M., Wallenberg, R., Deppert, K., Samuelson, L., 2012. Continous gas phase synthesis of nanowires with tunable properties. Nature 492, 90–94.
Himwas, C., Collin, S., Rale, P., Chauvin, N., Patriarche, G., Oheler, F., Julien, F.H., Travers, L., Harmand, J.-C., Tchernycheva, M., 2017. In situ passivation of GaAsP NWs. Nanotechnology 28, 495707.
Himwas, C., Den Hertog, M., Si Dang, L., Monroy, E., Songmuang, R., 2014. Alloy inhomogeneity and carrier localization in AlGaN sections and AlGaN/AlN Nanodisk in NWs with 240-350 nm emission. Appl. Phys. Lett. 105, 241908.
Holonyak, N., Bevacqua, S.F., 1962. Coherent (Visible) light emission from Ga(As1 − xPx) junctions. Appl. Phys. Lett. 1, 82–83. https://doi.org/10.1063/1.1753706.
Hortelano, V., Martínez, O., Cuscó, R., Artús, L., Jiménez, J., 2016. Cathodoluminescence study of Mg activation in non-polar and semi-polar faces of undoped/Mg-doped GaN core–shell nanorods. Nanotechnology 27, 095706. https://doi.org/10.1088/0957-4484/27/9/095706.
Hultin, O., Otnes, G., Borgström, M.T., Björk, M., Samuelson, L., Storm, K., 2016. Comparing hall effect and field effect measurements on the same single nanowire. Nano Lett. 16, 205–211. https://doi.org/10.1021/acs.nanolett.5b03496.

Hussain, L., Karimi, M., Berg, A., Jain, V., Borgström, M.T., Gustafsson, A., Samuelson, L., Pettersson, H., 2017. Defect-induced infrared electroluminescence from radial GaInP/AlGaInP quantum well nanowire array light- emitting diodes. Nanotechnology 28, 485205. https://doi.org/10.1088/1361-6528/aa913c.

Isheim, D., Kaszpurenko, J., Yu, D., Mao, Z., Seidman, D.N., Arslan, I., 2012. 3-D atomic-scale mapping of manganese dopants in lead sulfide nanowires. J. Phys. Chem. C 116, 6595–6600. https://doi.org/10.1021/jp300162t.

Iveland, J., Martinelli, L., Peretti, J., Speck, J.S., Weisbuch, C., 2013. Direct measurement of Auger electrons emitted from a semiconductor light-emitting diode under electrical injection: identification of the dominant mechanism for efficiency droop. Phys. Rev. Lett. 110, 177406.

Jahn, U., Lähnemann, J., Pfüller, C., Brandt, O., Breuer, S., Jenichen, B., Ramsteiner, M., Geelhaar, L., Riechert, H., 2012. Luminescence of GaAs nanowires consisting of wurtzite and zinc-blende segments. Phys. Rev. B 85, 045323. https://doi.org/10.1103/PhysRevB.85.045323.

Jain, S.C., McGregor, J.M., Roulston, D.J., 1990. Band-gap narrowing in novel III-V semiconductors. J. Appl. Phys. 68, 3747–3749. https://doi.org/10.1063/1.346291.

Jeganathan, K., Debnath, R.K., Meijers, R., Stoica, T., Calarco, R., Grützmacher, D., Lüth, H., 2009. Raman scattering of phonon-plasmon coupled modes in self-assembled GaN nanowires. J. Appl. Phys. 105, 123707. https://doi.org/10.1063/1.3148862.

Jeon, N., Loitsch, B., Morkoetter, S., Abstreiter, G., Finley, J., Krenner, H.J., Koblmueller, G., Lauhon, L.J., 2015. Alloy fluctuations act as quantum dot-like emitters in GaAs-AlGaAs core-shell nanowires. ACS Nano 9, 8335–8343. https://doi.org/10.1021/acsnano.5b04070.

Jiang, N., Gao, Q., Parkinson, P., Wong-Leung, J., Mokkapati, S., Breuer, S., Tan, H.H., Zheng, C.L., Etheridge, J., Jagadish, C., 2013. Enhanced minority carrier lifetimes in GaAs/AlGaAs core–shell nanowires through shell growth optimization. Nano Lett. 13, 5135–5140. https://doi.org/10.1021/nl4023385.

Joyce, H.J., Boland, J.L., Davies, C.L., Baig, S.A., Johnston, M.B., 2016. A review of the electrical properties of semiconductor nanowires: insights gained from terahertz conductivity spectroscopy. Semicond. Sci. Technol. 31, 103003. https://doi.org/10.1088/0268-1242/31/10/103003.

Joyce, H.J., Wong-Leung, J., Yong, C.-K., Docherty, C.J., Paiman, S., Gao, Q., Tan, H.H., Jagadish, C., Lloyd-Hughes, J., Herz, L.M., Johnston, M.B., 2012. Ultralow surface recombination velocity in InP nanowires probed by terahertz spectroscopy. Nano Lett. 12, 5325–5330. https://doi.org/10.1021/nl3026828.

Kayes, B.M., Atwater, H.A., Lewis, N.S., 2005. Comparison of the device physics principles of planar and radial p-n junction nanorod solar cells. J. Appl. Phys. 97, 114302. https://doi.org/10.1063/1.1901835.

Kayes, B.M., Nie, H., Twist, R., Spruytte, S.G., Reinhardt, F., Kizilyalli, I.C., Higashi, G.S., 2011. In: 27.6% Conversion efficiency, a new record for single-junction solar cells under 1 sun illumination. 2011 37th IEEE Photovoltaic Specialists Conference. Presented at the 2011 37th IEEE Photovoltaic Specialists Conference, pp. 000004–000008. https://doi.org/10.1109/PVSC.2011.6185831.

Ketterer, B., Heiss, M., Uccelli, E., Arbiol, J., Fontcuberta i Morral, A., 2011. Untangling the electronic band structure of wurtzite GaAs nanowires by resonant Raman spectroscopy. ACS Nano 5, 7585–7592. https://doi.org/10.1021/nn202585j.

Ketterer, B., Mikheev, E., Uccelli, E., Fontcuberta i Morral, A., 2010. Compensation mechanism in silicon-doped gallium arsenide nanowires. Appl. Phys. Lett. 97, 223103. https://doi.org/10.1063/1.3517254.

Khanal, D.R., Wu, J., 2007. Gate coupling and charge distribution in nanowire field effect transistors. Nano Lett. 7, 2778–2783. https://doi.org/10.1021/nl071330l.

Kim, H., Farrell, A.C., Senanayake, P., Lee, W.-J., Huffaker, D.L., 2016. Monolithically integrated InGaAs nanowires on 3D structured silicon-on-insulator as a new platform for full optical links. Nano Lett. 16, 1833–1839. https://doi.org/10.1021/acs.nanolett.5b04883.

Kishino, K., Yanagihara, A., Ikeda, K., Yamano, K., 2015. Monolithic integration of four-colour InGaN-based nanocolumn LEDs. Electron. Lett. 51, 852–854. https://doi.org/10.1049/el.2015.0770.

Kociak, M., Zagonel, L.F., 2017. Cathodoluminescence in the scanning transmission electron microscope. Ultramicroscopy 174, 50. https://doi.org/10.1016/j.ultramic.2016.11.018.

Koelling, S., Li, A., Cavalli, A., Assali, S., Car, D., Gazibegovic, S., Bakkers, E.P.A.M., Koenraad, P.M., 2017. Atom-by-atom analysis of semiconductor nanowires with parts per million sensitivity. Nano Lett. 17, 599–605. https://doi.org/10.1021/acs.nanolett.6b03109.

Koester, R., Hwang, J.-S., Salomon, D., Chen, X., Bougerol, C., Barnes, J.-P., Dang, D.L.S., Rigutti, L., de Luna Bugallo, A., Jacopin, G., Tchernycheva, M., Durand, C., Eymery, J., 2011. M-plane core-shell InGaN/GaN multiple-quantum-wells on GaN wires for electroluminescent devices. Nano Lett. 11, 4839–4845. https://doi.org/10.1021/nl202686n.

Kowalczewski, P., Redorici, L., Bozzola, A., Andreani, L.C., 2016. Silicon solar cells reaching the efficiency limits: from simple to complex modelling. J. Opt. 18, 054001. https://doi.org/10.1088/2040-8978/18/5/054001.

Krause, T., Hanke, M., Cheng, Z., Niehle, M., Trampert, A., Rosenthal, M., Burghammer, M., Ledig, J., Hartmann, J., Zhou, H., Wehmann, H.-H., Waag, A., 2016. Nanofocus x-ray diffraction and cathodoluminescence investigations into individual core–shell (In,Ga)N/GaN rod light-emitting diodes. Nanotechnology 27, 325707. https://doi.org/10.1088/0957-4484/27/32/325707.

Kreuzer, H.J., Nath, K., 1987. Field evaporation. Surf. Sci. 183, 591–608. https://doi.org/10.1016/S0039-6028(87)80230-3.

Kuang, Y., Vece, M.D., Rath, J.K., Dijk van, L., Schropp, R.E.I., 2013. Elongated nanostructures for radial junction solar cells. Rep. Prog. Phys. 76, 106502. https://doi.org/10.1088/0034-4885/76/10/106502.

LaPierre, R.R., 2011. Theoretical conversion efficiency of a two-junction III-V nanowire on Si solar cell. J. Appl. Phys. 110, 014310. https://doi.org/10.1063/1.3603029.

Lauhon, L.J., Adusumilli, P., Ronsheim, P., Flaitz, P.L., Lawrence, D., 2009. Atom-probe tomography of semiconductor materials and device structures. MRS Bull. 34, 738–743. https://doi.org/10.1557/mrs2009.248.

Lavenus, P., Messanvi, A., Rigutti, L., Bugallo, A.D.L., Zhang, H., Bayle, F., Julien, F.H., Eymery, J., Durand, C., Tchernycheva, M., 2014. Experimental and theoretical analysis of transport properties of core–shell wire light emitting diodes probed by electron beam induced current microscopy. Nanotechnology 25, 255201. https://doi.org/10.1088/0957-4484/25/25/255201.

Leamy, H.J., 1982. Charge collection scanning electron microscopy. J. Appl. Phys. 53, R51–R80. https://doi.org/10.1063/1.331667.

Lee, Y.H.J., Li, Z., Fu, L., Parkinson, P., Vora, K., Tan, H.H., Jagadish, C., 2012. In: Improved GaAs nanowire solar cells using AlGaAs for surface passivation. COMMAD 2012. Presented at the COMMAD 2012, pp. 131–132. https://doi.org/10.1109/COMMAD.2012.6472395.

Lefebvre, P., Gayral, B., 2008. Optical properties of GaN/AlN quantum dots. Comptes Rendus Phys. 9, 816–829. https://doi.org/10.1016/j.crhy.2008.10.008.

Lefebvre, W., Vurpillot, F., Sauvage, X., 2016. Atom Probe Tomography: Put Theory Into Practice. Academic Press.

Lewis, R.B., Nicolai, L., Küpers, H., Ramsteiner, M., Trampert, A., Geelhaar, L., 2017. Anomalous strain relaxation in core–shell nanowire heterostructures via simultaneous coherent and incoherent growth. Nano Lett. 17, 136–142. https://doi.org/10.1021/acs.nanolett.6b03681.

Li, S., Waag, A., 2012. GaN based nanorods for solid state lighting. J. Appl. Phys. 111, 071101. https://doi.org/10.1063/1.3694674.

Lim, S.K., Brewster, M., Qian, F., Li, Y., Lieber, C.M., Gradečak, S., 2009. Direct correlation between structural and optical properties of III – V nitride nanowire heterostructures with nanoscale resolution. Nano Lett. 9, 3940–3944. https://doi.org/10.1021/nl9025743.

Limbach, F., Hauswald, C., Lähnemann, J., Wölz, M., Brandt, O., Trampert, A., Hanke, M., Jahn, U., Calarco, R., Geelhaar, L., Riechert, H., 2012. Current path in light emitting diodes based on nanowire ensembles. Nanotechnology 23, 465301. https://doi.org/10.1088/0957-4484/23/46/465301.

Lindgren, D., Hultin, O., Heurlin, M., Storm, K., Borgström, M.T., Samuelson, L., Gustafsson, A., 2015. Study of carrier concentration in single InP nanowires by luminescence and Hall measurements. Nanotechnology 26, 045705. https://doi.org/10.1088/0957-4484/26/4/045705.

Liu, J.-L., Zhang, J.-L., Wang, G.-X., Mo, C.-L., Xu, L.-Q., Ding, J., Quan, Z.-J., Wang, X.-L., Pan, S., Zheng, C.-D., Wu, X.-M., Fang, W.-Q., Jiang, F.-Y., 2015. Status of GaN-based green light-emitting diodes. Chin. Phys. B 24, 067804. https://doi.org/10.1088/1674-1056/24/6/067804.

Loitsch, B., Jeon, N., Döblinger, M., Winnerl, J., Parzinger, E., Matich, S., Wurstbauer, U., Riedl, H., Abstreiter, G., Finley, J.J., Lauhon, L.J., Koblmüller, G., 2016b. Suppression of alloy fluctuations in GaAs-AlGaAs core-shell nanowires. Appl. Phys. Lett. 109, 093105. https://doi.org/10.1063/1.4962269.

Loitsch, B., Müller, M., Winnerl, J., Veit, P., Rudolph, D., Abstreiter, G., Finley, J.J., Bertram, F., Christen, J., Koblmüller, G., 2016a. Microscopic nature of crystal phase quantum dots in ultrathin GaAs nanowires by nanoscale luminescence characterization. New J. Phys. 18, 063009. https://doi.org/10.1088/1367-2630/18/6/063009.

Lorensen, W.E., Cline, H.E., 1987. Marching cubes: a high resolution 3D surface construction algorithm. Comput. Graph. 21, 163–169.

Mancini, L., Amirifar, N., Shinde, D., Blum, I., Gilbert, M., Vella, A., Vurpillot, F., Lefebvre, W., Lardé, R., Talbot, E., Pareige, P., Portier, X., Ziani, A., Davesnne, C., Durand, C., Eymery, J., Butté, R., Carlin, J.-F., Grandjean, N., Rigutti, L., 2014a. Composition of wide bandgap semiconductor materials and nanostructures measured by atom probe tomography and its dependence on the surface electric field. J. Phys. Chem. C 118, 24136–24151. https://doi.org/10.1021/jp5071264.

Mancini, L., Fontana, Y., Conesa-Boj, S., Blum, I., Vurpillot, F., Francaviglia, L., Russo-Averchi, E., Heiss, M., Arbiol, J., Morral, A.F.i., Rigutti, L., 2014b. Three-dimensional nanoscale study of Al segregation and quantum dot formation in GaAs/AlGaAs core-shell nanowires. Appl. Phys. Lett. 105, 243106. https://doi.org/10.1063/1.4904952.

Mancini, L., Hernández-Maldonado, D., Lefebvre, W., Houard, J., Blum, I., Vurpillot, F., Eymery, J., Durand, C., Tchernycheva, M., Rigutti, L., 2016. Multi-microscopy study of the influence of stacking faults and three-dimensional In distribution on the optical properties of m-plane InGaN quantum wells grown on microwire sidewalls. Appl. Phys. Lett. 108, 042102. https://doi.org/10.1063/1.4940748.

Mariani, G., Scofield, A.C., Hung, C.-H., Huffaker, D.L., 2013. GaAs nanopillar-array solar cells employing in situ surface passivation. Nat. Commun. 4, 1497. https://doi.org/10.1038/ncomms2509.

May, B.J., Sarwar, A.T.M.G., Myers, R.C., 2016. Nanowire LEDs grown directly on flexible metal foil. Appl. Phys. Lett. 108, 141103. https://doi.org/10.1063/1.4945419.

McMullen, E.R., Perdew, J.P., 1987. Theory of field evaporation of the surface layer in jellium and other metals. Phys. Rev. B 36, 2598–2606. https://doi.org/10.1103/PhysRevB.36.2598.

McMullen, E.R., Perdew, J.P., Rose, J.H., 1982. Effects of an intense electric field on metal surface geometry. Solid State Commun. 44, 945–949. https://doi.org/10.1016/0038-1098(82)90310-6.

Merano, M., Collin, S., Renucci, P., Gatri, M., Sonderegger, S., Crottini, A., Ganière, J.D., Deveaud, B., 2005b. High brightness picosecond electron gun. Rev. Sci. Instrum. 76, 085108. https://doi.org/10.1063/1.2008975.

Merano, M., Sonderegger, S., Crottini, A., Collin, S., Renucci, P., Pelucchi, E., Malko, A., Baier, M.H., Kapon, E., Deveaud, B., Ganière, J.-D., 2005a. Probing carrier dynamics in nanostructures by picosecond cathodoluminescence. Nature 438, 479. https://doi.org/10.1038/nature04298.

Metaferia, W., Persson, A.R., Mergenthaler, K., Yang, F., Zhang, W., Yartsev, A., Wallenberg, R., Pistol, M.-E., Deppert, K., Samuelson, L., Magnusson, M.H., 2016. GaAsP nanowires grown by aerotaxy. Nano Lett. 16, 5701–5707. https://doi.org/10.1021/acs.nanolett.6b02367.

Meuret, S., Zeijlemaker, H., Neu, E., Appel, P., Maletinsky, P., Polman, A., 2016. Time-resolved cathodoluminescence in a scanning electron microscope. In: European Microscopy Congress 2016: Proceedings. Wiley-VCH Verlag GmbH & Co. KGaA https://doi.org/10.1002/9783527808465.EMC2016.6725.

Miller, M.K., 2000. Atom Probe Tomography: Analysis at the Atomic Level. Springer, USA.

Miller, M.K., Russell, K.F., Thompson, K., Alvis, R., Larson, D.J., 2007. Review of atom probe FIB-based specimen preparation methods. Microsc. Microanal. 13, 428–436. https://doi.org/10.1017/S1431927607070845.

Mohajerani, M.S., Khachadorian, S., Schimpke, T., Nenstiel, C., Hartmann, J., Ledig, J., Avramescu, A., Strassburg, M., Hoffmann, A., Waag, A., 2016a. Evaluation of local free carrier concentrations in individual heavily-doped GaN:Si micro-rods by micro-Raman spectroscopy. Appl. Phys. Lett. 108, 091112. https://doi.org/10.1063/1.4943079.

Mohajerani, M.S., Müller, M., Hartmann, J., Zhou, H., Wehmann, H.-H., Veit, P., Bertram, F., Christen, J., Waag, A., 2016b. Direct correlations of structural and optical properties of three-dimensional GaN/InGaN core/shell micro-light emitting diodes. Jpn. J. Appl. Phys. 55, 05FJ09. https://doi.org/10.7567/JJAP.55.05FJ09.

Monemar, B., Ohlsson, B.J., Gardner, N.F., Samuelson, L., 2016. Nanowire-based visible light emitters, present status and outlook. In: Dayeh, S.A., Fontcuberta i Morral, A., Jagadish, C. (Eds.), Semiconductors and Semimetals, Semiconductor Nanowires II: Properties and Applications. Elsevier, pp. 227–271. https://doi.org/10.1016/bs.semsem.2015.10.002. Chapter 7.

Moody, M.P., Ceguerra, A.V., Breen, A.J., Cui, X.Y., Gault, B., Stephenson, L.T., Marceau, R.K.W., Powles, R.C., Ringer, S.P., 2014. Atomically resolved tomography to directly inform simulations for structure-property relationships. Nat. Commun. 5, 5501. https://doi.org/10.1038/ncomms6501.

Moratis, K., Tan, S.L., Germanis, S., Katsidis, C., Androulidaki, M., Tsagaraki, K., Hatzopoulos, Z., Donatini, F., Cibert, J., Niquet, Y.-M., Mariette, H., Pelekanos, N.T., 2016. Strained GaAs/InGaAs core-shell nanowires for photovoltaic applications. Nanoscale Res. Lett. 11, 176. https://doi.org/10.1186/s11671-016-1384-y.

Morkötter, S., Jeon, N., Rudolph, D., Loitsch, B., Spirkoska, D., Hoffmann, E., Döblinger, M., Matich, S., Finley, J.J., Lauhon, L.J., Abstreiter, G., Koblmüller, G., 2015. Demonstration of confined electron gas and steep-slope behavior in delta-doped GaAs-AlGaAs core-shell nanowire transistors. Nano Lett. 15, 3295–3302. https://doi.org/10.1021/acs.nanolett.5b00518.

Mounir, C., Schimpke, T., Rossbach, G., Avramescu, A., Strassburg, M., Schwarz, U.T., 2017. Polarization-resolved micro-photoluminescence investigation of InGaN/GaN core-shell microrods. J. Appl. Phys. 121, 025701. https://doi.org/10.1063/1.4973899.

Mounir, C., Schimpke, T., Rossbach, G., Avramescu, A., Strassburg, M., Schwarz, U.T., 2016. Optical properties and internal quantum efficiency of InGaN/GaN core-shell microrods for solid state lighting. J. Appl. Phys. 120, 155702. https://doi.org/10.1063/1.4964871.

Müller, E.W., Bahadur, K., 1956. Field ionization of gases at a metal surface and the resolution of the field ion microscope. Phys. Rev. 102, 624–631. https://doi.org/10.1103/PhysRev.102.624.

Müller, M., Schmidt, G., Metzner, S., Veit, P., Bertram, F., Krylyuk, S., Debnath, R., Ha, J.-Y., Wen, B., Blanchard, P., Motayed, A., King, M.R., Davydov, A.V., Christen, J., 2016. Structural and optical nanoscale analysis of GaN core–shell microrod arrays fabricated by combined top-down and bottom-up process on Si(111). Jpn. J. Appl. Phys. 55, 05FF02. https://doi.org/10.7567/JJAP.55.05FF02.

Nakai, E., Chen, M., Yoshimura, M., Tomioka, K., Fukui, T., 2015. InGaAs axial-junction nanowire-array solar cells. Jpn. J. Appl. Phys. 54, 015201. https://doi.org/10.7567/JJAP.54.015201.

Narchi, P., Neplokh, V., Piazza, V., Bearda, T., Bayle, F., Foldyna, M., Toccafondi, C., Prod'homme, P., Tchernycheva, M., i Cabarrocas, P.R., 2017. Surface potential investigation on interdigitated back contact solar cells by scanning electron microscopy and Kelvin probe force microscopy: effect of electrical bias. Sol. Energy Mater. Sol. Cells 161, 263–269.

Oberdorfer, C., Stender, P., Reinke, C., Schmitz, G., 2007. Laser-assisted atom probe tomography of oxide materials. Microsc. Microanal. 13, 342–346. https://doi.org/10.1017/S1431927607070274.

Pantzas, K., Patriarche, G., Troadec, D., Kociak, M., Cherkashin, N., Hÿtch, M., Barjon, J., Tanguy, C., Rivera, T., Suresh, S., Ougazzaden, A., 2015. Role of compositional fluctuations and their suppression on the strain and luminescence of InGaN alloys. J. Appl. Phys. 117, 055705. https://doi.org/10.1063/1.4907210.

Pasemann, L., 1981. A contribution to the theory of the EBIC contrast of lattice defects in semiconductors. Ultramicroscopy 6, 237–250.

Perea, D.E., Allen, J.E., May, S.J., Wessels, B.W., Seidman, D.N., Lauhon, L.J., 2006. Three-dimensional nanoscale composition mapping of semiconductor nanowires. Nano Lett. 6, 181–185. https://doi.org/10.1021/nl051602p.

Perovic, D.D., Castell, M.R., Howie, A., Lavoie, C., Tiedje, T., Cole, J.S.W., 1995. Field-emission SEM imaging of compositional and doping layer semiconductor superlattices. Ultramicroscopy 58, 104–113.

Pierret, A., Bougerol, C., Gayral, B., Kociak, M., Daudin, B., 2013a. Probing alloy composition gradient and nanometer-scale carrier localization in single AlGaN nanowires by nanocathodoluminescence. Nanotechnology 24, 305703. https://doi.org/10.1088/0957-4484/24/30/305703.

Pierret, A., Bougerol, C., den Hertog, M., Gayral, B., Kociak, M., Renevier, H., Daudin, B., 2013b. Structural and optical properties of Alx Ga1–xN nanowires. Phys. Status Solidi RRL 7, 868–873. https://doi.org/10.1002/pssr.201308009.

Progl, C.L., Parish, C.M., Vitarelli, J.P., Russell, P.E., 2008. Analysis of V defects in GaN-based light emitting diodes by scanning transmission electron microscopy and electron beam induced current. Appl. Phys. Lett. 92, 242103. https://doi.org/10.1063/1.2945232.

Prosa, T.J., Kostrna, S.L.P., Kelly, T.F., 2006. In: Laser atom probe tomography: application to polymers. 2006 19th International Vacuum Nanoelectronics Conference. Presented at the 2006 19th International Vacuum Nanoelectronics Conference, pp. 533–534. https://doi.org/10.1109/IVNC.2006.335331.

Qu, J., Wong, D., Du, S., Yang, L., Ringer, S., Zheng, R., 2015. Methodology exploration of specimen preparation for atom probe tomography from nanowires. In: 1st International Conference on Atom Probe Tomography & Microscopy. Ultramicroscopy, Vol. 159, pp. 427–431. https://doi.org/10.1016/j.ultramic.2015.06.003.

Reimer, L., 2013. Scanning Electron Microscopy: Physics of Image Formation and Microanalysis. Springer.

Ren, X., Riley, J.R., Koleske, D.D., Lauhon, L.J., 2015. Correlated high-resolution x-ray diffraction, photoluminescence, and atom probe tomography analysis of continuous and discontinuous InxGa1-xN quantum wells. Appl. Phys. Lett. 107, 022107. https://doi.org/10.1063/1.4926808.

Richter, A., Hermle, M., Glunz, S.W., 2013. Reassessment of the limiting efficiency for crystalline silicon solar cells. IEEE J. Photovoltaics. 3, 1184–1191. https://doi.org/10.1109/JPHOTOV.2013.2270351.

Rigutti, L., Blum, I., Shinde, D., Hernández-Maldonado, D., Lefebvre, W., Houard, J., Vurpillot, F., Vella, A., Tchernycheva, M., Durand, C., Eymery, J., Deconihout, B., 2014. Correlation of microphotoluminescence spectroscopy, scanning transmission electron microscopy, and atom probe tomography on a single nano-object containing an InGaN/GaN multiquantum well system. Nano Lett. 14, 107–114. https://doi.org/10.1021/nl4034768.

Riley, J.R., Detchprohm, T., Wetzel, C., Lauhon, L.J., 2014. On the reliable analysis of indium mole fraction within InxGa1-xN quantum wells using atom probe tomography. Appl. Phys. Lett. 104, 152102. https://doi.org/10.1063/1.4871510.

Riley, J.R., Padalkar, S., Li, Q., Lu, P., Koleske, D.D., Wierer, J.J., Wang, G.T., Lauhon, L.J., 2013. Three-dimensional mapping of quantum wells in a GaN/InGaN core-shell nanowire light-emitting diode array. Nano Lett. 13, 4317–4325. https://doi.org/10.1021/nl4021045.

Rühle, S., 2016. Tabulated values of the Shockley–Queisser limit for single junction solar cells. Sol. Energy 130, 139–147. https://doi.org/10.1016/j.solener.2016.02.015.

Savin, H., Repo, P., von Gastrow, G., Ortega, P., Calle, E., Garín, M., Alcubilla, R., 2015. Black silicon solar cells with interdigitated back-contacts achieve 22.1% efficiency. Nat. Nanotechnol. 10, 624–628. https://doi.org/10.1038/nnano.2015.89.

Schimpke, T., Mandl, M., Stoll, I., Pohl-Klein, B., Bichler, D., Zwaschka, F., Strube-Knyrim, J., Huckenbeck, B., Max, B., Müller, M., Veit, P., Bertram, F., Christen, J., Hartmann, J., Waag, A., Lugauer, H.-J., Strassburg, M., 2016. Phosphor-converted white light from blue-emitting InGaN microrod LEDs. Phys. Status Solidi A 213, 1577–1584. https://doi.org/10.1002/pssa.201532904.

Schulz, S., Tanner, D.P., O'Reilly, E.P., Caro, M.A., Martin, T.L., Bagot, P.A.J., Moody, M.P., Tang, F., Griffiths, J.T., Oehler, F., Kappers, M.J., Oliver, R.A., Humphreys, C.J., Sutherland, D., Davies, M.J., Dawson, P., 2015. Structural, electronic, and optical properties of m-plane InGaN/GaN quantum wells: insights from experiment and atomistic theory. Phys. Rev. B 92, 235419. https://doi.org/10.1103/PhysRevB.92.235419.

Sealy, C.P., Castell, M.R., Wilshaw, P.R., 2000. Mechanism for SE dopant contrast in SEM. J. Electron Microsc. (Tokyo) 49.

Shahmohammadi, M., Ganière, J.-D., Zhang, H., Ciechonski, R., Vescovi, G., Kryliouk, O., Tchernycheva, M., Jacopin, G., 2016. Excitonic diffusion in InGaN/GaN core–shell nanowires. Nano Lett. 16, 243–249. https://doi.org/10.1021/acs.nanolett.5b03611.

Shen, Y.R., 2003. The Principles of Nonlinear Optics. Wiley-Interscience.

Shin, J.C., Kim, K.H., Yu, K.J., Hu, H., Yin, L., Ning, C.-Z., Rogers, J.A., Zuo, J.-M., Li, X., 2011. In $_x$ Ga $_{1-x}$ as nanowires on silicon: one-dimensional heterogeneous epitaxy, bandgap engineering, and photovoltaics. Nano Lett. 11, 4831–4838. https://doi.org/10.1021/nl202676b.

Signorello, G., Lörtscher, E., Khomyakov, P.A., Karg, S., Dheeraj, D.L., Gotsmann, B., Weman, H., Riel, H., 2014. Inducing a direct-to-pseudodirect bandgap transition in wurtzite GaAs nanowires with uniaxial stress. Nat. Commun. 5, 3655. https://doi.org/10.1038/ncomms4655.

Silaeva, E.P., Arnoldi, L., Karahka, M.L., Deconihout, B., Menand, A., Kreuzer, H.J., Vella, A., 2014. Do dielectric nanostructures turn metallic in high-electric dc fields? Nano Lett. 14, 6066–6072. https://doi.org/10.1021/nl502715s.

Silaeva, E.P., Houard, J., Hideur, A., Martel, G., Vella, A., 2015. Field evaporation of semiconductors assisted by nonequilibrium phonon excitations. Phys. Rev. B 92, 195307. https://doi.org/10.1103/PhysRevB.92.195307.

Singh, R., Dailey, E.J., Drucker, J., Menéndez, J., 2011. Raman scattering from Ge-Si core-shell nanowires: validity of analytical strain models. J. Appl. Phys. 110, 124305. https://doi.org/10.1063/1.3667125.

Skromme, B.J., Sandroff, C.J., Yablonovitch, E., Gmitter, T., 1987. Effects of passivating ionic films on the photoluminescence properties of GaAs. Appl. Phys. Lett. 51, 2022–2024. https://doi.org/10.1063/1.98280.

Smith, L.M., Jackson, H.E., Yarrison-Rice, J.M., Jagadish, C., 2010 Insights into single semiconductor nanowire heterostructures using time-resolved photoluminescence. Semicond. Sci. Technol. 25, 024010. https://doi.org/10.1088/0268-1242/25/2/024010.

Spirkoska, D., Arbiol, J., Gustafsson, A., Conesa-Boj, S., Glas, F., Zardo, I., Heigoldt, M., Gass, M.H., Bleloch, A.L., Estrade, S., Kaniber, M., Rossler, J., Peiro, F., Morante, J.R., Abstreiter, G., Samuelson, L., Fontcuberta i Morral, A., 2009. Structural and optical properties of high quality zinc-blende/wurtzite GaAs nanowire heterostructures. Phys. Rev. B 80, 245325. https://doi.org/10.1103/PhysRevB.80.245325.

Stettner, T., Kostenbader, T., Ruhstorfer, D., Bissinger, J., Riedl H., Kaniber, M., Koblmüller, G., Finley, J.J., 2017. Direct coupling of coherent emission from site-selectively grown III–V nanowire lasers into proximal silicon waveguides. ACS Photonics 4, 2537–2543. https://doi.org/10.1021/acsphotonics.7b00805.

Storm, K., Halvardsson, F., Heurlin, M., Lindgren, D., Gustafsson, A., Wu, P.M., Monemar, B., Samuelson, L., 2012. Spatially resolved Hall effect measurement in a single semiconductor nanowire. Nat. Nanotechnol. 7, 718. https://doi.org/10.1038/nnano.2012.190.

Svensson, C.P.T., Mårtensson, T., Trägårdh, J., Larsson, C., Rask, M., Hessman, D., Samuelson, L., Ohlsson, J., 2008. Monolithic GaAs/InGaP nanowire light emitting diodes on silicon. Nanotechnology 19, 305201. https://doi.org/10.1088/0957-4484/19/30/305201.

Tchernycheva, M., Lavenus, P., Zhang, H., Babichev, A.V., Jacopin, G., Shahmohammadi, M., Julien, F.H., Ciechonski, R., Vescovi, G., Kryliouk, O., 2014b. InGaN/GaN core–shell single nanowire light emitting diodes with graphene-based P-contact. Nano Lett. 14, 2456–2465. https://doi.org/10.1021/nl5001295.

Tchernycheva, M., Messanvi, A., de Luna Bugallo, A., Jacopin, G., Lavenus, P., Rigutti, L., Zhang, H., Halioua, Y., Julien, F.H., Eymery, J., Durand, C., 2014a. Integrated photonic platform based on InGaN/GaN nanowire emitters and detectors. Nano Lett. 14, 3515–3520. https://doi.org/10.1021/nl501124s.

Tchernycheva, M., Neplokh, V., Zhang, H., Lavenus, P., Rigutti, L., Bayle, F., Julien, F.H., Babichev, A., Jacopin, G., Largeau, L., Ciechonski, R., Vescovi, G., Kryliouk, O., 2015.

Core–shell InGaN/GaN nanowire light emitting diodes analyzed by electron beam induced current microscopy and cathodoluminescence mapping. Nanoscale 7, 11692–11701. https://doi.org/10.1039/C5NR00623F.

Tchernycheva, M., Rigutti, L., Jacopin, G., Bugallo, A.d.L., Lavenus, P., Julien, F.H., Timofeeva, M., Bouravleuv, A.D., Cirlin, G.E., Dhaka, V., Lipsanen, H., Largeau, L., 2012. Photovoltaic properties of GaAsP core–shell nanowires on Si(001) substrate. Nanotechnology 23, 265402. https://doi.org/10.1088/0957-4484/23/26/265402.

Tchoulfian, P., Donatini, F., Levy, F., Dussaigne, A., Ferret, P., Pernot, J., 2014. Direct imaging of p–n junction in core–shell GaN wires. Nano Lett. 14, 3491–3498. https://doi.org/10.1021/nl5010493.

Tian, B., Zheng, X., Kempa, T.J., Fang, Y., Yu, N., Yu, G., Huang, J., Lieber, C.M., 2007. Coaxial silicon nanowires as solar cells and nanoelectronic power sources. Nature 449, 885–889. https://doi.org/10.1038/nature06181.

Tizei, L.H.G., Kociak, M., 2013. Spatially resolved quantum nano-optics of single photons using an electron microscope. Phys. Rev. Lett. 110, 153604. https://doi.org/10.1103/PhysRevLett.110.153604.

Tizei, L.H.G., Meuret, S., March, K., Hestroffer, K., Auzelle, T., Daudin, B., Kociak, M., 2014. A polarity-driven nanometric luminescence asymmetry in AlN/GaN heterostructures. Appl. Phys. Lett. 105, 143106. https://doi.org/10.1063/1.4897408.

Togonal, A.S., Foldyna, M., Chen, W., Wang, J.X., Neplokh, V., Tchernycheva, M., Nassar, J., Roca i Cabarrocas, P., Rusli, 2016. Core–shell heterojunction solar cells based on disordered silicon nanowire arrays. J. Phys. Chem. C 120, 2962–2972. https://doi.org/10.1021/acs.jpcc.5b09618.

Tomioka, K., Motohisa, J., Hara, S., Hiruma, K., Fukui, T., 2010. GaAs/AlGaAs core multishell nanowire-based light-emitting diodes on Si. Nano Lett. 10, 1639–1644. https://doi.org/10.1021/nl9041774.

Tourbot, G., Bougerol, C., Glas, F., Zagonel, L.F., Mahfoud, Z., Meuret, S., Gilet, P., Kociak, M., Gayral, B., Daudin, B., 2012. Growth mechanism and properties of InGaN insertions in GaN nanowires. Nanotechnology 23, 135703. https://doi.org/10.1088/0957-4484/23/13/135703.

Tsong, T.T., 1984. Pulsed-laser-stimulated field ion emission from metal and semiconductor surfaces: a time-of-flight study of the formation of atomic, molecular, and cluster ions. Phys. Rev. B 30, 4946–4961. https://doi.org/10.1103/PhysRevB.30.4946.

Tsong, T.T., Kinkus, T.J., 1984. Energy distributions of pulsed-laser field-desorbed gaseous ions and field-evaporated metal ions: a direct time-of-flight measurement. Phys. Rev. B 29, 529–542. https://doi.org/10.1103/PhysRevB.29.529.

Tsurumi, D., Hamada, K., 2013. Dopant mapping of semiconductors with scanning electron microscopy. SEI Tech. Rev. 141, 141–146.

Urban, A., Müller, M., Karbaum, C., Schmidt, G., Veit, P., Malindretos, J., Bertram, F., Christen, J., Rizzi, A., 2015. Optical emission of individual GaN nanocolumns analyzed with high spatial resolution. Nano Lett. 15, 5105–5109. https://doi.org/10.1021/acs.nanolett.5b01278.

van Dam, D., van Hoof, N.J.J., Cui, Y., van Veldhoven, P.J., Bakkers, E.P.A.M., Gómez Rivas, J., Haverkort, J.E.M., 2016. High-efficiency nanowire solar cells with omnidirectionally enhanced absorption due to self-aligned indium–tin–oxide Mie scatterers. ACS Nano 10, 11414–11419. https://doi.org/10.1021/acsnano.6b06874.

Vella, A., Vurpillot, F., Gault, B., Menand, A., Deconihout, B., 2006. Evidence of field evaporation assisted by nonlinear optical rectification induced by ultrafast laser. Phys. Rev. B 73, 165416. https://doi.org/10.1103/PhysRevB.73.165416.

Volotsenko, I., Molotskii, M., Barkay, Z., Marczewski, J., Grabiec, P., Jaroszewicz, B., Meshulam, G., Grunbaum, E., Rosenwaks, Y., 2010. Secondary electron doping

contrast: theory based on scanning electron microscope and Kelvin probe force microscopy measurements. J. Appl. Phys. 107, 014510.
Vurpillot, F., Cerezo, A., Blavette, D., Larson, D.j., 2004. Modeling image distortions in 3DAP. Microsc. Microanal. 384–390. https://doi.org/10.1017/S1431927604040486.
Waag, A., Wang, X., Fündling, S., Ledig, J., Erenburg, M., Neumann, R., Al Suleiman, M., Merzsch, S., Wei, J., Li, S., Wehmann, H.H., Bergbauer, W., Straßburg, M., Trampert, A., Jahn, U., Riechert, H., 2011. The nanorod approach: GaN NanoLEDs for solid state lighting. Phys. Status Solidi C 8, 2296–2301. https://doi.org/10.1002/pssc.201000989.
Wallentin, J., Anttu, N., Asoli, D., Huffman, M., Aberg, I., Magnusson, M.H., Siefer, G., Fuss-Kailuweit, P., Dimroth, F., Witzigmann, B., Xu, H.Q., Samuelson, L., Deppert, K., Borgstrom, M.T., 2013. InP nanowire array solar cells achieving 13.8% efficiency by exceeding the ray optics limit. Science 339, 1057–1060. https://doi.org/10.1126/science.1230969.
Wallentin, J., Borgström, M.T., 2011. Doping of semiconductor nanowires. J. Mater. Res. 26, 2142–2156. https://doi.org/10.1557/jmr.2011.214.
Wallentin, J., Wickert, P., Ek, M., Gustafsson, A., Reine Wallenberg, L., Magnusson, M.H., Samuelson, L., Deppert, K., Borgström, M.T., 2011. Degenerate p-doping of InP nanowires for large area tunnel diodes. Appl. Phys. Lett. 99, 253105. https://doi.org/10.1063/1.3669697.
Watson-Parris, D., Godfrey, M.J., Dawson, P., Oliver, R.A., Galtrey, M.J., Kappers, M.J., Humphreys, C.J., 2011. Carrier localization mechanisms in InGaN/GaN quantum wells. Phys. Rev. B 83, 115321. https://doi.org/10.1103/PhysRevB.83.115321.
Waugh, A.R., Boyes, E.D., Southon, M.J., 1976. Investigations of field evaporation with a field-desorption microscope. Surf. Sci. 61, 109–142. https://doi.org/10.1016/0039-6028(76)90411-8.
Wunnicke, O., 2006. Gate capacitance of back-gated nanowire field-effect transistors. Appl. Phys. Lett. 89, 083102. https://doi.org/10.1063/1.2337853.
Wurfel, P., 1982. The chemical potential of radiation. J. Phys. C Solid State Phys. 15, 3967. https://doi.org/10.1088/0022-3719/15/18/012.
Yacobi, B.G., Holt, D.B., 2013. Cathodoluminescence Microscopy of Inorganic Solids. Springer Science & Business Media.
Yang, F., Messing, M.E., Mergenthaler, K., Ghasemi, M., Johansson, J., Wallenberg, L.R., Pistol, M.-E., Deppert, K., Samuelson, L., Magnusson, M.H., 2015. Zn-doping of GaAs nanowires grown by Aerotaxy. J. Cryst. Growth 414, 181–186. https://doi.org/10.1016/j.jcrysgro.2014.09.051.
Yao, M., Huang, N., Cong, S., Chi, C.-Y., Seyedi, M.A., Lin, Y.-T., Cao, Y., Povinelli, M.L., Dapkus, P.D., Zhou, C., 2014. GaAs nanowire array solar cells with axial p-i-n junctions. Nano Lett. 14, 3293–3303. https://doi.org/10.1021/nl500704r.
Yoshimura, M., Nakai, E., Tomioka, K., Fukui, T., 2013. Indium phosphide core–shell nanowire array solar cells with lattice-mismatched window layer. Appl. Phys. Express 6, 052301. https://doi.org/10.7567/APEX.6.052301.
Yu, L., Rigutti, L., Tchernycheva, M., Misra, S., Foldyna, M., Picardi, G., Cabarrocas, P.R.i., 2013. Assessing individual radial junction solar cells over millions on VLS-grown silicon nanowires. Nanotechnology 24, 275401. https://doi.org/10.1088/0957-4484/24/27/275401.
Yu, P., Wu, J., Liu, S., Xiong, J., Jagadish, C., Wang, Z.M., 2016. Design and fabrication of silicon nanowires towards efficient solar cells. Nano Today 11, 704–737. https://doi.org/10.1016/j.nantod.2016.10.001.
Zagonel, L.F., Mazzucco, S., Tencé, M., March, K., Bernard, R., Laslier, B., Jacopin, G., Tchernycheva, M., Rigutti, L., Julien, F.H., Songmuang, R., Kociak, M., 2011. Nanometer scale spectral imaging of quantum emitters in nanowires and its correlation to their atomically resolved structure. Nano Lett. 11, 568–573. https://doi.org/10.1021/nl103549t.

Zagonel, L.F., Rigutti, L., Tchernycheva, M., Jacopin, G., Songmuang, R., Kociak, M., 2012. Visualizing highly localized luminescence in GaN/AlN heterostructures in nanowires. Nanotechnology 23, 455205. https://doi.org/10.1088/0957-4484/23/45/455205.

Zagonel, L.F., Tizei, L.H.G., Vitiello, G.Z., Jacopin, G., Rigutti, L., Tchernycheva, M., Julien, F.H., Songmuang, R., Ostasevicius, T., de la Peña, F., Ducati, C., Midgley, P.A., Kociak, M., 2016. Nanometer-scale monitoring of quantum-confined Stark effect and emission efficiency droop in multiple GaN/AlN quantum disks in nanowires. Phys. Rev. B 93, 205410. https://doi.org/10.1103/PhysRevB.93.205410.

Zhang, G., Tateno, K., Birowosuto, M.D., Notomi, M., Sogawa, T., Gotoh, H., 2015b. Controlled 1.1–1.6 μ m luminescence in gold-free multi-stacked InAs/InP heterostructure nanowires. Nanotechnology 26, 115704. https://doi.org/10.1088/0957-4484/26/11/115704.

Zhang, H., Dai, X., Guan, N., Messanvi, A., Neplokh, V., Piazza, V., Vallo, M., Bougerol, C., Julien, F.H., Babichev, A., Cavassilas, N., Bescond, M., Michelini, F., Foldyna, M., Gautier, E., Durand, C., Eymery, J., Tchernycheva, M., 2016. Flexible photodiodes based on nitride core/shell p–n junction nanowires. ACS Appl. Mater. Interfaces 8, 26198–26206. https://doi.org/10.1021/acsami.6b06414.

Zhang, H., Jacopin, G., Neplokh, V., Largeau, L., Julien, F.H., Kryliouk, O., Tchernycheva, M., 2015a. Color control of nanowire InGaN/GaN light emitting diodes by post-growth treatment. Nanotechnology 26, 465203. https://doi.org/10.1088/0957-4484/26/46/465203.

Zhang, Y., Wu, J., Aagesen, M., Holm, J., Hatch, S., Tang, M., Huo, S., Liu, H., 2014. Self-catalyzed ternary core–shell GaAsP nanowire arrays grown on patterned Si substrates by molecular beam epitaxy. Nano Lett. 14, 4542–4547. https://doi.org/10.1021/nl501565b.

Zhu, J., Yu, Z., Burkhard, G.F., Hsu, C.-M., Connor, S.T., Xu, Y., Wang, Q., McGehee, M., Fan, S., Cui, Y., 2009. Optical absorption enhancement in amorphous silicon nanowire and nanocone arrays. Nano Lett. 9, 279–282. https://doi.org/10.1021/nl802886y.

CHAPTER FIVE

Semiconductor Nanowires for Thermoelectric Generation

Gerard Gadea*, Alex Morata*, Albert Tarancon*,†,1
*Catalonia Institute for Energy Research (IREC), Barcelona, Barcelona, Spain
†Catalan Institution for Research and Advanced Studies (ICREA), Barcelona, Barcelona, Spain
[1]Corresponding author: e-mail address: atarancon@irec.cat

Contents

Semiconductors and Semimetals, Volume 98
ISSN 0080-8784
https://doi.org/10.1016/bs.semsem.2018.01.001

1. THERMOELECTRIC GENERATION

Thermoelectric generators (TEGs) make use of the thermoelectric effect for converting heat into electricity. This section introduces the basic principles of thermoelectricity and its use for thermoelectric generation.

1.1 The Seebeck Effect and Thermoelectric Generators

Thermoelectricity is the direct and thermodynamically reversible conversion of heat to electricity and vice versa. Thermoelectricity is encompassed by three differentiated effects that take name after the scientists who discovered them in the late XIX century: the Seebeck, Peltier and Thompson effects.

Among them, the Seebeck effect explains the transformation of heat into electricity and consists in the generation of a voltage difference along the ends of a material whose ends are kept at different temperatures. This Seebeck voltage V is proportional to the temperature difference ΔT and a constant characteristic of the material and the temperature of the measurement, namely the Seebeck coefficient S (Eq. 1):

$$V = S \cdot \Delta T \tag{1}$$

This effect is explained in terms of the temperature-driven random motion of charge carriers, namely electrons in metals and n-type semiconductors, and holes in p-type semiconductors. The carriers at the hot side move faster than those at the cold side, which yields to a net transport to the cold side, where they accumulate generating a voltage (Fig. 1A). This leads to a negative Seebeck coefficient for metals and n-type semiconductors and a positive Seebeck coefficient for p-type semiconductors (Eq. 1).

When the ends of a thermoelectric element subjected to a temperature difference are connected to a device, the Seebeck voltage generates a current flow in the closed loop (Fig. 1B). This current delivers a power P to the device—referred to as the load—which is proportional to the heat absorbed in the hot part Q_H and thus to the cross section A of the thermoelectric element—referred to as the leg. This is the working principle of a TEG. In TEGs a number of legs N are interconnected electrically in series and thermally in parallel, so the voltages add up in the same direction (Fig. 1C). In these multileg TEGs, P remains the same as it would be achieved with a single, thicker thermoelectric element of equal total cross section, but is delivered with higher voltage ($\propto N$) and lower current ($\propto 1/N$). This is

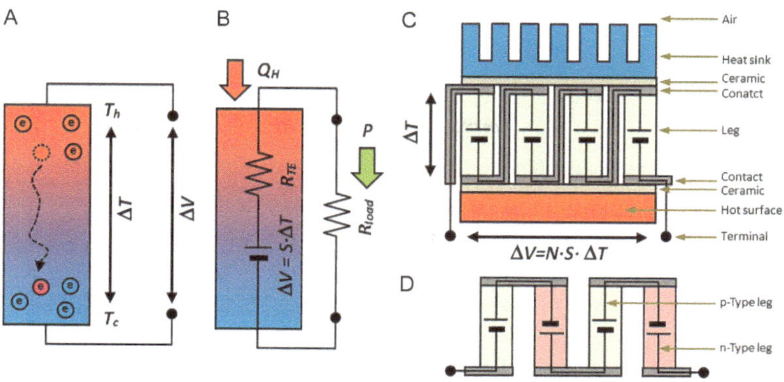

Fig. 1 (A) Seebeck effect in an electron conductor material (metal or n-type semiconductor). (B) Single-leg TEG connected to an electrical load. (C) Multiple leg TEG in unileg configuration, with voltage build-up flow and parts indicated. (D) Multiple leg TEG in unileg configuration, with voltage build-up indicated (only legs and contacts are shown).

done because Seebeck voltages of single legs are in the order of the mV range, while electronic devices have typically voltage requirements of few V. Two-leg configuration (Fig. 1D) using alternate p- and n-type semiconductors—with alternate sign of Seebeck coefficient—is commonly employed as it generally enables an easier fabrication approach.

In most heat recovery applications the hot source consists in a solid surface on which the TEG is placed, while the cold source is the surrounding air. In these cases a heat sink is incorporated on top of the TEG as in Fig. 1C and D, in order to increase air cooling by natural convection and thus achieve a higher ΔT.

1.2 Characteristics and Applications of Thermoelectric Generators

As in heat engines, the efficiency of thermal to electrical energy conversion $\eta = P/Q_H$ increases with temperature difference and is restrained to the Carnot limit. As ΔT goes from 100 to 700 K, η goes from 5% to 15% in TEGs, while for mechanical heat engines—Stirling machines, Rankine and Brayton turbines—it goes from 15% to 50% (Vining, 2009). These lower efficiencies translate to a higher overnight cost of TEG implementation—5–100 \$/W—when compared to that of steam turbines used in gas or coal generation plants—1–3 \$/W (LeBlanc et al., 2014). Though overnight cost comparison does not account for operation, maintenance, and replacement costs—for which TEGs are virtually exempt—it is clear that at the moment converting energy by means of the thermoelectric effect is more expensive than doing it with mechanical heat engines.

Despite their lower efficiency, TEGs have many advantages over heat engines in the field of thermal to electrical energy conversion. First, TEGs are scalable from the μm to the m range and are designed with a planar architecture. This means that they can cover a broad range of powers without economies-of-scale penalization and easily be fitted in confined spaces or portable devices. Second, their design is simple, with no movable parts or working fluid which implies a highly reliable and stable operation. Also their lack of parts that need periodical adjustment or replacement implies that they are maintenance-free, reducing costs, and enabling unmanned operation. For these reasons, TEGs are typically used in niche applications where access is hard, reliability is important, or maintenance is impossible. Some examples are the interior of cars for automotive energy recovery, powering remote monitoring stations in gas pipelines—converting the heat from a gas combustor—or in spacecrafts, converting the heat from a nuclear source within a radioisotope thermoelectric generator (Rowe, 2006).

1.3 Materials for Thermoelectric Conversion

The efficiency of an idealized TEG—with perfect electrical and thermal contacts and optimum dimensions for a specific connected device—is given by (Rowe, 2006)

$$\eta = \frac{P}{Q_H} = \left(1 - \frac{T_c}{T_h}\right) \cdot \frac{\sqrt{1 + ZT} - 1}{\sqrt{1 + ZT} - 1} \tag{2}$$

where T_c is the temperature at the cold side, T_h the temperature at the hot side, $T = (T_c + T_h)/2$ the average temperature, and ZT the so-called thermoelectric figure of merit, defined as

$$ZT = \frac{S^2 \sigma T}{k} \tag{3}$$

where S is the Seebeck coefficient, σ is the electrical conductivity, and k is the thermal conductivity. These are the so-called thermoelectric transport properties, as they relate the current and heat transport with thermal or electrical gradients applied. When a two-leg TEG is considered, average values of S, σ, and k are taken. As ZT increases so does η, approaching to the Carnot limit $1 - T_c/T_h$ when $ZT \rightarrow \infty$. Thus, good thermoelectric materials have high S (for having a high voltage), high σ (for avoiding power losses by Joule

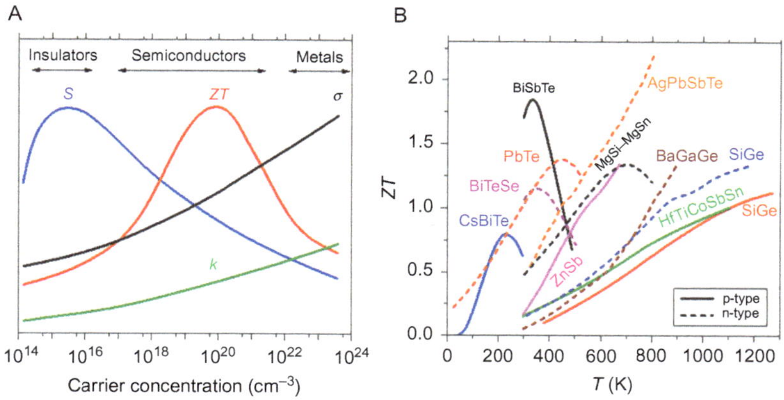

Fig. 2 (A) Illustrative plot of Seebeck coefficient S, electrical conductivity σ, thermal conductivity k, and thermoelectric figure of merit ZT with respect to carrier concentration. (B) ZT with respect to temperature for some semiconductor materials. *Data extracted from Rull-Bravo, M., et al., 2015. Skutterudites as thermoelectric materials: revisited. RSC Adv. 5 (52), 41653–41667. http://xlink.rsc.org/?DOI=C5RA03942H.*

effect), and low k (for being able to sustain high-temperature differences). Transport properties are mostly controlled by carrier concentration and are interrelated in such a manner that ZT is maximized for semiconductors, as in insulators σ is too low and in metals S is virtually null (Fig. 2A). In semiconductors carrier concentration is easily controlled by impurity doping. For this reason, semiconductors intended to be used in thermoelectric applications—insulators in their intrinsic state—need to be doped to their optimum value. In some cases, as in bismuth telluride alloys (Bi–Te), carrier concentration can also be controlled by tuning stoichiometry. Transport properties follow a similar tendency with temperature, so different semiconductors exhibit ZT peaks at different temperatures, which are optimum for their application (Fig. 2B). Typical ZT values of 1 for conventional thermoelectric materials (Bi_2Te_3, SiGe, PbTe) and the application of Eq. (3) yield to the aforementioned TEG efficiencies in the range of 5%–15% for commercially available thermoelectric materials. So far other materials with higher ZT of 1–2 shown in Fig. 2B have not reached the market.

1.4 Device Efficiency Considerations

Apart from the thermoelectric legs, TEGs are comprised by many other elements, as illustrated in Fig. 3A and B: metallic contact pieces joining legs and connecting the powered device (referred to as load), air between the legs and

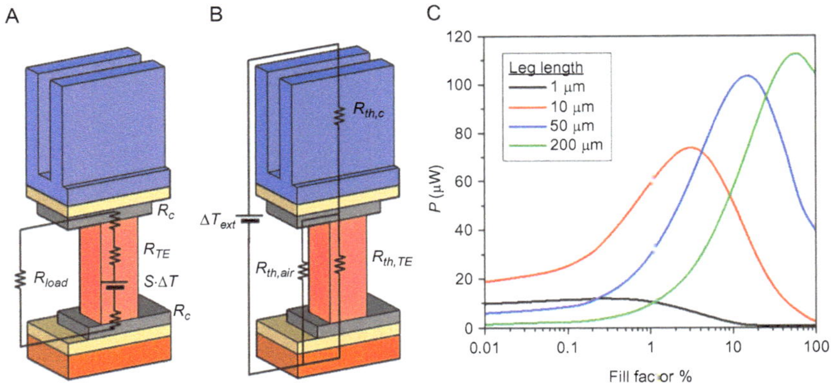

Fig. 3 (A) Section of a TEG leg and electrical model. (B) Section of a TEG leg and thermal model. ΔT is the temperature difference along the thermoelectric leg, while ΔT_{ext} stands for the temperature difference between the hot source and the cold air surroundings. (C) Example of thermal matching of the thermoelectric legs of a micro-TEG device. The power of a TEG of 1 mm^2 harvesting at $\Delta T = 100$ K is plotted for different leg lengths (*color of curves*) and fill factors (*x*-axis). *Data extracted from Santos, J.D., 2015. Simulation of Thermally Matched Unileg Microthermoelectric Generators Based on p-Type SiNWs. Barcelona.*

heat exchangers. While the figure of merit ZT indicates the potential for thermoelectric application in ideal conditions, the actual power per unit area (P/A) that a TEG is able to supply depends on other factors. These are indicated in Fig. 3A and B: (i) the thermal resistances in series and in parallel with that of the thermoelectric legs $R_{th,TE}$, which include those of air gaps surrounding the legs $R_{Th,air}$ and air convection through the heat exchanger in the cold side $R_{th,c}$; (ii) the metal–semiconductor electrical contact resistance of the legs with the contact stripes R_c; (iii) the electrical resistance of the load connected to the TEG R_L.

The maximization of the P/A is reached when the electrical matching of the load and thermal matching of the legs are attained. That implies changing the load, the leg length, and the fraction of TEG area covered by thermoelectric material (referred to as the fill factor) so that P/A is maximized for a determined material, and electrical/thermal parasitic resistances as illustrated in Fig. 3C. Electrical matching is attained when $R_L = 2R_c + R_{TE}$ (Rowe, 2006). In cases in which the temperature source is not thought to be affected by the TEG (e.g., when a small TEG is placed on top of a great, hot surface), $R_c \ll R_{TE}$ (negligible contact resistance) and $R_{th,air} \gg R_{th,TE}$ (negligible through-air heat bypass), then thermal optimization is attained when $R_{th,c} = R_{th,TE}$ (Leonov et al., 2011). This is why leg optimization is referred to as "thermal matching" as in this case the thermal resistance of the leg array $R_{th,TE}$ is matched to the thermal resistance of the heat sink $R_{th,c}$.

Another convenient metric for comparing materials is the overnight production cost in \$/W, which is calculated as the costs in \$ of thermoelectric material, heat exchangers, manufacturing, assembling, and the rest of components involved in the fabrication of a TEG able to convert 1 W at a specific external ΔT conditions (Yazawa and Shakouri, 2011; Yee et al., 2013)—"external" meaning that the difference is naturally achieved by using a heat sink and natural convection in air. \$/W metric allows assessing their suitability for application in a specific temperature scenario, indicating how good the material is at converting energy in these conditions and was calculated by LeBlanc et al. (2014) for several materials.

1.5 Microthermoelectric Generators for Energy Harvesting

Microthermoelectric generators (μTEGs) are TEGs with leg cross sections below 0.2 mm^2 which are fabricated with microfabrication techniques commonly used in the production of integrated circuits and microelectromechanical systems, as photolithography, thin-film deposition, etching, and pick-and-place tooling (Stark, 2015). μTEG microfabrication does not require expensive bulk processing as melting, milling, sintering, or metallization by soldering and enormously facilitate device parts interconnection by using micropatterning techniques. This leads to an area-based rather than volume-based cost dependency and allows for massive production of μTEGs and direct connection and integration with other microfabricated devices.

Micro-TEGs have typically sizes of 1–100 mm^2 and produce powers from 0.1 to 1000 μW when placed on top of hot surfaces. Combined with a power conditioning system and a small rechargeable battery—to cope with the intermittence of hot source availability—a single cm^2 of these μTEGs they can harvest 100–1000 mWh/year. This power is enough to feed small electronic devices such as microcontrollers, sensors, antennas, or LEDs for instance which in most devices have a similar average yearly consumption.

Reliability, small size, and integrability with electronic devices make μTEGs ideal candidates for covering the needs of wireless sensor network (WSN) nodes, rendering them autonomous whenever a heat source is available. This removes the necessity of battery replacement which is expensive and difficult, especially when large numbers of sensors are considered, often hard to access in industrial environments or remote stations. When heat source availability is restricted, setting the μTEG-powered devices in charge of the more energetically demanding tasks—send/receive to/from monitoring station—serves to increase network lifetime. The concept of extracting energy from ambient sources is known as energy harvesting. Energy harvesting has been attracting a great deal of interest lately due to the increase of the global number of sensors anticipating the forthcoming Internet of Things.

Another important remark of μTEGs is that so far they are the only devices that have achieved integration of thermoelectric nanomaterials as nanowires or thin films, which—as discussed in Section 2—present outstanding characteristics for exploitation of the thermoelectric effect. This is because—as discussed in Section 3—the fabrication methods for these nanomaterials are often such that not only their cross section but also their length is restrained to the μm scale, and thus μTEGs serve as an interface for bridging these nano- and microstructures to macroscopic thermal and electrical elements.

In the case of μTEGs, three things are considered relevant, namely, P/A in order to power the devices in a size-constrained area or from a size-constrained heat source, cost ($/W metric), and hot source temperature needed for operation.

2. SIZE EFFECTS IN NANOWIRES FOR THERMOELECTRIC APPLICATION

Nanomaterials are defined as materials with at least one dimension below a hundred nanometers. Thin films, nanowires, and quantum dots present one, two, and three dimensions constrained to this order. Nanomaterials have different properties from bulk materials due to the closeness of all their constituting atoms to material boundaries and to quantum mechanical effects that become significant in this size regime. As explained below, this affects their transport properties S, σ, and k, leading to a higher figure of merit ZT which leads in turn to an enhancement of thermoelectric performance.

Nanowires have two constrained dimensions—rather than one in thin films—allowing superior exploitation of size effects, and one free dimension—rather than none in quantum dots—allowing elongated structures able to withstand thermal gradients. Thus, nanowires are excellent candidates for application of the low size-driven improvement of thermoelectric transport properties. Due to their nanometric diameter and also their micrometric length—inherent to their fabrication routes—nanowires that are intended to exploit the thermoelectric effect must be integrated in μTEG devices. This implies that a series of geometric effects related to low dimensionality of μTEG legs must be taken into account. In this section, size effects, both in transport properties and in device aspects, are introduced and described.

2.1 Electronic Effects

The theoretical work of Hicks and Dresselhaus in 1993/1994 suggested that nanostructure thermoelectric materials should present an enhancement of ZT, triggering the research in this field. This was mostly attributed to quantum-mechanical effects that increase the thermoelectric power factor ($S^2\sigma$) with respect to bulk materials. Since then, much experimental and theoretical progress has been done. A comprehensive and updated description of these effects is given in the work of Pichanusakorn and Bandaru (2010a,b), recommended for the interested reader. An intuitive picture is given herein, based in the explanations of this and other recent works and energy band theory, which is extensively covered in several textbooks (Li, 2010; Rowe, 1995, 2006). Other effects significant at low dimensions—both positive and negative for thermoelectric application—and related to nanowires and other nanostructures are introduced herein as well.

2.1.1 Quantum Confinement

Differently from metals, in doped semiconductors thermal conductivity k does not change significantly with carrier concentration. This means that for reaching a maximum value of $ZT = S^2\sigma\, T/k$, an optimal carrier concentration n will be that which maximizes the thermoelectric power factor, defined as $S^2\sigma$, as illustrated in Fig. 2A. This optimal power factor can be further increased by using the quantum confinement effect, as explained below.

The electrical conductivity σ can be defined as

$$\sigma = n \cdot e \cdot \mu \tag{4}$$

where e is the charge of the carrier ($\pm$ the charge of the electron for a p-type or an n-type semiconductor), n is the carrier concentration, and μ the carrier mobility. Assuming constant μ, maximization of $S^2\sigma$ will be attained when S^2n is maximized. According to the derivation of Seebeck coefficient S by

Boltzmann transport equation with relaxation time approximation, this factor can be defined as

$$S^2 n = \frac{1}{(e \cdot T)^2} \left(\frac{\langle \tau \cdot E \rangle}{\langle \tau \rangle} - E_F \right)^2 n \tag{5}$$

where E_F is the Fermi level of the semiconductor, and $\langle \tau E \rangle / \langle \tau \rangle$ is the average energy of the electrons in the band weighed with the relaxation time τ. τ is the average time that an electron of energy E can travel freely before a scattering event randomized its momentum. If τ is constant with energy $\langle \tau E \rangle / \langle \tau \rangle = \langle E \rangle$. Thus $S^2 \sigma$ increases with the difference between the average energy of electrons $\langle E \rangle$ and the Fermi level E_F. Both parameters depend on how electrons fill the available energy levels in doped semiconductors, which is affected by low-dimensionality effects, as explained below.

The fraction of filled levels in a determined range of energy dE are given by $g(E) \cdot f(E)\ dE$, where $g(E)$ is the density of states (DOS) and $f(E)$ the Fermi distribution function (Li, 2010). The DOS function changes with energy and in semiconductors and insulators presents discontinuous regions referred to as bandgaps (Fig. 4A). $f(E)$ indicates the probability of an electron level of a certain energy E to be occupied, is centered in the Fermi energy E_F, and changes from zero to one in a range spanning a few kT energy units (Fig. 4B).

The shape and nature of the DOS functions are derived from quantum mechanics calculations applied to systems with many atoms. Isolated atoms have discrete energy levels. As they come close to other atoms—as when packed forming a solid—these energy levels are slightly pushed upward

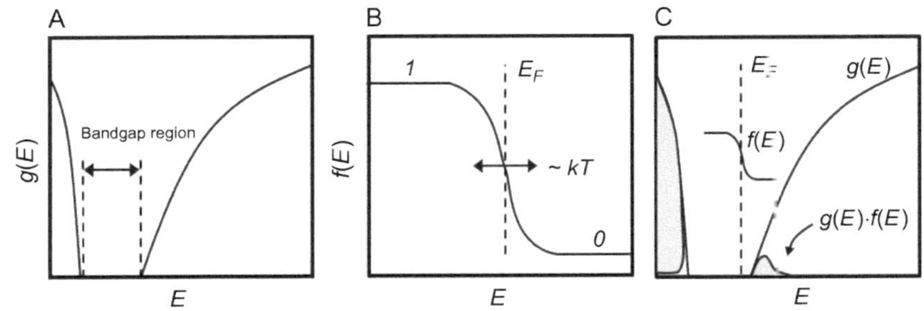

Fig. 4 (A) Density of states (DOS) function of a semiconductor material. (B) Fermi–Dirac distribution function along with the position of Fermi function E_F indicated. (C) Filling of the energy levels in a semiconductor material. *Shadowed area* represents levels filled with electrons.

or downward in order not to coincide with those of neighboring atoms and break the Pauli exclusion principle. This leads to the expansion of each level to an energy band, which can be thought of as a continuous range of allowed energy for electrons. The number of allowed energy levels (termed states) per unit volume per unit energy range at a certain energy E is the density of states function $g(E)$. When a large number of atoms are considered, the shape of $g(E)$ is dimension independent, as the interaction felt by electrons saturates with the distance. However, when any dimension is shrunk enough, $g(E)$ starts to show a saw pattern due to electrons perceiving that in one or more dimensions there is no longer a smooth transition between energy levels as illustrated in Fig. 5. As long as at least one of these dimensions is still large—thin films or nanowires, but not quantum dots—the DOS function will still be continuous.

As the diameter is reduced, subbands with sharp profile appear in few number and more definite, the DOS has a higher magnitude, and the minimum energy that electrons can have —termed as the ground energy E_0—is increased. The exact reasons and quantitative description of these changes can be found elsewhere (Kōmoto and Mori, 2013; Li, 2010). The magnitude of the DOS has been shown to increase as $1/d^2$ for nanowires, which is ultimately responsible for the increase of S^2n in thin films and nanowire structures (Pichanusakorn and Bandaru, 2010a,b).

Fig. 6 shows how the energy levels are filled in an extrinsic semiconductor with carrier concentration n, when it is bulk (Fig. 6B) and a thin nanowire ($d = 1$ nm, Fig. 6B). The carrier concentration n is fixed—determined by the concentration of dopant species which are considered to be fully ionized—and is represented by the shadowed area. As the semiconductor passes from bulk to nanowire, the increased magnitude of the density of states affects the electron distribution. The Fermi distribution f function

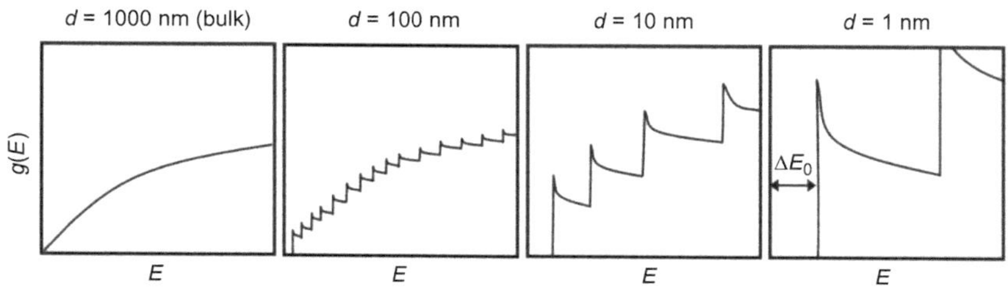

Fig. 5 Illustrative plot of density of states function $g(E)$ vs energy E for a nanowire whose diameter is reduced from bulk (represented by 1000 nm) to 1 nm. In the 1 nm case the increase of band ground energy E_0 is indicated.

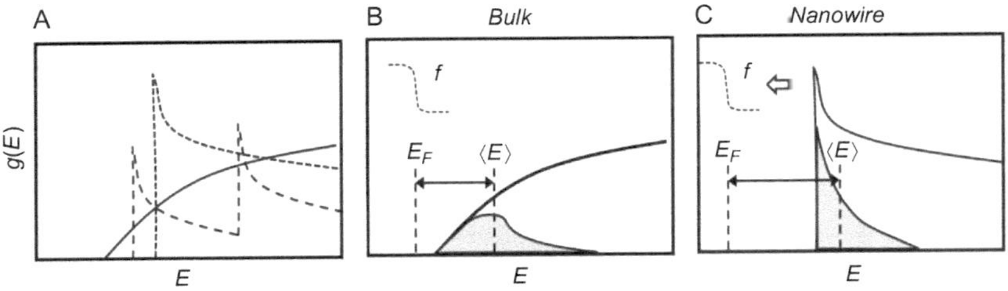

Fig. 6 (A) DOS function of a bulk semiconductor (*continuous line*), of a nanowire semiconductor of $d = 50$ nm (*dashed line*), and of a nanowire of $d = 1$ nm (*dotted line*). (B) Energy filling for a bulk semiconductor. (C) Energy filling for a semiconductor nanowire of $d = 1$ nm. In (B) and (C), values of Fermi level E_F and average energy $\langle E \rangle$ are indicated. *Shadowed areas* indicate levels occupied by electrons.

shifts toward lower E values by adopting a new E_F that makes that the concentration of electrons in this modified band—shadowed area in Fig. 6B and C—is still n. This implies that both $\langle E \rangle$ and E_F are decreased. Since the weight of the Fermi function $f(E)$ is higher than that of the density of states $g(E)$, these parameters change so that $\langle E \rangle - E_F$ is increased in the nanowire and thus so is S^2n (Eq. 5). This is done without a prejudice of σ—as n is fixed and determined by extrinsic doping—and μ is also treated as fixed. Low-dimensionality effects in μ are hard to predict. Some studies indicate that it does not change significantly with d, while others state that it might indeed be due to carrier scattering at interfaces, in such a way that it would overweigh the positive effect of S^2n, leading to a decrease of the power factor (Pichanusakorn and Bandaru, 2010a,b).

However, the aforementioned effect can be reverted in cases of high doping or intermediate nanowire diameter. As illustrated in Fig. 6A, the magnitude of the density of states can be lower than that of the bulk when higher n and/or d are considered. In these cases, all the argumentation presented before is applied in the opposite sense, leading to an upward shift of E_F and thus a lowering of S^2n. Cornett and Rabin (2011) predict that a reduction of up to 40% of the power factor can be expected in InSb nanowires of $d \approx 20$ nm due to this effect. Pichanusakorn and Bandaru (2010a,b) calculated the minimum diameters necessary for observation of positive quantum confinement effects for some typical semiconductor materials and found them to be in the range of 4–8 nm. Theoretically at smaller diameters the $1/d^2$ dependence of DOS magnitude may be observed and employed to improve the thermoelectric power factor. To the best of the authors' knowledge, experimental evidence of this effect has only been observed in the work of Wu et al. (2013), in which an enhancement of the power factor was measured for relatively thick (20–70 nm) indium arsenide nanowires.

2.1.2 Semimetal–Semiconductor Transition and Carrier Pocket Engineering

Thermoelectric materials are highly doped semiconductors, which implies that they possess mostly a single type of carrier, p-type (holes) or n-type (electrons). For this reason, so far, only a band of interest has been considered—the valence band where holes are located in p-type materials or the conduction band where conduction electrons are located in n-type materials. However, when both carriers are present in noticeable amounts, the Seebeck coefficient is obtained as

$$S = \frac{S_p \sigma_p + S_n \sigma_p}{\sigma_p + \sigma_n} \tag{6}$$

where S_p, S_n, σ_p, and σ_n, are the hole contribution to Seebeck coefficient, the electron contribution to Seebeck coefficient, the hole electrical conductivity, and the electron electrical conductivity, respectively. In highly doped semiconductors $\sigma_p \gg \sigma_n$ (or vice versa), S is mostly that of the majority carriers. However, when the temperature is high enough so that kT is in the order of the bandgap energy, then electrons are thermally promoted from the valence to the conduction band, leaving a hole behind. Soon carriers generated this way overcome those supplied by impurity doping and then $\sigma_p \approx \sigma_n$. Since $S_p \approx -S_n$ the average S coefficient rapidly drops (Eq. 6), and so does ZT as observable for any material after its optimum temperature (Fig. 2B). This is the so-called bipolar effect. For this reason a high bandgap is interesting for thermoelectric materials as it allows them to operate at a higher T where the efficiency of thermoelectric devices is higher (Eq. 2).

The bipolar effect is already present in undoped semiconductors, which possess the same number of holes and electrons—besides from a very low σ—and in semimetals. Semimetals do not have a bandgap as insulators or semiconductors—as the valence and conduction bands overlap—but do neither have a half-full band, as metals do. Thus their low Seebeck is due to a high bipolar effect, rather than to a high degeneracy.

Bismuth is a semimetal that possesses a high mobility μ (25 times that of silicon) and low lattice thermal conductivity $k_l \sim 1\,\mathrm{W/mK}$. Thus it would be a very good candidate for thermoelectric application if it were a semiconductor instead of a semimetal. This can be attained by

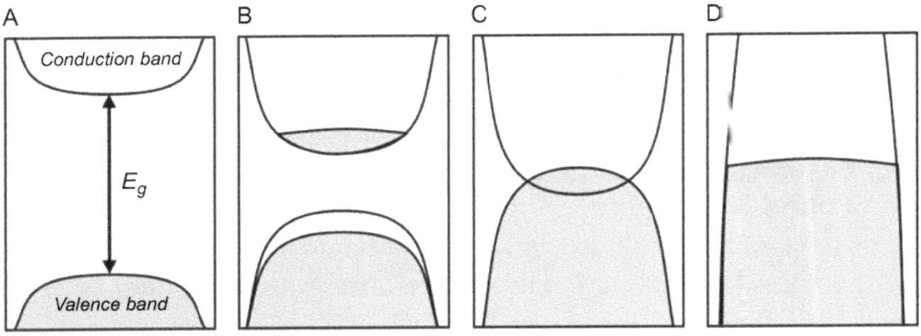

Fig. 7 Schematic representation of band structure (DOS and filled levels vs wave vector) for different types of materials. (A) Insulator, with bands and bandgap energy E_g indicated. (B) Semiconductor. (C) Semimetal. (D) Metal.

nanostructuring bismuth in the shape of nanowires. As can be seen in Fig. 5, one of the effects of low dimensionality in the density of states is the increase of ground energy E_0—i.e., the minimum energy of the conduction band. Similarly, the valence band is pushed downward when quantum confinement takes place. This implies that quantum confinement effect can be used to render bismuth a semiconductor instead of a semimetal (passing from Fig. 7C to B), enabling its use for thermoelectric application. Several works have experimentally proven this fact, with bismuth and bismuth antimonide nanowire arrays (Boukai et al., 2006; Lin et al., 2002). However, so far, there has not been a full *ZT* characterization or integration of bismuth-based nanowires in devices.

A similar approach is employed in carrier pocket engineering. In this case, the different shift of E_0 of different conduction/valence valleys located at different wave vectors is employed to bring them together at the same energy level. This way and by readjusting the doping level, one can use additional valleys for conduction and thus increase σ, as demonstrated by Rabin et al. (2001) for bismuth antimonide nanowires.

2.1.3 Detrimental Surface Effects

Some undesired effects may take place in nanowires regarding the effectiveness of their doping levels or carrier mobility, which could reduce σ. These are depletion of carriers due to the presence of surface charges, dopant deactivation due to the absence of a continuous bulk around the impurity atoms, and scattering of charge carriers at nanowire boundaries, as explained below.

Surface depletion of carriers in boron-doped silicon nanowires was investigated by Kimukin et al. (2006). These authors reported a depletion layer at the surface of the nanowire and attributed it to electrostatic repulsion of holes by H^+ ions adsorbed on the surface, a phenomenon naturally taking place at room temperature. According to that work, the depleted charge density at surface was 2.8×10^{12} cm^{-2}, which implied that, for their doping level of 2.1×10^{18} cm^{-3}, only NWs thinner than 80 nm presented a significant decrease in conductivity. In principle, this effect could be positive as well, if an accumulation rather than a depletion layer is formed. For instance, in the work of Pennelli et al. (2013), it is suggested that the formation of surface states at a Si/SiO_2 interface could be contributing to enhance σ in n-type silicon nanowires.

Regarding dopant deactivation, this effect was investigated for boron in silicon nanowires by Diarra et al. (2007) and Kimukin et al. (2006) being explained in terms of an increased ionization energy (E_i) in nanostructured materials with respect to bulk. E_i is of the same order as kT in bulk at room temperature, allowing full activation of the dopant atoms. This value is low enough in part due to the electrostatic screening provided by mobile charge carriers nearby the dopant center, which are abundant in the bulk material. However, as the size of the material is reduced and its ends are dragged closer to the dopant center, the number of carriers available for the screening becomes limited, consequently raising E_i. This in turn diminishes the fraction of B atoms ionized, and thus the effective doping level which determines S and σ. Kimukin et al. (2006) investigated this effect with phosphorus-doped n-type silicon nanowires and showed that it becomes significant at diameters below 30 nm.

The mean free path of charge carriers—which defines mobility μ and thus σ—is determined by the frequency of collisions with scattering centers as vibrating lattice atoms, crystalline defects, or ionized impurities and has typically values of 1–1000 nm depending on the material and temperature (Kittel, 2004). When the material size is reduced below this value—as in nanowires—important scattering of carriers at boundaries can take place (this effect is explained and illustrated in more detail for thermal transport in Section 2.2.1). In thermoelectric materials—heavily doped semiconductors—the dominant scattering mechanism is the ionized-impurity one, derived from the collisions of carriers with ionized dopant atoms, which present a large radius of interaction. In this case, the mean free path in bulk is in the order of 2–5 nm, implying that a significant mobility reduction takes place only when the diameter of the nanowire is reduced bellow this threshold. This issue has been theoretically discussed by Seong et al. (2012) and Shi (2012) and, so far, there is no experimental evidence of carrier scattering at nanowire boundaries.

2.1.4 Effects in Other Nanostructures

Other nanostructures and related effects are briefly introduced herein as often the former are combined with nanowires, as, for example, polycrystalline nanowires or superlattice nanowires.

The concept of energy filtering consists in increasing the Seebeck coefficient S by introducing scattering mechanisms that are stronger for low-energy electrons and weaker for high-energy electrons. This way a higher $\langle \tau E \rangle / \langle \tau \rangle$ will be obtained leading to the consequent increase in $S^2 n$ (Eq. 5). This can be attained by introducing crystal boundaries in the material, as for instance grain boundaries in polycrystalline materials. Grain boundaries have a thin charged layer due to abnormally coordinated atoms, which may serve as an energy filter: the carriers with small energies will not be able to surpass the barrier, being backscattered, whereas the high-energy ones will. This effect not only increases the S in polycrystalline materials but also decreases σ—as low-energy electrons cease to contribute to electric transport—and k—as phonons are scattered at these interfaces that disrupt crystal continuity. When the density of interfaces is high, as in nanograined bulk, the overall effect in ZT is positive (Bux et al., 2009; Kessler et al., 2013; Minnich et al., 2009; Suriano et al., 2015).

Another way of achieving energy filtering is by resonant doping (Rowe, 1995). In this case an impurity with energy below $\langle E \rangle$ is introduced in the material, so that mostly low-energy carriers interact with their states. The constant entering and exiting in these exotic states serve as an effective scattering mechanism, producing a delay in the transport of low-energy electrons along the material.

Superlattices are structures with many thin layers alternating a conductor—which is intended to be confined—and a barrier layer—an insulator. The superlattice configuration allows exploiting many effects, as for instance carrier pocket engineering (Koga et al., 1998), dopant modulation (Harman et al., 1996), or phonon coherent resonance (Chen, 1997), explained in detail elsewhere (Pichanusakorn and Bandaru, 2010a,b).

2.2 Thermal Effects

Another approach for increasing $ZT = S^2 \sigma T / k$ is decreasing the thermal conductivity of the semiconductor, k. Thermal conductivity is composed of a lattice component k_l and an electronic contribution k_e, which represent the contribution to heat transfer by atomic vibrations and by electrons, respectively:

$$k = k_l + k_e \tag{7}$$

Generally in semiconductors, the assumption of $k_e \ll k_l$ is considered. Moreover, k_e is linked to σ by means of the Wiedemann–Franz law, which states that $k_e \propto \sigma$, and thus the relation σ / k_e is invariant—a fact that cannot be overcome using a low-dimensional approach as with the S–σ relation, because electrons necessarily transport heat and charge at the same time. For these reasons, improvement of thermoelectric or thermal properties is dealt with by reducing k_l. Nanowires present mechanisms for reducing k_l, which are introduced and briefly described below. The formalism of heat transport by phonons and lattice dynamics, and explanations of low-dimensional effects taking place in nanowires can be found in several references (Pichanusakorn and Bandaru, 2010a,b; Rowe, 1995, 2006).

2.2.1 Phonon Boundary Scattering

Thermal conductivity in solids is usually explained in terms of phonons. Phonons are quasiparticles associated to the thermal vibrations with a frequency ω that transmit heat within a material. In a similar manner as electrons in Section 2.1.1, phonons can be treated as particles with a heat capacity C_v, a velocity v, and a relaxation time τ, which is the average time that takes for a phonon to encounter a scattering event that randomizes its momentum. The phonon mean free path (MFP $= v \cdot \tau$) is thus the average distance that a phonon can travel before it changes its direction, as illustrated in Fig. 8A for a bulk semiconductor. Lattice thermal conductivity k_l is defined in terms of the phonon properties as

$$k_L = \sum C_v \cdot v \cdot \text{MFP} \tag{8}$$

where the summation is made over all the phonon modes of all frequencies ω. C_v, v, and MFP depend on the frequency. The scattering events, which reduce the MFP, are caused by imperfections that disturb the crystalline periodicity of the lattice as ionized impurities, atoms of different mass, or boundaries of the material. Scattering is also produced by other particles or quasiparticles able to exchange energy and momentum as electrons or

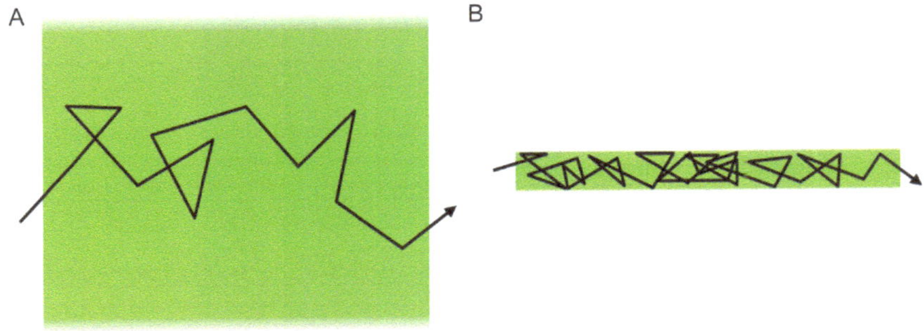

Fig. 8 Representation of heat transport in solid materials, in which heat is carried by phonons (*black arrow*). (A) Bulk material with MFP not limited by boundaries. (B) Nanowire material, with MFP limited by boundaries. The total distance traveled by the phonon in (B) is four times that of the phonon in (A) meaning that the thermal conductivity of (B) is reduced in the same factor.

other phonons. The contributions of different scattering mechanisms are added by Matthiessen's rule:

$$\frac{1}{\mathrm{MFP}} = \frac{1}{\mathrm{MFP}_{impurities}} + \frac{1}{\mathrm{MFP}_{boundaries}} + \frac{1}{\mathrm{MFP}_{phonon-phonon}} \tag{9}$$

but often one of these MFPs is much smaller than the rest and thus dominant.

An approach for reducing MFP and thus thermal conductivity of a semiconductor is forcing the increase of boundary events by diminishing the dimension of the material. As illustrated in Fig. 8B, this is accomplished in low-dimensional structures as nanowires.

The extent of this reduction can be quantified by the Casimir model. This model assumes that the nanowire has a constant cross section of diameter d and that the phonons scatter at boundaries with a certain degree of secularity represented by parameter p. Totally specular scattering is found for $p = 1$, while $p = 0$ means totally diffuse scattering. Specular scattering as in Fig. 9A does not change the momentum in the direction of heat propagation, meaning that it does not contribute to reduce the thermal conductivity. Conversely, in totally diffuse scattering the momentum is totally randomized at the boundary, meaning that phonons will scatter from there in all directions including backward. The mean free path due to boundary scattering is given by

$$\frac{1}{\mathrm{MFP}_{boundary}} = \frac{V(1-p)}{d} \tag{10}$$

where V is the volume per unit atom. A dominant $\mathrm{MFP}_{boundary}$ with totally diffusive scattering is the scenario, which contemplates a maximum

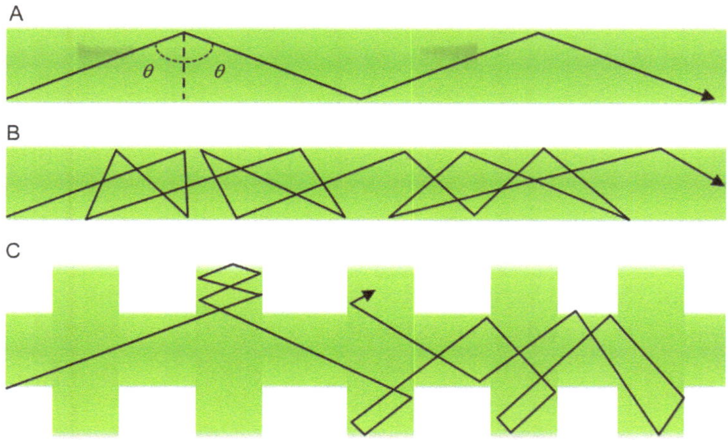

Fig. 9 Different types of phonon boundary scattering. (A) Casimir mode, purely specular ($p=1$). (B) Casimir model, purely diffusive ($p=0$, direction is randomized at each collision). (C) Scattering at a rough nanowire—idealized with a periodic structure—with $p\sim 1$, able to yield conductivities below the diffusive Casimir limit by means of phonon multireflection.

reduction of the thermal conductivity within this model, also known as the Casimir limit. In this case, the thermal conductivity reduction is given by

$$\frac{k_{NW}}{k_{bulk}}=\left(1+\frac{\mathrm{MFP}_{bulk}}{d}\right)^{-1} \tag{11}$$

where MFP_{bulk} is the mean free path that would be achieved in bulk derived from other mechanisms, in the absence of boundary scattering effect. Experimental k data of smooth silicon nanowires of different d fit well for this model with a scattering parameter of 0 (Li et al., 2003a). This indicates that even in smooth surfaces the effects of boundaries are contributing to reduce k. Conversely, at 4K a purely specular scattering is observed for silicon nanowires, as in Fig. 9A, which, combined with the long mean free path due to the reduced phonon–phonon interaction, results in thermal ballistic transport (Maire et al., 2017). This difference is due to p being proportional to e^{1/λ^2}, where $\lambda=v/\omega$ is the phonon wavelength. As the dominant wavelength increases from 2 to 300nm as the temperature is lowered from 300 to 4K, the boundary scattering loses its effectiveness becoming specular. Opposite, ballistic conduction was observed in Si–Ge nanowires at room temperature, at lengths as long as 8.3μm (Hsiao et al., 2013). This is due to the fact that alloy scattering induced by the random positioning of differently sized Si/Ge atoms within the lattice leads to a scattering of most high-frequency phonons. Thus the small fraction of low-frequency remaining phonons are the ones that carry the heat in Si–Ge alloys. Due to their long wavelength, they are not easily scattered by surface roughness and thus experiment ballistic conduction even at room temperature.

2.2.2 Rough Nanowires

An increased roughness has been shown to severely reduce the thermal conductivity in silicon nanowires (Hochbaum et al., 2008; Lim et al., 2012). Lim et al. (2012) found a good correlation between the parameters defining the roughness of the nanowire and the reduction of thermal conductivity, and also that this effect was more important than the diameter reduction. For thermal conductivity obtained this way reduction is below the Casimir limit. This is generally explained in terms of multiple reflections of phonons and localization of modes at the nanorecessions which constitute the rough surface boundary, as illustrated in Fig. 9C. Molecular dynamics simulations confirm this mechanism (Termentzidis et al., 2013), which is possible due to the enhanced roughness conferred by etching methods and also to the high surface-to-volume ratio conferred by the low diameter.

2.2.3 Thermal Quantum Effects in Confined Structures

There are other thermal effects that can contribute to k reduction in nanowire structures related to quantum, rather than classical effects. These effects are more likely to be observable in nanoheterostructures as core–shell or superlattice nanowires of materials with different acoustic impedances—typically Si–Ge. However, there is still no strong experimental confirmation.

Zou and Balandin (2001) calculated effects of localization of phonon modes. In the same manner as electron quantum confinement, the DOS of phonons is affected in small structures leading to exotic phenomena. The results suggested an important reduction of k for silicon when comparing 20 nm nanowires with bulk even supposing specular reflection. This additional, usually unaccounted, k-reducing effect could give an alternative explanation to the notorious fact that even smooth nanowires appear to behave as diffuse scatterers and that many materials go beyond the Casimir limit. Results from Wingert et al. (2011) in core–shell Ge–Si nanowires suggest a possible evidence of phonon confinement.

Chen et al. (2011) predicted a coherent phonon transport in thinner Ge/Si core/shell nanowires. In such a scenario, the waves would propagate in phase due to the acoustic confinement conferred by the silicon layer.

The interference of transverse (nonpropagating) and longitudinal (propagating) modes would lead to reduction in thermal conductivity by 45% at room temperature, if an optimum relation of shell-to-core thickness were employed.

2.3 Device Effects

2.3.1 High Weight of Contact Resistances

The total electrical resistance of a TEG leg is the sum of the thermoelectric material resistance R_{TE} (proportional to leg length L) and two times the contact resistance R_c (invariant with L), both of these terms being inversely proportional to leg cross section A (Eq. 12; Fig. 3A):

$$R = 2R_c + R_{TE} = 2\frac{\rho_c}{A} + \rho\frac{L}{A} \tag{12}$$

where ρ_c is the contact resistivity and ρ is the thermoelectric resistivity which is $1/\sigma$. The origin of R_c is the Schottky barrier formed between the semiconductor thermoelectric material and the contacts which may be metallic or semiconductor with a different level of doping. In semiconductor industry, ρ_c typically takes values from 10^{-4} to 10^{-7} Ω cm^2, depending on the height of the Schottky barrier (Scorzoni and Finetti, 1988).

The ratio of total resistance to thermoelectric material resistance, which ideally should be 1 for a perfect contact, is:

$$\frac{R}{R_{TE}} = 1 + \frac{\rho_c}{\rho L} \tag{13}$$

This ratio is directly related to the output power of a TEG: a value of 2 for instance means that the power of the TEG is half of the power that could be attained by removing the contact resistance. Fig. 10 shows a plot of R/R_c for typical σ and ρ_c values (100–1000 S/cm and 10^{-4}–10^{-7} Ω cm^2, respectively), and a range of lengths achievable for semiconductor nanowires (1–1000 μm). For nanowires of less than 100 μm, ρ_c in the range of 10^{-6}–10^{-5} Ω cm^2 is needed in order to avoid the detrimental effect of R_c, while for sizes below 10 μm a stricter condition of 10^{-7}–10^{-6} Ω cm^2 is naturally imposed. Above these values, contact resistances become significant, drastically reducing the power of the μTEG.

As explained in Section 2.3.1, ρ_c of nanowires takes values ranging from 6×10^{-7} to 5×10^{-4} Ω cm^2 depending on the method employed for contacting. Thus the contact formation is an important aspect that must be considered when conceiving nanowire-based μTEGs as it may be the

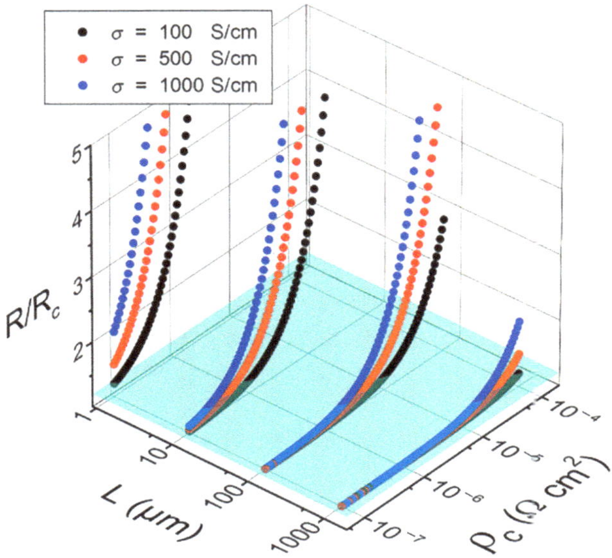

Fig. 10 Ratio of total resistance to thermoelectric material resistance R/R_c for different contact resistivity ρ_c, leg lengths L, and electrical conductivities σ (different curves). The *blue-shadowed plane* represents the level $R/R_{TE} = 1.2$, i.e., the contact resistance accounts for a 20% of the thermoelectric resistance.

decisive element determining whether the device will be able to harvest power from thermal sources or not.

A similar effect is to be expected with thermal contact resistance (sometimes referred to as interfacial thermal resistance). While its mechanisms and effects in thermoelectric devices have not been studied as much as its electrical counterpart, contact thermal resistances arise at metal–semiconductor interfaces due to different phonon properties of each media: density of modes, propagation speed, mean free path, and heat capacity. Although they may appear beneficial for thermoelectrics due to their contribution to lowering effective k, they actually reduce overall TEG power by diminishing the ΔT perceived by the thermoelectric material, which results in a lower Seebeck voltage. Thus they should be avoided and it is worth to note that their importance becomes higher as the legs are shortened, as in the case of nanowires.

2.3.2 Heat Bypass at Microscales

Heat bypass is the heat passing across the air gaps between the legs or the TEG (Fig. 3B, with thermal resistance R_{air}) and is deleterious for devices as thermalization of hot and cold parts will take place leading to a decrease of the Seebeck voltage. When legs have micrometric length—as nanowires in μTEG—some effects must be taken into account.

At very low distances (close to the mean free path of air molecules ~60 nm) the heat transport by conduction through air may enter into a ballistic/Knudsen regime: at this point, air molecules will pass from surface to surface without colliding with other molecules (Kaviany, 2013). This means that the heat transfer between two bodies will increase as the distance is shortened (in the diffusive regime), up to a maximum at which it will saturate (in the ballistic regime). This may be beneficial in the case of really short nanowires, or for avoiding transport in the outward direction of a dense array of nanowires. But it is unlikely that structures with such a short length/pitch are feasible or even useful for thermoelectric application.

Conversely, transport by radiation is enhanced at very short distances. According to the Stefan–Boltzmann law, in structures in the order of 1–100 μm radiative heat transport—scaling with $\Delta(T^4)$—should become significant with respect to conduction heat transport—scaling with ΔT—at temperatures higher than ≈300 K. This is the argumentation line usually given when stating the limits of accuracy of thermal conductivity measurements performed with suspended platforms separated by a few μm (Shi et al., 2003). However, apart from far-field radiation considered by Stefan–Boltzmann law, near-field radiative heat transport must also be taken into account. Fig. 11A shows radiative thermal conductance between two planes

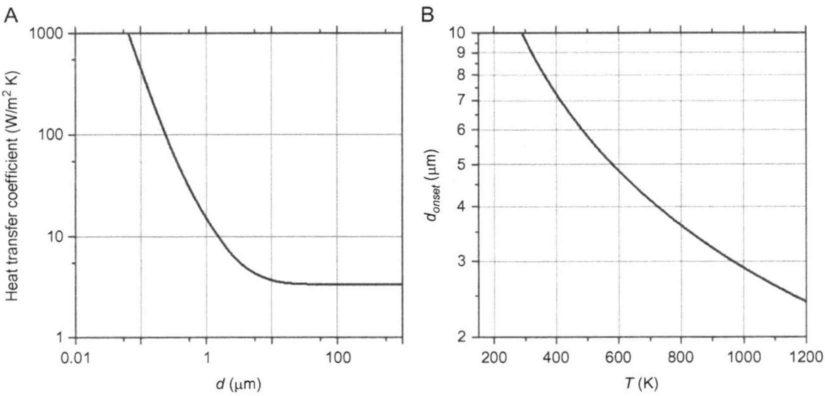

Fig. 11 (A) Radiative heat transfer coefficient as a function of distance for two plates of *z*-cut sapphire. (B) Onset distance at which near-field radiative heat transfer becomes important, calculated from the Wien's law for dominant wavelength of blackbody emission. *Panel (A): Data extracted from Ottens, R.S., et al., 2011. Near-field radiative heat transfer between macroscopic planar surfaces. Phys. Rev. Lett. 107 (1), 014301.*

as a function of distance d, with data extracted from Ottens et al. (2011). A constant value of conductance is found at high distances, as expected by far-field radiation, which is distance independent. However, a clear increase is observed when $d < 1$–$10\,\mu m$, due to near-field radiative heat transfer. Near-field heat transfer is explained in terms of the interaction of evanescent electromagnetic waves among the two surfaces. These waves arise from the constant processes of internal reflection of electromagnetic waves within the two materials, whose amplitudes of the magnetic/electric fields span a few times the length of their wavelengths away from the surface.

The onset distance at which this effect becomes significant can be approximated as one to three times the dominant radiation wavelength given by the Wien's law and thus is inversely proportional to absolute temperature T (Kittel, 2004). This distance is plotted in Fig. 11B. This effect should be carefully considered when designing μTEG and nanowire characterization microstructures as it may lead to a very high flow of bypass heat. For example, in the case of Fig. 11A for sapphire, transport by radiation is enhanced by a factor of 120 when the distance is 100 nm, and by 300 when the distance is 80 nm.

2.3.3 Limited Dimensions and Fill Factor

As described in Section 1.4, the length and the fill factor of the leg lengths should be adapted for maximizing the targeted TEG metrics—normally P/A for μTEG devices integrating nanowires. The optimum values depend mostly in nanowire ZT and heat sink thermal resistance (Yazawa and Shakouri, 2011). Typical optimum leg lengths and fill factors of TEG devices are in the order of 1–100 mm and 20%–50%, respectively (LeBlanc et al., 2014). However, the length of the nanowires is limited to 0.1 mm in the best of cases, due to fabrication constraints (Section 3). While this effect might normally be compensated by increasing the fill factor, this property is also limited in the case of nanowires. The densest arrays obtained by metal-assisted chemical etching (MACE; Section 3.4.2) present an approximated fill factor of 30%. However, it must be taken into account that heat sinking is generally worse in μTEGs, meaning that lower lengths and fill factors are required for attaining thermal matching.

2.3.4 Mechanical and Heat Sinking Constraints

Due to their high aspect ratio, nanowire arrays with low density are relatively fragile and thus unable to withstand the pressures that are normally applied to macroscopic—and even thin film—TEG elements. The main implication in μTEG devices is that a heat sink or contacts cannot be pressed on top of a vertical array nor they can be pressed in general above suspended microstructures that normally constitute μTEG devices. In general, this implies a poor heat sinking when compared to bulk devices, with the consequent decrease in power (Briand et al., 2015).

While mechanical robustness can be conferred to the array by embedding it in an electrical and thermal insulator matrix, this approach conveys an enhancement of the effective thermal conductivity of the array (Li et al., 2012b). This enhancement may be significant, as the thermal conductivity of the matrix solid—however low—is much higher than that of the air and the nanowire arrays present, in general, a low fill factor (Dunham et al., 2015).

While the abovementioned effects are important for most nanowire arrays and configurations, it is worth to note that vertical arrays of densely packed nanowires present surprising mechanical properties which prevent them from suffering these drawbacks. For example, Xu et al. (2012, 2013) were able to safely apply a pressure of the order of 10 bar in dense arrays of nanowire grown by the MACE technique (Section 3.4.2).

3. FABRICATION AND INTEGRATION OF THERMOELECTRIC NANOWIRES

Differently from the legs of a bulk thermoelectric material, nanowires cannot be fabricated, defined, and arranged in TEGs by conventional bulk methods used for processing thermoelectric materials, such as zone melting, dicing, and soldering (Rowe, 1995). Instead, they are obtained by means of microfabrication techniques such as photolithography, etching, electrochemical deposition (ECD), and chemical vapor deposition. In this section, the methods available for nanowire fabrication are introduced and explained, showing examples of the most relevant cases and making an emphasis on their scalability toward integration in μTEG devices. Important steps including contact formation and doping are discussed separately. The possible application of nanowires as a bulk within TEGs is briefly discussed in the last subsection.

3.1 Scalability Considerations

Nanowires offer excellent properties for exploitation of the thermoelectric effect (Section 2) and may indeed be fabricated with a wide range of diameters, crystalline forms, and compositions with the methods explained in the following sections. However, they will only be useful for thermoelectric application in the measure in which their fabrication and integration in μTEG devices are feasible—by means of a low electrical contact resistance—and scalable. Scalability is based in three aspects. First, nanowires must be able to cover a surface with a high fill factor—i.e. many nanowires connected in parallel and densely packed—in order for μTEGs to have a low area normalized resistance. Second, the synthesis of a power unit/leg—a nanowire here—must be fast. And third, the cost of producing a power unit must be as low as possible. While this is hard to assess, it can be inferred by the microfabrication equipment used in the process flow, the efficiency of the precursor used, and the temperature budget of the most relevant steps.

3.2 Bottom-Up and Top-Down Strategies

A first classification can be made depending on using bottom-up or top-down approaches. Bottom-up methods consist in controlling reaction and precipitation of semiconductor species so that they directly form solid nanowire structures (Fig. 12, from middle to left). Top-down methods consist in performing thinning and etching processes to bulk semiconductors at the nanoscale so that the resulting structures are nanowires (Fig. 12, from middle to right).

Etching processes necessarily require the disposal of part of the semiconductor, implying that top-down etching-based methods make a more ineffective use of precursors than bottom-up, additive-based ones. Moreover, top-down methods define nanowires in a specific direction—often literally top to down—implying that the arrays grown in each etch process must be within the plane of the etched layer, as in Fig. 12 right. However, in bottom-up methods nanowires can theoretically grow in any plane and direction, as in Fig. 12 left, permitting to obtain arrays extending in the three dimensions in a same process, increasing their throughput and thus their scalability.

Top-down methods are more reproducible than currently existing bottom-up methods. This arises from the fact that crystallinity, doping, and composition can be finely controlled in bulk materials and these properties are directly transferred to the resulting top-down nanowires once etched. While both groups employ microfabrication techniques, top-down etching methods are more mainstream and generally they can be easily chosen so that they are compatible with CMOS technology. Conversely, bottom-up methods often require the use of non-CMOS elements such as metallic nanoparticles or porous templates (Sections 3.3.1 and 3.3.3),

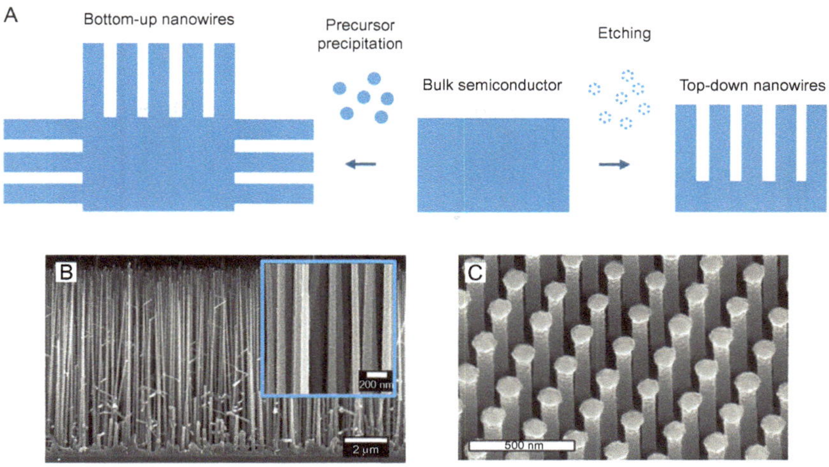

Fig. 12 (A) *Top row*: Schematic of bottom-up vs top-down nanowire growth methods. From *middle* to *left*: bottom-up, with direct formation of nanowire structures by precipitation of semiconductors in a nanowire shape. From *middle* to *right*: top-down, with etching of nanowires from a bulk semiconductor. (B) SEM image of a silicon–germanium nanowire array produced by a bottom-up method (vapor–liquid–solid growth). (C) SEM image of silicon nanowire array grown by a top-down method (laser interference nanolithography). *Panel (B): From Gadea, G., et al., 2017. Integration of Si/Si-Ge Nanostructures in Micro-Thermoelectric Generators. University of Barcelona. http://ggd.a2hosted.com/dissertation/2017_Gadea_thesis.pdf. Panel (C): From Curtin, B.M., Fang, E.W., Bowers, J.E., 2012. Highly ordered vertical silicon nanowire array composite thin films for thermoelectric devices. J. Electron. Mater. 41 (5), 887–894. http://link.springer.com/10.1007/s11664-012-1904-1.*

which lowers their compatibility hindering to some extent their potential for massive fabrication and implementation using mainstream technology.

As discussed above, both families present advantages and disadvantages regarding scalability and nanowire quality. So far, the growth of nanowire arrays for thermoelectric application has been implemented successfully in devices by means of both top-down and bottom-up methods (Section 6).

3.3 Bottom-Up Growth

3.3.1 Metal Nanoparticle-Based Methods

Metal nanoparticle-based nanowire growth methods follow the mechanism illustrated in Fig. 13A. A substrate is covered by metallic nanoparticles of diameter $d \sim 100\,\text{nm}$, brought to reaction/deposition conditions—temperature and pressure—and exposed to precursors. These precursors selectively react in the metallic nanoparticle—which acts as catalyst—and

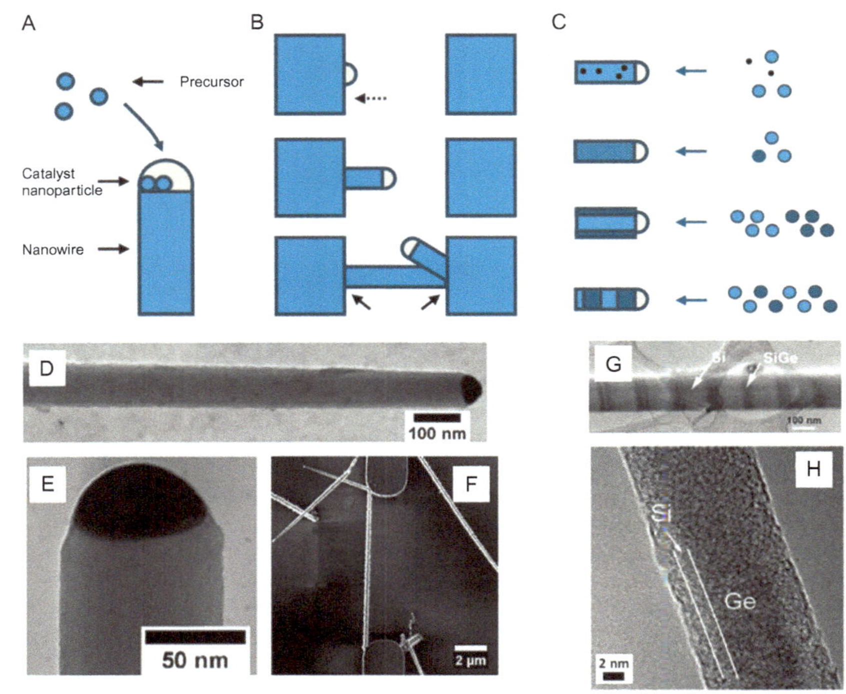

Fig. 13 (A) Metallic nanoparticle catalyst-driven nanowire formation mechanism. (B) Epitaxial growth of nanowire from a crystalline plane (*dotted arrow*) and formation of monolithic connection (*solid arrows*). (C) From *top* to *down*: formation of doped nanowire, alloy nanowire, core–shell nanowire, and superlattice nanowire. (D–F) TEM and SEM images of silicon nanowires obtained by vapor–liquid–solid (VLS) growth. (G) TEM image of silicon/silicon–germanium superlattice nanowire grown by laser-assisted VLS. (H) TEM image of germanium/silicon core/shell nanowire grown by VLS. *Panels (D)–(F): From Gadea, G., et al., 2015. Towards a full integration of vertically aligned silicon nanowires in MEMS using silane as a precursor. Nanotechnology 26 (19), 195302. http://stacks.iop.org/0957-4484/26/i=19/a=195302?key=crossref.f9bc9ad33e1c0a3bf6a9d5335c9e829c. Panel (G): From Li, D., Wu, Y., Fan, R., et al., 2003. Thermal conductivity of Si/SiGe superlattice nanowires. Appl. Phys. Lett. 83 (15), 3186–3188. Panel (H): From Wingert, M.C., et al., 2011. Thermal conductivity of Ge and Ge-Si core-shell nanowires in the phonon confinement regime. Nano Lett. 11 (12), 5507–5513.*

incorporate within it the dissolved atoms of the semiconductor species of interest. Reaction and incorporation go on until a critical supersaturation is reached and the semiconductor atoms precipitate in the most favorable interface available, which is the nanoparticle–substrate interface. Continuous incorporation of the semiconductor in the metal nanoparticle and precipitation at nanoparticle–substrate interface leads to growth of a semiconductor structure whose diameter is limited by the size of the nanoparticle, i.e., a nanowire.

While nanowire placement, diameter, and density (NWs/μm^2) are chiefly controlled by the nanoparticles, length L in the order of 1–100 μm is controlled by the reaction conditions—which define a growth rate in the order of nm/min—and the growth time—the time of exposition to precursors. Appropriate choice of the substrate, catalyst, and reaction conditions yields to ordered precipitation following underlying substrate substructure, i.e., epitaxially. This allows for growth of single-crystalline nanowires aligned in preferential directions and monolithically connected to the substrate, which is very convenient for integration in thermoelectric and other electronic devices (Fig. 13B and F; Dávila et al., 2012). By exposing nanoparticles to controlled amounts of precursors of different semiconductors or dopants, one can tailor composition and substructure of the resulting nanowires (Schmidt et al., 2010): simultaneous exposure to semiconductor and dopant leads to in situ doped semiconductor nanowires (necessary for increasing σ), simultaneous exposure to two semiconductors leads to alloy nanowires (convenient for reducing k), and alternate exposure to different semiconductors leads to heterostructures (superlattice and core–shell nanowires, convenient for reducing k/increasing $S^2\sigma$) (Fig. 13C, G, and H).

Different types of metal nanoparticle-based nanowire growth are considered in the next according to the phases that host the semiconductor species prior to their incorporation into the solid nanowire. For example, in the vapor–liquid–solid growth, the first of these methods discovered by Wagner and Ellis (1964), the semiconductor passes from a vapor phase (within a precursor molecule) to a liquid phase (within a metal–semiconductor eutectic alloy droplet) and later to a solid phase (constituting the nanowire). This method and many others discovered more recently sharing the same general mechanism (depicted in Fig. 13A) have been developed and implemented for nanowire growth (Dasgupta et al., 2014). A brief description of each one and some considerations regarding employment of catalyst nanoparticles is presented in the following sections.

3.3.1.1 Vapor–Liquid–Solid Growth

Vapor–liquid–solid (VLS) growth is the most widely used method for growing semiconductor nanowires owing to its flexibility, scalability, and capacity of producing crystalline nanowires of controlled dimensions. In this method the nanoparticle-seeded substrate is loaded into a chemical vapor deposition (CVD) reactor, brought to reaction conditions and exposed to precursor gases as for instance silane/disilane (Si precursors), germane (Ge precursor), diborane (p-type doping of Si/Ge), and phosphine (n-type dopant of Si/Ge). Choice of the semiconductor and metal nanoparticles must be such that they form a eutectic alloy during growth conditions, presenting a reduced melting point with respect to the pure species. This way a liquid rather than a solid catalyst nanoparticle will participate in the process, offering a high capacity for dissolving the semiconductor species and thus allowing for a high growth rate in the order of 100–1000 nm/min (Schmidt et al., 2010).

Growth temperature varies widely from 300°C to 800°C depending on the precursors employed. Higher temperatures lead to a higher surface diffusivity of the semiconductor at droplet–substrate interface promoting epitaxial growth. Also higher temperatures permit the nucleation of smallest wires, which otherwise would not be able to grow due to a thermodynamic hindering arising from the high surface energy of nanowire lateral facets (Gibbs–Thomson effect in nanowires (Schmidt et al., 2010)). However, high temperatures lead also to migration of metal catalyst to the inner and lateral parts of the wire, which leads to a limited L and a tapered and thicker cross section, detrimental for k reduction. High temperature leads also to a faster reaction rate both within the drop (which produces out-of-interface crystallite precipitation giving rise to kinks) and uncatalyzed at the sidewalls (which further contributes to a tapered and thick cross section) (Gadea et al., 2015). Introduction of HCl mitigates fast reaction and metal migration to some extent, allowing for high alignment, length, and constant and small diameters when combined with a high growth temperature (Oehler et al., 2009). Pressure employed varies in a wide range, 0.1 mbar (low-pressure CVD) to atmospheric pressure. The lower the partial pressure of the precursor—also attainable by dilution of precursor with a carrier gas as hydrogen or argon—the higher the alignment but the lower the growth rate, which may become impractical for integration (Gadea et al., 2017). Nanoparticle size also influences the preferred growth direction, apparently due to the different surface energies of the crystal planes that nanowires expose at sidewalls and at top—interfacing with the eutectic droplet. The relative weight of these surfaces varies with nanowire diameter, explaining why silicon nanowires below 20 nm grow reverently in the 211 direction, while above 50 nm they do it in the 111 direction, as proposed by Schmidt et al. (2005).

There are many examples of doped semiconductor nanowires for thermoelectric applications grown by the VLS growth, including silicon, silicon–germanium, germanium/silicon core–shell, zinc oxide, indium selenide, gallium nitride, gallium arsenide, and indium phosphide among others (Kolasinski, 2006).

The main advantages of VLS method are its compatibility with silicon technology—as it is based in CVD, a mainstream microfabrication technique—and the ease of controlling doping and composition—as diffusion problems are not important in the gas phase which means that compositions are readily controlled by the relative partial pressure of precursors during growth. VLS is the most studied and well known of nanoparticle-based methods. Moreover it is of the few that has been able to attain array integration in devices, in a horizontal fashion and within trenches as depicted in Fig. 13B (Chaudhry et al., 2007; Dávila et al., 2012).

Some variations of this method employ vapor transport (VT)—rather than CVD—in order to achieve precursor precipitation. Pulsed laser deposition (PLD) for instance can be implemented to VLS growth using VT. In this setup a pulsed laser evaporates precursors from a target and those are transported by a carrier gas to a zone containing catalysts, so VLS growth can take place. This approach with a flow of silane and a target of germane allowed the formation of superlattice nanowires with sharp interfaces (Si/Si–Ge with interfaces of few Å, g) and very thin nanowires ($d < 4\,\text{nm}$ for Si) (Li et al., 2003b; Morales, 1998; Schmidt et al., 2010). VT can also be employed to grow metal oxides with an in situ formation of the catalyst particles (Comini, 2006). Typically, a metal oxide powder is evaporated in a tubular furnace and transported to a colder zone where it condenses. A proper control of conditions leads to the formation of metal droplets, which become liquid and act as catalysts for the VLS growth of metal oxide nanowires.

3.3.1.2 Molecular Beam Epitaxy

Using the same approach as in VLS but employing a molecular beam epitaxy (MBE) equipment instead of a CVD, one can obtain aligned and crystalline nanowires that grow by means of a resembling mechanism, which was described by Zakharov et al. (2006). In this case, pure semiconductor atoms ejected by a Knudsen cell cross a high vacuum chamber and adsorb onto the substrate. These adsorbed atoms are referred to as adatoms. Fast surface supersaturation and surface diffusion lead to adatom migration, from the nanowire neighboring substrate (a few μm radius), up to the nanowire sidewalls, to the catalyst nanoparticle and finally to the catalyst–nanowire planar interface. The driving force of this process is the surface energy minimization for the adatom incorporating at the nanoparticle–nanowire interface with respect to the nanowire vacuum–substrate interface

Though not so well studied, this method produces similar results to VLS growth, so it has a same set of advantages and disadvantages. MBE nanowire growth method is attractive for the perfect crystallinity of the as-grown nanowires and the facility for changing the source making it suitable for heterostructure growth. Alas, it needs utilization of expensive MBE equipment and it is rather slow compared to other techniques with a growth rate in the order of 0.01 nm/min.

3.3.1.3 Vapor–Solid–Solid Growth

Vapor–solid–solid (VSS) growth works similar to VLS except that the catalytic nanoparticle does not form an eutectic with the semiconductor and thus remains a solid solution during growth, which is attained by choosing a different metal or lowering growth temperature. For example, Si nanowires can be synthesized by VSS growth using a Cu catalyst at temperatures as low as 400°C, while the eutectic temperature of $Cu–Si_3$ is above 800°C (Dasgupta et al., 2014).

Since solubility within a solid is much lower than within a liquid, the supersaturation within the catalyst is reached at lower concentration and so is the precipitation rate. Thus, typical growth rates of VSS are in the order of 1–10 nm/min. However, the lower growth rate means that they can work also at higher temperature before fast precipitation starts, which allows obtaining highly crystalline, low-tapered structures. The low solubility is also a requirement for forming sharp interfaces between different materials in nanowire heterostructures. For instance, VLS requires complex setups involving a CVD reactor coupled to a PLD equipment to form Si/Ge nanowire superlattices, while these could be readily obtained in a CVD reactor by VSS growth (Dasgupta et al., 2014).

3.3.1.4 Solution/Supercritical Fluid–Liquid–Solid Growth

In solution–liquid–solid (SLS) growth stabilized metallic nanocolloids are loaded in a flask together with an organic solution containing a surfactant and dissolved semiconductor precursor species. Then, the flask is heated up to temperatures lower than in VLS/VSS growth (<350°C) so that reaction, dissolution, and precipitation take place in the catalyst nanoparticle, leading to nanowire growth as in Fig. 13A. The presence of surfactant avoids agglomeration of the nanowires as they are formed, restricting the growth to the axial dimension and allowing the formation of nanowires up to several hundreds of μm in length. Then, the solvent is removed and a powder composed of randomly oriented nanowires is obtained. After growth, the surfactant needs to be removed as it covalently bonds to nanowire surfaces compromising their electrical properties (Dasgupta et al., 2014).

Clear advantages of this method are the lower growth temperature and the fact that growth takes place in nanoparticles dispersed through all the volume of the flask, allowing to obtain several mg/mL at a fast rate in a few minutes process—differently from gas-based methods as VLS/VSS, in which growth takes place only in substrate surfaces containing nanoparticles. However, it is more limited with regard to available reactant variety as in VLS and all possibilities of growing aligned and monolithically connected nanowires are lost. In principle, there is no incompatibility for using this method with nanoparticles adsorbed onto a substrate as in VLS, which allows obtaining aligned and integrated nanowires. However, if done, its main advantage—high throughput—is missing. Besides, whichever the approach, critical point drying (CPD) will be needed anyways to prevent agglomeration of the wires due to solvent surface tension forces (Passi et al., 2013). Thus, nanowires grown this way are most likely to be applied in composites or in a sintered form, as described in Section 3.6.

A variation of this method follows a microfluidics-based approach, in which a flow containing precursors is injected into a microreactor containing the catalyst particles. This allows sudden changes in exposition to precursors, as in VLS/VSS growth, allowing the growth of nanowire heterostructures (Laocharoensuk et al., 2013). In the so-called supercritical fluid–liquid–solid (SFLS) growth, the phase containing the precursors is brought to a supercritical state by increasing the pressure in a special reactor, above 200 bar. Supercritical conditions enhance the solubility of precursors in nanoparticles, allowing growth at 300°C below the eutectic point needed in VLS growth using the same metal and semiconductor combination (Tuan et al., 2005).

3.3.1.5 Catalyst Nanoparticles—Controlled Deposition and Removal

In the methods described above the nanoparticle size and density control the nanowire diameter, density (nanowires/μm^2 in vapor-phase and nanowires/mL in liquid-phase methods), and location of the particles within the substrate (vapor-phase methods) (Xia et al., 2003). While diameter and density are not exactly those of catalyst nanoparticles employed, they are clearly controlled by the latter (Gadea et al., 2015; Gao et al., 2005; Schmidt et al., 2010).

For solution-based nanowire growth methods—SLS and SFLS—surfactant-stabilized metal nanoparticles can be obtained commercially or fabricated by means of colloidal synthesis or other suitable approaches (Johnston, 2012). The availability of monodisperse sizes and compositions allows finely controlling the morphology of the resulting nanowires.

In vapor-phase methods—VLS and VSS, nanoparticles are adsorbed onto a surface, which can be attained by electrostatic colloid deposition of thin-film dewetting (Fig. 14).

In electrostatic colloid deposition, the target surface is functionalized with a positively charged polymer layer and subsequently exposed to a colloidal dispersion of metal nanoparticles stabilized with negatively charged surfactants. Electrostatic attraction leads to irreversible nanoparticle deposition, with a density of 1–10 nanoparticles/μm^2 controlled by colloid concentration in the dispersion and exposure time (Fig. 14B). By masking the deposition of the polymer layer with photolithography methods, one can restrict the growth of nanowire arrays to the targeted surfaces forming patterns (Hochbaum et al., 2005).

In thin-film dewetting, nanoparticles are obtained by deposition of a metallic thin film followed by an annealing, typically at 400°C. Solid/eutectic state dewetting of the metal thin film leads to the formation of an array of metallic nanoparticles, presenting a broader size distribution and higher density—thousands of nanoparticles/μm^2—than those obtained by colloid deposition (Gadea et al., 2015, 2017).

Metallic thin films for this may be deposited by physical methods (evaporation, sputtering) or chemical ones (electro and electroless deposition—also known as galvanic displacement). Chemical methods present two advantages. First, they can cover horizontal surfaces as microtrenches allowing for nanowire integration (Fig. 13B). Second, they can selectively cover target surfaces, since only those exposing conductors and semiconductors susceptible to oxidation will participate in the redox process which brings about thin-film deposition. While μTEGs do generally need to maximize nanowire density, limiting nanowire growth to target areas is convenient for avoiding short circuit of critical device parts. In this sense, patterned colloid deposition and chemical-based thin-film dewetting constitute very advantageous methods for device integration—the former having more size control and the latter exhibiting higher densities potential due to its intrinsic

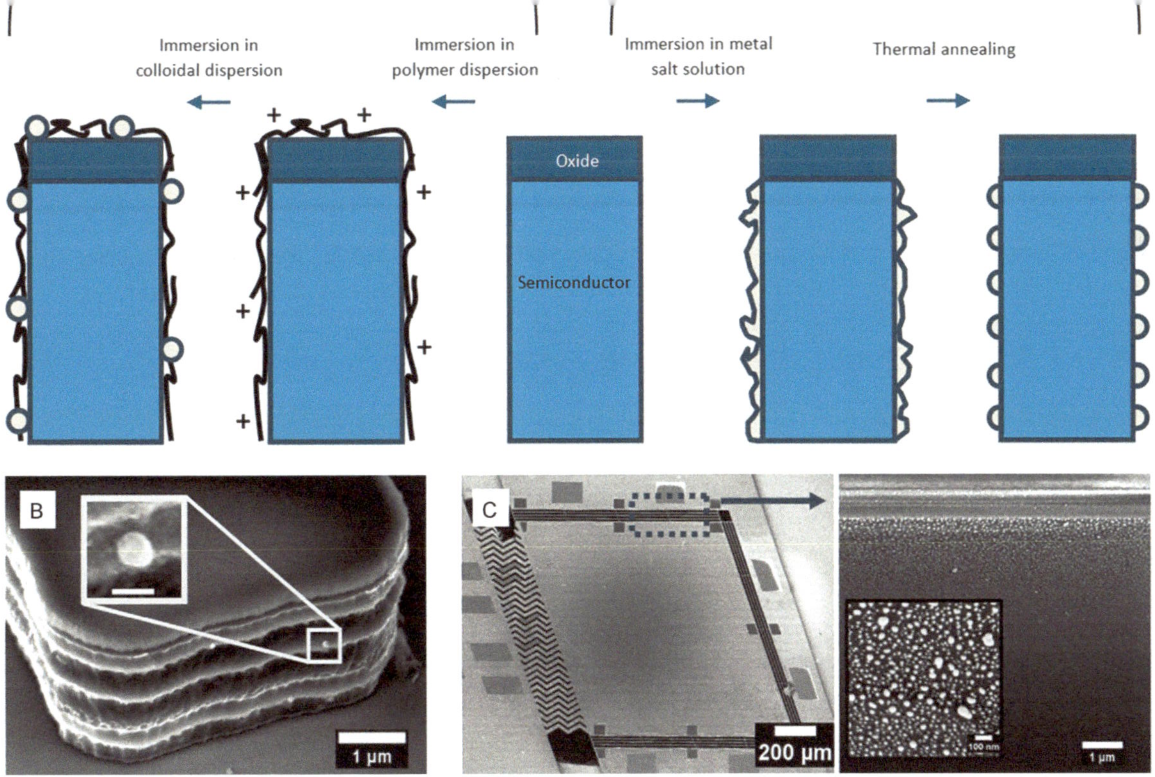

Fig. 14 (A) *Top row*: Schematic of deposition of metal catalyst nanoparticles for subsequent nanowire growth on top of an oxide-passivated semiconductor. From *middle* to *left*: Colloid deposition, involving the exposition to a positively charged polymer and negatively charged metal nanoparticle colloids. From *middle* to *right*: Thin-film dewetting including a selective deposition by galvanic displacement and an annealing process in order to produce the dewetting. (B, C) SEM image of catalyst nanoparticles deposited in vertical device walls by electrostatic deposition (low density and nonselective, in B) and galvanic displacing (high density and selective in silicon parts, in C). *From Gadea, G., et al., 2017. Integration of Si/Si-Ge Nanostructures in Micro-Thermoelectric Generators. University of Barcelona. http://ggd.a2hosted.com/dissertation/2017_Gadea_thesis.pdf.*

nature and the higher diffusion of dissolved precursors, rather than colloids, within trenches (Fig. 14C).

It is worth to note that, while high density is generally an advantage for achieving higher density of nanowires, it actually may be a drawback when it implies a large presence of small nanoparticles. In VLS processes, small nanoparticles may not produce nucleation and besides grow in nonintended crystalline directions giving rise to kinked nanowires that interfere with the normal growth of aligned nanowires. Thus, adding very small particles (<30 nm) to an array of moderately sized particles (<50 nm) may actually turn into a lower density of aligned nanowires after VLS growth (Gadea et al., 2017). For this reason, it is often convenient to use masking elements that restrict thin-film deposition to holes with homogeneous separation and size. The use of self-assembling polymeric nanospheres or block copolymers for instance has been combined with galvanic displacement obtaining controlled depositions of gold particles, which have later been used for the growth of highly aligned and monodisperse nanowires by gold-catalyzed MBE growth (Doerk et al., 2011; Fridman, 2005).

The catalyst metallic nanoparticles and their rests can be removed after nanowire growth with selective wet etching, although their presence does not prejudice the thermoelectric behavior of semiconductor nanowires (Franssila, 2010). Conversely, in some cases they have shown to serve as dopants leading to an increase of $S^2\sigma$ (Pichanusakorn and Bandaru, 2010a,b). On the other hand, photovoltaic effect and field-effect switching response are severely hampered by the presence of metals with ionization energies lying in the middle of the bandgap, due to their ability to induce excess carrier recombination (Dasgupta et al., 2014). For these applications, other methods that do not use nanoparticles are more interesting such as crystal seed-based growth (Greene et al., 2005), screw dislocation-based growth (Feng et al., 2008), thin-film oxidative growth (Xu et al., 2004), and selective area epitaxial deposition (Joyce and Baldrey, 1962). However, for thermoelectric applications these methods do not offer additional advantages over catalyst-based ones, which remain faster and thus more convenient for implementation in TEG devices, and thus they are not presented herein.

3.3.2 Solvothermal Synthesis

Solvothermal synthesis of nanowires is a high-pressure solution-based nanowire growth method that does not employ any catalyst. An organic solution of semiconductor precursors and surfactants is heated up above the boiling point of the solvent (typically <250°C) in a reactor able to withstand high pressures, as in SFLS. The increase of pressure in the liquid phase, due to the partial evaporation, can go up to hundreds of bars, therefore favoring the conditions for precursor decomposition and crystal nucleation. Then, the growth of nanowires proceeds at high rates by crystal growth in preferred directions with surfactant adsorbed onto sidewalls preventing agglomeration. If the solvent employed is water, it is referred to as hydrothermal synthesis, which involves higher temperature and pressure, resulting in a safer manipulation due to the absence of a toxic/flammable solvent (Park et al., 2015; Shi et al., 2008).

The advantages and drawbacks of this method are those of SLS and SFLS. Additionally, its simplicity and low-temperature budget are very suitable for massive production. However, it is only applicable to materials with very anisotropic crystal structures exhibiting a clearly preferred growth direction, which luckily comprise some classical thermoelectric materials such as PbTe, SbTe, or Bi_2Te_3 (Xia et al., 2003; Zhang et al., 2012).

3.3.3 Porous Template-Assisted Methods

A porous alumina template, several tens of microns long and comprising aligned cylindrical pores with diameters ranging 10–100 nm and density of ~100 pores/μm^2, is purchased or fabricated by means of anodization of an aluminum foil following well-established processes (Martín et al., 2014). In a different approach, pores may be obtained by producing channels in a polymeric material by means of ion track technology (Koukharenko et al., 2008; Li et al., 2009). Then, the pores are filled with the desired thermoelectric material by one of the approaches described below. This way, arrays of aligned nanowires are formed with diameters, densities, and lengths controlled by the template pore size, density, and duration of the deposition, respectively (Fig. 15). While the alumina template confers mechanical stability to the array, it may compromise the effective k of the template-nanowire composite, as it has a higher thermal conductance than the void/air that would be occupying its place in case of its absence. Thus, the template can be removed by means of subsequent wet etching, revealing free-standing nanowires. However, a CPD-drying step must be employed in order to prevent surface tension-derived stresses hampering the nanowire electronic properties (Section 3.5.4).

Pore filling is generally attained by ECD, as depicted in Fig. 15. First, a metal electrode is evaporated at one side of the template. Then, it is contacted and immersed in a plating bath containing the targeted semiconductor

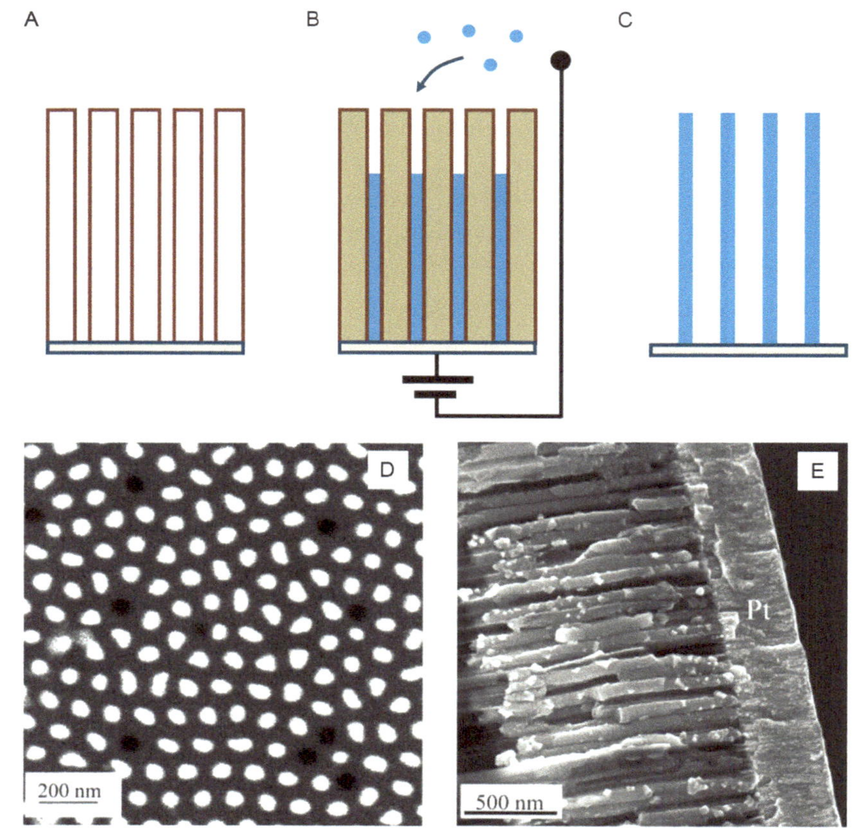

Fig. 15 (A–C) *Top row*: Bottom-down fabrication of nanowires by alumina-template-assisted electrochemical deposition (ECD). (A) A metal is evaporated in one side of the template. (B) The template is immersed in an electroplating bath loaded with semiconductor precursors. An electric potential is applied between the electrode and the bath, leading to the growth of nanowires within the channels. (C) After the ECD the alumina template is removed by selective wet etching. (D, E) SEM images of bismuth telluride nanowires grown by alumina-templated ECD growth. (D) *Top view*, before template removal. (E) Cross section of the nanowire array and the platinum (Pt) electrode, after membrane removal by wet etching. *Reprinted with permission of Martín-González, M., et al., 2003. Direct electrodeposition of highly dense 50 nm $Bi_2Te_{3-y}Se_y$ nanowire arrays. Nano Lett. 3 (7) 973–977. https://doi.org/10.1021/nl034079s. Copyright (2003) American Chemical Society.*

precursors. Upon application of a potential difference between the electrolyte and the electrode, nanowires are grown from the electrode side to the free side. Voltage modulation and pulsed deposition may be used to switch between single-crystal and polycrystalline growth and force partial pore coverage, which leads to diameter control (Li et al., 2005). Moreover, a supersaturated alloy can be electrodeposited and subsequently thermally annealed in order to produce heterostructures, as demonstrated by Wang et al. (2007) who obtained Bi_2T_3/Te superlattice nanowires. In a second approach, pores are filled by high-pressure injection, using a reactor able to sustain vacuum, heating, and high pressures up to 350 bar (Zhang et al., 1999). Target semiconductor–dopant powder with the template on top is loaded into the reactor. Then, air is evacuated to moderate vacuum and heat is applied, so the semiconductor is molten and the template immersed in it. Then, high pressure is applied in order to fill the pores and the system is cooled slowly or fast in order to achieve single-crystal or polycrystalline nanowires, respectively. A third approach consisting in the use of metal nanoparticle nanowire growth can be used as well (Shimizu et al., 2007). In this case, the template is placed on top of a crystalline substrate exhibiting a specific crystalline plane and metal nanoparticles are deposited at the pore bottom by depositing a thin film from top (Section 3.3.1). Then, a VLS growth is performed, so single-crystal nanowires grow following the underlying plane direction, insensitive to some extent to preferred growth direction at these specific growth conditions.

These methods have a very high potential, allowing the growth of dense arrays of aligned nanowires with fine control of their properties. The availability of several methods for the filling allows deposition of many diverse materials. Besides, they offer a higher crystalline control and lower possibility of intercrossing of kinked nanowires, as compared to VLS/VSS growth. Yet, the length of the arrays as in metal nanoparticle-mediated growth is $<100\,\mu m$, which implies that they have to somehow be integrated in μTEG devices, similar to the abovementioned bottom-up methods. Differently from nanoparticle-mediated nanowire growth methods, there is no strategy taking into account simultaneous direct growth and device integration of nanowires by this method, as in VLS growth (Fig. 13B). So far, these methods need a subsequent step of cleaving and metal deposition—with the consequent risk of formation of Schottky contacts—in order to be able to generate μTEG structures (Wang et al., 2005). Also, when ECD is used for filling the pores, diffusion issues due to the different size of the pores lead to a limitation of the maximum length and fill factor that one can attain, as fastest, wider channels grow mushroom-like structures that prevent filling of neighboring smaller pores. In the case of filling by pressurized injection, a cleaving process needs to be done after formation in order to release the nanowire-containing membrane and to recover the great portion of the nonused precursor—a process so far nonautomatable and thus of complex implementation.

3.3.4 Electrospinning of Semiconductor Nanofibers

A solution-containing gel precursors or nanocrystals is blended with a dissolved polymer and injected into an electrospinning equipment. The as-electrospun fibers are dried at mild temperatures (typically 60°C) and annealed at high temperatures (typically >750°C), in order to remove the polymer and decompose the gel precursors or even sinter the nanocrystals. The final form consists of aligned or unaligned polycrystalline fibers with diameters in the order of 100 nm piled one up another forming a dense fabric. This novel route seems promising for upscaling the production of nanowire (long nanofibers in this case) materials, as it is based in electrospinning—a technology well established in industry—and can cover arbitrarily large areas in a porous format, which helps to mitigate internal heat bypass by blocking air convection. These fabrics are the most close to bulk nanowire structures achieved so far, potentially able to work as macroscopic TEGs. However, fabrics are hard to scale in the cross-plane direction, which implies that their application must be with heat also flowing in this direction, which is perpendicular to that of fiber alignment, surely compromising electrical transport. Besides, fibers are deposited in a way that they must necessarily contact one another and only polycrystalline nanofibers have been attained so far, meaning that the material may behave more like a polycrystalline bulk—for which other methods already exist for producing denser forms. Moreover, the electrospinning approach requires to add and burn a mass of polymer equivalent to that of the employed precursor, implying additional material and processing costs. Some examples of these fibers are oxide thermoelectrics such as $NaCo_2O_4$, $CaCo_4O_9$, and LaSrCoO (Greiner and Wendorff, 2007; Ma et al., 2010).

3.4 Top-Down Growth

3.4.1 Nanolithography Methods

Fabrication of nanowires by nanolithography comprises three general steps (Fig. 16). First, a flat surface of the targeted semiconductor bulk is covered with another material referred to as the mask—typically a polymer—defining features with dimensions in the order of 100 nm. Second, an anisotropic etch process is performed, from top to down, which is blocked at the regions covered by the mask—typically a deep reactive ion etching (DRIE).

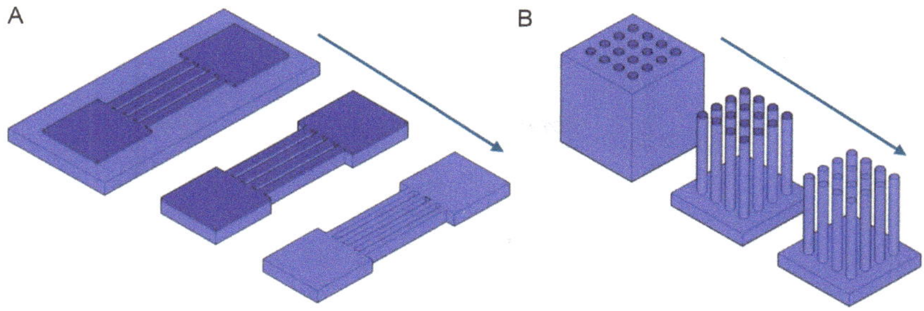

Fig. 16 Top-down, nanolithography-based process for obtaining nanowires. (A) Horizontal and (B) vertical configurations. First, a mask is deposited on top of the semiconductor. Then, top-down etching process is performed leading to the definition of the wires. Then, the mask is removed and free-standing nanowires are obtained.

And third, the mask is removed by means of a material selective etching. This way, semiconductor structures are obtained, with a cross section and a depth controlled by the shape of the mask and the duration of the etching process, respectively (Fig. 16). The use of thin and supported semiconductor substrates, stripe-like masks, and short etch times leads to horizontally aligned nanowires (Fig. 16A), while the use of thick semiconductor substrates, dot-like masks, and longer etching times leads to vertically aligned nanowires (Fig. 16B). Both configurations present different advantages and drawbacks, discussed below.

Two approaches may be adopted in order to define the masking particles that will determine nanowire dimensions. The first consists in the direct use of nanolithography techniques able to engrave features in the sub-100 nm regime. Some of them, which are not discussed thoroughly herein, are electron beam lithography (Heron et al., 2009), X-ray lithography, nanosphere lithography (Fridman, 2005), nanoimprint lithography (Yan et al., 2005), interference lithography (Curtin et al., 2012), superlattice nanowire array transfer lithography (SNAP) (Melosh, 2003), sidewall transfer lithography (Hållstedt et al., 2008), and atomic force microscope (AFM) oxidation lithography (Martínez et al., 2010). The second approach is based in the postprocessing sacrificial oxidation technique (Section 3.5.3), which allows working with mask dimensions higher than those intended for the nanowires, thus enabling the use of common optical lithography techniques (Li et al., 2011). In this case, masking particles with dimensions >100 nm are deposited leading to the formation of thick nanowires upon etching. Afterward, the diameter of these nanowires is reduced with an oxidation-etching process.

3.4.1.1 Horizontal Configuration

Planar arrays of horizontally aligned nanowires can be defined and monolithically integrated in μTEG devices in a simple nanolithography process, by defining thicker regions for the current collection, as in Fig. 16A. This is so far the more effective and reproducible manner of contacting nanowire arrays. Subsequent nanowire suspension by selective substrate etching and CPD is convenient in order to avoid trough-substrate parallel thermal conductance—which compromises effective *k*. Some examples of this fabrication are the horizontal and monolithically integrated nanowires characterized in different works (Boukai et al., 2008; Jang et al., 2012; Totaro et al., 2012), which were obtained by means of techniques such as SNAP, e-beam, or sacrificial oxide-based nanolithography. In the work of Farrell et al. (2012), large arrays of horizontally aligned nanowires were fabricated using optical interference lithography and block copolymers.

However, only a single layer can be attained in the simplest form of this method, implying a very poor fill factor when normalized to μTEG planar surface (Section 5). In order to solve this issue, two approaches have been recently implemented to horizontal nanolithography techniques. In one of them, developed by Ferri et al. (2011), a nitride/oxide superlattice is first fabricated by thin-film deposition and subsequent optical lithography. Then, a selective wet etching creates nanometric recessions in the resulting structures, which are filled by CVD-deposited polysilicon. Then, a last etching process defines the NWs (Fig. 17A and C). Indeed, this is an excellent method for extending the nanowire array to the third dimension and one of the few that enable the integration of n-type and p-type polysilicon nanowire arrays in μTEG devices fully fabricated by bottom mainstream microfabrication techniques. It must be said that the horizontal pitch is still limited by optical lithography and that the scalability of the superlattice deposition limits the scalability of the whole process—as each oxide layer defines one 2D nanowire array stage. Also, it is likely that the presence of nitride-supporting structures with wide cross sections needs to be dealt with after nanowire definition in order to prevent parasitic thermal conductance. Still, this approach has a great potential as improvement of these issues is technologically feasible. In a second approach, the scalloping of a DRIE process—normally an unwanted effect—is controlled and smartly harnessed for defining several stages of horizontal suspended nanowires, as depicted in Fig. 17B and D (Tasdemir et al., 2016). This method allows for monolithic integration—actually, it is in need of it to prevent collapse—and indeed presents a great potential due to its simplicity. However, reproducibility and

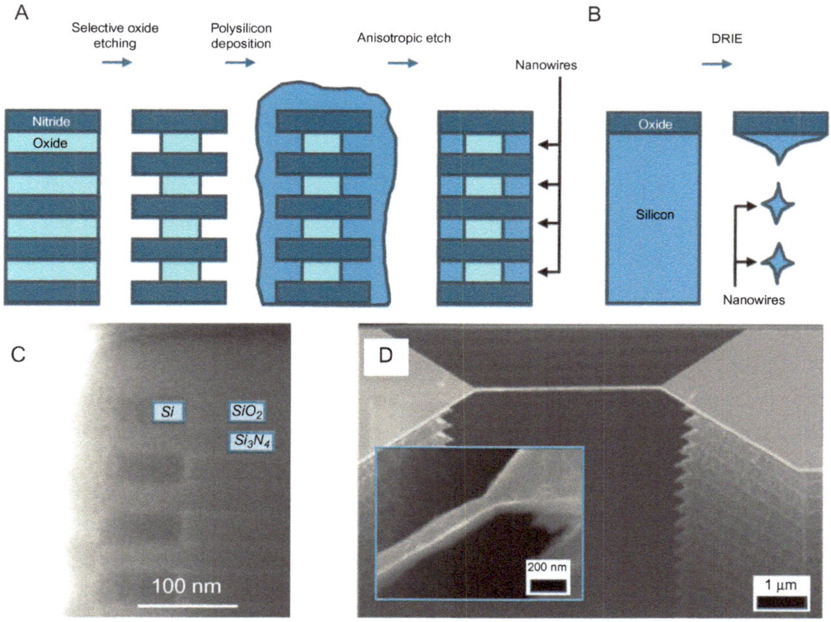

Fig. 17 Horizontal nanolithography-based strategies for extending horizontal nanowire arrays in the third dimension, downward the wafer. The schematics of (A, B) represent cross sections of wafers, which after processing lead to cross section of nanowires—which would extend normally to the paper plane. (A) Nanorecession method, with SEM images of silicon nanowires shown in (C). (B) Deep reactive ion etching (DRIE) scalloping control method, with SEM images of silicon nanowire shown in (D). *Panel (A): Ferri, M., et al., 2011. Ultradense silicon nanowire arrays produced via top-down planar technology. Microelectron. Eng. 88 (6), 877–881. http://linkinghub.elsevier.com/retrieve/pii/S0167931710004557. Panel (B): From Nasr Esfahani, M., et al., 2017. Monolithic technology for silicon nanowires in high-topography architectures. Microelectron. Eng. 183–184, 42–47.*

control seem very difficult with the shape of the scalloping affecting both the vertical pitch and the nanowire diameter and being influenced by DRIE parameters and nanowire surroundings. Overall, it is a promising method, with a very low yield so far but with a lot of room for improvement.

3.4.1.2 Vertical Configuration

Definition of vertically aligned nanowire arrays by nanolithography yields very high densities, as in the case of bottom-up methods, ensuring a high fill factor and thus a high potential for μTEG power (Fig. 16B). However, unlike in horizontal nanolithography or VLS methods, there is not a direct solution for forming electrical contacts in the growth process.

Thus postprocessing steps are needed, as in template-based synthesis, including matrix embedding, polishing, patterned metal deposition on top, thermal formation of nonrectifying contacts (Li et al., 2011), and removal of the supportive embedding matrix (if parasitic thermal conductance has to be avoided).

Li et al. (2011) used this approach to define a CMOS-compatible process to obtain arrays of vertically aligned silicon nanowires integrated in μTEG devices. In that work, deep ultraviolet lithography (DUV) was employed for defining the masking silicon oxide particles, with diameters >100 nm, and subsequent sacrificial oxidation was carried out to reduce the diameter down to 80 nm. In the same line, Curtin et al. (2012) integrated vertically aligned silicon nanowires in full-fledged μTEG devices defined by a nanolithography process. In this case, the silicon oxide masks were defined by laser interference.

3.4.2 Metal-Assisted Chemical Etching

MACE starts with the deposition of a noble metal thin film—few nm—on top of a doped semiconductor bulk of interest (Fig. 18A). Then, it is immersed in a solution containing an acid—typically hydrofluoric acid, HF—and an oxidant substance—typically hydrogen peroxide, H_2O_2. At this point, a redox reaction occurs triggered by the presence of the metal, in which the semiconductor is oxidized and dissolved into the solution, leaving gaps behind. This kind of etching reactions is often aided by imposing an electrostatic potential. Since here no electric potential is applied to promote the reaction—a metal is used in its place—these reactions are referred to as electroless etching (EE). Eq. (14) shows the EE process for dissolution of silicon in HF and H_2O_2, supported by the presence of silver (Ag):

$$\begin{aligned} Si + 4Ag^+ + 6F \rightarrow \; & 4Ag + SiF_6{}^{6-} \\ 2Ag + H_2O_2 + H^+ \rightarrow \; & 2Ag^+ + H_2O \end{aligned} \tag{14}$$

The initial dissolution leads to the rupture of the metallic film in nanoparticles similarly as in a dewetting process (Fig. 18B). Then, the etching proceeds at the nanoparticle–semiconductor interface, defining channels into the semiconductor surfaces (Fig. 18C). The channels elongate with the oxidant and the dissolved semiconductor flowing through them to and from the semiconductor–nanoparticle interface, respectively. While in the whole scheme the metal or its cations are not consumed (Eq. 14), the particles are not technically "catalysts" as they participate continuously in the reactions, dissolving and precipitating, as indicated by Huang et al. (2011). When the

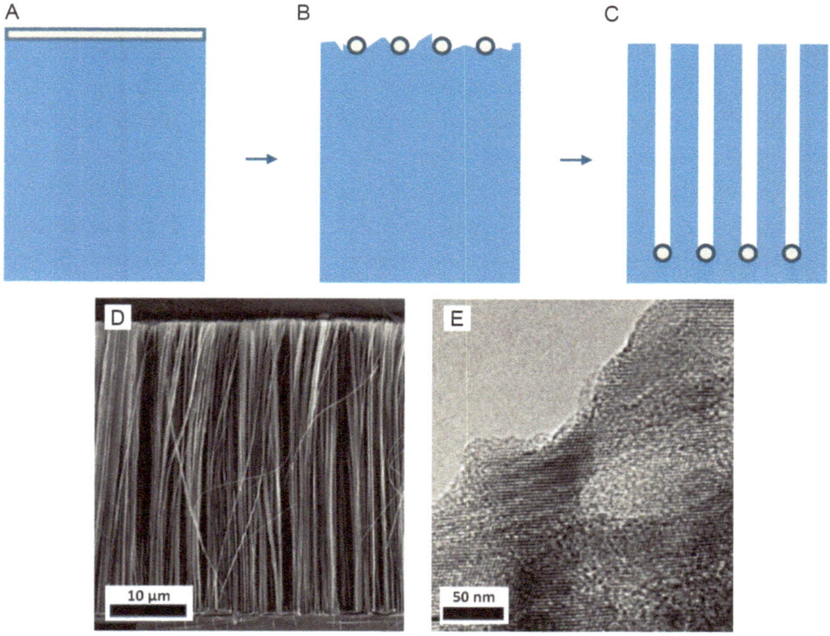

Fig. 18 (A–C) Top-down formation of nanowires by metal-assisted chemical etching (MACE). (A) A noble metal thin film is deposited on top of a semiconductor. (B) The metal is immersed in a solution containing an acid and an oxidant. In a first stage of the process, the metal supports the irregular etch of the top surface, splitting into nanoparticles. (C) As the process is continued, the etch takes place at the metal–semiconductor interface defining channels that lead to the formation of nanowires. (D, E) SEM and TEM images, respectively, of rough silicon nanowires grown by MACE. The high-resolution TEM detail in (E) shows how roughness is formed without compromising the crystalline continuity. *Reprinted with permission of Hochbaum, A.I., et al., 2009. Single crystalline mesoporous silicon nanowires. Nano Lett. 9 (10), 3550–3554. Copyright (2009) American Chemical Society.*

density of channels is high enough, the resulting structures are aligned arrays of nanowires (Fig. 18C and D). This method is often referred to as EE in the literature, as MACE is indeed a specific variant of EE.

Length of the channels—and thus of the nanowire arrays—can be controlled by the immersion time—typically 15–60 min—in the etchant solution and has typical values of 1–100 μm. As in VLS growth, metal losses during growth limit the maximum length of the nanowires. As appreciated in Fig. 18C, the cross section of the nanowires is ideally that of noncovered portion of semiconductor top face. Thus the top metal pattern ultimately determines nanowire diameter, which can vary widely from 5 to

1000 nm (Huang et al., 2011). If as described above a simple thin-film deposition is used, a rather random process of nanoparticle formation is ruling the formation of top metal pattern at the early stage of the dissolution. This can be controlled to some extent by adjusting the film thickness in a similar manner as in thin-film dewetting. However, broad dispersions will be obtained, which may cause analogue problems to those in dewetted thin films for VLS nanocatalyst formation, as discussed above. Thus, for a thinner control a metal thin-film deposition accompanied by auxiliary metal pattering methods as those introduced in Section 3.3.1 may be employed, allowing precise determination of nanowire diameter and pitch Some examples comprise the works of Puyoo et al. (2011) and Curtin et al. (2012), who used nanosphere lithography and laser interference lithography, combined with MACE to obtain highly aligned silicon nanowires.

Different from the methods described above, the as-formed nanowires present a very rough surface (Fig. 18E), which has proven to be very effective in reducing thermal conductivity (Hochbaum et al., 2008; Lim et al., 2012). Moreover, this roughness can be controlled by process parameters and also by changing the doping type and degree, as Lim et al. (2012) demonstrated for silicon nanowires. Actually, very high doping levels lead to a too pronounced roughness and dissolution rate, which turns out to impede the formation of continuous nanowire structures. For this reason, if highly doped and rough nanowires are targeted, these must be etched out by MACE from an intermediately doped semiconductor and then doped in another subsequent process (Xu et al., 2013).

Similar to porous template-assisted methods, MACE allows to grow very dense arrays—with the fill factor of ~30%—of vertically aligned nanowires. Moreover, being a top-down method, the properties of the underlying substrate—crystallinity, composition, and doping—are easily transferred to the as-etched nanowires. They are monolithically connected to the underlying substrate by one side, solving partly the connection issue but, as in vertical nanolithography or pore-assisted methods, top connection must be formed by additional processing. The most interesting aspect of these arrays, however, is that their synthesis incorporates mechanisms to confer them a controlled surface roughness that is proven to be very effective in the reduction of the thermal conductivity. While EE-mediated surface roughening has been shown to be generally applicable to all nanowires grown by means of postgrowth etching (Section 3.5.3), it is a great advantage to be able to integrate it in a route which encompasses growth, doping, and roughening of the as-synthesized nanowires—almost complete except for the electrical connection. It is worth to mention also that the method is relatively inexpensive, as it is simple and can take place at room conditions.

3.5 Other Processing Steps

3.5.1 Contact Formation

Contact resistance ρ_c of nanowires discussed in Section 2.3.1 depends on the methods employed for forming the electrical connections.

Xu et al. (2012) directly contacted large MACE grown silicon and silicon–germanium arrays by sandwiching between two pressurized copper plates. However, the high contact resistance obtained this way severely diminishing the power of the TEG devices. While in a later version increased doping of the upper part of the nanowires—the nonmonolithically contacted one—reduced a great deal this resistance, it still influenced the electrical behavior of the device (Xu and Fobelets, 2014; Xu et al., 2013).

Contacts might be formed monolithically within the definition process of the nanowire by nanolithography, in one side if nanowires are defined vertically or in both if they are horizontal (Fig. 16). In a second approach, contacts may be formed at the end of a VLS/VSS growth process, as indicated in Section 3.3.1, also in a monolithic fashion. Monolithic integration of nanowires in highly doped silicon structures by means of CVD-VLS growth method yielded 3–5 $\times 10^{-6}\ \Omega\,\text{cm}^2$ for silicon nanowires (Chaudhry et al., 2007).

Contacts might also be formed by evaporation of a metal in a separate process after growth. Metal evaporation yielded ρ_c of $5 \times 10^{-4}\ \Omega\,\text{cm}^2$ for Ti/Au contacts in silicon nanowires (Mohney et al., 2005) and 10^{-1}–$10^{1}\ \Omega\,\text{cm}^2$ for Ag, Au, and Pd contacts in low-doped Ge nanowires—which presented a high Schottky barrier (Koleśnik-Gray et al., 2013). Chemically cleaning the surface with ammonium polysulfide before depositing and Au–Cr contact layer led to ρ_c values as low as $1.4 \times 10^{-7}\ \Omega\,\text{cm}^2$ for InP nanowires (Sourribes et al., 2013). Ti/Au deposition followed by rapid annealing led to $3 \times 10^{-6}\ \Omega\,\text{cm}^2$ for Ga nanowires. Silicidation of contacts, achieved by masked deposition of Ti, Ni, and Pt contacts followed by rapid thermal annealing, led to low ρ_c values for silicon nanowires (Hong et al., 2011; Lin et al., 2008). A combination of VLS growth method for one side and Ni deposition and thermal annealing for the other leads to 6–10 $\times 10^{-7}\ \Omega\,\text{cm}^2$ for InAs nanowires (Berg et al., 2015). Finally, while metal contacts are more versatile—being able to be applied to nanowires with any final form—monolithic and alloy-based contacts (as the silicide-based) offer lower ρ_c values, often being necessary for effective integration of nanowires into devices.

3.5.2 Doping Control

The doping level is the concentration of p- or n-type doping impurities in a semiconductor, having typical values in the order of 10^{16}–10^{21} atoms/cm^3. This parameter is extremely influential in thermoelectric transport properties, as it determines the carrier concentration—being the former and the latter approximately equal if proper dopant is chosen such that ionization is complete (Kittel, 2004). As introduced in Section 1.3, S, σ, and k_e are correlated in a way that an optimum carrier concentration exists for $ZT = S\sigma T/(k_e + k_l)$ maximization. Thus a fine control of doping level is needed to reach electronic optimization of the as-grown nanowires.

Many growth methods described in this section provide a mechanism to achieve an in situ doping process. In bottom-up methods, this is achieved by tuning the amount of precursors. For example, in VLS-grown silicon nanowire the ratio of dopant to precursor atoms in the vapor phase controls the proportion of dopant in the resulting nanowires. Conductivities exceeding 1000 S/cm were obtained by this method, which is enough for optimization of electric properties (Schmidt et al., 2010). In template-based ECD of bismuth telluride nanowires using out-of-stoichiometric proportions of bismuth/telluride, one can alter the carrier concentration, which is not technically doping but achieves the same goal of controlling electronic properties of the resulting nanowires (Rowe, 1995). In top-down methods, the doping level of the substrate is directly transferred to the resulting nanowires after etching.

However in some cases in situ doping does not yield high enough impurity concentration for electronic optimization or it does it in a non-homogeneous fashion that may be detrimental for a high σ. For instance, radial and axial doping gradients are sometimes observed in top-down, in situ doped VLS nanowires (Allen et al., 2009; Du et al., 2013). Also, the in situ doping process might influence the morphology of the as-grown nanowires or hinder it totally. In top-down MACE grown nanowires, for instance, using substrate wafer with high doping levels leads to a fast etching rate, which does not allow nanowire formation, limiting to some extent the doping level directly attainable by this method (Huang et al., 2011).

In the aforementioned cases, the use of an additional ex situ doping/drive-in process seems convenient. Kanungo et al. (2009) researched the effects of ion implantation on undoped silicon nanowires grown by VLS growth, achieving in the range of 10^{18}–10^{19} dopant atoms/cm^3, for p- and n-type doping. Doerk et al. (2010) achieved ex situ doping of silicon nanowires in the vapor phase by exposing them to BBr_3 at temperatures ranging 700–800°C, reaching ~10^{18} cm^{-3}. Ingole et al. (2008) used a proximity-diffusion spin-on-dopant (SOD)-based approach. In a first step, nanowires were exposed to SOD vapors (B_2O_3) that led to formation of a thin boron-rich layer surrounded by a thin oxide layer in the outmost part. After removal of the oxide, a drive-in process at ~1000°C led to a uniform dopant distribution with concentrations up to 3×10^{20} B atoms/cm^3. Xu and Fobelets (2014) used a resembling postgrowth SOD approach—spinning, oxide removal, and drive-in at 900°C—for improving the electrical properties of MACE-grown silicon nanowires. In the well-known work of Hochbaum et al. (2008), highly doped silicon nanowires were obtained by ex situ doping by ion implantation of MACE nanowires grown from a low-doped substrate wafer.

3.5.3 Postgrowth Etching

Nanowire surface can be partially etched after growth in order to enhance phonon boundary scattering effect, which diminishes k, improving the thermoelectric behavior.

In the sacrificial oxidation technique, the outer layer of the nanowires is thermally oxidized in a dry process and subsequently removed by an etching process, in this case wet and selective for the semiconductor oxide. Li et al. (2011) used this method to reduce the diameter of their nanowires defined by optical lithography from 100 to 80 nm, for instance.

Also, an EE reaction as the one used in the MACE synthesis of nanowires can be employed for roughening nanowires after their growth, thus conferring more effective means to enhance phonon boundary scattering (Section 2.2.2). This approach was used by Lee et al. (2016) and Lim et al. (2012) for roughening VLS-grown silicon nanowires for the characterization of roughness effects. The resulting roughness parameters—RMS and correlation lengths—could be controlled by limiting the etching time and concentration of acids/salts in the etchant solution.

These strategies can be applied to all kinds of nanowires after growth to improve their thermal behavior—as long they are not yet embedded or sintered (Section 3.6). However, it does not modify the pitch, meaning that the diameter reduction comes with the necessary reduction of array fill factor. Also, attaining a controlled and homogeneous etching rate becomes challenging when dense arrays are considered, as diffusion–reaction effects may lead to inefficient/null etching of inner nanowires and excessive/total etching of outer nanowires.

3.5.4 Drying of Nanowire Arrays

Some routes for growing nanowire arrays involve a wet process after nanowire growth, for example, MACE growth or EE-based postgrowth roughening treatments. If an array is dried by simple evaporation, then capillary forces lead to stick of the nanowires. This effect—nanostructure collapse upon drying (Namatsu et al., 1999)—may be detrimental for two reasons: first the surface-tension forces can be high enough to mechanically damage the nanowires so their σ is affected, and second because nanowires stick together and may not be as effective as free nanowires in terms of k reduction (Section 3.6).

This effect can be avoided by using CPD (Passi et al., 2013). In a CPD process, the liquid containing the nanowires is brought to high pressure and then exchanged by liquid CO_2. Then CO_2 is brought to a supercritical state, and by means of applying high T/P, forced to circumvent the triple point of the phase diagram without crossing the vapor–liquid line. When the sample is brought back to room conditions, CO_2 is released as a gas in a smooth transition. This way the nanowire array can be dried without experiencing strong capillary forces. Hill et al. (2010) developed another approach in which capillary stiction due to evaporation drying is significantly decreased by applying an electric field to the sample so that nanowires are forced to stay aligned during the process.

3.6 Bulk-Like Application

In Sections 3.1–3.5 the integration has been mostly considered for isolated nanowires, that is, nanowires that form aligned arrays and noncolliding with each other, which implies the necessity of integration in μTEG devices. However in some cases—solvothermal/laser-assisted synthesis for instance—alignment and connection are unfeasible, as nanowires are obtained as a disordered powder. In this case, nanowires are often included as inclusions within another thermoelectric material or sintered (e.g., by spark plasma), targeting to apply them as a nanostructured bulk. Then nanowire-based bulk material can be then integrated in a TEG by means of conventional macroscopic techniques—dicing, assembling, and contact soldering. This approach is actually applicable to all nanowires, synthesized by any method, and is indeed very practical, removing the problem of a limited fill factor. However, it necessarily requires additional steps for contact formation.

It is worth to note also that it is doubtful that positive effects for thermoelectric application remain the same when the nanowires are embedded in an electrically and thermally continuous three-dimensional medium. The properties of randomly aligned, sintered nanowires would reasonably resemble those of the nanostructured bulk, which is indeed appropriate for thermoelectric applications but to a lower extent and obtainable by other more suitable fabrication approaches (Kessler et al., 2013; Minnich et al., 2009). Analogously, the properties of a randomly aligned nanowire-containing composite should present properties resembling to those of composites including nanoinclusions. While $S^2\sigma$ of the matrix could be enhanced by exploiting modulation doping, and phonon scattering would take place at nanowires (Dresselhaus et al., 2007), no additional boundary scattering or quantum effects could take place in the matrix, implying that the low dimensionality-derived improvement of thermoelectric performance would be lower than that achieved in isolated nanowires.

4. THERMOELECTRIC CHARACTERIZATION TECHNIQUES FOR NANOWIRES

Dimensional, structural, and microstructural characterization of nanomaterials such as nanowires is typically carried out with a variety of advanced and reliable nanoscopy techniques. However, measuring their electrical and thermal transport properties (S, σ, and k) represents an important challenge due to the reduced dimensions of these nanostructures and the corresponding small response under voltage or thermal gradients. This small response forces the use of microstructures integrating the nanostructures and specific local probe methods—typically used for single nanowire measurements—or effective medium approximations, complex modeling, and thin-film/bulk methods—applicable to large nanowire array measurements. In this section, the characterization techniques used for single and nanowire arrays are introduced and described. Additionally, a small section introduces the means so far available for the determination of dopant concentration in nanowires, which is a key parameter influencing thermoelectric transport properties, as well as for the determination of the composition of thermoelectric alloys.

4.1 Single Nanowires

4.1.1 Seebeck Coefficient (S)—Heated Microstructures

Seebeck measurements are attained by forcing and measuring a temperature difference across the nanowire and measuring an open-circuit voltage, which allows the calculation of S as $S = \Delta V/\Delta T$ (Eq. 1). Since, in these measurements, the current passing is virtually zero, there is no voltage drop due to the contact resistance which means that two-probe measurements are enough to determine *S*. The measurements are dimension independent so for knowing *S* there is no need of fine determination of *L* and/or *d*.

Currently developed thermal characterization of microstructures (Fig. 19) typically allows: (i) imposing a ΔT using zigzag heaters; (ii) measuring the ΔT by reading the resistance increment in stripe sensors; (iii) measuring the ΔV generated at the wire ends by using the same stripe for determining open-circuit voltage at the ends of the nanowire (Rojo et al., 2013). The use of two heaters allows forcing very small ΔT and controlling overall measurement temperature. The principle by which the temperature

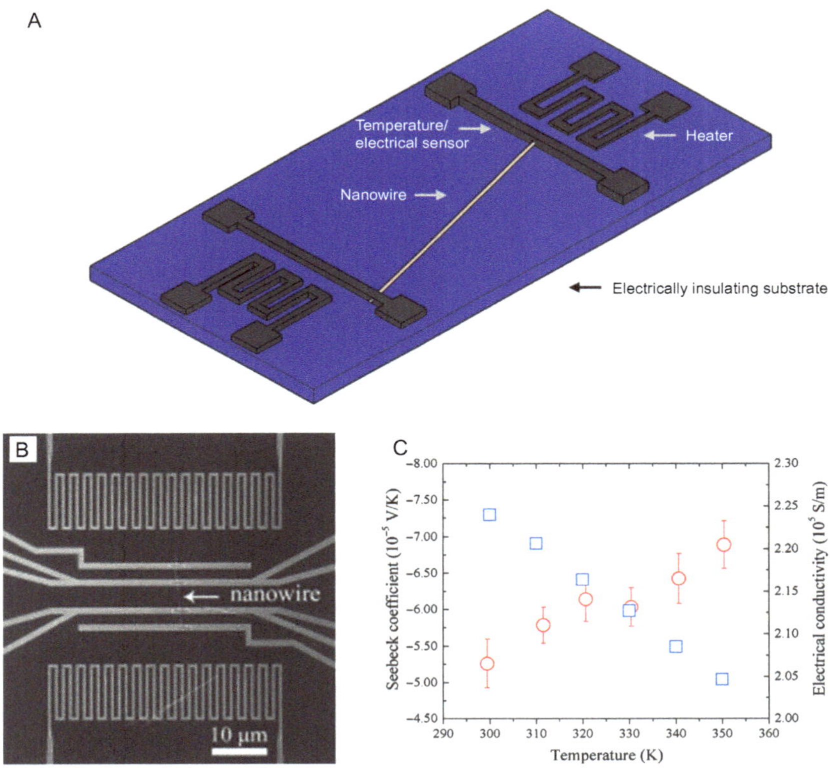

Fig. 19 (A) Planar-heated microstructure used for determination of Seebeck coefficient of nanowires. (B) SEM image of bismuth telluride nanowire integrated in a planar-heated microstructure for *S* and σ measurement, with results shown in (C). *From Shin, H.S., et al., 2014. Thermopower detection of single nanowire using a MEMS device. Measurement 51 (1), 470–475.*

is read is the linear increase of resistance of the metallic stripes with temperature. Thus a separate experiment for determining the temperature coefficient of resistance (TCR) of the stripes must also be performed. Heated microstructures are essentially planar insulating substrates with metallic thin films patterned on top—typically platinum with a few hundreds of nm. Zigzagging highly resistive structures serve as heaters, while straight strips serve as electrical contacts.

Two distinct manners exist for the integration of nanowires in these characterization structures. In a first approach, nanowires directly grow horizontally integrated in devices containing the contacts and heaters, through nanolithography methods (Section 3.4.1) or VLS monolithic integration (Section 3.3.1). In second, nanowires grown by these or any other methods are first dispersed in an organic solvent and then drop casted on top of an insulating chip. After solvent evaporation deposited nanowires are localized by, for instance, SEM and then contacts and heaters are patterned by means of electron beam-induced or ion beam-induced deposition (EBID and IBID, respectively). Since in S measurements electrical contact resistance has no effect—provided that a minimum stable electrical conductance can be achieved—the metal chosen for the contact formation is not very relevant.

To the best knowledge of the authors, this is the only method reported so far for the determination of the Seebeck coefficient of single nanowires. Some examples of its use has been reported elsewhere (Chen et al., 2010a; Jang et al., 2010; Lee et al., 2009; Noroozi et al., 2014; Wu et al., 2013). The fact that S measurements are not dependent on the electrical contact resistance and that the thermal contact resistance influence is avoided by using separate tracks for forcing and measuring current makes this configuration reasonably accurate.

4.1.2 Electrical Conductivity (σ)

For a fairly constant cross-sectional area structure such as in a nanowire, the electrical conductivity σ (i.e., the inverse of the electrical resistivity ρ in Eq. 12) is given by

$$\sigma = \frac{1}{A \cdot \frac{dR}{dL}} \tag{15}$$

Therefore, conductivity measurements should be able to determine the I–V relationship of the nanowire by forcing voltage and measuring current (or vice versa) in order to obtain $R = dV/dI$. Moreover, a precise determination of the nanowire dimensions, length L and cross section A, is also required.

4.1.2.1 Contact Resistance and Uniformity Considerations

When the contact resistance R_c can be neglected ($\rho_c < 10^{-6}\ \Omega\,\text{cm}^2$ for 10 μm nanowires for instance, see Fig. 10), dR/dL may be approximated as R/L, but when it is not the case, then R_c must be decoupled from the measurement. This can be done by performing a four-probe measurement as illustrated in Fig. 20A or by using a transmission line model (TLM). This TLM approach involves measuring wires of different lengths in the two-probe configuration as in Fig. 20B (Reeves and Harrison, 1982). In the first case, dR/dL is given by R_{4w}/L and the contact resistance as $R_c = (R_{2w} - R_{4w})/2$, while, in the second, dR/dL is the slope divided by A and R_c the intercept of the R_{2w} vs L plot divided by $2A$.

Eq. (15) and methods depicted in Fig. 20 assume that the resistance of the nanowire decreases proportionally with its cross section and this cross section constantly along the nanowire. These assumptions may be wrong if, for example, the nanowire presents tapering as in the case of some VLS-grown nanowires or if the effects of surface charges are nonnegligible as in the case of thin and lightly doped nanowires (Kimukin et al., 2006). Other models developed by Mohney et al. (2005) and Chaudhry et al. (2007) deal with

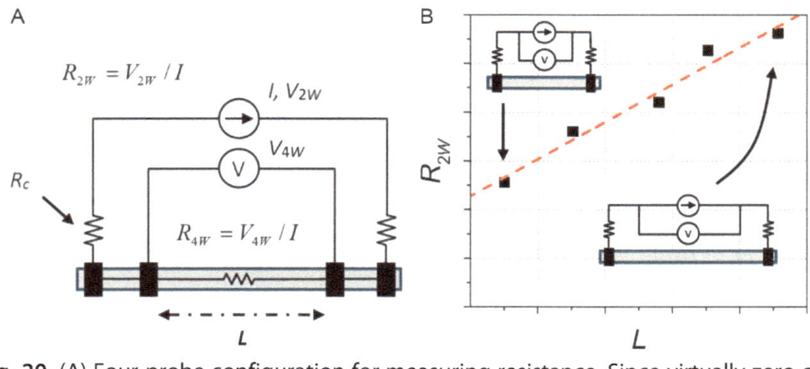

Fig. 20 (A) Four-probe configuration for measuring resistance. Since virtually zero current flows through the current sensing probes the influence in the contact resistances is avoided. (B) Illustrative plot of a transmission line method (TLM) for finding the resistivity (slope of the curve) and the contact resistance (half the *Y*-intercept) from a plot of *R* measured in the two-wire configuration in samples with different lengths.

these nonuniformities and allow for σ and R_c calculations by using the same principle of measuring R of different nanowires.

The determination of the contact resistivity ($\rho_c = R_c/A$) is interesting regarding integration as it may explain why a nanowire-based μTEG is or not performing well. In this sense, the measurement of ρ_c should be done in the same conditions as in which nanowires are to be integrated in the final μTEG device.

4.1.2.2 Four- and Two-Probe Microstructures

Microstructures with insulating substrates and four-probe contacts allow measuring electrical resistance as $R = dV/dI$ and calculating electrical conductivity from Eq. (15). The same microstructures used in Seebeck measurements serve for this purpose as long as they have four probes for avoiding the influence of the contact resistance, as shown in Fig. 19B. Two-probe microstructures may also be used provided measurements of nanowires of different lengths are available in order to perform a TLM-like method, as described above. The fabrication and nanowire integration methods are the same in Section 4.1.1.

Some examples of these measurements include the work of Chaudhry et al. (2007) and Koleśnik-Gray et al. (2013) who measured the electrical properties of semiconductor nanowires with a TLM method and with four probes, respectively. It must be noted that the characterization by TLM, though more complex—requiring the integration of nanowires with different lengths—is much more precise and allows extraction of other parameters such as surface charge. After all, the calculation made by comparing resistances measured by two and four wires is equivalent to using a TLM method with only two points.

4.1.2.3 Scanning Probe Microscopy

A nanowire is contacted as in the methods described above, so that electrical current or voltage difference can be forced/measured across its ends. Then, the nanowire is scanned along its axial position by means of a scanning probe microscopy (SPM) mode able to locally deliver and measure a current or to measure a voltage V, with a spatial resolution the order of 100 nm.

Three SPM techniques that can be used so far for this purpose are scanning tunneling microscopy (STM), conductive atomic force microscopy (C-AFM), and Kelvin force scanning probe microscopy (KPFM). The first two are able to force and measure currents that flow from them to the underlying substrate, only difference being the magnitude and the origin of the current—nA tunneling current in STM and μA drift current in C-AFM. KPFM is able to measure the surface potential of the samples, and so the voltage differences among different positions in a homogeneous material.

In current-sensing probes such as STM and C-AFM, a fixed voltage difference V is set between the tip and one of the nanowire ends and the nanowire is scanned along its axial direction, while the current that flows from tip to wire I is recorded with respect to the position x. Resistance at each point is found as $R = V/I$ as a function of x. These data together with nanowire dimensions—given by SEM/TEM rather than AFM topography, as the latter usually presents more artifacts—can be used as in the TLM method introduced above to obtain the σ and ρ_c, from dR/dx and $R|_{x=0}$, respectively.

In KPFM the voltage V rather than the current is sensed. In this case, the contacts to force a fix current I through the nanowire, and the KPFM probe is scanned along its axial direction, recording V vs x. The corresponding $R(x)$ is calculated and employed to determine σ and ρ_c as described above.

Apart from needing the integration of nanowires in contacting microstructures as in four-probe microstructures, these methods need to perform a scan using special SPM equipment—despite only two contacts are needed, rather than four. However, these techniques are interesting as they enable measurements in a local fashion: variations of σ along the wire may be determined by changes in the $R(x)$ slope, and different contact resistances may be measured at both sides. For example, Koren et al. (2009) employed KPFM to infer the gradual change of doping of a nanowire along its axial and radial directions. Assumptions as employed in using four- or two-probe structures are not needed and the exact contributions of contact effects are obtained. Some examples of the use of SPM methods for determination of nanowire σ are the works of Ma et al. (2001), Birjukovs et al. (2008), and Vinaji et al. (2012) who used STM, C-AFM, and KPFM, respectively.

4.1.3 Thermal Conductivity (k)

Thermal conductivity k measurements are based on forcing and measuring a heat flow Q through the nanowire, measuring the temperature difference ΔT developed at its ends, and determining its cross section A. This allows obtaining k through its definition (Eq. 16) or analogous versions adapted to the employed measurement:

$$k = \frac{Q \cdot L}{A \cdot \Delta T} \tag{16}$$

While dimensions L and d leading to A calculation are commonly obtained by SEM/TEM, Q and ΔT are measured by means of different methods introduced in the following sections. Heat flow at such small scales is a magnitude especially challenging to determine and must be inferred through some assumptions that imply using a high vacuum—to avoid transport by conduction—and low temperature—to avoid transport by radiation.

4.1.3.1 Suspended Heated Microstructures

The use of two suspended heated structures was first employed by Shi et al. (2003), and since then, it has been widely used for the determination of thermal conductivity of nanowires.

Fig. 21 shows two suspended microstructures of an electrically insulating material incorporating four-probe connected zigzag heaters with a

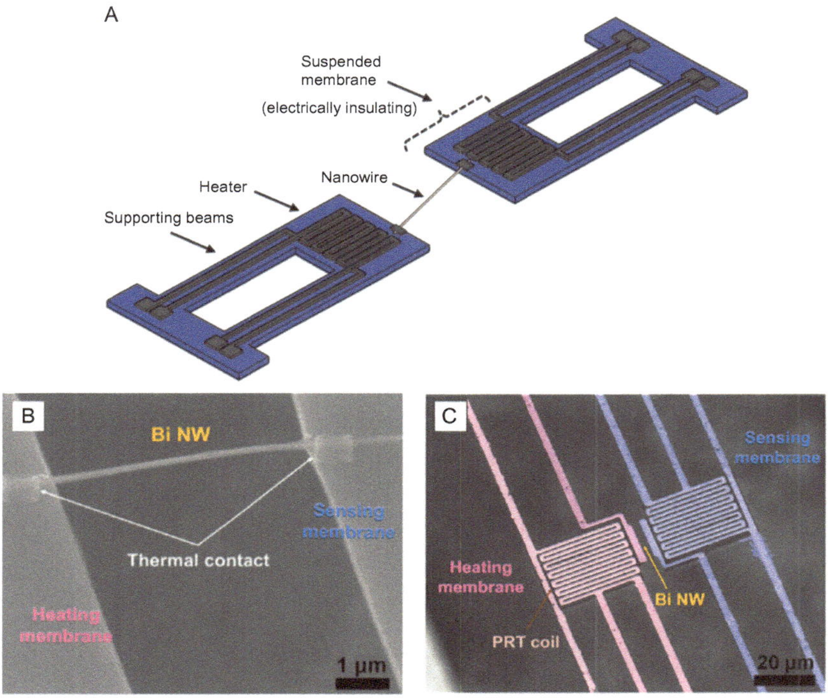

Fig. 21 (A) Suspended heated structures for the determination of thermal conductivity of nanowires. (B, C) SEM images of bismuth silicon nanowire integrated in suspended microstructures for k measurement. *Reprinted with permission of Roh, J.W., et al., 2011. Observation of anisotropy in thermal conductivity of individual single-crystalline bismuth nanowires. ACS Nano 5 (5), 3954–3960. Copyright (2011) American Chemical Society.*

suspended nanowire in between. In a typical measurement a high enough current to produce noticeable Joule heat dissipation is forced through one of the heaters, while a small current—only for resistance measurement purposes—is applied to the other. Part of the dissipated heat is going from the high current (hot platform) to the low current (cold platform) through the nanowire, while another part is going through the support legs of both platforms to the external frame of the chip, kept at constant temperature.

The resistances of both heaters serve for the determination of the temperature of each one of the suspended structures, as the metal sensors in Seebeck measurements, through calibration of their TCR (Section 4.1.1). The increase of temperature (with respect to a no current scenario) of the hot platform is defined as ΔT_h and the temperature increase of the cold platform is defined as ΔT_c. The heat dissipated in the hot platform is obtained as $Q_h = R_h \cdot I^2$, R_h being the four-probe resistance of the hot side heater. Heat also dissipates in the metal stripes that connect the heater to the external frame. This lead heat is obtained as $Q_L = R_L \cdot I^2$, R_L being the resistance of one beam, which can be obtained as well due to the four contacts fabricated in the device.

Then, applying a linear model for thermal transport and assuming that all heat is transferred by conduction, the thermal conductivity of the suspended nanowire can be obtained as (Shi et al., 2003)

$$k = \frac{L}{A} \frac{Q_s + Q_L}{\Delta T_h + \Delta T_s} \frac{\Delta T_s}{\Delta T_h - \Delta T_s} \tag{17}$$

Microstructures are fabricated by conventional microfabrication techniques, including a selective etch to suspend the platforms and the surrounding supports. Integration of the nanowire in such structures could be reached by the methods described above (monolithic or drop casting and contact evaporation). However, since platforms must be suspended by means of DRIE, when using drop-casting deposition microstructure definition cannot be conveniently performed after nanowire growth, as with S and σ setups. Instead, high-density drop casting is performed in a chip containing many microstructures, hoping that statistically some wire will be suspended between two structures and so it can be fixed and contacted by means of EBID or IBID. Another approach uses a micromanipulator tip to electrostatically attach and release a nanowire in the target structure prior to the metal deposition, ensuring the success of the integration (Lee et al., 2010).

There is no manner of implementing suspended temperature sensors in the middle of the wire—and if it was done, there would not be any way to prevent heat loss across them. For these reasons, a four-probe-like temperature measurement cannot be satisfactorily performed—while it can in S measurements (Fig. 19). This means that thermal contact resistance cannot be decoupled from the measurement and it must be either reduced or determined by means of a TLM-like method (see Section 4.1.2). For instance in the work from Hochbaum et al. (2008), thermal contact resistance is reduced by depositing wide platinum patches, while in Maire et al. (2017) it is measured with a TLM method, by doing thermal measurements of nanowires with multiple lengths.

In order to keep the assumptions of the method true, one must work at $T < 300\,\mathrm{K}$ to avoid transport by radiation and pressure below 10^{-4} mbar to avoid transport by through-air conduction (Moore and Shi, 2011). Nonlinear distributions of temperature due to far-field radiation lead to error estimations up to 25% if not considered. The introduction of a radiation shield in these measurements has recently allowed to characterize thermal conductivity of Si NWs up to 700 K (Lee et al., 2016). However, the works that study the accuracy limits of this method due to radiation effects do not often consider enhanced near-field radiative transport, which given the scale and temperature of the measurements might be influent (Section 2.3.2; Joulain, 2008).

Leaving aside heat leak considerations, the accuracy of the method is limited by the noise fluctuations in the four-probe measurements of heater resistances. Nanowires with thermal conductance $G < 1\,\mathrm{nW/K}$ will not be able to be safely measured (Rojo et al., 2013). In order to overcome this issue, a Wheatstone bridge may be implemented to reduce the noise while lowering the nanowire conductance resolution to $\sim 10\,\mathrm{pW/K}$ (Wingert et al., 2012). For instance, nanowires interesting for thermoelectric application, e.g., of $10\,\mu\mathrm{m}$ in length and k of $10\,\mathrm{W/m\,K}$, must be measured using a differential bridge for $d < 35\,\mathrm{nm}$. When a smaller k of $1\,\mathrm{W/m\,K}$ is considered, the use of a bridge becomes necessary for all $d < 100\,\mathrm{nm}$ and, even so, the measurement is limited to NWs with $d > 15\,\mathrm{nm}$. While shortening the nanowire can increase its conductance, it must be taken into account that the weight of the thermal contact resistance increases, which may lead to inference of an erroneously low value of k if not properly determined (Section 2.3.1).

This is the most popular method employed for the determination of the thermal conductivity of nanowires because of different reasons. First, thermal conductivities of many materials can be measured including insulators. Second, it is agreed that this method is more trustable than alternative ones (despite many assumptions are required). And third, because it allows full characterization in the case of thermoelectric nanowires, i.e., by incorporating additional electrical contacts, the suspended heated microstructures serve as well for the determination of S, σ, and ρ_c and therefore ZT in the same nanowire. However, one must take into account that any thermal contact resistance will have an effect in the Seebeck coefficient in this setup—while it is not like this in nonsuspended structures with additional sensors—and should be determined or minimized.

Most measurements of thermal conductivity of semiconductor nanowires were performed with this approach, as, for example, VLS and MACE silicon nanowires by Boukai et al. (2008) and Hochbaum et al. (2008).

4.1.3.2 Self-Heating of Suspended Nanowires

In this method, a current is applied though a nanowire suspended in between two contacting, nonsuspended structures with feature contacts for electrical access (Fig. 22, and Fig. 13F for a silicon nanowire).

The self-heating caused by Joule effect, i.e., the dissipation of a certain power $P = RI^2$, raises the mean temperature of the nanowire, which results in a variation of its resistance, R. Assuming that heat is transported by conduction and knowing the variation of R with T, one can infer the thermal conductivity through (Karg et al., 2013)

$$k = \frac{L}{12A}\frac{Q}{\Delta T} \tag{18}$$

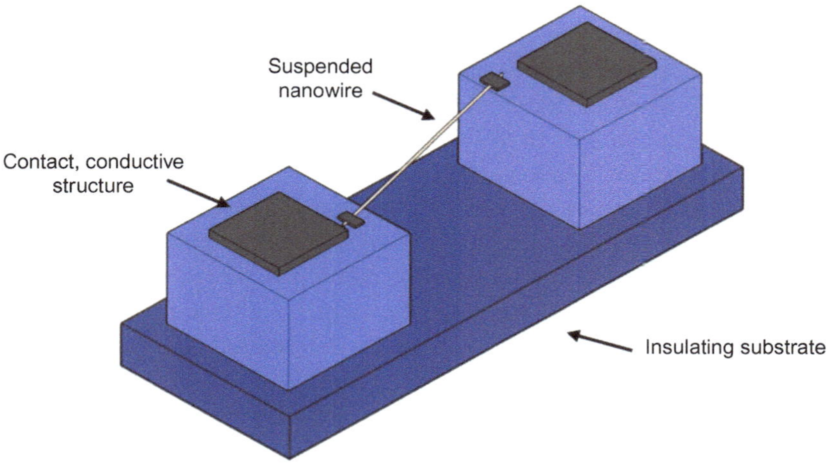

Fig. 22 Bridge structures for determination of thermal and electrical conductivity of nanowires.

In order for this method to work, the nanowires must present ohmic behavior—which implies a high enough doping level—and a small contact resistance. An important contact resistance, even if measured, will produce heating at nanowire ends compromising the assumptions of the employed thermal model. Like in suspended microstructures described above, the fabrication and nanowire integration can be attained monolithically or by drop casting and the same microstructure may be used for determining σ and ρ_c. It is worth noting again that the impossibility of using suspended sensing thermometers implies that the effects of thermal contact resistance cannot be avoided and thus it should be minimized or determined.

This method can be applied using direct current (DC) (Karg et al., 2014; Völklein et al., 2009) or alternating current (AC) (Heron et al., 2009; Lu et al., 2002). In the AC version, a 3ω approach can be used for increasing the accuracy in the measurement. In this case, a current of frequency ω and peak amplitude I_o is forced through the wire, while a voltage at frequency 3ω is measured in a lock-in amplifier. The magnitude of the 3ω voltage $V_{3\omega}$ is related to the thermal conductivity by

$$V_{3\omega} = \frac{4 \cdot I_0{}^3 \cdot L \cdot R \cdot (dR/dT)}{\pi^4 \cdot k \cdot A\sqrt{1 + (2 \cdot \tau \cdot \omega)^2}} \tag{19}$$

where $\tau = \rho \cdot C_P \cdot L^2/\pi^2 k$, with ρ being the density and C_P the specific heat capacity. The main advantages of 3ω compared to DC self-heating are: (i) it presents a much higher signal-to-noise ratio since many voltage readings are integrated over several AC cycles by the lock-in amplifier; (ii) it reduces the electromigration by forcing current in both senses; and (iii) it allows to simultaneously find k and C_P. One can first obtain k by measuring at the low-frequency limit in Eq. (19) and later find C_P by measuring at higher frequencies in which a $V_{3\omega}$ depends on ω.

4.1.3.3 Scanning Thermal Microscopy

A few SPM methods based in scanning thermal microscopy (SThM) may be employed for the characterization of the thermal conductivity of the nanowire. AFM probes with a nanothermometer—either a resistor or a thermocouple—integrated in the tip are used to locally heat and sense temperature, in an area with a radius of typically ~100 nm.

Puyoo et al. (2011) described a 3ω-based method for determining the thermal conductivity of vertically aligned nanowires embedded in a silica matrix. In this approach, the polished sample was scanned from top, being able to swiftly obtain thermal information of all the nanowires of a $5 \times 5\,\mu m^2$ region, by contacting their cross section from the top. Thermal conductivity was assessed by fitting the data—and previous data regarding tip and convection derived effects—to a complex model involving the heat spreading from the tip to the composite matrix, including several parasitic resistances. Results in good agreement with the literature were obtained in Si, Si–Ge, and Bi_2Te_3 nanowires using this method. The SThM–3ω method was also applied by Stranz et al. (2011) for thermal characterization of vertically aligned—nonembedded—silicon nanowires of 250 nm in diameter.

Another approach based in SThM but with a nanowire lying in a silica substrate was employed by Soudi et al. (2011). In this case, a gallium nitride nanowire was laterally scanned under vacuum conditions. The heat equation was solved by this specific configuration, accounting for the effects of the underlying substrate and the contact resistances involved in the heat transfer. Parameters related to the substrate were obtained by Raman thermometry, allowing to fit the data from the thermal scan to the model and thus obtaining the nanowire thermal conductivity.

In the work of Menges et al. (2016), the temperature profile of a self-heated and suspended indium phosphide nanowire was obtained by SThM. The 2ω-based scanning thermometry method employed therein allowed obtaining the $T(x)$ profile without influence of the tip-nanowire interfacial contact resistance. In principle, this information together with the power dissipated in the AC self-heating can be used to determine the thermal conductivity of the suspended nanowire by fitting the data to the parabolic heat equation solution—if conventional measures are taken, as using a temperature below 300 K and high vacuum.

Recently, Park et al. demonstrated the potential of scanning wave thermal microscopy for determining the thermal properties of nanowires (Kwon et al., 2003; Park et al., 2016). The approach consists in self-heating a suspended nanowire with an AC source and measuring the phase lag between thermal and electrical signals with the SThM tip with respect to its position on the nanowire. The gradient of this lag along the axial position (termed phase lag gradient, PLG) is directly related to the thermal conductivity of the nanowire. Choosing the proper frequency range the effect of thermal contact resistances can be decoupled, in resembling approach employed for thin film 3ω measurements—based in the limited reach of the heat wave, depending on its frequency. Reasonable results were obtained in the measurement of Bi_2T_3 nanowires—close to bulk values for a $d > 250$ nm nanowire.

SThM-based methods are very promising as they allow obtaining the whole profile of temperatures, giving insight of the variations in the local thermal properties and the effects of the nanowire-to-substrate contact resistances. However, there are important tip-related parasitic thermal conductances/resistances that may affect these measurements when so small structures are targeted. Thermal conductance from the tip to the rest of the probe/surroundings is much higher than the conductance of the nanowire. Alternative heat paths may be decoupled as in many cases they can be considered constant or modeled. However, since the thermal contact resistance between the tip and the nanowire is of the same order (or even higher) than that of the nanowire and of unpredictable nature—arising from the random contacting of the tip and sample nanofacets—it is necessary a proper determination of this contact resistance at each point. Novel SThM methods and models described above tackle this issue representing a promising family of techniques for the determination of k in nanowires.

4.1.3.4 Microoptical Methods

In micro-Raman thermometry, a suspended nanowire at vacuum conditions is heated with a microlaser spot and its optical response in the infrared is read with micro-Raman spectrometry. The Raman shift changes linearly with temperature, allowing inferring temperature at the position x (once this variation is calibrated in a separate experiment). The amount of heat absorbed is a function of the laser shape and size and nanowire diameter and absorption. Taking into account all the parameters and the nanowire dimensions, one can infer the thermal conductivity of the nanowire by fitting the data to the solution of the heat equation assuming no radiative/convective losses (Soini et al., 2010):

$$\Delta T(x) = \frac{P_{abs}}{kA}\left(\frac{x^2}{L} - \frac{L}{4}\right) \tag{20}$$

where P_{abs} is the power absorbed by the nanowire.

A great advantage of this method is the possibility to decouple contact resistance effects, as in SThM ones. As the temperature profile is obtained, one can deduce its effect from the ΔT in the nanowire close to the heat sinks (as in SPM methods for σ, at Section 4.1.2). A very positive point when compared to SThM is not having to deal with the tip-to-sample contact resistance. Moreover, if the suspended nanowire is electrically connected, a full thermoelectric characterization can be performed, determining also

S and σ from the open-circuit voltage and slope, respectively, of I–V curves measured at different laser positions (Yazji et al., 2015). However, it must be noted that the calculation of the absorbed power requires assumptions, modeling, and knowledge of absorptivity, and that the spatial resolution is still in the range of 0.5–1 μm, meaning that safe determination of k is only possible in long enough wires.

Microphotoluminescence technique uses a similar approach to that described above for micro-Raman. A suspended nanowire is heated by a laser pulse and its visible optical florescence signal is recorded as a function of the position. This signal is also a function of the temperature, implying that the same above described methodology can be used to determine thermal properties of the nanowire.

4.2 Nanowire Arrays

In thermoelectric characterization of nanowire arrays, it is convenient that their spatial distribution and dimensions are homogeneous to a certain degree or an average value will be obtained, which might arise from the contribution of very different values of σ and k depending on the diameter of each case (Section 2.2.1). If not, special attention must be taken when determining the dimensions by imaging, which might also require some specific statistical processing and modeling.

4.2.1 Single Nanowire Methods

In the case of aligned arrays of free nanowires or meshes, properly reengineered structures for accommodating wide arrays of nanowires and adapted models for accounting for their multiplicity can be acceptable—except for the SPM or optical cases in which it is not possible to access to subjacent nanowires in a three-dimensional arrangement. Some examples are the full characterization (S, σ, and k) of 2D Si nanowire arrays done by Boukai et al. (2008), the S measurements performed in arrays by Dávila et al. (2012) for Si NWs, and the k measurements performed by Pennelli et al. (2014) in Si NWs.

It is worth noting that in large and dense arrays, most of the nanowires are surrounded by other nanowires at the same temperature and thus shielded from the radial thermal transport by radiation or air conduction. This is very positive as it helps to reinforce the assumption of heat transport being only by means of axial solid conduction in k measurements, raising the temperature limit of these measurements.

4.2.2 Bulk and Thin-Film Methods

In the so-called effective medium approximations, nanowires can be treated as composite bulk/thin-film material composed by the nanowires and a matrix. This matrix can be air or vacuum in the case of free nanowire arrays or a thermally and electrically insulating material in the case of embedded nanowires (Curtin and Bowers, 2012; Li et al., 2011). This last approach is often adopted to give mechanical stability to nanowire arrays, which are obtained without connections, and need some pressure.

By modeling these nanowire composites assuming homogeneity in dimensions and spatial distribution, some methods applicable to bulk and thin films can be used as detailed below.

4.2.2.1 Conventional Methods (S, σ, and k)

Conventional setups for measuring transport properties of thermoelectric materials adapted for thin samples can be used with dense packs of aligned nanowires or nanowires embedded in a matrix, as far as they present enough robustness to resist pressure of external contacting elements. For example, Chen et al. (2010a) characterized as-fabricated bismuth telluride nanowire arrays embedded in alumina templates, using standard equipment for macroscopic samples, i.e., a four-probe thermal measurement for S and a flash thermal diffusivity measurement for k. Abramson et al. (2004) employed the well-established thin-film 3ω technique for measuring k of an array of vertically aligned silicon nanowires embedded in a polymeric matrix, with a model accounting for sample anisotropy. Xu et al. (2013) measured S for free-standing silicon-germanium nanowires using a macroscopic setup. Similarly, Chen et al. (2010b) employed the hot-disk transient method used in bulk samples for measuring k of dense, free-standing silicon nanowire arrays.

However, these measurements are significantly affected by electrical and thermal contact resistances, which affect σ, and S and k, respectively. While electrical contact resistance can be reduced by alloying (for instance Ni silicidation in silicon nanowires), there is not an established solution for the interfacial thermal resistance. Also, the thermal conductivity measurement will be affected by the parallel heat flow through the matrix, which always presents a higher fill factor and thus a nondespicable thermal conductance.

4.2.2.2 Photoacoustic Technique (k)

A laser is concentrated in a micrometric spot on top of a large nanowire array sample. Then the laser is suddenly interrupted. The sharp temperature transition generates a heat wave that propagates through the composite, generating a corresponding volume wave—due to thermal expansion—that is eventually translated into a pressure—or sound—wave. The sound wave propagates to the underlying substrate and through it to the air where it is transported to a microphone. Complex models involving thermal and mechanical transport relate the power of each sound harmonic to the thermal conductivity of the top material initially heated by the laser. While this method was originally devised for thin-film samples, it can be applied to nanowires assuming a homogeneous medium with anisotropic thermal conductivity.

Vertically aligned silicon nanowires could be measured by this method. Values in the order of that of silicon nanowires of the literature were obtained for cross-plane k, while values close to k of air were obtained in the in-plane directions, which was reasonable taking into account the vertical orientation of the wires.

4.2.2.3 Time-Domain Thermoreflectance (k)

An array of embedded nanowires exhibiting a top polished face is exposed to two pulsed lasers of different energies. The so-called heating pulses, with high energy, produce heat waves that propagate to the underlying cold substrate. The probing pulses, with low energy, are used to measure the change of reflectance of the sample with the ΔT produced by the heating pulses.

In its simplest approximation, the thermal transient behavior taking place after the laser impact spans a lapse of time of the order of $\tau = L^2/\alpha$, where $\alpha = k/\rho \cdot C_P$ is the thermal diffusivity of the sample. τ is typically in the range of 0.1–10 μs for nanowires. Determination of τ leads to the determination of α, which in turn leads to the determination of k of an effective medium. As in the photoacoustic technique, a homogeneous and anisotropic model must be applied in order to account for the different properties of the nanowires and the embedding matrix (Rojo et al., 2013).

4.3 Determination of Doping Level and Composition

The doping level is indicative of the degree of electronic optimization of the as-grown nanowires and thus its determination—or at least a qualitative inference of its order—is necessary (Section 3.5.2). The classical techniques used in bulk—secondary ion mass spectroscopy (SIMS), spreading resistance profiling, and junction staining (Sze, 2006)—are typically incompatible with nanowires, as they need to destroy a high volume of material in order to be applied. A differential SIMS approach—comparing signal of substrates with/without nanowires—can be employed to obtain an order estimate, but the detection limit of this method seems to be very low (Ingole et al., 2008). Thus, other strategies have been developed and employed for determining of doping in nanowires, which are briefly introduced below. A recommended review is presented in Wallentin and Borgström (2011). Some of the techniques even allow determination of doping as a function of the position, revealing possible inhomogeneities associated to the employed fabrication technique, as for instance axial and radial dopant gradation in some VLS-grown nanowires (Agrawal et al., 2011; Imamura et al., 2008). Knowing these defects at the local level is interesting as they may hint ways to improve the thermoelectric properties of the as-grown nanowires.

Atom probe tomography is an evolution of the SIMS technique for nanoscales, able to detect the position of each atom of a small nanostructure in the nm range. It works by field-evaporating one by one the atoms of the specimen as it is scanned in the three dimensions, while approaching the evaporating electrode (Agrawal et al., 2011; Miller and Forbes, 2009). After evaporation, the atoms are directed to a SIMS detector. With this method, local doping level of phosphorus-doped germanium nanowires could be determined, indicating a significant radial segregation related to the in situ VLS-mediated doping/growth process (Perea et al., 2009; Zhang et al., 2009). Sun et al. (2016) determined a lower radial gradation when ex situ doping is employed. The resolution of this method, i.e., the lowest composition that can be detected, is said to be in the part-per-million range, which would correspond to $5–10^{16}$ atoms/cm^3 in Si or Ge.

In another approach the Fano resonance of p-type dopants can be measured in a micro-Raman setup. This is the resonance between the backscattered Raman signals of the silicon with that of the dopant atoms. This technique is nondestructive and has micrometric precision, allowing determination of doping gradients along the axial direction of a suspended nanowire in a similar setup as in Section 4.1.3 (Imamura et al., 2008). However, only p-type doping levels above 10^{19} cm^{-3} can be measured. Scanning photocurrent microscopy is an electrooptical-based method. In this case, a concentrated laser is focused in a section of a wire with micrometric resolution and generates excess carriers—i.e., a photovoltaic-derived current. This small photocurrent is measured, and with a model, the doping level as a position in the nanowire is inferred (Allen et al., 2009; Richter et al., 2008).

Other methods use inference by measurement of electric properties. This would imply using variants of the contact-based or SPM-based setups described in Section 4.1.2 for determining σ and by making some assumptions determining the doping level—the simplest of which assumes that the doping–conductivity relation is the same for bulk and for nanowires (Ingole et al., 2008). However, this has the same issues related to σ measurements: influence of contact resistance and homogeneity assumption, often ignoring the effects of surface carriers or dopant deactivation, more likeable to take place in nanowires (Section 2.1.3). In this sense, Kimukin et al. (2006) performed σ measurements in silicon nanowires after successive etch-based thinning steps, being able to infer radial distribution of dopant and surface charges. Also, Garnett et al. (2009) demonstrated that cyclic capacitance–voltage (C–V) measurements could be employed to determine nanowire doping, even giving insight on radial dopant distribution. Finally, the theoretical work of Pichanusakorn and Bandaru (2010a,b) shows that for any thermoelectric bulk/nanostructure at any temperature, the optimum doping level is such that Seebeck coefficient is in the range of 130–200 μV/K. Thus, measuring S might actually serve as a way to determine whether doping optimization has been achieved or not. This is very practical as S measurement is the most trustable one among those of thermoelectric transport properties, not being influenced by contacts. Xu and Fobelets (2014) compared S measurements of equally doped nanowire and bulk samples in order to estimate the doping level. Pennelli et al. (2013) determined the doping level of large arrays of interconnected n-type silicon nanowires by S and σ measurements, leading to similar values of 3×10^{19} and 4×10^{19} cm^{-3}, respectively.

In some cases and especially materials, electron beam-based methods—energy-dispersive X-ray spectroscopy (EDX) and electron energy loss spectroscopy (EELS)—can be used as well, although generally dopant signals are rather weak in these systems (Wallentin and Borgström, 2011).

Similar to dopant concentration evaluation, it is important to accurately determine the stoichiometry and distribution of elements in thermoelectric alloys. For example, in Si–Ge compositions, the alloying effect induces phonon scattering at a maximum level when the concentration is 50% Si–50% Ge (although it is drastically reduced yet for compositions ranging from 20 to 80%). Also in Bi_2Te_3 nanowires, the carrier concentration is controlled by means of tuning the Bi and Te proportions, rather than doping. EDX and EELS are generally employed for these purposes, as they are generally accessible, practical, and have enough resolution.

5. SEMICONDUCTOR NANOWIRE-BASED μTEG DEVICES

An exhaustive review of the available data on nanowire-based μTEG devices has been carried out for this work (see Table 1). The table includes data on nanowire, device, and performance properties together with data of bulk/thick film silicon μTEGs included for comparison.

Fig. 23 shows two nanowire-based μTEGs, with planar (A, B) and vertical (C, D) architectures, from Dávila et al. (2012) and Curtin et al. (2012), respectively.

Although being an exhaustive list of works, the number of studies is still rather low due to the challenging nature of growing and integrating nanowire structures in devices. Moreover, most of them are made of silicon or Si-based nanowires (Si NWs), with specific works on bismuth telluride (Bi_2Te_3), copper and zinc oxide (CuO and ZnO), and polymer-embedded tellurium nanowires (Te). Silicon–germanium (Si–Ge) NWs were also recently integrated in devices by Xu et al. (2012) and Gadea et al. (2017). Silicon has a high thermal conductivity in its bulk shape, and thus its thermoelectric efficiency fully relays on an effective nanostructuration (Schierning, 2014). Despite this fact, the number of silicon nanowire-based μTEGs is the highest, which is clearly due to the well-known integration capabilities of the silicon-based microelectronics industry. All of the silicon nanowire-based devices make use of top-down methods except for the work of Dávila et al. (2012), which employed CVD-based VLS growth. Regardless of top-down or bottom-up, the techniques employed in these works are based in mainstream microfabrication techniques.

Bi_2Te_3 is the champion thermoelectric material at an industrial level in bulk. However, so far it has not been successfully integrated in nanowire-based devices able to generate thermoelectric power. Unfortunately, since the work of Wang et al. (2005), where a device architecture was presented and the integration issues and solutions were discussed, there has been no notice of further progress. On the other hand, the metal-oxide-nanowire-based μTEG fabricated by Zappa et al. (2014) seems still in a very early phase. The very low power density might be attributed to its design, which is perhaps more focussed on the optimization of its functionality as a sensor. CuO nanowires were grown in thin layers on top of thin-film substrates, while the ZnO ones grew in a disordered manner forming porous thin long stripes. Probably this is the cause of the high resistance reported ($R \sim 9\,M\Omega$), which is clearly hampering the device performance. Despite this low

Table 1 Compilation of Data From the Literature on Micro-TEGs Based on Nanowires

References	Material	Ar.[a]	NW Growth Method	Contacts	NW Surroundings	d (nm)[b]	L (μm)[b]	S (μV/K)[b]	σ (S/cm)[b]	$S^2\sigma$ (μW/cm K^2)[b]	k (W/m K)[b]	ZT @ RT[b]	R_{tot}/R_{NW}[b,c]	Doping (cm^{-3})[b]
Wang et al. (2005)	n-Type Bi_2Te_3 p-Type Bi_2Te_3	V	ECD in alumina template	Au sputtering	Embedded in Al_2O_3	50	100	260 −188	—	—	—	—	—	—
Li et al. (2011)	n-Type Si p-Type Si	V	Nanolithography (DUV + sacrificial oxidation)	Evaporation of Ni–Si	Embedded in SiO_2	80	1	±39	385	0.6	—	—	2.2	5.00E + 19
Li et al. (2012a)	n-Type Si p-Type Si	V	Nanolithography (DUV + sacrificial oxidation)	Evaporation of Ni–Si	Embedded in polyimide	80	1	±80	2200	14	9	0.047	—	5.00E + 19
Xu et al. (2012)	p-Type Si	V	MACE	Cu pressure contacting	Free standing	50–250	19	<u>559</u>[d]	0.18	0.05	—	—	<u>2.5E ± 05</u>[d]	—
	p-Type Si–Ge	V	CVD (for Si–Ge thick film) + MACE	Cu pressure contacting	Free standing	50–250	19	<u>531</u>[d]	—	—	—	—	—	—
Xu et al. (2013)	n/p-Type Si	V	SOD (n-type doping) + MACE	Cu pressure contacting	Free standing	—	—	—	—	—	—	—	—	—
Xu and Fobelets (2014)	p-Type Si	V	MACE + SOD (p-type doping)	Cu pressure contacting	Free standing	50–250	20	240	~263	~16	—	—	2.3	~2E + 19[d]
Curtin and Bowers (2012)	n-Type Si	V	Nanolithography (laser interference)	Evaporation of Ni/Au	Embedded in SiO_2	80–90	1	−284	<u>21</u>[d]	1.7	7.5	<u>0.007</u>[d]	—	2.50E + 19
Dávila et al. (2012)	p-Type Si	P	VLS	Monolithic VLS	Free standing	100	10 (90)[d]	<u>160</u>[d]	—	—	—	—	—	—

Gadea et al. (2017)	p-Type Si	P	VLS	Monolithic VLS	Free standing	80–140	10 (40)[d]	175	393	9	60	0.006	~1	~5×10^{19}[d]
	p-Type Si	P	VLS	Monolithic VLS	Free standing	80–140	10 (40)[d]	175	393	9	60	0.006	~1	~5×10^{19}[d]
	p-Type Si–Ge	P	VLS	Monolithic VLS	Free standing	60–75	10 (40)[d]	141	18	0.4	—	—	—	~$8–20 \times 10^{19}$[d]
Ferri et al. (2011) Data extracted from Fonseca et al. (2016)	n-Type pSi p-Type pSi	P	Nanolithography (nanorecession)	Monolithic etch (poly)	Embedded in SiO_2/ Si_3N_4	70–100	950	−160 240	500 250	13 14	—	—	—	—
Pennelli et al. (2013)	n-Type Si	P	Nanolithography (EBL + sacrificial oxidation)	Monolithic etch	Laying on thin SiO_2	60	3 (1000)[d]	−240	540	31	—	—	—	4×10^{19}
Hashimoto et al. (2017)	n-Type Si	P	Nanolithography (EBL + sacrificial oxidation)	Monolithic etch	Laying on bulk Si	45 × 36	10	−206	148	6	—	—	—	—
Yang et al. (2012)	p-Type Te	P	Solvothermal	Ag paste	Embedded in polymer	—	—	—	—	—	—	—	—	—
Zappa et al. (2014)	n-Type ZnO p-Type CuO	P	VLS/VT Thin-film oxidation	Cu sputtering	Mat-like stripes, on top of thin film	~20–100[d]	~1[d]	−110 +430	—	—	—	—	—	—
Glosch et al. (1999)	p-Type BULK Si	P	Optical lithography	Al	Free standing	$10 \times 7\ \mu m^2$	100	—	250	—	—	—	—	—

Continued

Table 1 Compilation of Data From the Literature on Micro-TEGs Based on Nanowires—cont'd

References	Power (μW)[e]	OCV (mV)[f]	I_{SC} (mA)	*R* (Ohm)	Δ*T* NW (K)	Δ*T* Achieved by[g]	Dev. Area (mm²)[h]	#TC[i]	#NW[i]	Dens. (NW/cm²)[h]	FF (%)	*P*/*A* (μW/cm²)[h,j]	*P*/*A*·Δ*T*² (μW/cm² K²)[h,j]
Wang et al. (2005)	—	—	—	—	—	—	0.5	144	—	1.0E+09[d]	—	—	—
Li et al. (2011)	0.0015	1.5	0.0038	396[d]	0.12	Heater	25	164	9.6E±07[d]	3.8E±08[d]	60	0.006[d]	0.4
Li et al. (2012a)	0.4700	27.9	0.067	416[d]	2.1[d]	Heater	25	164	9.6E±07[d]	3.8E±08[d]	60	1.9[d]	0.4[d]
Xu et al. (2012)	0.00034	1.9	0.0008[d]	2400	3.4	Heater	225	1	2–10E±09[d]	1–5 E±09[d]	50	0.0002[d]	0.00001[d]
	0.00108	2.6	0.0014[d]	1800	4.9	Heater	225	1	2–10E±09[d]	1–5 E±09[d]	50	0.0005[d]	0.00002[d]
Xu et al. (2013)	0.00112	2.1[d]	0.0021[d]	1000	—	Heater	150	1	–	—	—	0.0007[d]	—
Xu and Fobelets (2014)	0.65000	2.2[d]	1.2	1.8	8	Heater	400	1	2–10E±09[d]	6–30 E±08[d]	30	0.16[d]	0.003[d]
Curtin and Bowers (2012)	30	17	7	2.4	56	Heater	0.0025	1	6.3E±07[d]	2.50E±08[d]	15	1,200,000[d]	380[d]
Dávila et al. (2012)	9.6	58	1	58[d]	300	Heater	1.5[d]	1	—	—	—	960	0.011[d]
Gadea et al. (2017)	0.0011	0.84	0.010[d]	85	2.6	Harvesting @ 200°C	2	1	1.0E±09[d]	5.0E±06[d]	0.05	0.11	0.016[d]
	0.0440	1.93	0.136[d]	14.2	4.7	Harvesting @ 300°C	1.5	1	4.9E±08[d]	3.3E±06[d]	0.03	4.4	0.2[d]
	0.0490	2.8	0.104[d]	27	8.7	Harvesting @ 300°C	1.5	1	4.9E±08[d]	3.3E±06[d]	0.01	4.9	0.07[d]
Ferri et al. (2011) Data extracted from Fonseca et al. (2016)	—	500	—	—	—	Harvesting @ 50°C	100	264	2.5E+05	2.5E±05[d]	0.001	—	—

Pennelli et al. (2013)	—	0.7–1.2[d]	—	3.2E+04	3–5	Heater	2.4	1	5.5E+04	2.30E±06[d]	0.006	—	—
Hashimoto et al. (2017)	0.000003	5.00E−02	0.00024[d]	208[d]	—	—	>0.005[d]	—	4.0E+02	<8E±06[d]	<0.01	<0.06[d]	—
Yang et al. (2012)	0.003[d]	38	0.00032	1.2E±05[d]	55	Heater	10	1	—	—	—	0.03[d]	0.00001[d]
Zappa et al. (2014)	0.000016	24[d]	0.000003	9E+06	6	Heater	400	5	—	—	—	0.000004	0.0000001[d]
Glosch et al. (1999)	5	5200	0.0058	9E+05	20	Heater	16.5	1000	1000 (TCs)	6E±01 (TCs)[d]	0.4	30[d]	0.08[d]

[a]"Ar." stands for architecture (planar or vertical arrays of nanowires).
[b]Thermoelectric properties (S, σ, k, $S^2\sigma$, ZT) correspond to room temperature values for the nanowires, not the devices (i.e., not including contact effects).
[c]R_{tot}/R_{NW} is Eq. (13), relating total resistance including contact one, with that of the nanowires, and was calculated from R_c and R_{tot} values given by the references.
[d]*Underlined values* are estimations calculated with data from the references. Values preceded by a swung dash (~) are said to be approximated in references. *NW length values* in *parenthesis* refer to length of nanowire arrays connected in series.
[e]"Power" is the maximum power P—electrically matched—attained by the μTEG, when the nanowires are set to a temperature difference indicated by ΔT.
[f]OCV and I_{SC} are the open-circuit voltage and the short-circuit current that can be attained in these conditions, respectively, and R is the resistance. P, OCV, I_{SC}, and R are values corresponding to the device rather than the nanowires and in the calculations performed herein are related by:

$$P = \frac{OCV \cdot I_{SC}}{4} \tag{21}$$

$$R = \frac{OCV}{I_{SC}} \tag{22}$$

[g]"ΔT achieved by" indicates how was the temperature difference which allowed generation of P achieved: if it was forced by using a heater or if it was generated by placing the μTEG device on top of a hot surface.
[h]The device area is considered as the total device area as observed from top, regardless of the device architecture. All area-normalized parameters are normalized to this area. Thus the fill factor for instance is considered as "Area of all cross sections of all nanowires/total device area seen from top," and density (NWs/cm^2) is analogously so.
[i]"#TC" and "#NW" are the number of thermocouples and nanowires, respectively.
[j]P/A is the power density and $P/(A\Delta T^2)$ the efficiency factor defined in Briand et al. (2015), which accounts for the fact that P/A scales with ΔT^2 in μTEG devices.

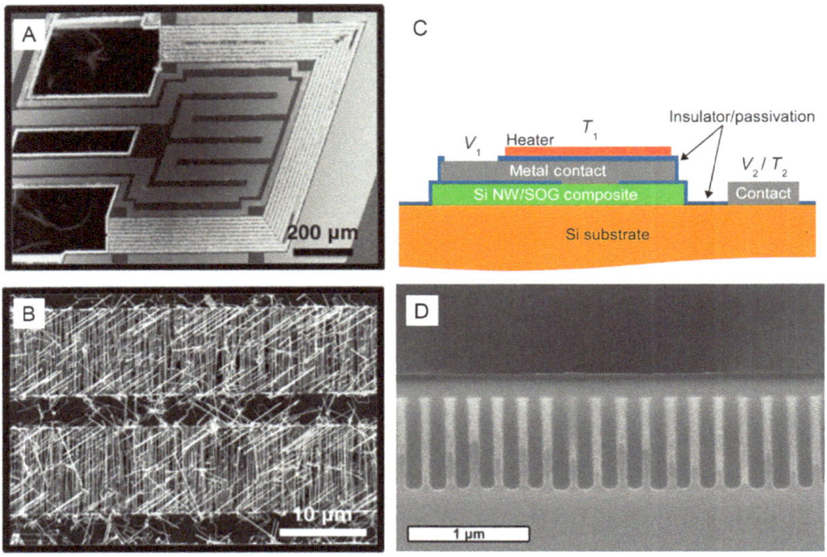

Fig. 23 Two silicon nanowire-based silicon microthermoelectric generators (μTEGs). (A, B) SEM top views of the device devised by Dávila et al. with a planar architecture (Dávila et al., 2012). Bottom-up aligned nanowires monolithically integrated in microtrenches are magnified in (B). (C, D) The device of Curtin et al., comprising vertically defined top-down nanowires (Curtin et al., 2012). (C) A schematic cross section of the device, with the nanowire array colored in *green*. An actual SEM cross section of the array of nanowires embedded in spin-on-glass (SOG) is shown in (D).

performance, it is a pioneering work on self-powering sensing devices, in this case, a temperature sensor. Besides, the device from Zappa et al. is one of the few devices which managed to create a multileg device, with five thermocouples. The same considerations can be applied to the work of Yang et al. (2012). Its recent irruption with an alternative material and a disordered disposition of the nanowires conforming a thin sheet represents an interesting breakthrough, although the measured 100 kΩ of resistance is a high value for a single thermocouple.

Regarding the Si–Ge nanowire-based μTEG reported by Xu et al. (2012), the synthesis of the NWs was carried out by first growing a thick film of poly Si–Ge by CVD on top of a silicon wafer and then performing a MACE growth. However, the contact resistance of the μTEG—determined by the contact obtained by sandwiching the array in between two Cu blocks—was so high that it substantially limited the device power. In a more recent work (Xu and Fobelets, 2014), the contact was improved to some extent for Si–Ge nanowires although a final solution is still pending.

In the same work, Si NWs were treated with an SOD-based doping step that substantially diminished their contact resistance. In the work of Gadea et al. (2017), Si–Ge nanowire arrays were integrated for the first time in planar-based μTEG devices. In this case, a low doping level resulted in a low power factor. Nonetheless, a power of 5 μW/cm^2 was obtained when harvesting on top of surfaces at 300°C, due to the improved thermal properties of the Si–Ge nanowire array.

Three of the silicon nanowire listed devices present vertical architectures, defined by the top-down methods nanolithography and MACE, while four of them present planar architectures, one based in VLS growth and the rest in nanolithography. The planar-based devices present a much lower fill factor <1%, in comparison to that of their vertical counterparts going from 15% to 60%. This is due to the fact that, despite having a very high density in the lateral dimension (few NW/μm), the amount of layers in the cross-plane dimension is limited: for instance, there is a single layer in the case of Hashimoto et al. (2017) and an extension of 15 μm depth in the works of Dávila et al. (2012) and Gadea et al. (2017). Besides, these devices need additional contacts and supportive areas, which reduce the available number of nanowires per unit area. Conversely, these problems are nonexisting in vertical architecture μTEGs in which the nanowires can be grown at lengths only limited by the method and contacting structures can be placed over or under the array. On the other hand, in the planar designs (Dávila et al., 2012; Ferri et al., 2011; Gadea et al., 2017; Pennelli et al., 2013; Watanabe et al., 2017), all the connections are monolithic and obtained in situ during the growth process, reducing substantially the contact resistance and simplifying the integration route. The contact resistance was estimated to be reduced in a factor of 2 and 10,000 in the early works of Li et al. (2011) and Xu et al. (2012), respectively, an issue that was not even present in the work of Gadea et al. (2017).

The planar architecture adopted by different authors (Dávila et al., 2012; Ferri et al., 2011; Gadea et al., 2017)—also observed in thin-film μTEGs (Perez-Marín et al., 2014)—makes use of a suspended platform to thermally isolate the cold from the hot parts. The thermoelectric legs—the nanowires—are integrated in between the platform and the surrounding frame, allowing to generate a temperature difference and thus a Seebeck voltage. The device used in the work of Gadea et al. (2017) is a second generation of the early design employed by Dávila et al. (2012) presenting improved electrical and thermal interfacing, as introduced in Calaza et al. (2016), which led to an increase of the efficiency factor $P/(A \cdot \Delta T^2)$ by a factor of 2. The performance of this device was measured in the harvesting mode, i.e., directly placing it on top of a hot surface with the cold part being cooled by natural convection, as it would be in a real application. Promising results with power densities up to 4.5 μW/cm^2 were obtained when the Si NW-based μTEG is put on top of a heat source at 300°C. Opposite to the unileg device employed by Gadea et al., Ferri et al. (2011) defined two-leg planar architectures based on n/p-type polycrystalline Si nanowires. The high number of thermocouples (264) together with the long nanowire length (~950 μm) allowed the generation of a very high OCV of 500 mV when harvesting on top of a surface at a temperature as low as 50°C. So far, the authors did not publish more information about this interesting μTEG and thus values of R or P/A are not yet available. Similarly, the work of Pennelli et al. (2013) does not report any value of power since it is focused on characterization rather than on nanowire-based μTEGs. However, it shows the suitability of planar networks of thin Si nanowires for implementation in μTEGs, as the former can be extended to very long lengths and the transversal interconnection leads to high reliability in case of nanowire break. It is worth noting that these thin nanowires (60 nm) are suspended presenting a low fill factor in the buried oxide film of the device layer, which is much thicker. Thus, the presence of a nonnegligible parallel thermal conductance is expected, which could lead to power reduction by thermalization of the hot and cold parts. In this sense, the case presented by Watanabe et al. (2017) is more extreme showing a single layer of nanowires laying on top of bulk silicon, across which the ΔT is established in a very short length. The authors claim the use of the "evanescent thermal field," arguing that using short lengths might justify the approach (at first sight seems to lead to a thermal short circuit of the hot and cold parts, thus avoiding the existence of a usable temperature difference). Further results in this novel work should be carefully followed since this approach will enable a rather simple fabrication route.

Looking at the top-down-based works with vertical architecture, a clear evolution is observed with noticeable improvement of the last generation of the devices from Li et al. (2012a) and Xu and Fobelets (2014). The first one shows an enhanced performance due to high thermal resistance, filling while the second takes advantage of the implementation of the SOD doping process after growth that allows reducing the contact and the intrinsic resistance of the as-grown MACE nanowires. Reasonable powers of ~1 μW/cm^2 were obtained when ΔTs of 2–8 K were used. As explained in Section 1.5, power densities in the range of 10–100 μW/cm^2 are enough for most energy harvesting common scenarios. Regarding the work of Curtin and Bowers (2012), a surprisingly high power density of 1.2 W/cm^2 was obtained when working at a ΔT of 56 K. This value is slightly below that of fully optimized commercial state of the art TAGS bulk modules when operating with a ΔT of 415 K (TECTEC, 2017). In an attempt to normalize the results with respect to the ΔT employed, the TEC efficiency factor $P/(A \cdot \Delta T^2)$ was calculated (Briand et al., 2015). Despite this normalization, a huge discrepancy still exists since the calculations of the efficiency factor of the μTEG reported by Curtin and Bowers (2012) based on silicon (with a ZT of 0.007) are clearly superior than the commercial module based in TAGS (with a ZT of 1.5). However, this difference is expected taking into account the manner in which material and device parameters influence in the efficiency coefficient, $P/(A \cdot \Delta T^2)$. Using Eqs. (1), (15), (21), and (22) leads to

$$\frac{P}{A \cdot \Delta T^2} = \frac{V^2}{4 \cdot R \cdot A \cdot \Delta T^2} = \frac{S^2 \cdot \sigma \cdot FF}{4 \cdot L} \tag{23}$$

with P/A being the same multiplied by ΔT^2. This implies that in experiments in which ΔT is fixed, the power density is expected to increase with the power factor, the fill factor, and the inverse of the nanowire length. Thus, this metric might serve to reveal the degree of electronic optimization, as includes the power factor calculated with device—rather than nanowire—parameters, which are effective S and effective σ. However, it does not give insight on thermal behavior as ΔT is forced by means of a heater and it scales with $1/L$ which might be arbitrarily short. For instance, a μTEG that achieves a high $P/(A \cdot \Delta T^2)$ by using short leg lengths is prone to possess a poor thermal behavior—i.e., a low thermal resistance—for the same reason. A measure that would be more complete comparing full thermoelectric performance is the harvested power at a fixed substrate temperature and cooling conditions. For instance, the power harvested on top of hot surfaces when cooling by natural convection in air, as given by some authors of Table 1 (Ferri et al., 2011; Fonseca et al., 2016; Gadea et al., 2017). Unfortunately, these data are not often provided in these terms in the literature, not even in the bulk or thick-film generators as the one employed here for comparison purposes (Glosch et al., 1999). All in all, it is difficult to assess and compare the full thermoelectric performance of the devices presented in Table 1 and even confirm noticeable progress with respect to the bulk.

6. SUMMARY AND CONCLUSIONS

Semiconductor nanowires are excellent candidates for thermoelectric generation applications due to low-size effects affecting the transport properties S, σ, and k, which lead to an enhancement of the thermoelectric figure of merit $ZT = S^2\sigma T/k$. Their elongated shape and outstanding properties together with a certain compatibility with known microfabrication processes (especially in the case of silicon-based nanowires) make them ideal for being implemented in μTEG devices dedicated to energy harvesting applications as powering WSN nodes.

It seems clear that low-dimensional classical effects take place in semiconductor nanowires. For nanowires of $d < 100\,\text{nm}$, an increased surface-to-volume ratio combined with phonon boundary scattering leads to a significant reduction of k, which has been directly measured and theoretically explained. It is important to mention that the determination of the thermal transport properties of nanowires at high temperatures still remains a challenge due to the interference of radiation in the measurements.

Electronic quantum effects are not so well supported. While the claimed semimetal–semiconductor transition has been achieved in few cases, experimental observation of positive quantum confinement effects in nanowires is still pending. Theoretical studies predict its occurrence in extremely thin nanowires ($d < 5\,\text{nm}$) which, so far, have not been fabricated and integrated in characterization structures.

More generally, there is still room for improvement in the electrical optimization of the nanowires as in most cases, despite achieving a drastic reduction of k, they possess a low power factor $S^2\sigma$ which ultimately renders their ZT lower than in optimized bulk. A higher control of the doping and the surface charges seems necessary for future developments.

Regarding the fabrication and integration of these nanowires in power devices, there are many routes able to grow dense arrays of aligned and doped semiconductor nanowires potentially applicable for thermoelectric generation. While most techniques have the potential for fabricating nanowires of many different semiconductors, they have been mainly applied to bismuth telluride, silicon–germanium, and, mostly, silicon (due to its compatibility with mainstream microfabrication technology). Not all fabrication methods are scalable and only a few of them are compatible with the integration of nanowires in μTEG devices, which requires connecting them as legs in a TEG with a low contact resistance. Due to this fact, so far only few devices achieved the integration of nanowires into fully fledged μTEG devices exhibiting powers in the order of 1–10 μW/cm^2. These values should be considered the beginning of the story since, differently from their bulk—and even some thin film—counterparts, the integration of nanowires in μTEGs is still in an early stage and there is plenty of room for improvement by optimization of their electrical and thermal transport properties, fill factors, achieved ΔT, and reduction of contact resistances.

REFERENCES

Abramson, A.R., et al., 2004. Fabrication and characterization of a nanowire/polymer-based nanocomposite for a prototype thermoelectric device. J. Microelectromech. Syst. 13 (3), 505–513. Available at: http://ieeexplore.ieee.org/document/1303629/.

Agrawal, R., et al., 2011. Characterizing atomic composition and dopant distribution in wide band gap semiconductor nanowires using laser-assisted atom probe tomography. J. Phys. Chem. C 115 (36), 17688–17694.

Allen, J.E., et al., 2009. Nonuniform nanowire doping profiles revealed by quantitative scanning photocurrent microscopy. Adv. Mater. 21 (30), 3067–3072.

Berg, M., et al., 2015. A transmission line method for evaluation of vertical InAs nanowire contacts. Appl. Phys. Lett. 107 (23), 232102.

Birjukovs, P., et al., 2008. Electrical characterization of bismuth sulfide nanowire arrays by conductive atomic force microscopy. J. Phys. Chem. C 112 (49), 19680–19685. Available at: http://pubs.acs.org/doi/abs/10.1021/jp805422k.

Boukai, A.I., Xu, K., Heath, J.R., 2006. Size-dependent transport and thermoelectric properties of individual polycrystalline bismuth nanowires. Adv. Mater. 18 (7), 864–869.

Boukai, A.I., et al., 2008. Silicon nanowires as efficient thermoelectric materials. Nature 451 (7175), 168–171. Available at: http://www.ncbi.nlm.nih.gov/pubmed/18185583. Accessed 30 April 2014.

Briand, D., Yeatman, E., Roundy, S., 2015. Micro Energy Harvesting. Wiley-VCH Verlag GmbH & Co. KGaA.

Bux, S.K., et al., 2009. Nanostructured bulk silicon as an effective thermoelectric material. Adv. Funct. Mater. 19 (15), 2445–2452.

Calaza, C., et al., 2016. Thermal test of an improved platform for silicon nanowire-based thermoelectric micro-generators. J. Electron. Mater. 45 (3), 1689–1694. Available at: http://link.springer.com/10.1007/s11664-015-4168-8.

Chaudhry, A., et al., 2007. Ultra-low contact resistance of epitaxially interfaced bridged silicon nanowires. Nano Lett. 7 (6), 1536–1541. Available at: http://pubs.acs.org/doi/abs/10.1021/nl070325e. Accessed 18 August 2014.

Chen, G., 1997. Size and Interface effects on thermal conductivity of superlattices and periodic thin-film structures. J. Heat Transfer 119, 220.

Chen, C.L., et al., 2010a. Fabrication and characterization of electrodeposited bismuth telluride films and nanowires. J. Phys. Chem. C 114 (8), 3385–3389.

Chen, T.-G., et al., 2010b. Phonon thermal conductivity suppression of bulk silicon nanowire composites for efficient thermoelectric conversion. Opt. Express 18 (S3), A467. Available at: https://www.osapublishing.org/oe/abstract.cfm?uri=oe-18-S3-A467.

Chen, J., Zhang, G., Li, B., 2011. Phonon coherent resonance and its effect on thermal transport in core-shell nanowires. J. Chem. Phys. 135 (10), 104508. Available at: http://adsabs.harvard.edu/cgi-bin/nph-data_query?bibcode=2011JChPh.135j4508C&link_type=EJOURNAL%5Cnpapers3://publication/doi/10.1063/1.3637044.

Comini, E., 2006. Metal oxide nano-crystals for gas sensing. Anal. Chim. Acta 568 (1–2), 28–40.
Cornett, J.E., Rabin, O., 2011. Thermoelectric figure of merit calculations for semiconducting nanowires. Appl. Phys. Lett. 98 (18), 182104.
Curtin, B.M., Bowers, J.E., 2012. Thermoelectric properties of silicon nanowire array and spin-on glass composites fabricated with CMOS-compatible techniques. MRS Proc. 1408 (x), 1–6. p.mrsf11-1408-bb11-02. Available at: http://journals.cambridge.org/abstract_S1946427412001947.
Curtin, B.M., Fang, E.W., Bowers, J.E., 2012. Highly ordered vertical silicon nanowire array composite thin films for thermoelectric devices. J. Electron. Mater. 41 (5), 887–894. Available at: http://link.springer.com/10.1007/s11664-012-1904-1.
Dasgupta, N.P., et al., 2014. 25th anniversary article: semiconductor nanowires—synthesis, characterization, and applications. Adv. Mater. 26 (14), 2137–2183.
Dávila, D., et al., 2012. Monolithically integrated thermoelectric energy harvester based on silicon nanowire arrays for powering micro/nanodevices. Nano Energy 1 (6), 812–819. Available at: http://linkinghub.elsevier.com/retrieve/pii/S2211285512001401. Accessed 23 May 2014.
Diarra, M., et al., 2007. Ionization energy of donor and acceptor impurities in semiconductor nanowires: importance of dielectric confinement. Phys. Rev. B Condens. Matter Mater. Phys. 75 (4), 045301.
Doerk, G.S., et al., 2010. Ex situ vapor phase boron doping of silicon nanowires using BBr3. Nanoscale 2 (7), 1165–1170. Available at: http://www.ncbi.nlm.nih.gov/pubmed/20648329%5Cnhttp://www.ncbi.nlm.nih.gov/pubmed/20648344.
Doerk, G.S., et al., 2011. Micellar block copolymer templated galvanic displacement for epitaxial nanowire device integration. J. Mater. Chem. 21 (24), 8807. Available at: http://xlink.rsc.org/?DOI=c1jm10693g.
Dresselhaus, M.S., et al., 2007. New directions for low-dimensional thermoelectric materials. Adv. Mater. 19 (8), 1043–1053.
Du, S., et al., 2013. Quantitative dopant distributions in GaAs nanowires using atom probe tomography. Ultramicroscopy 132, 186–192.
Dunham, M.T., et al., 2015. Power density optimization for micro thermoelectric generators. Energy 93, 2006–2017.
Farrell, R.A., et al., 2012. Large-scale parallel arrays of silicon nanowires via block copolymer directed self-assembly. Nanoscale 4 (10), 3228. Available at: http://xlink.rsc.org/?DOI=c2nr00018k.
Feng, X., et al., 2008. Vertically aligned single crystal TiO2 nanowire arrays grown directly on transparent conducting oxide coated glass: synthesis details and applications. Nano Lett. 8 (11), 3781–3786.
Ferri, M., et al., 2011. Ultradense silicon nanowire arrays produced via top-down planar technology. Microelectron. Eng. 88 (6), 877–881. Available at http://linkinghub.elsevier.com/retrieve/pii/S0167931710004557.
Fonseca, L., et al., 2016. Smart integration of silicon nanowire arrays in all-silicon thermoelectric micro-nanogenerators. Semicond. Sci. Technol. 31 (8), 084001.
Franssila, S., 2010. Introduction to Microfabrication. John Wiley & Sons, Ltd.
Fridman, 2005. Ordered arrays of silicon nanowires produced by nanosphere lithography and molecular beam epitaxy. Nano Lett. 5, 2524–2527.
Gadea, G., et al., 2015. Towards a full integration of vertically aligned silicon nanowires in MEMS using silane as a precursor. Nanotechnology 26 (19), 195302. Available at: http://stacks.iop.org/0957-4484/26/i=19/a=195302?key=crossref.f9bc9ad33e1c0a3bf6a9d5335c9e829c.
Gadea, G., et al., 2017. Integration of Si/Si-Ge Nanostructures in Micro-Thermoelectric Generators. University of Barcelona. Available at: http://ggd.a2hosted.com/dissertation/2017_Gadea_thesis.pdf.

Gao, D., et al., 2005. Selective growth of Si nanowire arrays via galvanic displacement processes in water-in-oil microemulsions. J. Am. Chem. Soc. 127 (13), 4574–4575. Available at: http://www.ncbi.nlm.nih.gov/pubmed/15796513. Accessed 15 October 2014.

Garnett, E.C., et al., 2009. Dopant profiling and surface analysis of silicon nanowires using capacitance-voltage measurements. Nat. Nanotechnol. 4 (5), 311–314.

Glosch, H., et al., 1999. Thermoelectric converter for energy supply. Sens. Actuators A 74 (1), 246–250.

Greene, L.E., et al., 2005. General route to vertical ZnO nanowire arrays using textured ZnO seeds. Nano Lett. 5 (7), 1231–1236.

Greiner, A., Wendorff, J.H., 2007. Electrospinning: a fascinating method for the preparation of ultrathin fibers. Angew. Chem. Int. Ed. 46 (30), 5670–5703.

Hållstedt, J., Hellström, P.E., Radamson, H.H., 2008. Sidewall transfer lithography for reliable fabrication of nanowires and deca-nanometer MOSFETs. Thin Solid Films 517 (1), 117–120.

Harman, T.C., Spears, D.L., Manfra, M.J., 1996. High thermoelectric figures of merit in PbTe quantum wells. J. Electron. Mater. 25 (7), 1121–1127. Available at: http://link.springer.com/10.1007/BF02659913.

Hashimoto, S., et al., 2017. Anomalous Seebeck coefficient observed in silicon nanowire micro thermoelectric generator. Appl. Phys. Lett. 111 (2), 023105.

Heron, J.S., et al., 2009. Mesoscopic size effects on the thermal conductance of silicon nanowire. Nano Lett. 9 (5), 1861–1865.

Hill, J.J., et al., 2010. Eliminating capillary coalescence of nanowire arrays with applied electric fields. ACS Appl. Mater. Interfaces 2 (7), 1992–1998.

Hochbaum, A.I., et al., 2005. Controlled growth of Si nanowire arrays for device integration. Nano Lett. 5 (3), 457–460.

Hochbaum, A.I., et al., 2008. Enhanced thermoelectric performance of rough silicon nanowires. Nature 451 (7175), 163–167. Available at: http://www.ncbi.nlm.nih.gov/pubmed/18185582. Accessed 23 May 2014.

Hong, S.H., et al., 2011. Electrical characteristics of nickel silicide-silicon heterojunction in suspended silicon nanowires. Solid-State Electron. 56 (1), 130–134.

Hsiao, T.K., et al., 2013. Observation of room-temperature ballistic thermal conduction persisting over 8.3 μm in SiGe nanowires. Nat. Nanotechnol. 8 (7), 534–538.

Huang, Z., et al., 2011. Metal-assisted chemical etching of silicon: a review. Adv. Mater. 23 (2), 285–308.

Imamura, G., et al., 2008. Distribution of active impurities in single silicon nanowires. Nano Lett. 8 (9), 2620–2624.

Ingole, S., et al., 2008. Ex situ doping of silicon nanowires with boron. J. Appl. Phys. 103 (10), 104302.

Jang, M., et al., 2010. The characteristics of Seebeck coefficient in silicon nanowires manufactured by CMOS compatible process. Nanoscale Res. Lett. 5 (10), 1654–1657. Available at: http://www.nanoscalereslett.com/content/5/10/1654.

Jang, M., et al., 2012. Top-down processed silicon nanowires for thermoelectric applications. J. Nanosci. Nanotechnol. 12 (4), 3552–3554. Available at: http://www.ncbi.nlm.nih.gov/pubmed/22849166.

Johnston, R.L., 2012. Metal nanoparticles and nanoalloys. Front. Nanosci. 3 (1), 1–42.

Joulain, K., 2008. Near-field heat transfer: a radiative interpretation of thermal conduction. J. Quant. Spectrosc. Radiat. Transf. 109 (2), 294–304.

Joyce, B.D., Baldrey, J.A., 1962. Selective epitaxial deposition of silicon. Nature 195 (4840), 485–486.

Kanungo, P.D., et al., 2009. Ex situ n and p doping of vertical epitaxial short silicon nanowires by ion implantation. Nanotechnology 20 (16), 165706. Available at: http://stacks.iop.org/0957-4484/20/i=16/a=165706.

Karg, S.F., et al., 2013. Measurement of thermoelectric properties of single semiconductor nanowires. Journal of Electronic Materials 42 (7), 2409–2414. Available at: http://link.springer.com/10.1007/s11664-012-2409-7. Accessed 12 January 2015.

Karg, S.F., et al., 2014. Full thermoelectric characterization of InAs nanowires using MEMS heater/sensors. Nanotechnology 25 (30), 305702. Available at: http://www.ncbi.nlm.nih.gov/pubmed/25004861.

Kaviany, M., 2013. Heat Transfer Physics. Cambridge University Press.

Kessler, V., et al., 2013. Thermoelectric properties of nanocrystalline silicon from a scaled-up synthesis plant. Adv. Eng. Mater. 15 (5), 379–385. Available at: http://doi.wiley.com/10.1002/adem.201200233. Accessed 1 December 2017.

Kimukin, I., Islam, M.S., Williams, R.S., 2006. Surface depletion thickness of p-doped silicon nanowires grown using metal-catalysed chemical vapour deposition. Nanotechnology 17 (11), S240–S245.

Kittel, C., 2004. Introduction to Solid State Physics. Wiley.

Koga, T., et al., 1998. Carrier pocket engineering to design superior thermoelectric materials using GaAs/AlAs superlattices. Appl. Phys. Lett. 73 (20), 2950–2952.

Kolasinski, K.W., 2006. Catalytic growth of nanowires: vapor-liquid-solid, vapor-solid-solid, solution-liquid-solid and solid-liquid-solid growth. Curr. Opin. Solid State Mater. Sci. 10 (3–4), 182–191.

Koleśnik-Gray, M.M., et al., 2013. Contact resistivity and suppression of Fermi level pinning in side-contacted germanium nanowires. Appl. Phys. Lett. 103 (15), 153101.

Kōmoto, K., Mori, T., 2013. Thermoelectric Nanomaterials : Materials Design and Applications. Springer. Available at: https://books.google.fr/books?id=tGu4BAAAQBAJ&dq=thermoelectric+nanomaterials&hl=es&source=gbs_navlinks_s. Accessed 1 December 2017.

Koren, E., et al., 2009. Nonuniform doping distribution along silicon nanowires measured by Kelvin probe force microscopy and scanning photocurrent microscopy. Appl. Phys. Lett. 95 (9), 92105. Available at: http://aip.scitation.org/doi/10.1063/1.3207887.

Koukharenko, E., et al., 2008. Towards a nanostructured thermoelectric generator using ion-track lithography. J. Micromech. Microeng. 18 (10), 104015.

Kwon, O., Shi, L., Majumdar, A., 2003. Scanning thermal wave microscopy (STWM). J. Heat Transfer 125 (1), 156. Available at: http://heattransfer.asmedigitalcollection.asme.org/article.aspx?articleid=1446270.

Laocharoensuk, R., et al., 2013. Flow-based solution-liquid-solid nanowire synthesis. Nat. Nanotechnol. 8 (9), 660–666.

LeBlanc, S., et al., 2014. Material and manufacturing cost considerations for thermoelectrics. Renew. Sustain. Energy Rev. 32, 313–327. Available at: https://doi.org/10.1016/j.rser.2013.12.030.

Lee, C.-H., et al., 2009. Thermoelectric power measurements of wide band gap semiconducting nanowires. Appl. Phys. Lett. 94 (2), 22106. Available at: http://aip.scitation.org/doi/10.1063/1.3067868. Accessed 23 October 2014.

Lee, K.M., et al., 2010. Focused ion beam-assisted manipulation of single and double beta-SiC nanowires and their thermal conductivity measurements by the four-point-probe 3-omega method. Nanotechnology 21 (12), 125301.

Lee, J., et al., 2016. Thermal transport in silicon nanowires at high temperature up to 700 K. Nano Lett. 16 (7), 4133–4140. Available at: http://pubs.acs.org/doi/abs/10.1021/acs.nanolett.6b00956.

Leonov, V., Fiorini, P., Vullers, R.J.M., 2011. Theory and simulation of a thermally matched micromachined thermopile in a wearable energy harvester. Microelectron. J. 42 (4), 579–584. Available at: https://doi.org/10.1016/j.mejo.2010.08.002

Li, S.S., 2010. Semiconductor Physical Electronics. Springer, New York.

Li, D., Wu, Y., Kim, P., et al., 2003a. Thermal conductivity of individual silicon nanowires. Appl. Phys. Lett. 83 (14), 2934–2936.
Li, D., Wu, Y., Fan, R., et al., 2003b. Thermal conductivity of Si/SiGe superlattice nanowires. Appl. Phys. Lett. 83 (15), 3186–3188.
Li, L., et al., 2005. Synthetic control of large-area, ordered bismuth nanowire arrays. Mater. Lett. 59 (10), 1223–1226.
Li, X., et al., 2009. High density p-type Bi0.5Sb1.5Te3 nanowires by electrochemical templating through ion-track lithography. Phys. Chem. Chem. Phys. 11 (18), 3584–3590.
Li, Y., et al., 2011. Chip-level thermoelectric power generators based on high-density silicon nanowire array prepared with top-down CMOS technology. IEEE Electron Device Lett. 32 (5), 674–676. Available at: http://ieeexplore.ieee.org/lpdocs/epic03/wrapper.htm?arnumber=5735183.
Li, Y., Buddharaju, K., Tinh, B.C., et al., 2012a. Improved vertical silicon nanowire based thermoelectric power generator with polyimide filling. IEEE Electron Device Lett. 33 (5), 715–717.
Li, Y., Buddharaju, K., Singh, N., et al., 2012b. Top-down silicon nanowire-based thermoelectric generator: design and characterization. J. Electron. Mater. 41 (6), 989–992. Available at: http://link.springer.com/10.1007/s11664-012-1901-4. Accessed 18 December 2013.
Lim, J., et al., 2012. Quantifying surface roughness effects on phonon transport in silicon nanowires. Nano Lett. 12 (5), 2475–2482. Available at: http://pubs.acs.org/doi/abs/10.1021/nl3005868.
Lin, Y.M., et al., 2002. Semimetal-semiconductor transition in Bi1-xSbx alloy nanowires and their thermoelectric properties. Appl. Phys. Lett. 81 (13), 2403–2405.
Lin, Y.C., et al., 2008. Single crystalline PtSi nanowires, PtSi/Si/PtSi nanowire heterostructures, and nanodevices. Nano Lett. 8 (3), 913–918.
Lu, L., Yi, W., Zhang, D.L., 2002. 3 omega method for specific heat and thermal conductivity measurements. Rev. Sci. Instrum. 72 (7), 2996–3003. Available at: http://aip.scitation.org/doi/10.1063/1.1378340.
Ma, D.D.D., Lee, C.S., Lee, S.T., 2001. Scanning tunneling microscopic study of boron-doped silicon nanowires. Appl. Phys. Lett. 79 (15), 2468–2470. Available at: http://aip.scitation.org/doi/10.1063/1.1409276.
Ma, F., et al., 2010. Nanocrystalline structure and thermoelectric properties of electrospun NaCo2O4 nanofibers. J. Phys. Chem. C 114 (50), 22038–22043.
Maire, J., Anufriev, R., Nomura, M., 2017. Ballistic thermal transport in silicon nanowires. Sci. Rep. 7, 41794. Available at: http://www.nature.com/articles/srep41794.
Martín, J., et al., 2014. Ordered three-dimensional interconnected nanoarchitectures in anodic porous alumina. Nat. Commun. 5, 5130.
Martínez, R.V., Martínez, J., Garcia, R., 2010. Silicon nanowire circuits fabricated by AFM oxidation nanolithography. Nanotechnology 21 (24), 245301.
Melosh, N.A., 2003. Ultrahigh-density nanowire lattices and circuits. Science 300 (5616), 112–115. Available at: http://www.sciencemag.org/cgi/doi/10.1126/science.1081940.
Menges, F., et al., 2016. Temperature mapping of operating nanoscale devices by scanning probe thermometry. Nat. Commun. 7, 10874.
Miller, M.K., Forbes, R.G., 2009. Atom probe tomography. Mater Charact 60 (6), 461–469. Available at: https://doi.org/10.1016/j.matchar.2009.02.007.
Minnich, A.J., et al., 2009. Bulk nanostructured thermoelectric materials: current research and future prospects. Energ. Environ. Sci. 2 (5), 466. Available at: http://xlink.rsc.org/?DOI=b822664b.
Mohney, S.E., et al., 2005. Measuring the specific contact resistance of contacts to semiconductor nanowires. Solid State Electron. 49 (2), 227–232.

Moore, A.L., Shi, L., 2011. On errors in thermal conductivity measurements of suspended and supported nanowires using micro-thermometer devices from low to high temperatures. Meas. Sci. Technol. 22 (1), 015103.
Morales, A.M., 1998. A laser ablation method for the synthesis of crystalline semiconductor nanowires. Science 279 (5348), 208–211. Available at: http://www.sciencemag.org/cgi/doi/10.1126/science.279.5348.208.
Namatsu, H., Yamazaki, K., Kurihara, K., 1999. Supercritical drying for nanostructure fabrication without pattern collapse. Microelectron. Eng. 46 (1), 129–132.
Noroozi, M., et al., 2014. In: Fabrication and thermoelectric characterization of GeSn nanowires. ULIS 2014—2014 15th International Conference on Ultimate Integration on Silicon, pp. 125–128.
Oehler, F., et al., 2009. The effects of HCl on silicon nanowire growth: surface chlorination and existence of a "diffusion-limited minimum diameter" Nanotechnology 20, 475307. Available at: http://www.ncbi.nlm.nih.gov/pubmed/19875870. Accessed 1 August 2014.
Ottens, R.S., et al., 2011. Near-field radiative heat transfer between macroscopic planar surfaces. Phys. Rev. Lett. 107 (1), 014301.
Park, H., et al., 2015. Aqueous chemical synthesis of tellurium nanowires using a polymeric template for thermoelectric materials. CrstEngComm 17 (5), 1092–1097. Available at: http://xlink.rsc.org/?DOI=C4CE02222J.
Park, K., et al., 2016. Measurement of thermal conductivity of Bi2Te3 nanowire using high-vacuum scanning thermal wave microscopy. Appl. Phys. Lett. 108 (7), 71907. Available at: https://doi.org/10.1063/1.4942098%5Cnhttp://aip.scitation.org/doi/10.1063/1.4942098.
Passi, V., et al., 2013. Design guidelines for releasing silicon nanowire arrays by liquid and vapor phase hydrofluoric acid. Microelectron. Eng. 103, 57–65.
Pennelli, G., et al., 2013. Seebeck coefficient of nanowires interconnected into large area networks. Nano Lett. 13 (6), 2592–2597.
Pennelli, G., Nannini, A., Macucci, M., 2014. Indirect measurement of thermal conductivity in silicon nanowires. J. Appl. Phys. 115 (8), 084507.
Perea, D.E., et al., 2009. Direct measurement of dopant distribution in an individual vapour-liquid-solid nanowire. Nat. Nanotechnol. 4 (5), 315–319.
Perez-Marín, A.P., et al., 2014. Micropower thermoelectric generator from thin Si membranes. Nano Energy 4, 73–80.
Pichanusakorn, P., Bandaru, P., 2010a. Nanostructured thermoelectrics. Mater. Sci. Eng. R 67 (2–4), 19–63.
Pichanusakorn, P., Bandaru, P.R., 2010b. Minimum length scales for enhancement of the power factor in thermoelectric nanostructures. J. Appl. Phys. 107 (7), 074304.
Puyoo, E., et al., 2011. Scanning thermal microscopy of individual silicon nanowires. J. Appl. Phys. 109 (2), 24302. Available at: http://scitation.aip.org/content/aip/journal/jap/109/2/10.1063/1.3524223.
Rabin, O., Lin, Y.M., Dresselhaus, M.S., 2001. Anomalously high thermoelectric figure of merit in Bi1-x-Sbx nanowires by carrier pocket alignment. Appl. Phys. Lett. 79 (1), 81–83.
Reeves, G.K., Harrison, H.B., 1982. Obtaining the specific contact resistance from transmission line model measurements. IEEE Electron Device Lett. 3 (5), 111–113.
Richter, T., et al., 2008. Doping concentration of GaN nanowires determined by opto-electrical measurements. Nano Lett. 8 (9), 3056–3059.
Rojo, M.M., et al., 2013. Review on measurement techniques of transport properties of nanowires. Nanoscale 5 (23), 11526. Available at: http://xlink.rsc.org/?DOI=c3nr03242f.
Rowe, D.M., 1995. CRC Handbook of Thermoelectrics. CRC Press.

Rowe, D.M., 2006. Thermoelectrics Handbook—Macro to Nano. CRC Press.
Schierning, G., 2014. Silicon nanostructures for thermoelectric devices: A review of the current state of the art. Physica Status Solidi A. 211 (6), 1235–1249. Available at: http://doi.wiley.com/10.1002/pssa.201300408. Accessed 17 July 2014.
Schmidt, V., Senz, S., Gösele, U., 2005. Diameter-dependent growth direction of epitaxial silicon nanowires. Nano Lett. 5 (5), 931–935.
Schmidt, V., Wittemann, J.V., Gösele, U., 2010. Growth, thermodynamics, and electrical properties of silicon nanowires. Chem. Rev. 110 (1), 361–388. https://doi.org/10.1021/cr900141g. Available at:.
Scorzoni, A., Finetti, M., 1988. Metal/semiconductor contact resistivity and its determination from contact resistance measurements. Mater. Sci. Rep. 3 (2), 79–137.
Seong, M., et al., 2012. Modeling and theoretical efficiency of a silicon nanowire based thermoelectric junction with area enhancement. J. Appl. Phys. 111 (12), 124319.
Shi, L., 2012. Thermal and thermoelectric transport in nanostructures and low-dimensional systems. Nanoscale Microscale Thermophys. Eng. 16 (2), 79–116.
Shi, L., et al., 2003. Measuring thermal and thermoelectric properties of one-dimensional nanostructures using a microfabricated device. J. Heat Transfer 125 (5), 881. Available at: http://heattransfer.asmedigitalcollection.asme.org/article.aspx?articleid=1446905.
Shi, W., et al., 2008. Hydrothermal synthesis and thermoelectric transport properties of impurity-free antimony telluride hexagonal nanoplates. Adv. Mater. 20 (10), 1892–1897.
Shimizu, T., et al., 2007. Synthesis of vertical high-density epitaxial Si(100) nanowire arrays on a Si(100) substrate using an anodic aluminum oxide template. Adv. Mater. 19 (7), 917–920.
Soini, M., et al., 2010. Thermal conductivity of GaAs nanowires studied by micro-Raman spectroscopy combined with laser heating. Appl. Phys. Lett. 97 (26), 245708.
Soudi, A., Dawson, R.D., Gu, Y., 2011. Quantitative heat dissipation characteristics in current-carrying GaN nanowires probed by combining scanning thermal microscopy and spatially resolved Raman spectroscopy. ACS Nano 5 (1), 255–262.
Sourribes, M.J.L., et al., 2013. Minimization of the contact resistance between InAs nanowires and metallic contacts. Nanotechnology 24 (4), 045703.
Stark, I., 2015. Micro thermoelectric generators. In: Briand, D., Yeatman, E., Roundy, S., Brand, O., Fedder, G.K., Hierold, C., Korvink, J.G., Tabata, O. (Eds.), Micro Energy Harvesting. John Wiley & Sons, pp. 245–269.
Stranz, A., Waag, A., Peiner, E., 2011. Thermal characterization of vertical silicon nanowires. J. Mater. Res. 26 (15), 1958–1962. Available at: http://www.journals.cambridge.org/abstract_S0884291411000604.
Sun, Z., et al., 2016. Dopant diffusion and activation in silicon nanowires fabricated by ex situ doping: a correlative study via atom-probe tomography and scanning tunneling spectroscopy. Nano Lett. 16 (7), 4490–4500.
Suriano, F., et al., 2015. Influence of grain size on the thermoelectric properties of polycrystalline silicon nanowires. J. Electron. Mater. 44 (1), 371–376.
Sze, S.M., 2006. Semiconductor Devices: Physics and Technology. Wiley.
Tasdemir, Z., et al., 2016. A deep etching mechanism for trench-bridging silicon nanowires. Nanotechnology 27 (9), 95303. Available at: http://stacks.iop.org/0957-4484/27/i=9/a=095303?key=crossref.d9451ef0b58376b2cd342ee57243f6ac.
TECTEC, 2017. PBTAGS-200.004A Thermoelectric Module Specification Sheet. Available at: http://thermoelectric-generator.com/wp-content/uploads/2016/05/TEG1-PBTAGS-220.004A-SPEC-SHEET.pdf.
Termentzidis, K., et al., 2013. Modulated SiC nanowires: molecular dynamics study of their thermal properties. Phys. Rev. B Condens. Matter Mater. Phys. 87 (12), 125410. Available at: https://link.aps.org/doi/10.1103/PhysRevB.87.125410.

Totaro, M., Bruschi, P., Pennelli, G., 2012. Top down fabricated silicon nanowire networks for thermoelectric applications. Microelectron. Eng. 97, 157–161. Available at: https://doi.org/10.1016/j.mee.2012.04.007.
Tuan, H.Y., et al., 2005. Catalytic solid-phase seeding of silicon nanowires by nickel nanocrystals in organic solvents. Nano Lett. 5 (4), 681–684.
Vinaji, S., Bacher, G., Mertin, W., 2012. Material and doping contrast in III/V nanowires probed by Kelvin probe force microscopy. In: Lorke, A. et al., (Ed.), Nanoparticles From the Gasphase: Formation, Structure, Properties. Springer Berlin Heidelberg, Berlin, Heidelberg, pp. 185–206. Available at: https://doi.org/10.1007/978-3-642-28546-2_8.
Vining, C.B., 2009. An inconvenient truth about thermoelectrics. Nat. Mater. 8 (2), 83–85. Available at: http://www.nature.com/doifinder/10.1038/nmat2361.
Völklein, F., et al., 2009. The experimental investigation of thermal conductivity and the Wiedemann-Franz law for single metallic nanowires. Nanotechnology 20 (32), 325706. Available at: http://www.ncbi.nlm.nih.gov/pubmed/19620755.
Wagner, R.S., Ellis, W.C., 1964. Vapor-liquid-solid mechanism of single crystal growth. Appl. Phys. Lett. 4 (5), 89–90.
Wallentin, J., Borgström, M.T., 2011. Doping of semiconductor nanowires. J. Mater. Res. 26 (17), 2142–2156. Available at: http://www.journals.cambridge.org/abstract_S0884291411002147.
Wang, W., et al., 2005. A new type of low power thermoelectric micro-generator fabricated by nanowire array thermoelectric material. Microelectron. Eng. 77 (3–4), 223–229. Available at: http://www.sciencedirect.com/science/article/pii/S0167931704005143. Accessed 15 August 2014.
Wang, W., et al., 2007. Bi2Te3/Te multiple heterostructure nanowire arrays formed by confined precipitation. J. Am. Chem. Soc. 129 (21), 6702–6703.
Watanabe, T., et al., 2017. In: A scalable Si-based micro thermoelectric generator. IEEE Electron Devices Technology and Manufacturing Conference Proceedings of Technical Papers.
Wingert, M.C., et al., 2011. Thermal conductivity of Ge and Ge-Si core-shell nanowires in the phonon confinement regime. Nano Lett. 11 (12), 5507–5513.
Wingert, M.C., et al., 2012. Ultra-sensitive thermal conductance measurement of one-dimensional nanostructures enhanced by differential bridge. Rev. Sci. Instrum. 83 (2) 024901.
Wu, P.M., et al., 2013. Large thermoelectric power factor enhancement observed in InAs nanowires. Nano Lett. 13 (9), 4080–4086.
Xia, Y., et al., 2003. One-dimensional nanostructures: synthesis, characterization, and applications. Adv. Mater. 15 (5), 353–389. Available at: https://doi.org/10.1002/adma.200390087.
Xu, B., Fobelets, K., 2014. Spin-on-doping for output power improvement of silicon nanowire array based thermoelectric power generators. J. Appl. Phys. 115 (21), 214306. Available at: http://aip.scitation.org/doi/10.1063/1.4881781.
Xu, C.H., Woo, C.H., Shi, S.Q., 2004. Formation of CuO nanowires on Cu foil. Chem. Phys. Lett. 399 (1–3), 62–66.
Xu, B., et al., 2012. Thermoelectric performance of $Si_{0.8}Ge_{0.2}$ nanowire arrays. IEEE Trans. Electron Devices 59 (12), 3193–3198. Available at: http://ieeexplore.ieee.org/document/6313899/.
Xu, B., et al., 2013. n-Si–p-Si1 − xGex nanowire arrays for thermoelectric power generation. Solid-State Electron. 83, 107–112. Available at: http://linkinghub.elsevier.com/retrieve/pii/S0038110113000518.
Yan, X.M., et al., 2005. Fabrication of large number density platinum nanowire arrays by size reduction lithography and nanoimprint lithography. Nano Lett. 5 (4), 745–748. Available at: https://doi.org/10.1021/nl050228q%5Cnhttp://pubs.acs.org/doi/full/10.1021/nl050228q%5Cnhttp://pubs.acs.org/doi/pdf/10.1021/nl050228q.

Yang, Y., et al., 2012. Nanowire-composite based flexible thermoelectric nanogenerators and self-powered temperature sensors. Nano Res. 5 (12), 888–895. Available at: http://link.springer.com/10.1007/s12274-012-0272-8.

Yazawa, K., Shakouri, A., 2011. Cost-efficiency trade-off and the design of thermoelectric power generators. Environ. Sci. Technol. 45 (17), 7548–7553.

Yazji, S., et al., 2015. Complete thermoelectric benchmarking of individual InSb nanowires using combined micro-Raman and electric transport analysis. Nano Res. 8 (12), 4048–4060.

Yee, S.K., et al., 2013. $ per W metrics for thermoelectric power generation: beyond ZT. Energ. Environ. Sci. 6 (9), 2561–2571. Available at: http://xlink.rsc.org/?DOI=C3EE41504J.

Zakharov, N., et al., 2006. Growth phenomena of Si and Si/Ge nanowires on Si (111) by molecular beam epitaxy. J. Cryst. Growth 290 (1), 6–10. Available at: http://linkinghub.elsevier.com/retrieve/pii/S0022024805016076.

Zappa, D., et al., 2014. Integration of ZnO and CuO nanowires into a thermoelectric module. Beilstein J. Nanotechnol. 5, 927–936. Available at: http://www.beilstein-journals.org/bjnano/content/5/1/106.

Zhang, Z.B., et al., 1999. Processing and characterization of single-crystalline ultrafine bismuth nanowires. Chem. Mater. 11 (12), 1659–1665.

Zhang, S., et al., 2009. Relative influence of surface states and bulk impurities on the electrical properties of Ge nanowires. Nano Lett. 9 (9), 3268–3274.

Zhang, G., et al., 2012. Rational synthesis of ultrathin n-type Bi_2Te_3 nanowires with enhanced thermoelectric properties. Nano Lett. 12 (1), 56–60. Available at: http://pubs.acs.org/doi/abs/10.1021/nl202935k.

Zou, J., Balandin, A., 2001. Phonon heat conduction in a semiconductor nanowire. J. Appl. Phys. 89 (5), 2932–2938.

CHAPTER SIX

Measurement of the Thermoelectric Properties of Individual Nanostructures

Francesco Rossella*, Giovanni Pennelli†, Stefano Roddaro*,†,1
*NEST, Scuola Normale Superiore and Istituto Nanoscienze—CNR, Pisa, Italy
†Università di Pisa, Pisa, Italy
[1]Corresponding author: e-mail address: stefano.roddaro@unipi.it

Contents

Semiconductors and Semimetals, Volume 98
ISSN 0080-8784
https://doi.org/10.1016/bs.semsem.2018.02.001

1. INTRODUCTION

Thermoelectric (TE) effects and the existence of an intrinsic connection between charge and heat transport phenomena in materials have been well known since the discoveries of Thomas Seebeck in the early 19th century. When a material is subject to a temperature gradient, high-energy carriers residing in hot regions tend to diffuse to the cold ones, producing a charge build-up and an electrostatic potential difference. In open-circuit conditions, such a process continues until drift and diffusion currents balance and a Seebeck voltage is established. TE effects provide a very interesting way to convert waste heat into useful energy using a solid-state converter: this possibility has been widely explored since the 1950s, but in practice TE devices have been relegated so far to a small group of niche scientific, medical, or military applications. The key reason for such a limited diffusion of this technology fundamentally lies in the relatively low thermodynamic efficiency of existing TE devices. After a long period of slow progress, research on TE physics has been renovated during the past two decades thanks to the advent of nanoscience and nanotechnology. Nanowires (NWs) have played an important role in this new research direction and have been crucial in the demonstration of various strategies put forward to improve the efficiency of TE energy conversion. As a result, recent years have seen the development of novel TE nanomaterials with properties that have been impossible to achieve so far in any known conventional bulk material and have revived hopes for a more widespread application of TE technology.

The chapter is organized as follows: after a brief introduction about the problem of thermodynamic efficiency of TE heat engines (Section 2), the rest of the chapter focuses on the challenges and opportunities related to the measurement of the TE parameters of nanomaterials in general and of individual nanostructures in particular; thermal transport techniques based on suspended microstructures are quickly reviewed in Section 3; the importance of the field-effect control is discussed in Section 4 and its combination with full TE characterization methods in Section 5; finally, perspectives and opportunities concerning the investigation of the low-temperature TE response of nanostructures are addressed in Section 6.

2. TE EFFECTS AND NANOSTRUCTURES

2.1 The Problem of Conversion Efficiency

The first large-scale efforts to exploit TE effects for energy conversion date back to the 1950s, when it was discovered that heavily doped semiconductors feature large Seebeck coefficients and can be used to implement useful solid-state heat engines (Mahan et al., 1997). Devices based on TE effects indeed display a set of unique properties that are very attractive for applications. First of all, they are typically very reliable since they do not contain any moving part: TE generators coupled to radioactive heat sources have been in operation for a total combined time of over 300 years, in the harsh environment of deep space probes, with no failures and obviously no maintenance (Dudzinski, 2013). In addition, different from the case of other heat conversion technologies, TE generators and refrigerators can be downscaled and integrated in very small devices and, for instance, used to harvest energy from a variety of existing devices that produce waste heat during their normal operation (Snyder and Toberer, 2008; Vining, 2009). Unfortunately, TE devices also have one crucial drawback: their thermodynamic efficiency is disappointingly low and significantly smaller than the one offered by alternative technologies. Indeed, an ideal efficient TE material needs to comply with a set of conflicting requirements that are in practice very challenging to meet: it should display large electrical conductivity σ and Seebeck coefficient S, and a small thermal conductivity $\kappa = \kappa_e + \kappa_{ph}$, where κ_e and κ_{ph} are the conductivity components due to electrons and phonons. While each of these parameters can be strongly modified by tuning the concentration of free carriers n (see Fig. 1), it is extremely challenging to optimize all of them at the same time and maximize the overall TE performance, for a variety of reasons (Snyder and Toberer, 2008). First of all, in typical bulk materials σ and κ_e follow the Wiedemann–Franz relation which locks their ratio. In addition, large Seebeck coefficients are typically only obtained for poorly conductive materials. The problem of efficiency maximization can be more precisely quantified by means of the dimensionless figure of merit

$$ZT = \frac{\sigma S^2 T}{\kappa}$$

where T is the absolute temperature. The value of ZT is in a monotonic correspondence with the maximum attainable thermodynamic efficiency η according to the relation

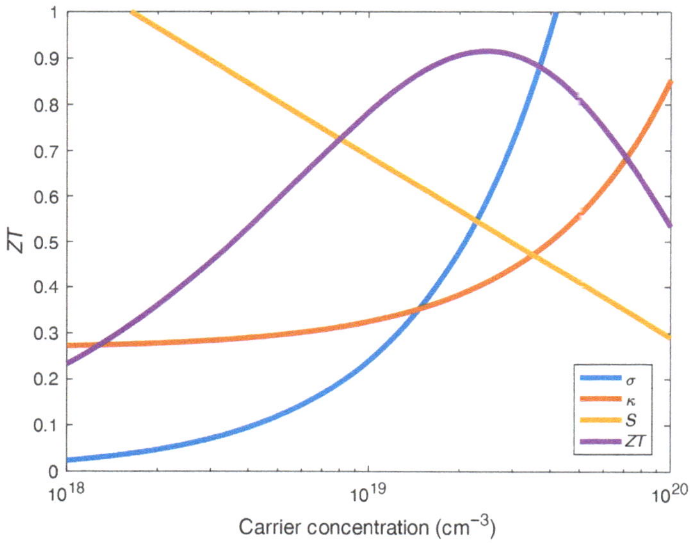

Fig. 1 Qualitative dependence of the TE parameters on the carrier concentration *n*. *Adapted from Snyder, G.J., Toberer, E.S., 2008. Nat. Mater. 7, 105–114.*

$$\frac{\eta}{\eta_C} = \frac{\sqrt{Z\bar{T}+1}-1}{\sqrt{Z\bar{T}+1}+T_C/T_H}$$

where T_C and T_H are the temperature of the cold and hot reservoirs, $\bar{T}$ is the average temperature, and $\eta_C = (T_H - T_C)/T_H$ is the Carnot efficiency (Snyder and Toberer, 2008). The resulting efficiency as a function of ZT in the limit of small thermal bias is reported in Fig. 2: a large ZT is clearly required to obtain a large efficiency. Research activities in the 1950s established the foundations of TE device physics and allowed the identification of a class of "good" TE materials—for instance Be_2Te_3 at around room temperature or SiGe in the 700–1000°C range—which allow reaching $ZT \cong 1$, corresponding to a unappealing performance of less than 15% of Carnot efficiency, in the best possible electrical and thermal matching conditions. As highlighted by a number of authors, $ZT > 3$ is the generally accepted condition for making some large-scale application of TE devices economically appealing (Bell, 2008; Majumdar, 2004; Vineis et al., 2010; Vining, 2009). Such a condition has been completely out of reach for decades.

2.2 The Advent of Nanomaterials

Since the 1990s, the advent of new nanomaterials has strongly revived research activities in TE physics. The resulting paradigm shift has been

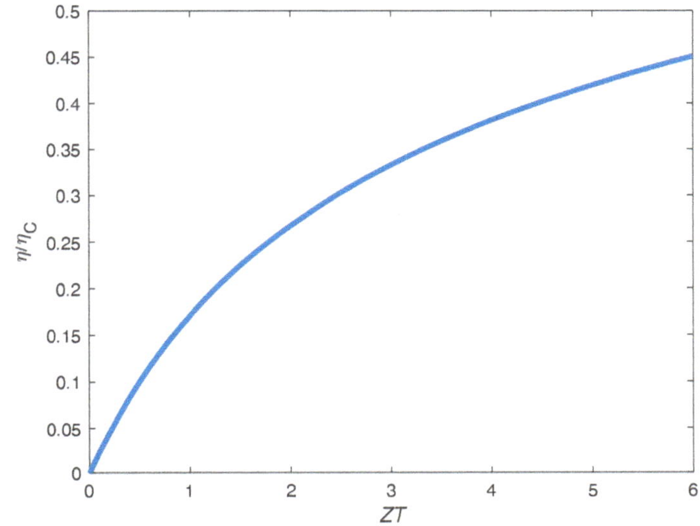

Fig. 2 Correspondence between efficiency and the figure of merit *ZT* in the limit of a small thermal bias.

driven by a set of theoretical predictions about the possible strategies for engineering an improvement of the value of ZT (Baxter et al., 2009; Chen et al., 2012; Dresselhaus et al., 2007; Heremans et al., 2013; Hicks and Dresselhaus, 1993a,b; Majumdar, 2004; Vineis et al., 2010). Since phonon heat conduction constitutes a pure loss for what concerns TE energy conversion,[a] a first crucial idea consists in trying to *reduce the crystal thermal conductivity* κ_{ph}. A central strategy, typically referred to as the "phonon glass/electric crystal" concept (Bell, 2008), consists in manipulating the nanoscale texture of a material and introducing a number of random or controlled interfaces that are able to induce a strong phonon boundary scattering while not disturbing too much electronic transport. This can be achieved by taking advantage of the different length scales associated with phonon and electron transport in solids. A further exploration direction consists in the manipulation of the phononic density of states (Khitun et al., 1999; Kwon et al., 2016), which can become relevant nanoscale conductors or in finely heterostructured systems. A second class of novel ideas to increment efficiency rather aims at boosting the value of the so-called power factor σS^2, which appears at the numerator of the expression for ZT. Various proposals have been put forward in this context and most of them typically take

[a] This statement neglects the impact of phonon drag phenomena, which might become relevant in particular in the context of one-dimensional conductors (Boukai et al., 2008; Herring, 1954).

advantage of the engineering of quantum confinement and density of states (Dresselhaus et al., 2007; Heremans et al., 2013; Nakpathomkun et al., 2010), or of the clever design of energy filtering to allow the diffusion of carriers only within a well-controlled energy window (Humphrey and Linke, 2005). The manipulation and reduction of κ_{ph} is now an established fact, even if the magnitude of the effect and its exact control are still debated. Semiconductor NWs have provided a formidable platform for experiments in this context (Ali et al., 2017; Caballero-Calero and Martín-González, 2016; Yang, 2010): about 10 years ago experiments performed on rough silicon NWs (Boukai et al., 2008; Hochbaum et al., 2008) have in fact strikingly confirmed prediction by showing evidence for a hundredfold reduction of the thermal conductivity with respect to bulk values for silicon. Thanks to the progress on the reduction of κ_{ph}, various laboratory prototypes of materials with a claimed ZT well above 2 have been developed, bringing new hopes for a wider diffusion of TE technology. Research efforts for the control and maximization of the power factor σS^2 are still ongoing and few interesting demonstrations have been performed in recent years, in particular in prototypes operating at cryogenic temperatures (Nakpathomkun et al., 2010; Wu et al., 2013).

2.3 TE Properties of an Individual Nanostructure

The rapid growth of the research field of nanoscale TE goes hand in hand with the development and refinement of experimental methods for the determination of the TE parameters. Open issues remaining in the strategies for the improvement of the parameters entering the value of ZT often require an investigation at the level of the individual nanostructure, possibly as a function of various control parameters such as the free carrier density. On the other hand, the direct measurement of the TE response of a single nanostructure is intrinsically more challenging than the one of a macroscopic material for a variety of reasons: it is harder to create a significant temperature bias over a small spatial scale (Roddaro et al., 2014); due to the sheer scale of the interfaces, measurements are much more affected by systematic effects due to thermal and electrical contact resistances (Yazji et al., 2016); nanostructures often display a sizeable variability due to mesoscopic effects as well as to the simple fact that they can be hard to fabricate in a completely reproducible and predictable way. These difficulties also imply that it is particularly relevant to be able to determine *all* the TE parameters on the same nanostructure (Yazji et al., 2015), in order to avoid possible wrong conclusions drawn from the fact that nominally identical nanostructures can in fact behave in a relatively different way.

It is important to notice that the emergence of new tools for the study of TE effects at the scale of an individual nanostructure is also an interesting development in itself, since it provides novel tools for the fundamental investigation of the physics of nanostructures and of condensed matter. The TE response yields in fact an information which is not equivalent to the one that can be derived from conventional charge transport experiments. TE measurements have been in fact proposed to provide new insight on the properties of nanostructures (Gluschke et al., 2015; Lim et al., 2014); to provide new independent evidences of exotic electronic states (Khim et al., 2015; Leijnse, 2014); or even to obtain an independent estimate of fundamental transport parameters such as free carrier mobility and distinguish between different experimental sources of systematic errors (Roddaro et al., 2013; Schmidt et al., 2014).

2.4 Perspectives Beyond Energy Conversion

The full comprehension of the mechanisms ruling thermodynamics at the meso- and nanoscale and its connection with the macroscopic limit remains today a great intellectual challenge. The investigation of TE physics of nanomaterials and of individual nanostructures has in fact a relevance that goes beyond the context of energy conversion and of the development of novel high-ZT materials, and research activities can be framed in the much wider context of heat management and thermodynamics at the nanoscale (Chen, 2005). Today's nanoelectronics is indeed mostly limited by heat dissipation constraints (Moore and Shi, 2014; Schelling et al., 2005) and future quantum computing technology will also likely require a careful control of heat transport and the impact of statistical and quantum fluctuation. A number of interesting further investigation directions have emerged in recent years in connection with the thermodynamics at the nanoscale, in particular concerning the investigation of quantum thermodynamics (Pekola, 2015), the nonlinear TE response of systems in strongly out-of-equilibrium conditions (Erdman et al., 2017; Svensson et al., 2013; Talbo et al., 2017), and the feeble boundary between heat and mechanical work in physical systems that are small with respect to the relevant length scale such as the electron and phonon mean free path (Luckyanova et al., 2012; Luo and Chen, 2013; Nardi et al., 2011).

3. DETECTION OF HEAT TRANSPORT WITH SUSPENDED MICROSTRUCTURES

3.1 The Challenge of Heat Measurement

Measuring thermal properties such as thermal conductivity κ or heat capacitance C is generally significantly more challenging than assessing conventional charge transport characteristics. First of all, virtually any material is characterized by a finite thermal conductivity; thus it is hard to force heat conduction to follow a specific path and it is not always completely obvious to predict the impact of parallel heat leaks in the solid-state regions that surround the measured material. Second, heat flows cannot be directly detected but rather have to be indirectly deduced from the read out of local thermometers and from the temperature behavior of known and predictable physical systems. These challenges are even further exacerbated when thermal measurements have to be performed on an individual nanostructure: a small heat conductance or heat capacity is obviously harder to detect on top of background signals due to stray heat leak paths; thermal contact resistances can easily become nonnegligible due to small contact areas and again to the small heat conductance of the target materials (Yazji et al., 2016; Yu et al., 2006); sample-to-sample variability is not uncommon in nanostructures, due to the above-mentioned reasons. Obviously, the thermal properties of a nanostructured material can also be investigated by addressing the ensemble behavior, for instance using bundles of NWs or fiber that are surely easier to manipulate and measure with respect to an individual nanoscale object. Nevertheless, the measurement of an individual nanostructure has some important added values, in particular in the context of fundamental research in TE: first of all the intrinsic variability at this scale implies that ensemble averages cannot be particularly meaningful of the behavior of an individual NW; in addition, wire-to-wire thermal coupling is generally largely unknown and makes it hard to estimate and investigate the fundamental thermal properties of the target nanostructure.

3.2 The Suspended Microstructure Approach

Suspended microstructures have been widely used for the investigation of thermal conductivity of individual nanostructures such as nanotube, NWs, and nanosheets since early 2000 (Kim et al., 2001; Li et al., 2003b; Shi et al., 2003). The technique makes it possible to extremely limit heat leak paths and allowed the first heat transport measurements over an individual nanostructure. This constituted an important step forward with respect to technique addressing the thermal properties of nanostructure assemblies since it allowed sorting out the intrinsic nanostructure conductivity from the uncontrolled thermal barrier due at the interface between nearby nanostructures in the assembly (Kim et al., 2001). Two suspended heater/thermometer platforms are typically fabricated by selective etching of a 100-nm-thick SiN layer and are connected to the bulk Si substrate by a set of high aspect ratio (typically about 100 μm long and a fraction of square micron in cross-sectional area) beams, which thus have a very low thermal conductance (on the order of few tens of nW/K). As visible in the panels of Fig. 3, each platform typically integrates a resistive heater and a resistive thermometer, usually fabricated by depositing a thin Pt layer. The device is operated by heating one of the platforms by Joule effect (heating membrane, temperature T_h) and by measuring the resulting temperature increase in the second platform (sensing membrane, temperature T_s), which is sensitive to the thermal connection provided by the nanostructure and to the mild thermal sinking provided by the suspension beams. The device is typically operated in vacuum to avoid convective heating and at low thermal bias to minimize radiative effects. The technique is capable of detecting a thermal conductance of 0.1–1 nW/K, depending on the

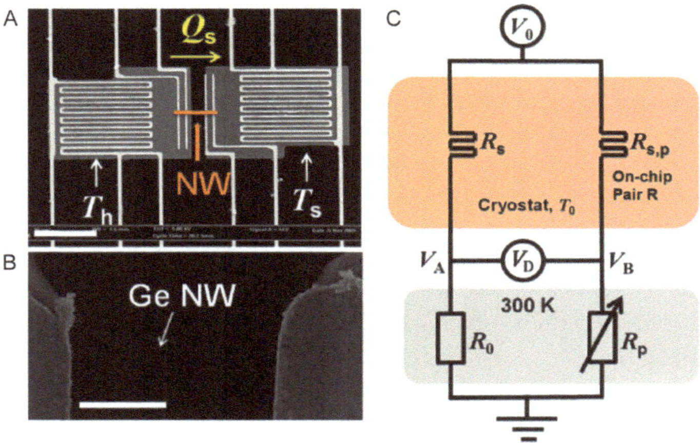

Fig. 3 Suspended microheaters for the measurement of the thermal conductivity of individual NWs: scanning electron image of the device and of the NW (A, B); Wheatstone bridge setup for the measurement of the thermoresistance on the sensing membrane (C). *Reproduced with permission from Wingert, M.C., Chen, Z.C.Y., Dechaumphai, E., Moon, J., Kim, J.H., Xiang, J., Chen, R., 2011. Nano Lett. 11, 5507–5513. Copyright 2011 American Chemical Society.*

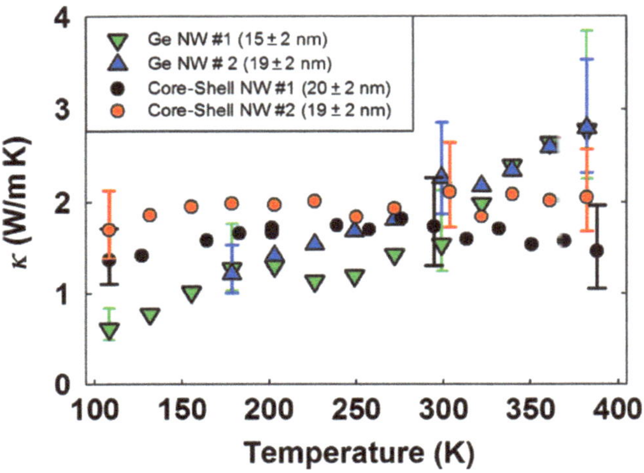

Fig. 4 Comparison of thermal conductivity of Ge and Ge–Si core–shell NWs with different diameters. *Reproduced with permission from Wingert, M.C., Chen, Z.C.Y., Dechaumphai, E., Moon, J., Kim, J.H., Xiang, J., Chen, R., 2011. Nano Lett. 11, 5507–5513. Copyright 2011 American Chemical Society.*

measurement temperature (Shi et al., 2003), and can be used to determine the conductance of individual NWs and the impact of nanostructuring on the conductivity κ (see Fig. 4).

Suspended microstructures have provided the first important evidences of thermal conductance reduction at the scale of the individual nanostructure. These include in particular the first demonstration of a strong suppression of the thermal conductivity in small diameter Si NWs (Boukai et al., 2008; Hochbaum et al., 2008). Since their introduction, suspended micromembranes have been used to characterize a wide range of nanomaterials and have shed new light on the effects of nanostructuring on the thermal conductivity of a material (Ko et al., 2016; Kwon et al., 2016; Lee et al., 2012; Mavrokefalos et al., 2007, 2009; Pettes and Shi, 2009; Swinkels et al., 2015; Weathers et al., 2013; Wingert et al., 2011; Zhou et al., 2005, 2011a,b).

3.3 Challenges, Perspectives, and Limitations

The discussed method poses a set of critical technical challenges in particular for what concerns the positioning and contacting of the nanostructure, since the suspended structures are very fragile and obviously incompatible with conventional lithographic techniques. Individual NWs are typically added using in situ micromanipulators or drop casting, while thermal and electrical contacts can be fabricated by beam-induced deposition of amorphous carbon (only for thermal contact) or of metallic layers such as Pt. This technology is also clearly not compatible with conventional wet surface passivation techniques which are typically used to obtain reliable Ohmic contacts. As a consequence, as also highlighted by various authors (Yazji et al., 2016; Yu et al., 2006), the role of thermal contact resistances has to be carefully taken into account. The problem becomes particularly relevant in the limit of small nanostructures which are characterized by contacts with extremely small interface areas.

Thanks to its exceptional sensitivity, the suspended microstructure method has also been used to estimate the thermal parameters of individual nanostructures. Recent experiments have explored the possibility to detect the heat capacity C of a single nanofiber or NW, which can be relevant in the dynamical response of a TE material and which can yield useful information about the phonon density of states in the nanostructure. Heat capacity measurements can also be performed using the so-called 3ω technique (Lu et al., 2001), which will be discussed in detail in Section 5. A variant was recently demonstrated also in the case of suspended microstructures for the case of insulating nanostructures, which do not allow exploiting self-heating by Joule effect (Chen et al., 2008; Zheng et al., 2016).

While suspended microstructures have provided an exceptional tool for the study of the thermal conductivity κ of an individual nanostructure and have allowed a crucial step forward with respect to experiments using nanostructure assemblies, they are not ideal for the study of the enhancement of the power factor σS^2 in nanoscale conductors. In fact, beyond the intrinsic technical complexity they entail, they lack a fundamental possibility which is the focus on the next section: the possibility to control and tune the nanostructure electronic configuration by field effect.

4. FIELD-EFFECT CONTROL OF THE TE RESPONSE

4.1 The Importance of Field-Effect Control

Each of the parameters entering the TE response—i.e., the conductivities σ and κ, and the Seebeck coefficient S—is first of all strongly dependent on the density of free carriers n or, equivalently, on the band filling. Thus, the maximization of the figure of merit ZT requires first and foremost achieving an optimal doping (see Fig. 1). While the investigation of the TE response of an individual nanostructure poses various technical challenges, a unique advantage is offered by the possibility to tune them by field effect. Differently from bulk materials, nanostructures are in fact typically small on the scale of the screening length and thus their band filling can be *tuned* during the experiment by simple gating. This provides a crucial advantage that in part compensates the relatively poorer reproducibility of nanomaterials and basically allows "simulating" all the possible doping levels on the same nanostructure. Obviously, field-effect control is also crucial to control the effects of quantum confinement or to exactly tune the configuration of energy filters built into the device structure. In both these cases, which are particularly relevant in the context of the enhancement of the power factor σS^2, the correct device operation in fact typically requires a precise positioning of the chemical potential within the energy levels of the device. The importance of, and the possibilities offered by, field-effect control was only recently recognized and exploited in a set of experimental works on individual semiconductor single crystals with high aspect ratios, or NWs. This pioneering strategy was first demonstrated in lead chalcogenides-based materials: using a field-effect gated device, Liang and coworkers were able to tune the Seebeck coefficient of individual PbSe NWs by a factor three (Liang et al., 2009). Many further experiments have exploited the same technique on a variety of nanostructures (Bejenari et al., 2010; Brovman et al., 2016; Curtin et al., 2013; Moon et al., 2013; Neophytou and Kosina, 2014; Persson et al., 2009a,b; Roddaro et al., 2013; Saito et al., 2016; Tian et al., 2012; Wu et al., 2013; Zhou et al., 2011c).

4.2 Semiconductor Nanowire Synthesis

Semiconductor NWs are typically grown using epitaxial techniques such as the molecular or chemical beam epitaxy (MBE or CBE, respectively). During the growth, the NWs self-assemble starting from "seeds" of catalysts, typically constituted by Au nanoparticles, deposited on a suitable substrate (see Fig. 5). The control of the growth conditions and of the III–V precursors (Dubrovskii et al., 2016; Gomes et al., 2015) allows obtaining a variety of material and geometry combinations: for example, it is possible to obtain axial heterostructures (Ercolani et al., 2009; Pitanti et al., 2011; Romeo et al., 2012; Rossella et al., 2014; Zannier et al., 2016) or, taking advantage of the more efficient relaxation of the strain in NWs, radial heterostructures between materials with either small or even relatively large lattice mismatch (Jiang et al., 2007; Rocci et al., 2016; Roddaro et al., 2008; Xiang et al., 2006). This is illustrated in Fig. 6 for the relevant cases of an InAs/InP axial heterostructure NW and an InAs–GaSb radial heterostructure NW.

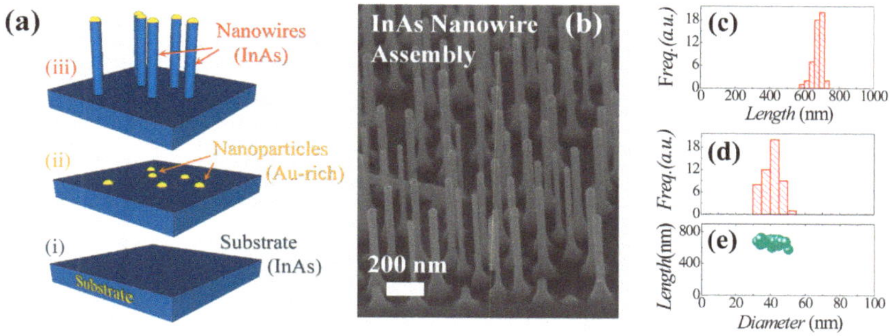

Fig. 5 (A) Sketch of the growth procedure: Au nanoparticles on InAs (111) B substrates are obtained from colloids or thin film dewetting and used as growth catalysts. The diameter of the NWs can be controlled either by changing the film thickness (panel (B), 20 min annealing at 510°C, overlays are Gaussian fits of the data) or by choosing colloidal nanoparticles of different diameters (panel (C), 20 min annealing at 465°C). *Adapted from Floris, F., Fornasari, L., Marini, A., Bellani, V., Banfi, F., Roddaro, S., Ercolani, D., Rocci, M., Beltram, F., Cecchini, M., Sorba, L., Rossella, F., 2017. Nanomaterials 7, 400.*

Furthermore, it is worth mentioning that the arrangement of the NWs on the growth substrate can be controlled by exploiting nanolithography techniques to define the pattern of metallic nanoparticles that catalyze the growth of the nanocrystals (Persson et al., 2009a,b; Vukajlovic-Plestina et al., 2017; Wu et al., 2002). Among the possible different materials that can be combined in NWs, III–V semiconductors allow to realize nanostructures that exhibit excellent electrical transport properties, offer the ability to (i) control thermal properties thanks to the appropriate choice of geometry, dimensions, and elements, and to (ii) exploit the field effect for the control of the band filling. For TE studies, the NWs are usually removed from the growth substrate and transferred onto a substrate (typically SiO_2/Sip++) suitable for the fabrication of nanodevices and compatible with the field-effect control. Standard e-beam lithography is used to manufacture the devices.

The ability to synthesize advanced heterostructures using NWs offers a number of advantages in the context of the fundamental investigation of the TE response of nanoscale systems. While the development of large-scale NW-based nanostructured materials is still a long-term objective today, heterostructure NWs offer the chance to address various fundamental issues in nanoscale TE. For instance, thermal conductivity suppression has been pursued not only using rough NWs but also axial heterostructures (Li et al., 2003a) or even homostructures (Caroff et al., 2009; Zhou et al., 2011a) that

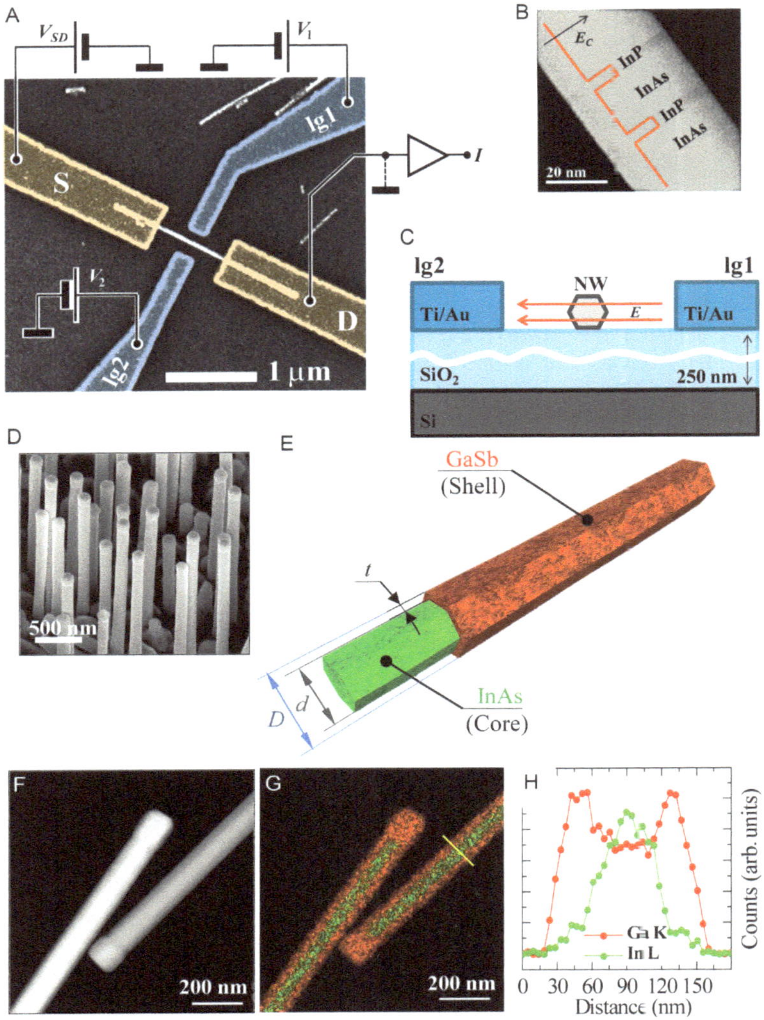

Fig. 6 InAs/InP axial heterostructure NWs: scanning (A) and transmission (B) electron picture of a device embedding axial barriers; (C) sketch of the device structure. InAs–GaSb core–shell heterostructure NWs: (D) scanning electron micrograph, (E) schematic of a single NW, (F) scanning TEM micrograph, and (G) EDX element map of two core–shell NWs transferred onto a standard TEM grid (Ga, *red-dotted area*; In, *green-dotted area*); (H) cross-section line profile (*yellow cut* in panel G): indium (Gallium) peak in correspondence of the NW core (shell) is observed. *Reproduced with permission from Rocci, M., Rossella, F., Gomes, U.P., Zannier, V., Rossi, F., Ercolani, D., Sorba, L., Beltram, F., Roddaro, S., 2016. Nano Lett. 16, 7950–7955; Roddaro, S., Pescaglini, A., Ercolani, D., Sorba, L., Beltram, F., 2011. Nano Lett. 11, 1695–1699. Copyright 2011 and 2016 American Chemical Society.*

are intended to suppress the diffusion of phonons. In addition, the use of axial nanostructure has been exploited to investigate how quantum confinement, interactions, and energy filtering can affect the TE response (Svensson et al., 2013). Radial heterostructures have been instead predicted to allow a modulation of the thermal conductivity by various authors (Bi et al., 2012; Lü, 2009; Yang et al., 2015). Interesting possibilities also involve the engineering of the interaction between radial thermocouples, which might be used to tune the TE response of a coaxial thermal pair (Abergel, 2015; Wu et al., 2014). Such core–shell nanostructures could be for instance developed starting from the broken-gap InAs/GaSb combination (Gluschke et al., 2015; Rocci et al., 2016), which has been recently investigated in different contexts.

4.3 Field-Effect Seebeck Platform

A recent example of field-effect Seebeck platform derived from the pioneering works by Liang et al. (2009) is shown in Fig. 7A (Roddaro et al., 2013, 2014). Devices are built starting from InAs NWs deposited over the SiO_2/Si n++ substrate by drop casting. Source (S) and drain (D) electrodes are obtained by aligned e-beam lithography, chemical passivation, and Ti/Au evaporation (Roddaro et al., 2013). The contacts operate both as electrical leads and resistive thermometers, thanks to their multicontact arrangement. This allows sampling the temperatures T_S and T_D at the two ends of the field-effect transistor and thus the thermal bias $\Delta T = T_D - T_S$. The four-wire thermometers are operated using AC current biases of few mA and a phase-locked measurement technique. The transport characteristics of the NW are simultaneously monitored by imposing a DC bias V_{DS} and measuring the resulting DC current I_{DS}. A thermal bias of few kelvin per micron can be induced by biasing a heater integrated in the doped Si substrate, as discussed further in the next paragraph. In the presence of a thermal bias the IV characteristics of the NW remain linear but are shifted according to

$$RI_{DS} = V_{DS} + V_{th},$$

where R is the NW resistance and $V_{th} = (S - S_{Au})\Delta T$ is the resulting Seebeck voltage, which depends both on the NW Seebeck coefficient and on one of the Au electrodes, which yields a negligible TE response with respect to the NW (Roddaro et al., 2013). The substrate can also be operated as a backgate and thus allows measuring the TE response as a function of the

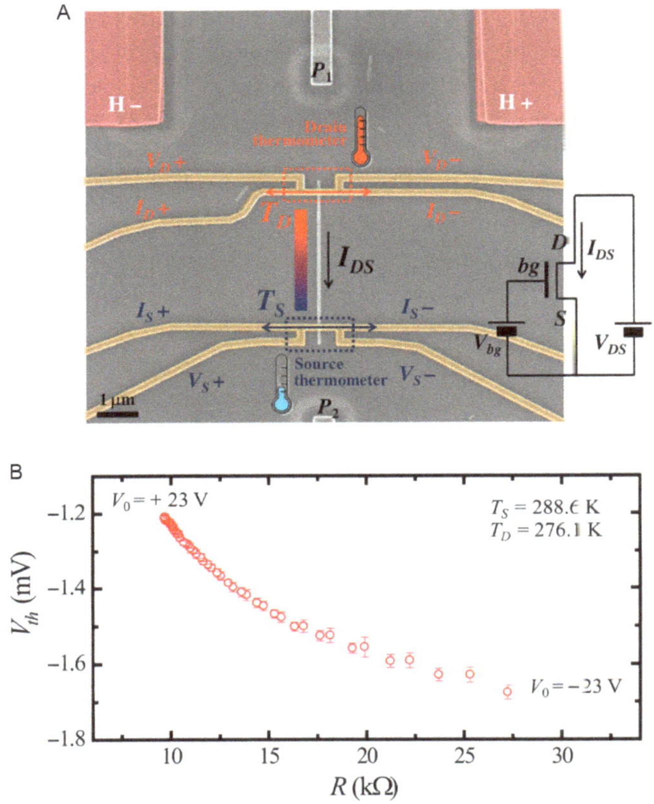

Fig. 7 (A) Scanning electron micrograph of a single NW field-effect transistor for the modulation of the thermoelectric parameters. The measurement is set up for electronic transport measurements visible in overlay. (B) A large thermovoltage can be obtained and controlled by the field effect, as a result of the change of resistance in the NW. *Reproduced with permission from Roddaro, S., Ercolani, D., Safeen, M.A., Suomalainen, S., Rossella, F., Giazotto, F., Sorba, L., Beltram, F., 2013. Nano Lett. 13, 3638–3642. Copyright 2013 American Chemical Society.*

NW carrier density. Both the thermovoltage and the NW resistance depend on the substrate bias V_0 and an example of the resulting $V_{th}(R)$ parametric curve is reported in Fig. 7B. The technique allows a complete characterization, as seen in the panels of Fig. 8. Thanks to the direct measurement of temperatures at source and drain electrodes and to the large ΔT attainable, a very precise mapping of the power factor σS^2 and of its factors as a function of the temperature (between 300 and 77 K) can be obtained as shown in Fig. 8. As discussed in Section 2, each of the parameters can be

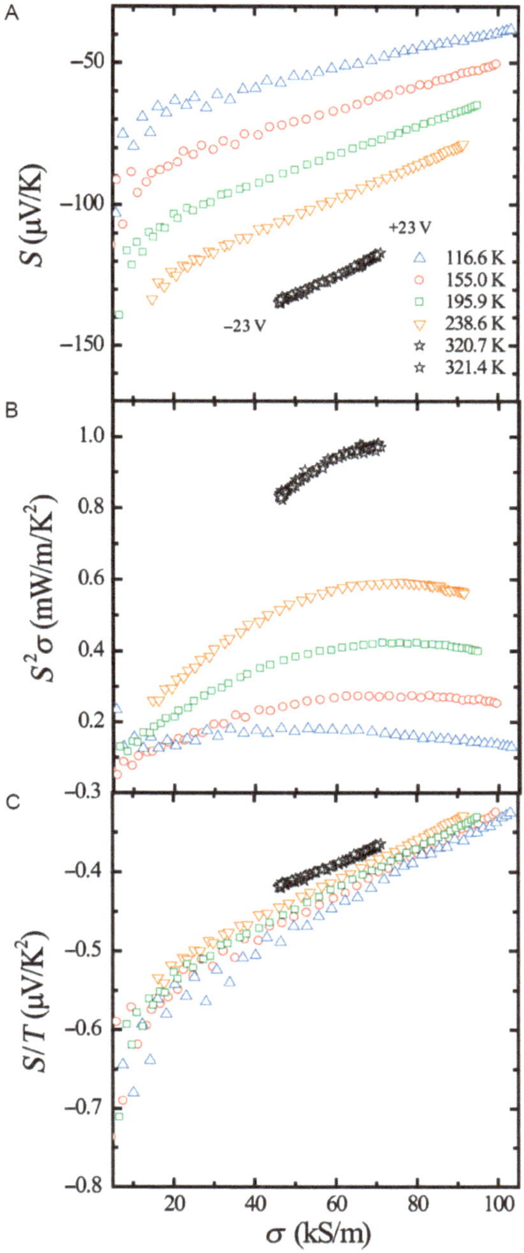

Fig. 8 Field-effect control can be used to strongly modulate both the Seebeck coefficient and the conductivity (see panel (A)). The corresponding power factor σS^2 and S/T are visible in panels (B) and (C). *Reproduced with permission from Roddaro, S., Ercolani, D., Safeen, M.A., Suomalainen, S., Rossella, F., Giazotto, F., Sorba, L., Beltram, F., 2013. Nano Lett. 13, 3638–3642. Copyright 2013 American Chemical Society.*

widely modulated by field effect and a maximal σS^2 is obtained for a specific optimal conductivity.

Since the measured nanostructures are typically only few (or even one) micron long, it is crucial to be able to induce a sufficiently large thermal gradient in order to obtain a sizeable thermovoltage. A standard heating architecture consists in realizing a metallic surface heater on top of the SiO_2/Si n++ substrate, in close proximity to the nanostructure (Liang et al., 2009). In this case the metallic strip is heated by Joule effect and the formation of a thermal gradient relies on heat diffusion through the oxide and into the substrate below the nanostructure (see top part of Fig. 9). This scheme allows the control of the nanostructure behavior by field effect using the Si n++ substrate, which is electrically decoupled from the NW and the heater. This approach however typically fails to establish temperature

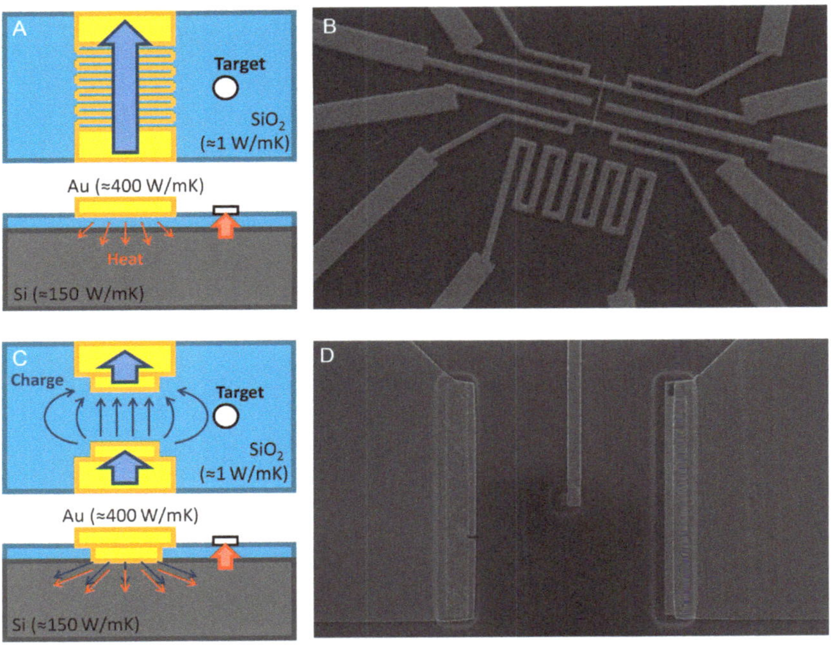

Fig. 9 Comparison between a surface and buried heater architecture used to induce a thermal bias on an individual nanostructure. In a surface heater (sketch in panel (A) and electron microscope picture of typical device in panel (B)) the thermal gradient is established due to heat diffusion across the oxide and into the Si substrate. Differently, the buried heater (sketch in panel (C) and electron microscope picture of typical device in panel (D)) relies on the graded Joule heating derived from the current diffusion in the substrate.

gradients larger than few tenths of kelvin per micron (Liang et al., 2009; Small, 2004; Small et al., 2003; Tian et al., 2012). The alternative approach used in the experiment of Fig. 7 consists of a buried heater scheme that exploits the graded Joule heating derived from the diffusion in the conductive bulk of a SiO_2/Si n++ of the current injected from the H_+ and H_- contacts (see bottom part of Fig. 9). The architecture is compatible with a field-effect control of the nanostructure, which can be induced by the average voltage $V_0 = (V_{H+} + V_{H-})/2$, while the heater current is controlled by the bias $V_{H+} - V_{H-}$. The different working principle of the heater yields uniform thermal gradients up to about 10 K/μm, i.e., one order of magnitude beyond values typically reported in the literature for top heaters. Beside outperforming top heaters reported so far in the literature, it opens new possibilities for the investigation of TE effects on individual nanoscale active elements, still retaining the possibility to be fabricated in any position and orientation on a Si/SiO_2 substrate, thus being applicable to the investigation of the effect of a thermal bias on nanostructures transferred by drop casting or similar methods, i.e., typically with a random position and orientation. Using this device architecture, record-high thermovoltage values in excess of 1 mV were demonstrated in devices comprising a single NW as active element, as well as the modulation of these values by a field effect (Fig. 14B).

Noticeably, the InAs NW represents a model system of particular relevance for the investigation of strongly confined low-dimensional systems (Fuhrer et al., 2007; Rossella et al., 2014; Thelander et al., 2003), which in turn may provide intriguing heat transport and TE properties (Hoffmann et al., 2009; Zhang et al., 2011). It is worth to notice that the scheme is of general interest for the investigation of a large class of nanomaterials with potential TE applications. Exploiting this approach, one can precisely map the evolution of the Seebeck coefficient S in individual NWs as a function of temperature (from room temperature down to low temperature) and—thanks to the gate tunability—as a function of σ modulated by the field effect.

The investigation of the gate-dependent TE response and the comparison with TE models can be useful to extract useful information about the transport parameters of the NW, which are not directly accessible in standard charge transport experiments. For example, as seen in Fig. 10, the field-effect evolution of both V_{th} and R typically displays a strong hysteresis as a consequence of the slow traps charging dynamics of localized states in the oxide and on the surface of the NW, which screen the action of the gate. Differently from conventional charge transport, comparison with TE measurements allows to isolate the variations due to the mobile carriers in the NW.

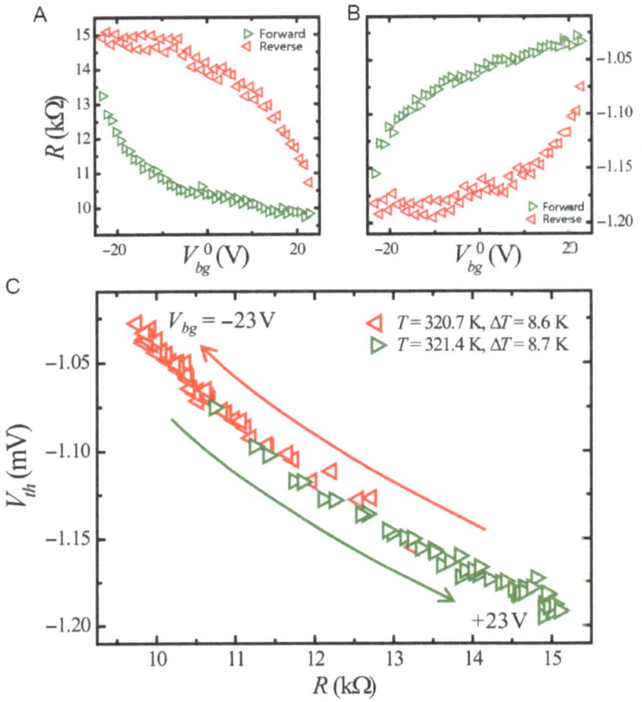

Fig. 10 Room temperature evolution of the resistance (A) and thermovoltage (B) as a function of the gate voltage. A very large hysteresis between the forward (*green*) and backward (*red*) sweeps is observed as a consequence of surface/oxide trap charging dynamics. The parametric evolution in panel (C) is largely independent from the sweep direction since it is only caused by mobile carriers and is almost insensitive to surface charge. *Reproduced with permission from Roddaro, S., Ercolani, D., Safeen, M.A., Suomalainen, S., Rossella, F., Giazotto, F., Sorba, L., Beltram, F., 2013. Nano Lett. 13, 3638–3642. Copyright 2013 American Chemical Society.*

In fact, the parametric evolution $V_{th}(R)$ displays *no* hysteresis since both the thermovoltage and the resistance of the NW only depend on the amount of mobile charges. The analysis of this kind of data can thus be used to extract an alternative and more intrinsic estimate of transport parameters of free carriers in the NW. The most important one entering in the S vs σ relation is the electron mobility μ. The obtained values are typically significantly larger with respect to ones resulting by standard field effect-based mobility estimates, in the same nanostructure (Roddaro et al., 2013). The discrepancy is consistent with the action of surface states' screening, which typically leads to an overestimate of the carrier density and to an underestimate of mobility (Dayeh et al., 2007; Halpern et al., 2012; Storm et al., 2012).

5. FIELD-EFFECT AND COMPLETE TE CHARACTERIZATION

5.1 Thermal Transport Using Micro-Raman

While semiconductor NWs represent ideal nanostructures for investigating the effects of low-dimensionality and thermal conductivity suppression on the TE response, it is very challenging to accurately and simultaneously measure the temperature gradients and the heat flow in the same nanostructure. Among the different strategies proposed to face this issue, in addition to the one using suspended microthermometer devices (Shi et al., 2003), a very powerful one consists in the combined use of spatially resolved optical spectroscopies and transport measurements within the framework of *suspended* NW device architectures that can allow to determine all the TE parameters in single NWs and to quantify the *ZT* figure of merit of the devices. Micro-Raman spectroscopy—traditionally applied to address thermal properties in different nanomaterials (Faugeras et al., 2010; Rossella et al., 2012)—was proved to represent a very powerful tool in this context and, most importantly, it is compatible with standard lithography and is thus more reliable for what concerns electrical/thermal contacts (Doerk et al., 2010; Soini et al., 2010; Yazji et al., 2016). This technique was recently used by Yazji and coworkers to laser-heat single NWs, while simultaneously measuring the associated electrical response (Yazji et al., 2015), see a sketch of the method in Fig. 11.

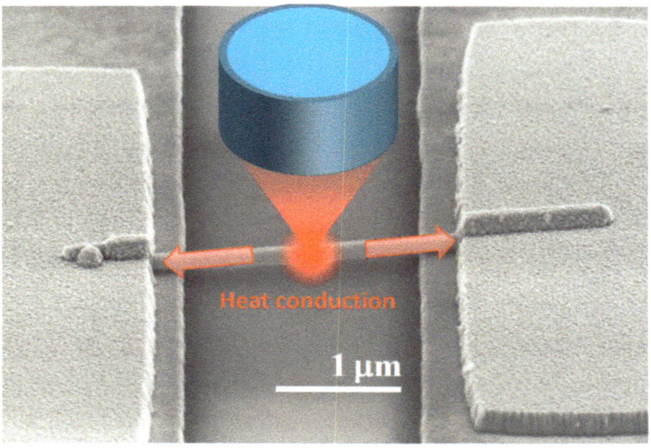

Fig. 11 Scanning electron micrograph of the typical sample configuration including an individual NW suspended between two gold pads and a schematic of the micro-Raman heating approach.

Temperature and electrical data were rationalized within a one-dimensional heat equation model including Gaussian heat source and provided for the first time reliable values of κ in nanoscale InSb, that were discussed in comparison to their bulk counterpart (Bowers et al., 1959; Nakwaski, 1988; Yamaguchi et al., 2005). In addition, the thermal contact resistances at the source and drain electrodes of the NW were shown to play a crucial role, so they cannot be neglected in such mesoscopic devices. For the case of InSb NWs, the obtained TE parameters were compared with those obtained by field-effect transistor Seebeck measurements, reporting a very good agreement between the two independently measured set of parameters. Noticeably, the combination of laser heating, Raman spectroscopy, and simple electrical measurements also allow the estimate of the thermal contact resistances at the nanostructure-source/drain contacts, crucial parameters that can drastically affect the operation of real TE devices at the nanoscale.

5.2 Perspectives for Suspended Nanowire Devices

The experimental scheme proposed by Yazji and coworkers has not implemented so far gate tunability and thus does not exploit the possibility to explore the dependence of the TE parameters upon changing the carrier density. In absence of field effect, n and thus σ are fixed to the pristine value of the as-grown nanostructure under test. In principle, however the architecture is compatible with gate control and these optical and electrical techniques can be combined to extract the values of S and κ as a function of n. To implement this kind of advanced functionalities and exploit the field-effect conventional substrate gating or even suspended gates can be coupled to suspended NW architectures. This scenario is illustrated in Fig. 12, where the scanning electron micrographs of multigate and multicontact suspended NW devices are reported. These represent good candidates for a new generation of single nanostructure-based devices allowing the simultaneous measurement of each TE parameter in combination with the gate control of the electrical conductivity. In the specific case depicted in figure, the measurement of κ can be achieved exploiting the 3ω method, which will be discussed in the next section.

Noticeably, suspended NW device architectures can allow the investigation of new thermal–electrical transport regimes and to study the feasibility of exotic phenomena such as thermal rectification, i.e., the thermal resistance is a function of the sign of the applied thermal gradient. Actually, since

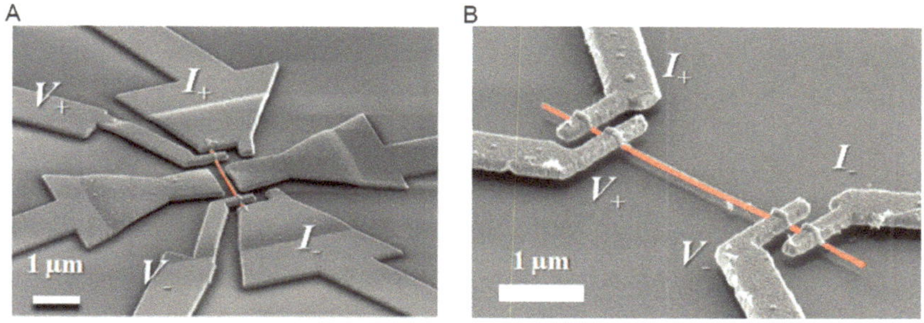

Fig. 12 (A) Scanning electron microscopy micrograph of a suspended NW device. Electrode pairs for current injection and voltage measurement are indicated, including wide source and drain current electrodes for an improved thermal anchoring. The device is also equipped with side gate electrodes for the modulation of the carrier density in the NW. (B) Detail of a suspended NW device not equipped with side gates.

the seminal experimental paper by Chang et al. demonstrating nonnegligible thermal rectification in asymmetrically mass-loaded carbon nanotubes (Casati, 2007; Chang et al., 2006), directed thermal transport in nanostructured materials has been attracting significant attention (Roberts and Walker, 2011). A solid-state thermal rectifier at the micro/nanoscale represents in fact an ideal component for thermal logic circuits, a potential paradigm shift for thermal management and energy applications. A similar approach was proposed for different types of high aspect ratio nanostructures including graded III–V semiconductor core–shell NWs (Liu et al., 2014), saw tooth-faceted silicon NWs (Ross et al., 2005), and telescopic Si NWs (Cartoixà et al., 2015).

5.3 Measuring Individual Nanostructures Using the 3ω Technique

As argued in the previous sections, a key ingredient for the development of innovative nanostructured materials for TE applications is the improvement of techniques and procedures which allow the reliable measurement of the thermal conductivity κ on a nanometric scale. Heaters and temperature sensors—required for the measurement of κ—cannot be very small with respect to the characteristic dimensions of the nanostructure to be characterized. The assumption that they are negligible—made by conventional data analysis—can lead to misleading results. Hence, both advanced measurement techniques and accurate modeling for data reduction represent crucial requirements. In this section, we present the use of the 3ω technique (Cahill, 1990) for the measurement of the thermal conductivity in NWs and nanostructures. One of the advantages of this method is that, upon certain conditions, the same structure under test—e.g., a NW—can be used both for heating and for measuring the temperature.

The 3ω technique was originally developed for the measurement of the thermal conductivity in thin films deposited on a substrate. A metal strip (resistor) is fabricated on top of the film (see Fig. 11A) and is used both for the generation of heat (via Joule effect) and for the measurement of the temperature (Cahill, 1990; Wang and Sen, 2009). The latter depends on the capability of the film to dissipate the generated heat; hence, it depends on the thermal characteristics of the film. In the configuration shown in Fig. 13A, the thermal transport (red arrows) is dominant in the direction perpendicular to the film plane. Fig. 13B shows the setup for the measurement of the κ in the film plane. As the film is deposited on a suspended membrane (Jain and Goodson, 2008), the Joule heating generated by the metal track is dissipated along a direction laying on the film itself. This concept (see Fig. 13C) can also be applied to the thermal characterization of suspended nanostructures, as silicon nanomembranes (Pennelli and Macucci, 2016; Pennelli et al., 2017a,b).

In the mentioned cases, the resistor is biased by a sinusoidal current

$$I(t) = I_P \sin(\omega t)$$

while the heat P generated by the metal track via Joule effect can be expressed as $P(t) = R(T)I^2(T)$, where R is the electrical resistance. The heat being dissipated, the temperature $T_R(t)$—driven by the instantaneous power $P(t)$—depends on the geometry of the resistor and on the geometry, thermal conductivity, and thermal capacity C_V of the nanostructure. Thus, the measurement of the temperature of the resistor $T_R(t)$ allows to determine the thermal parameters κ and C_V. The same resistor, used as heater, also provides an indirect measurement of the temperature through its resistance value R. If the temperature T_R has a reasonably small range of variation, one can assume a linear dependence of R upon T: $R(T) = R_0(1 + \alpha(T - T_0))$, where T_0 is a

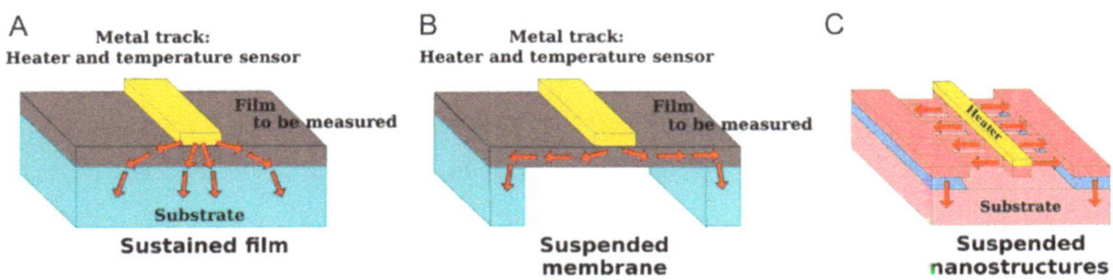

Fig. 13 Devices based on a metal resistor for the 3ω measurement.

reference temperature, R_0 is the resistance at T_0, and α is the temperature coefficient of the resistance ($\alpha = R_0^{-1} \cdot \partial R/\partial T$) which depends on the material used for the fabrication of the resistor/heater. The instantaneous power

$$P(t) = R(t) I_p^2 \cdot (1 + \cos 2\omega t)/2$$

oscillates at frequency 2ω, and the resistor temperature $T_R(t)$ and the resistance value $R(t) = R(T_R(t))$ oscillate at the same frequency. Since the current varies with a frequency ω, then the voltage drop across the heater (typically measured using a four-probe configuration) has a component at frequency 3ω, given by

$$V(t) = V_0 + V_{1\omega} \sin(\omega t + \theta_{1\omega}) + V_{3\omega} \sin(3\omega t + \theta_{3\omega}).$$

The amplitude $V_{3\omega}$ and phase $\theta_{3\omega}$ of the third harmonic allow to determine κ and C_V, if the geometrical factors are known, and generally both $V_{3\omega}$ and $\theta_{3\omega}$ depend on ω. However, for $\omega < \omega_t$ (cut-off frequency), $V_{3\omega}$ is frequency independent and its phase is π, i.e., opposite in sign with respect to the first harmonic. The value of ω_t is proportional to $1/\tau_t$, where $\tau_t = L^2/D_t$ is a time constant which depends on the thermal diffusivity coefficient $D_t = \kappa/C_V$ and on the characteristic length L of the nanostructure (e.g., the film thickness or the length of the suspended nanostructure).

Here, L represents the penetration depth of a heat wave with frequency ω_t. For frequencies $\omega \ll 2\pi/\tau_t$, we can assume that the heat wave is in phase with the local temperature; hence, the third harmonic is in phase (opposite in sign) with the biasing current. In this case, the local temperature—and hence the amplitude of the third harmonic—depends only on the biasing current and on the thermal conductivity. In this frame, a key point is the development of suitable models to relate the amplitude of the third harmonic with the thermal conductivity.

Fig. 14 shows a different approach for 3ω measurement. The structure to be characterized is used at the same time for the generation of heat and for the measurement of the temperature via the output voltage (self-heating self-measuring approach). The structure must be suspended so that the heat which is locally generated by Joule effect, due to the biasing current $I(t)$, flows through the structure and is dissipated at both ends (Fig. 14, right panel). Hence, the temperature depends on the thermal characteristics of the structure, and in particular on its capability of dissipating. From the fabrication point of view, this approach is easier because it does not require any additional metal resistor. However, the electrical properties of

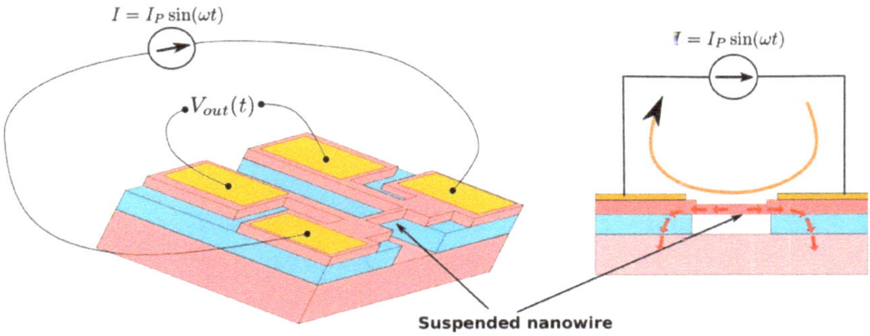

Fig. 14 A NW is suspended between contacts. This device is suitable for 3ω measurements with the self-heating self-measuring approach.

the material the nanostructure is made of need to be precisely known, in particular the dependence of the electrical resistivity ρ on temperature. Hence, the full electrical characterization as a function of temperature must be performed before using the 3ω method. Conversely, the resistor-based approach allows to use well-known materials such as metals for the fabrication of the resistor.

The self-heating self-measuring approach is very suitable for nanostructures such as carbon nanotubes (Choi et al., 2006; Yi et al., 1999) and semiconducting NWs (Bosseboeuf et al., 2015; Pennelli et al., 2014). The SEM micrograph reported in Fig. 15 (Pennelli, 2014) shows a suspended silicon NW, with four electrodes for the electrical and thermal characterization with the 3ω method. Fig. 16 reports schematics of different measurement setups. The sinusoidal current signal, fed between the two current probes, is obtained from a sinusoidal voltage source combined with a voltage to current converter (transconductance amplifier). The output voltage is collected through the other two contacts and measured by means of a lock-in amplifier, or by a spectrum analyzer, which allows to determine the amplitude $V_{3\omega}$ and phase $\theta_{3\omega}$ of the third harmonic. Other experimental setups for 3ω measurements make use of a Wheatstone bridge (Fig. 14, left panel), yielding to a reduction of the first harmonic component of the output signal (Chirtoc and Henry, 2008). However this configuration, although relevant for the measurement of the temperature of the tip in thermal scanning microscopy applications (Chirtoc and Henry, 2008; Puyoo et al., 2011), cannot be used with four-contact devices.

A key point of the 3ω technique is the development of suitable models to relate the amplitude of the third harmonic with the thermal conductivity of the nanostructure under test. Robust analytical models were developed for

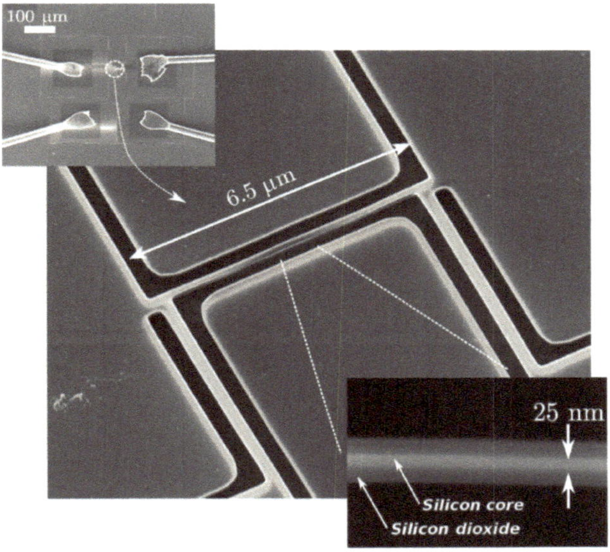

Fig. 15 A silicon NW suspended between leads for four-contact characterization. *From Pennelli, G., 2014. Beilstein J. Nanotechnol. 5, 1268–1284.*

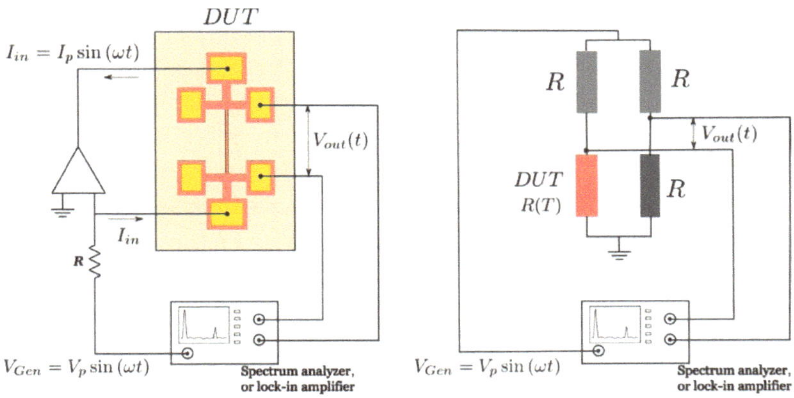

Fig. 16 Two different experimental setups for the 3ω measurement.

the case of supported thin films or suspended membranes (Fig. 13, sketches in panel A and B) exploiting the resistor approach. These models are based on the analytical solution of the heat transport equation, which can only be found at the price of three main approximations. First, the electrical power to be dissipated is evaluated assuming constant resistance value, $P(t) = R(t)$ $I^2(t) \simeq R_0 I^2(t)$. Second, the heater (resistor), which sets the boundary conditions for the solution of the heat transport equation, is treated as a one-dimensional heat source. Third, the electrical conductivity of the material to be measured is neglected, limiting the model applicability to insulating or semi-insulating materials. For the particular case of nanoscale structures (Fig. 13, sketch C), the width of the resistor cannot be much smaller with respect to the nanostructure itself. To overcome all these limitations, a numerical solution of the thermal and electrical transport equations (Pennelli and Macucci, 2013; Pennelli et al., 2017a,b) needs to be implemented for the correct analysis of the experimental data.

Analytical models for the self-heating self-measuring approach were developed (Lu et al., 2001) for suspended structures where the transverse dimension is much smaller with respect to the length L (high aspect ratio). For this case, in the low-frequency regime $\omega \ll 2\pi/\tau_t = 2\pi D_t/L^2$ and assuming a linear dependence of the resistance on temperature, $V_{3\omega}$ is proportional to the third power of the current peak amplitude and to the reciprocal of the thermal conductivity, i.e.,

$$V_{3\omega} \simeq \frac{I_p^3}{\kappa} \alpha \frac{4R_0^2 L}{\pi^4 S},$$

where S is the cross-sectional area. In the case of semiconductor NWs, the linear dependence of the resistance with temperature, taken into account with the parameter α, can be assumed only for high doping values and for low variation of the temperature of the NW. In general, the thermal and electrical transport equations need to be solved with numerical techniques, assuming as boundary conditions a fixed temperature at the ends of the NW and a given current density in the NW. Fig. 17 (Pennelli et al., 2014)

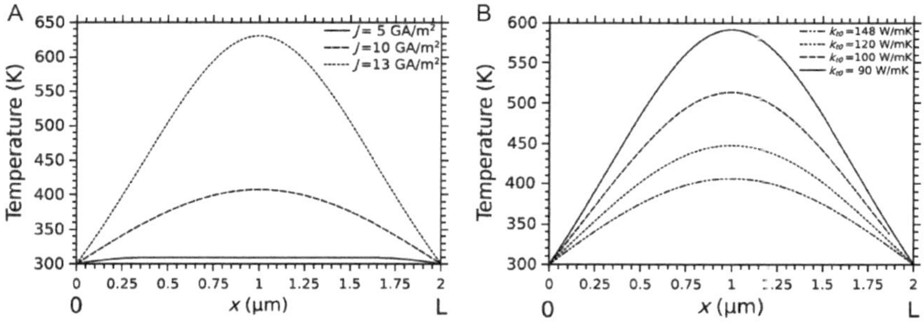

Fig. 17 Temperature distribution along a NW with a length $L = 2\,\mu m$. *On the left*: temperature for different values of the current density. *On the right*: temperature for different values of the thermal conductivity. *From Pennelli, G., 2014. Beilstein J. Nanotechnol. 5, 1268–1284 with permission of AIP.*

shows the temperature distribution in a silicon NW for different values of current density (see the left panel) and for different values of thermal conductivity. Silicon offers the advantage that its electrical characteristics, also as a function of temperature, are very well known for its prevalence in the fabrication of integrated circuits.

6. PERSPECTIVES AND LOW-TEMPERATURE TEs

A number of fundamental issues in TE physics are more easily studied at low temperatures. While these kinds of experiments have less direct connections with practical applications of TE effects to energy conversion, they can provide a very valuable insight on the impact of quantum and interaction effects, which become more easily observable at temperatures of the order of 1 K or less. Experimental platforms discussed so far typically exploited resistive thermometry and resistive heating. Both of these become ineffective at low temperatures because of the saturation of thermoresistances and because of the decoupling between the electrons and phonons at low temperatures. Techniques that are able to directly control the electronic distribution and measure the electronic temperatures are necessary in this case. In this section, we review recent efforts toward the development of novel techniques for the control and detection of the electron distribution in NWs. A very popular method is based on the use of quantum dots, which can act as selective energy filters (Nakpathomkun et al., 2010). The two further strategies here quickly discussed exploit hybrid superconductor-NW devices and noise spectroscopy.

6.1 Hybrid Super-Nanowire Devices

Hybrid architectures combining superconductive elements with normal metals represent a refined technology for the local measurement and manipulation of heat at low temperatures and have been the subject of extensive research efforts (Giazotto et al., 2006; Muhonen et al., 2012). In particular, devices integrating a set of normal-insulator-superconductor (NIS) tunnel junctions between a normal metal (N) and Al (Pekola et al., 2004) or other superconductors (S) (Nevala et al., 2012; Quaranta et al., 2011) have exhibited significant nanorefrigeration in the millikelvin regime, much larger than the one achievable using alternative approaches such as those based on quantum dots. The transfer of hybrid superconductor–metal technology to semiconductor NWs has progressed significantly in the past decade with the demonstration of NW-based Josephson junctions using a variety of superconductors (Doh et al., 2005; Frielinghaus et al., 2010; Paajaste et al., 2015; Spathis et al., 2011; Tiira et al., 2017) and further exotic devices (Giazotto et al., 2011; Mourik et al., 2012; Roddaro et al., 2011).

A first demonstration of hybrid NIS junctions integrated on an InAs NW has been realized only in very recent times and is visible in Fig. 18 (Mastomäki et al., 2017). The device architecture shown in panel (A) exploits a conventional SiO_2/Si n++ substrate where contacts are fabricated using a metallic multilayer. The NW is first contacted using a Ti/AlMn layer (5/50 nm), where the AlMn alloy contains sufficient Mn impurities (nominally, 2%) to quench superconductivity. The alloy is then controllably oxidized in situ to obtain a well-controlled tunnel barrier with a thickness of 1–2 nm. The contact is then completed by the evaporation of a superconductive Al layer (50 nm). An alternative approach could entail the use of epitaxial core–shell barriers where tunnel barrier transparency—which is crucial for the correct operation of the NIS junctions (Giazotto et al., 2006)—can in principle be finely controlled during the growth. The same contact technology can be used to implant either a tunnel thermometer or a heating/refrigeration junction. In the experiment the central junctions were operated as thermometry, while the outer ones provided the

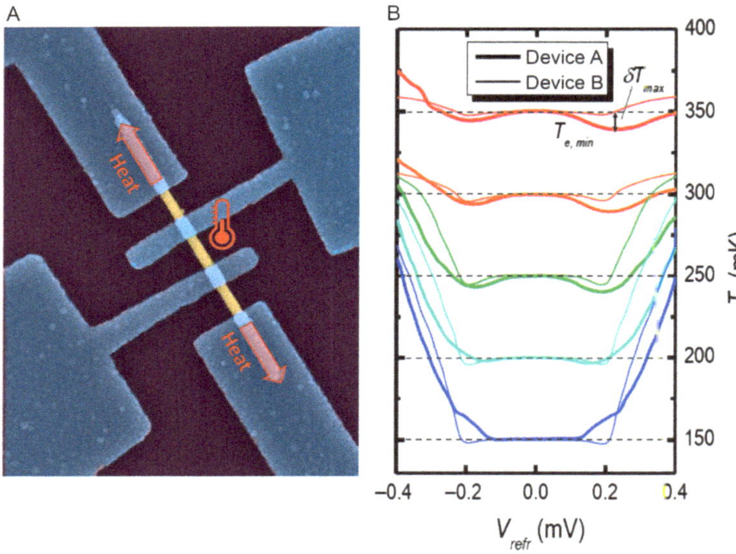

Fig. 18 (A) Scanning electron picture of the hybrid superconductor tunnel device. (B) Refrigeration effect as a function of the bath temperature and of the refrigeration bias.

cooling. The electronic temperature T_e as a function of the bath temperature T_{bath} and of the refrigerator bias V_{refr} is visible in Fig. 18B and a top cooling of about $\delta T_{\text{max}} \approx 10$ mK could be obtained in this proof-of-principle device starting from a bath temperature of 250 mK.

6.2 Shot Noise Thermometry

An interesting alternative approach to directly address the temperature and distribution of the electron system of a NW consists in exploiting the thermal, or Johnson-Nyquist, and shot noise which derived from the statistical fluctuations of the transport signal as a consequence of the finite temperature and of the quantization of the charge. The power spectral density of thermal noise is equal to

$$S_I(\omega = 0) = 4Gk_BT$$

where G is the conductance and T is the temperature, and thus can be used to implement a primary electron thermometer. At a large bias current I shot noise is expected, with a power spectral density $S_I = 2FqI$ where q is the carriers' charge and F the Fano factor, as visible in Fig. 19 for an InAs NW (Tikhonov et al., 2016a). Noise thermometry was exploited to perform

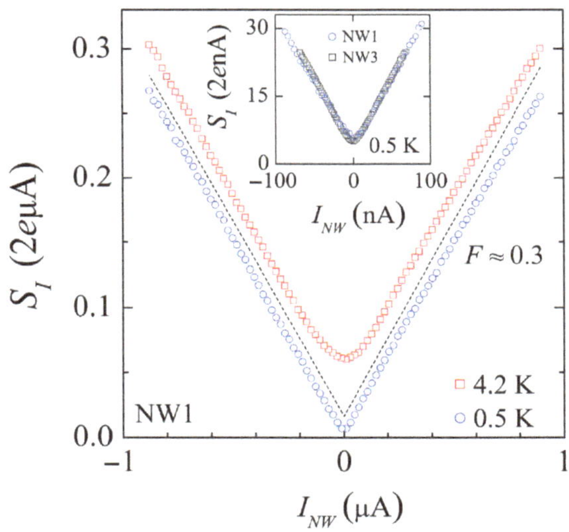

Fig. 19 Current noise spectral density as function of the bias current measured on a NW at a bath temperature of 0.5 and 4.2 K. The Fano factor $F = 0.3$ is compatible with a diffusive elastic transport regime. *From Tikhonov, E.S., Shovkun, D. V., Ercolani, D., Rossella, F., Rocci, M., Sorba, L., Roddaro, S., Khrapai, V.S., 2016. Sci. Rep. 6, 30621.*

TE measurements (Tikhonov et al., 2016b) and recent results have shown that it can be also used to obtain a *local* measurement of the electronic properties (Tikhonov et al., 2016a).

REFERENCES

Abergel, D.S.L., 2015. Appl. Phys. Lett. 106, 213103.
Ali, A., Chen, Y., Vasiraju, V., Vaddiraju, S., 2017. Nanotechnology 28, 282001.
Baxter, J., Bian, Z., Chen, G., Danielson, D., Dresselhaus, M.S., Fedorov, A.G., Fisher, T.S., Jones, C.W., Maginn, E., Kortshagen, U., Manthiram, A., Nozik, A., Rolison, D.R., Sands, T., Shi, L., Sholl, D., Wu, Y., 2009. Energy Environ. Sci. 2, 559.
Bejenari, I., Kantser, V., Balandin, A.A., 2010. Phys. Rev. B Condens. Matter Mater. Phys. 81, 1–14.
Bell, L.E., 2008. Science 321, 1457–1461.
Bi, K., Wang, J., Wang, Y., Sha, J., Wang, Z., Chen, M., Chen, Y., 2012. Phys. Lett. A Gen. At. Solid State Phys. 376, 2668–2671.
Bosseboeuf, A., Allain, P.E., Parrain, F., Le Roux, X., Isac, N., Jacob, S., Poizat, A., Coste, P., Maaroufi, S., Walther, A., 2015. Adv. Nat. Sci. Nanosci. Nanotechnol. 6, 25001.
Boukai, A.I., Bunimovich, Y., Tahir-Kheli, J., Yu, J.-K., Goddard III, W.A., Heath, J.R., 2008. Nature 451, 168–171.
Bowers, R., Ure, R.W., Bauerle, J.E., Cornish, A.J., 1959. J. Appl. Phys. 30, 930–934.
Brovman, Y.M., Small, J.P., Hu, Y., Fang, Y., Lieber, C.M., Kim, P., 2016. J. Appl. Phys. 119, 234304.
Caballero-Calero, O., Martín-González, M., 2016. Scr. Mater. 111, 54–57.
Cahill, D.G., 1990. Rev. Sci. Instrum. 61, 802–808.
Caroff, P., Dick, K.A., Johansson, J., Messing, M.E., Deppert, K., Samuelson, L., 2009. Nat. Nanotechnol. 4, 50–55.
Cartoixà, X., Colombo, L., Rurali, R., 2015. Nano Lett. 15, 8255–8259.
Casati, G., 2007. Nat. Nanotechnol. 2, 23.
Chang, C.W., Okawa, D., Majumdar, A., Zettl, A., 2006. Science 314, 1211.
Chen, G., 2005. Nanoscale Energy Transport and Conversion. Oxford University Press.
Chen, R., Hochbaum, A.I., Murphy, P., Moore, J., Yang, P., Majumdar, A., 2008. Phys. Rev. Lett. 101, 1–4.
Chen, Z.G., Hana, G., Yanga, L., Cheng, L., Zou, J., 2012. Prog. Nat. Sci. Mater. Int. 22, 535–549.
Chirtoc, M., Henry, J.F., 2008. Eur. Phys. J. Spec. Top. 153, 343–348.
Choi, T.-Y., Poulikakos, D., Tharian, J., Sennhauser, U., 2006. Nano Lett. 6, 1589–1593.
Curtin, B.M., Codecido, E.A., Kra, S., Bowers, J.E., 2013. Nano Lett. 13, 5502–5508.
Dayeh, S.A., Soci, C., Yu, P.K.L., Yu, E.T., Wang, D., 2007. J. Vac. Sci. Technol. B Microelectron. Nanometer Struct. Process. Meas. Phenom. 25, 1432.
Doerk, G.S., Carraro, C., Maboudian, R., 2010. ACS Nano 4, 4908–4914.
Doh, Y.J., Van Dam, J.A., Roest, A.L., Bakkers, E.P.A.M., Kouwenhoven, L.P., De Franceschi, S., 2005. Science 309, 272–275.
Dresselhaus, M.S., Chen, G., Tang, M.Y., Yang, R., Lee, H., Wang, D., Ren, Z., Fleurial, J.P., Gogna, P., 2007. Adv. Mater. 19, 1043–1053.
Dubrovskii, V.G., Sibirev, N.V., Berdnikov, Y., Gomes, U.P., Ercolani, D., Zannier, V., Sorba, L., 2016. Nanotechnology 27, 375602.
Dudzinski, L., 2013. Radioisotope Power Systems: Radioisotope Thermoelectric Generator (RTG). NASA.

Ercolani, D., Rossi, F., Li, A., Roddaro, S., Grillo, V., Salviati, G., Beltram, F., Sorba, L., 2009. Nanotechnology 20, 505605.
Erdman, P.A., Mazza, F., Bosisio, R., Benenti, G., Fazio, R., Taddei, F., 2017. Phys. Rev. B 95, 1–26.
Faugeras, C., Faugeras, B., Orlita, M., Potemski, M., Nair, R.R., Geim, A.K., 2010. ACS Nano 4, 1889–1892.
Frielinghaus, R., Batov, I.E., Weides, M., Kohlstedt, H., Calarco, R., Schäpers, T., 2010. Appl. Phys. Lett. 96, 1–4.
Fuhrer, A., Fröberg, L.E., Pedersen, J.N., Larsson, M.W., Wacker, A., Pistol, M.-E., Samuelson, L., 2007. Nano Lett. 7, 243–246.
Giazotto, F., Heikkilä, T.T., Luukanen, A., Savin, A.M., Pekola, J.P., 2006. Rev. Mod. Phys. 78, 217–274.
Giazotto, F., Spathis, P., Roddaro, S., Biswas, S., Taddei, F., Governale, M., Sorba, L., 2011. Nat. Phys. 7, 857–861.
Gluschke, J.G., Leijnse, M., Ganjipour, B., Dick, K.A., Linke, H., Thelander, C., 2015. ACS Nano 9, 7033–7040.
Gomes, U.P., Ercolani, D., Zannier, V., Beltram, F., Sorba, L., 2015. Semicond. Sci. Technol. 30, 115012.
Halpern, E., Elias, G., Kretinin, A.V., Shtrikman, H., Rosenwaks, Y., Halpern, E., Elias, G., Kretinin, A.V., Shtrikman, H., Rosenwaks, Y., 2012. Appl. Phys. Lett. 262105, 1–5.
Heremans, J.P., Dresselhaus, M.S., Bell, L.E., Morelli, D.T., 2013. Nat. Nanotechnol. 8, 471–473.
Herring, C., 1954. Phys. Rev. 96, 1163.
Hicks, L.D., Dresselhaus, M.S., 1993a. Phys. Rev. B 47, 16631–16634.
Hicks, L.D., Dresselhaus, M.S., 1993b. Phys. Rev. B 47, 12727–12731.
Hochbaum, A.I., Chen, R., Delgado, R.D., Liang, W., Garnett, E.C., Najarian, M., Majumdar, A., Yang, P., 2008. Nature 451, 163–167.
Hoffmann, E.A., Nilsson, H.A., Matthews, J.E., Nakpathomkun, N., Persson, A.I., Samuelson, L., Linke, H., 2009. Nano Lett. 9, 779–783.
Humphrey, T.E., Linke, H., 2005. Phys. Rev. Lett. 94, 3–6.
Jain, A., Goodson, K.E., 2008. J. Heat Transfer 130, 102402.
Jiang, X., Xiong, Q., Nam, S., Qian, F., Li, Y., Lieber, C.M., 2007. Nano Lett. 7, 3214–3218.
Khim, H., López, R., Lim, J.S., Lee, M., 2015. Eur. Phys. J. B 88, 151.
Khitun, A., Balandin, A., Wang, K.L., 1999. Superlattices Microstruct. 26, 181.
Kim, P., Shi, L., Majumdar, A., McEuen, P.L., 2001. Phys. Rev. Lett. 87, 215502.
Ko, T.Y., Shellaiah, M., Sun, K.W., 2016. Sci. Rep. 6, 1–9.
Kwon, S., Wingert, M.C., Zheng, J., Xiang, J., Chen, R., 2016. Nanoscale 8, 13155–13167.
Lee, E.K., Yin, L., Lee, Y., Lee, J.W., Lee, S.J., Lee, J., Cha, S.N., Whang, D., Hwang, G.S., Hippalgaonkar, K., Majumdar, A., Yu, C., Choi, B.L., Kim, J.M., Kim, K., 2012. Nano Lett. 12, 2918–2923.
Leijnse, M., 2014. New J. Phys. 16, 15003.
Li, D., Wu, Y., Fan, R., Yang, P., Majumdar, A., 2003a. Appl. Phys. Lett. 83, 3186–3188.
Li, D., Wu, Y., Kim, P., Shi, L., Yang, P., Majumdar, A., 2003b. Appl. Phys. Lett. 83, 2934–2936.
Liang, W., Hochbaum, A.I., Fardy, M., Rabin, O., Zhang, M., Yang, P., 2009. Nano Lett. 9, 1689–1693.
Lim, J.S., López, R., Sánchez, D., 2014. New J. Phys. 16, 015003.
Liu, Y.Y., Zhou, W.X., Tang, L.M., Chen, K.Q., 2014. Appl. Phys. Lett. 105, 203111.
Lü, X., 2009. J. Appl. Phys. 106, 064305.
Lu, L., Yi, W., Zhang, D.L., 2001. Rev. Sci. Instrum. 72, 2996–3003.

Luckyanova, M.N., Garg, J., Esfarjani, K., Jandl, A., Bulsara, M.T., Schmidt, A.J., Minnich, A.J., Chen, S., Dresselhaus, M.S., Ren, Z., Fitzgerald, E.A., Chen, G., 2012. Science 338, 936–939.
Luo, T., Chen, G., 2013. Phys. Chem. Chem. Phys. 15, 3389.
Mahan, G., Sales, B., Sharp, J., 1997. Phys. Today 50, 42–47.
Majumdar, A., 2004. Science 303, 777–778.
Mastomäki, J., Roddaro, S., Rocci, M., Zannier, V., Ercolani, D., Sorba, L., Maasilta, I.J., Ligato, N., Fornieri, A., Strambini, E., Giazotto, F., 2017. Nano Res. 10, 3468–3475.
Mavrokefalos, A., Pettes, M.T., Zhou, F., Shi, L., 2007. Rev. Sci. Instrum. 78, 34901.
Mavrokefalos, A., Moore, A.L., Pettes, M.T., Shi, L., Wang, W., Li, X. 2009. J. Appl. Phys. 105, 104318.
Moon, J., Kim, J.H., Chen, Z.C.Y., Xiang, J., Chen, R., 2013. Nano Lett. 13, 1196–1202.
Moore, A.L., Shi, L., 2014. Mater. Today 17, 163–174.
Mourik, V., Zuo, K., Frolov, S.M., Plissard, S.R., Bakkers, E.P.A.M., Kouwenhoven, L.P., 2012. Science 336, 1003.
Muhonen, J.T., Meschke, M., Pekola, J.P., 2012. Rep. Prog. Phys. 75, 046501.
Nakpathomkun, N., Xu, H.Q., Linke, H., 2010. Phys. Rev. B Condens. Matter Mater. Phys. 82, 235428.
Nakwaski, W., 1988. J. Appl. Phys. 64, 159–166.
Nardi, D., Travagliati, M., Siemens, M.E., Li, Q., Murnane, M.M., Kapteyn, H.C., Ferrini, G., Parmigiani, F., Banfi, F., 2011. Nano Lett. 11, 4126–4133.
Neophytou, N., Kosina, H., 2014. Appl. Phys. Lett. 105, 1–6.
Nevala, M.R., Chaudhuri, S., Halkosaari, J., Karvonen, J.T., Maasilta, I.J , 2012. Appl. Phys. Lett. 101, 112601.
Paajaste, J., Amado, M., Roddaro, S., Bergeret, F.S., Ercolani, D., Sorba, L., Giazotto, F., 2015. Nano Lett. 15, 1803–1808.
Pekola, J.P., 2015. Nat. Phys. 11, 118–123.
Pekola, J.P., Heikkilä, T.T., Savin, A.M., Flyktman, J.T., Giazotto, F., Hekking, F.W.J., 2004. Phys. Rev. Lett. 92, 56804.
Pennelli, G., 2014. Beilstein J. Nanotechnol. 5, 1268–1284.
Pennelli, G., Macucci, M., 2013. J. Appl. Phys. 114, 214507.
Pennelli, G., Macucci, M., 2016. Semicond. Sci. Technol. 31, 54001.
Pennelli, G., Nannini, A., Macucci, M., 2014. J. Appl. Phys. 115, 11–16.
Pennelli, G., Dimaggio, E., Macucci, M., 2017a. arXiv. https://arxiv.org/abs/1708.03142.
Pennelli, G., Dimaggio, E., Macucci, M., 2017b. arXiv. https://arxiv.org/abs/1710.04403.
Persson, A.I., Fröberg, L.E., Samuelson, L., Linke, H., 2009a. Nanotechnology 20, 225304.
Persson, A.I., Koh, Y.K., Cahill, D.G., Samuelson, L., Linke, H., 2009b. Nano Lett. 9, 4484–4488.
Pettes, M.T., Shi, L., 2009. Adv. Funct. Mater. 19, 3918–3925.
Pitanti, A., Ercolani, D., Sorba, L., Roddaro, S., Beltram, F., Nasi, L., Salviati, G., Tredicucci, A., 2011. Phys. Rev. X 1, 1–7.
Puyoo, E., Grauby, S., Rampnoux, J.M., Rouvìre, E., Dilhaire, S., 2011. J. Appl. Phys. 109, 24302.
Quaranta, O., Spathis, P., Beltram, F., Giazotto, F., 2011. Appl. Phys. Lett 98, 2009–2012.
Roberts, N.A., Walker, D.G., 2011. Int. J. Therm. Sci. 50, 648–662.
Rocci, M., Rossella, F., Gomes, U.P., Zannier, V., Rossi, F., Ercolani, D., Sorba, L., Beltram, F., Roddaro, S., 2016. Nano Lett. 16, 7950–7955.
Roddaro, S., Fuhrer, A., Brusheim, P., Fasth, C., Xu, H.Q., Samuelson, L., Xiang, J., Lieber, C.M., 2008. Phys. Rev. Lett. 101, 1–4.
Roddaro, S., Pescaglini, A., Ercolani, D., Sorba, L., Giazotto, F., Beltram, F., 2011. Nano Res. 4, 259–265.

Roddaro, S., Ercolani, D., Safeen, M.A., Suomalainen, S., Rossella, F., Giazotto, F., Sorba, L., Beltram, F., 2013. Nano Lett. 13, 3638–3642.
Roddaro, S., Ercolani, D., Safeen, M.A., Rossella, F., Piazza, V., Giazotto, F., Sorba, L., Beltram, F., 2014. Nano Res. 7, 1–9.
Romeo, L., Roddaro, S., Pitanti, A., Ercolani, D., Sorba, L., Beltram, F., 2012. Nano Lett. 12, 4490–4494.
Ross, F.M., Tersoff, J., Reuter, M.C., 2005. Phys. Rev. Lett. 95, 1–4.
Rossella, F., Soldano, C., Bellani, V., Tommasini, M., 2012. Adv. Mater. 24, 2453–2458.
Rossella, F., Bertoni, A., Ercolani, D., Rontani, M., Sorba, L., Beltram, F., Roddaro, S., 2014. Nat. Nanotechnol. 9, 997–1001.
Saito, Y., Iizuka, T., Koretsune, T., Arita, R., Shimizu, S., Iwasa, Y., 2016. Nano Lett. 16, 4819–4824.
Schelling, P.K., Shi, L., Goodson, K.E., 2005. Mater. Today 8, 30–35.
Schmidt, V., Mensch, P.F.J., Karg, S.F., Gotsmann, B., Das Kanungo, P., Schmid, H., Riel, H., 2014. Appl. Phys. Lett. 104, 0–4.
Shi, L., Li, D., Yu, C., Jang, W., Kim, D., Yao, Z., Kim, P., Majumdar, A., 2003. J. Heat Transfer 125, 881.
Small, J., 2004. Microscale Thermophys. Eng. 8, 1–5.
Small, J.P., Perez, K.M., Kim, P., 2003. Phys. Rev. Lett. 91, 256801.
Snyder, G.J., Toberer, E.S., 2008. Nat. Mater. 7, 105–114.
Soini, M., Zardo, I., Uccelli, E., Funk, S., Koblmüller, G., Fontcuberta i Morral, A., Abstreiter, G., 2010. Appl. Phys. Lett. 97, 2008–2011.
Spathis, P., Biswas, S., Roddaro, S., Sorba, L., Giazotto, F., Beltram, F., 2011. Nanotechnology 22, 105201.
Storm, K., Halvardsson, F., Heurlin, M., Lindgren, D., Gustafsson, A., Wu, P.M., Monemar, B., Samuelson, L., 2012. Nat. Nanotechnol. 7, 718–722.
Svensson, S.F., Hoffmann, E.A., Nakpathomkun, N., Wu, P.M., Xu, H.Q., Nilsson, H.A., Sánchez, D., Kashcheyevs, V., Linke, H., 2013. New J. Phys. 15, 105011.
Swinkels, M.Y., Van Delft, M.R., Oliveira, D.S., Cavalli, A., Zardo, I., Van Der Heijden, R.W., Bakkers, E.P.A.M., 2015. Nanotechnology 26, 385401.
Talbo, V., Saint-Martin, J., Retailleau, S., Dollfus, P., 2017. Sci. Rep. 7, 1–13.
Thelander, C., Mårtensson, T., Björk, M.T., Ohlsson, B.J., Larsson, M.W., Wallenberg, L.R., Samuelson, L., 2003. Appl. Phys. Lett. 83, 2052–2054.
Tian, Y., Sakr, M.R., Kinder, J.M., Liang, D., MacDonald, M.J., Qiu, R.L.J., Gao, H.J., Gao, X.P.A., 2012. Nano Lett. 12, 6492–6497.
Tiira, J., Strambini, E., Amado, M., Roddaro, S., San-Jose, P., Aguado, R., Bergeret, F.S., Ercolani, D., Sorba, L., Giazotto, F., 2017. Nat. Commun. 8, 1–9.
Tikhonov, E.S., Shovkun, D.V., Ercolani, D., Rossella, F., Rocci, M., Sorba, L., Roddaro, S., Khrapai, V.S., 2016a. Sci. Rep. 6, 30621.
Tikhonov, E.S., Shovkun, D.V., Ercolani, D., Rossella, F., Rocci, M., Sorba, L., Roddaro, S., Khrapai, V.S., 2016b. Semicond. Sci. Technol. 31, 104001.
Vineis, C.J., Shakouri, A., Majumdar, A., Kanatzidis, M.G., 2010. Adv. Mater. 22, 3970–3980.
Vining, C.B., 2009. Nat. Mater. 8, 83–85.
Vukajlovic-Plestina, J., Kim, W., Dubrovski, V.G., Tütüncüoğlu, G., Lagier, M., Potts, H., Friedl, M., Fontcuberta i Morral, A., 2017. Nano Lett. 17, 4101–4108.
Wang, H., Sen, M., 2009. Int. J. Heat Mass Transf. 52, 2102–2109.
Weathers, A., Bi, K., Pettes, M.T., Shi, L., 2013. Rev. Sci. Instrum. 84, 84903.
Wingert, M.C., Chen, Z.C.Y., Dechaumphai, E., Moon, J., Kim, J.H., Xiang, J., Chen, R., 2011. Nano Lett. 11, 5507–5513.
Wu, Z.H., Mei, X.Y., Kim, D., Blumin, M., Ruda, H.E., 2002. Appl. Phys. Lett. 81, 5177–5179.

Wu, P.M., Gooth, J., Zianni, X., Svensson, S.F., Gluschke, J.G., Dick, K.A., Thelander, C., Nielsch, K., Linke, H., 2013. Nano Lett. 13, 4080–4086.
Wu, K., Rademaker, L., Zaanen, J., 2014. Phys. Rev. Appl. 2, 1–8.
Xiang, J., Vidan, A., Tinkham, M., Westervelt, R.M., Lieber, C.M., 2006. Nat. Nanotechnol. 1, 208–213.
Yamaguchi, S., Matsumoto, T., Yamazaki, J., Kaiwa, N., Yamamoto, A., 2005. Appl. Phys. Lett. 87, 1–3.
Yang, P., 2010. 2010 IEEE 3rd International Nanoelectronics Conference (INEC), 49–49.
Yang, K., Cantarero, A., Rubio, A., D'Agosta, R., 2015. Nano Res. 8, 2611–2619.
Yazji, S., Hoffman, E.A., Ercolani, D., Rossella, F., Pitanti, A., Cavalli, A., Roddaro, S., Abstreiter, G., Sorba, L., Zardo, I., 2015. Nano Res. 8, 4048–4060.
Yazji, S., Swinkels, M.Y., De Luca, M., Hoffmann, E.A., Ercolani, D., 2016. Semicond. Sci. Technol. 31, 1–8.
Yi, W., Lu, L., Dian-lin, Z., Pan, Z., Xie, S., 1999. Phys. Rev. B 59, R9015–R9018.
Yu, C., Saha, S., Zhou, J., Shi, L., Cassell, A.M., Cruden, B.A., Ngo, Q., Li, J., 2006. J. Heat Transfer 128, 234.
Zannier, V., Ercolani, D., Gomes, U.P., David, J., Gemmi, M., Dubrovskii, V.G., Sorba, L., 2016. Nano Lett. 16, 7183–7190.
Zhang, Y., Dresselhaus, M.S., Shi, Y., Ren, Z., Chen, G., 2011. Nano Lett. 11, 1166–1170.
Zheng, J., Wingert, M.C., Moon, J., Chen, R., 2016. Semicond. Sci. Technol. 31, 1–6.
Zhou, J., Jin, C., Seol, J.H., Li, X., Shi, L., 2005. Appl. Phys. Lett. 87, 1–3.
Zhou, F., Moore, A.L., Bolinsson, J., Persson, A., Fröberg, L., Pettes, M.T., Kong, H., Rabenberg, L., Caroff, P., Stewart, D.A., Mingo, N., Dick, K.A., Samuelson, L., Linke, H., Shi, L., 2011a. Phys. Rev. B Condens. Matter Mater. Phys. 83, 1–10.
Zhou, F., Persson, A., Samuelson, L., Linke, H., Shi, L., 2011b. Appl. Phys. Lett. 99, 17–20.
Zhou, G., Li, L., Li, G.H., 2011c. J. Appl. Phys. 109, 114311.

CHAPTER SEVEN

Piezoelectric Semiconducting Nanowires

Yonatan Calahorra, Canlin Ou, Chess Boughey, Sohini Kar-Narayan[1]
University of Cambridge, Cambridge, United Kingdom
[1]Corresponding author: e-mail address: sk568@cam.ac.uk

Contents

1. INTRODUCTION

Semiconductor nanowires have been at the forefront of nanotechnology, bringing forth advances in electronics, optoelectronics, and basic physics, to name a few. More recently, there has been increasing interest in the existence of *piezoelectricity* in some of these materials, whereby they are capable of inter-converting mechanical and electrical energy. Piezoelectricity has been found to manifest more strongly in nanowires of some semiconductor

Semiconductors and Semimetals, Volume 98
ISSN 0080-8784
https://doi.org/10.1016/bs.semsem.2018.02.002

materials as compared to the bulk, which makes nanowires particularly attractive for applications in strain sensing and mechanical energy harvesting. At the same time, the geometry and unclamped nature of nanowires render them sensitive to small forces/deflections, and hence there has been an ongoing effort to incorporate them into so-called piezoelectric *nanogenerators* that are capable of harvesting energy from ambient vibrations to power autonomous devices, such as ubiquitous wireless sensor nodes. Furthermore, semiconducting piezoelectric materials have been driving the emerging fields of piezotronics and piezo-phototronics, where the development of a piezo-potential in response to stress or strain can be used to tune the electronic and/or optical properties of the material by altering the band structure of the semiconductor. There are many potential sensing and energy-based applications that can arise from the development of this area of research, but only a few candidate semiconducting materials systems have been explored thus far for their piezotronic and photo-piezotronic properties. Further development of the field requires investigation and development of a wide range of piezoelectric semiconducting nanowires. However, the direct measurement of piezoelectric properties of these nanowires is challenging, and thus rarely reported. This is due to generic difficulties associated with measurements of piezoelectric properties of nanoscale objects using conventional scanning probe microscopy techniques, which present key characterization challenges in this field. Recent advances in nanoscale characterization of piezoelectric semiconducting nanowires, such as the development of nondestructive piezoresponse force microscopy (ND-PFM), are opening up new avenues into the detailed investigation of nanoscale piezoelectricity in these materials. This chapter provides an overview of piezoelectric semiconducting nanowires, with a focus on the nature of the effect itself, how it manifests in different materials, how it is measured at the nanoscale, and finally different types of piezoelectric energy-related devices that can be achieved using these nanowires. A greater understanding of the underlying physics that governs piezoelectricity in semiconducting nanowires will serve to underpin further developments of their applications in energy devices.

2. BACKGROUND

2.1 Piezoelectricity

The piezoelectric effect results from relative displacement of ions or polar components within a piezoelectric material due to an applied stress, resulting in a change in net polarization. Piezoelectricity, first described by the Curie brothers (Curie and Curie, 1882), refers to the linear interrelation between the electrical and mechanical states of certain materials. Piezoelectricity is manifested in 20 out of the 21 noncentrosymmetric crystal classes, with the exception being the 432 cubic system. Centrosymmetric crystal classes do not exhibit piezoelectricity. Piezoelectricity is manifested as both a "direct" effect, where polarization is created in response to an applied stress, and a "converse" effect, where an applied electric field creates a strain within the material. Piezoelectricity is described mathematically with a set of complementary equations given below:

$$D = \epsilon_T E + dT \quad (1)$$

$$S = d^t E + s_E T \quad (2)$$

where T and S represent stress and strain, respectively, which are second rank tensors related to an applied force and the normal to the area over which it acts, D and E represent the electric displacement and electric field vectors, ϵ is the dielectric permittivity of the material, where subscript T represents constant stress, d is the piezoelectric tensor (third rank), and d_T is the transpose of d. s_E is the elastic compliance at constant electric field. These terms are second, third, and fourth rank tensors, respectively, and due to symmetry considerations Voigt notation is often used for convenience to describe the tensors. For the piezoelectric tensor this takes the form of a 3×6 matrix. There are four types of piezoelectric coefficients

$$d_{ij} = \left(\frac{\partial D_i}{\partial T_j}\right)^E \quad (3)$$

$$e_{ij} = \left(\frac{\partial D_i}{\partial S_j}\right)^E \quad (4)$$

$$g_{ij} = -\left(\frac{\partial E_i}{\partial T_j}\right)^D \quad (5)$$

$$h_{ij} = -\left(\frac{\partial E_i}{\partial S_j}\right)^D \quad (6)$$

where the superscript indicates differentiation under constant or zero field. d_{ij} and e_{ij} are the most commonly used coefficients, often coined the piezoelectric strain coefficients (d) and charge coefficients (e). i ranges from 1 to 3 and j from 1 to 6 (Voigt), where 1, 2, 3 are direction axes and the 3-axis typically denotes the polarization direction, while 4, 5, 6 denote shear deformation about 1, 2, 3 axes, correspondingly. d_{ij} therefore describes polarization arising in the i direction in response to stress along the j direction, or alternatively strain in the j direction in response to an electric field along the i direction.

2.2 Piezoelectric Semiconductors

The most common semiconductors, with the exception of silicon and germanium, are piezoelectric; these include III–Vs and II–VIs. The wurtzite GaN family of materials, and ZnO are best known for their piezoelectric properties. Piezoelectricity in semiconductor materials can bring about both advantages and disadvantages. Unlike cubic structured GaAs/AlGaAs high electron mobility transistor (HEMT) technology which relies on the separation of dopants from the charge carriers (Hu et al., 1997), GaN/AlGaN HEMT technology is based on the piezoelectricity and spontaneous polarization discontinuities in III-N heterostructures and does not rely on doping to achieve conductance (Fig. 1) (Ambacher et al., 1999). Interestingly, in the more established GaN LED lighting technology, the existence of piezoelectricity and spontaneous polarization is undesired, reducing emission efficiency (Kim et al., 2007). This particular issue is directing research efforts toward "non-polar" III-N technology, and polarization-matched layers. For an exhaustive discussion on polarization in semiconductors, consult Ref. Wood and Jena (2007).

Semiconductors include bound and free charges, and the relation between the charges and external and internal potentials is dominated by Poisson's equation and Fermi–Dirac distribution (Sze and Ng, 2006), taking the following form:

$$-\nabla^2\phi = \nabla\cdot D = \rho \tag{7}$$

$$\rho = q\cdot\left(N_D^+ - N_A^- + p - n\right) \tag{8}$$

$$n = \frac{2}{\sqrt{\pi}} N_c F_{1/2}\left(q\phi/k\tilde{T}\right) \tag{9}$$

where q is the electron charge, ϕ is the electric potential (related to the conduction band and Fermi level through $-q\phi = E_C - E_F$), and ρ is the free charge distribution, determined by the ionized donor and acceptor concentrations (N_D^+, N_A^-), and electron and hole concentrations (n, p). The free carrier concentrations are then related to the potential through the effective density of states (N_c) and Fermi integral ($F_{1/2}$). k is the Boltzmann constant, and here $\tilde{T}$ is temperature.

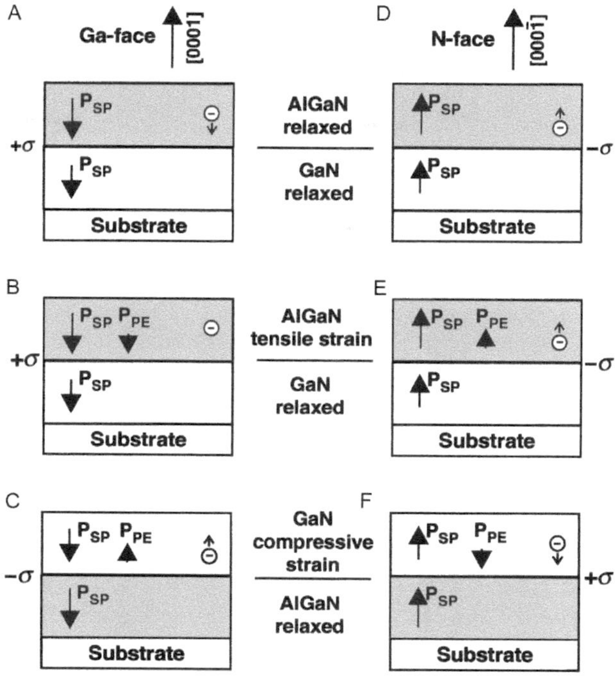

Fig. 1 Schematic of the interplay between spontaneous polarization, piezoelectricity, and crystal orientation in the GaN/AlGaN heterostructure system, leading to the creation of a two-dimensional electron gas at the interface. (A–C) Ga-face heterostructure growth with relaxed AlGaN on GaN, strained AlGaN on GaN, compressed GaN on AlGaN, correspondingly; (D–F) N-face heterstructure with relaxed AlGaN on GaN, strained AlGaN on GaN, compressed GaN on AlGaN, correspondingly. *Reproduced from Ambacher, O., Smart, J., Shealy, J.R., Weimann, N.G., Chu, K., Murphy, M., Schaff, W.J., Eastman, L.F., Dimitrov, R., Wittmer, L., et al., 1999. Two-dimensional electron gases induced by spontaneous and piezoelectric polarization charges in N-and Ga-face AlGaN/GaN heterostructures. J. Appl. Phys. 85 (6), 3222–3233 with permission from AIP.*

For a piezoelectric semiconductor, Eqs. (1), (2), and (7)–(9) are coupled to describe the electro-mechanics of the material. It is important to note that observation of piezoelectric effects in semiconductors relies on the existence of a depletion region, where the semiconductor is not neutral, and therefore is related to junctions and surfaces. Initial observations of coupled behavior were studied in the early and late 1970s reporting photomechanical coupling in piezoelectric semiconductors (Łagowski and Gatos, 1972, 1974) and mechanical modulation of metal-semiconductor Schottky contact barrier height (Kusaka, 1978; Kusaka et al., 1974), later coined as the piezotronic

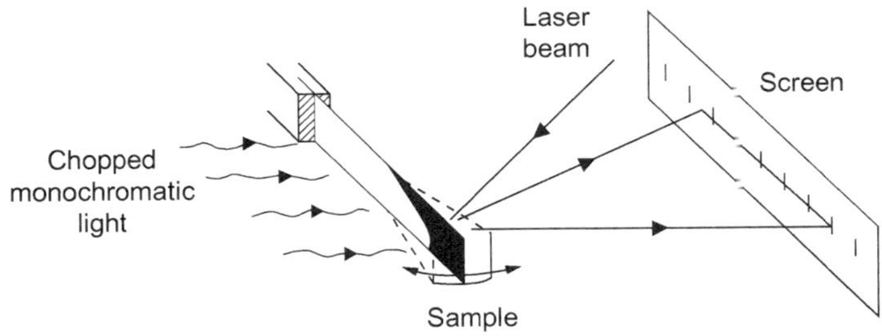

Fig. 2 Schematic of the optical setup to measure the photomechanical coupling in piezoelectric semiconducting cantilevers. *Reproduced from Lagowski, J., Gatos, H.C., 1974. Photomechanical vibration of thin crystals of polar semiconductors. Surface Sci. 45 (2), 353–370 with permission from Elsevier.*

effect (Wang, 2007; Zhou et al., 2008). Łagowski and Gatos (1972, 1974) described the deformation of CdS and GaAs cantilevers in response to light absorption (Fig. 2). They ascribe the observation to generation of surface voltage and converse piezoelectric deformation. This effect was exhibited only when the polar surfaces, {0001} in CdS and {111} in GaAs, were examined (the wide faces of the cantilever). No measurable effect was found for samples with non-polar surfaces, indicating that this effect is related to piezoelectricity. Furthermore, no effect was measured when GaP samples were examined.

Kusaka et al. examined the direct piezoelectric effect, where the Schottky barrier height was measured during bending of metal-semiconductor junctions of CdS and GaP (Kusaka, 1978; Kusaka et al., 1974). The Schottky barrier change was monitored through current measurements, and it was found that the polarity of the change corresponded to the crystal face of the Schottky contact, and the material (II–VI and III–V semiconductors' piezoelectric coefficients have opposite signs)—indicating that the effect originated in piezoelectricity. It is interesting to note that GaP was examined in both experiments; however, it only exhibited a piezoelectric related effect in the latter—where a device was fabricated. As mentioned by Łagowski and Gatos (1974), this is most likely due to the presence of a surface accumulation layer in free GaP surfaces, shortening the generated surface photovoltage. This layer is depleted when forming a Schottky contact, allowing for the piezoelectric effect to modify the barrier height.

By solving the piezoelectric equations coupled to Poisson's equation (Eqs. 1, 2, 7–9) an expression for the piezotronic effect, the change in barrier height with strain, was presented

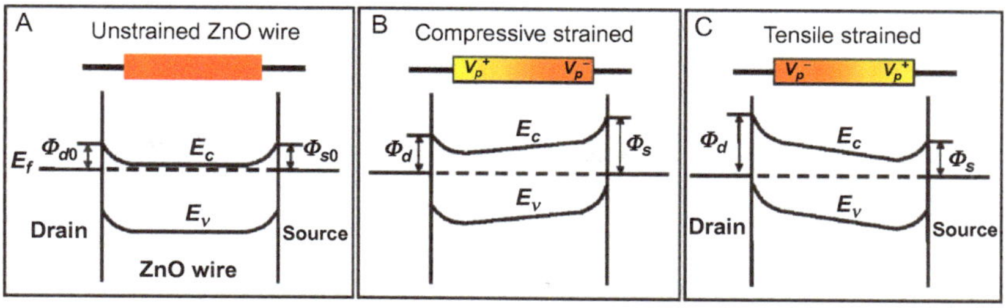

Fig. 3 The fields developing within a ZnO NW and near the Schottky contacts for (A) unstrained; (B) compressed; (C) stretched NWs. *Reproduced from Zhou, J., Fei, P., Gu, Y., Mai, W., Gao, Y., Yang, R., Bao, G., Wang, Z.L., 2008. Piezoelectric-potential-controlled polarity-reversible Schottky diodes and switches of ZnO wires. Nano Lett. 8 (11), 3973–3977 with permission from ACS.*

$$\Delta\phi_B = \pm\frac{d_{31}}{S_{11}} \cdot \frac{V_{bi}}{qN_D} \cdot \frac{1}{R} \tag{10}$$

where V_{bi} is the built-in voltage, and $1/R$ the curvature of the substrate the piezotronic Schottky diode was mounted on. Fig. 3 shows a schematic representation of the piezotronic effect in NWs, as was extensively studied in work from Z.L. Wang's group. Importantly, direct stretching and compressing along the polar axis are more accessible in NWs compared to the bulk.

3. WHY ARE PIEZOELECTRIC SEMICONDUCTING NANOWIRES SPECIAL?

This section explores early studies that capture the essence of piezoelectric semiconductor coupling: interesting phenomena arise in junction regions, where the unique behavior of semiconductors comes into play. When it comes to semiconductor NWs, their inherent electromechanical characteristics may contribute to an enhanced piezoelectric response, as explained in the following section.

3.1 Mechanical Properties of Piezoelectric Semiconducting Nanowires

Nanowires have unique mechanical properties due to their size and slender shape. Nanowires exhibit increased failure stress, compared to known bulk properties. The main reason is the smaller likelihood of structural defects in smaller sizes (Gordon et al., 2009; Wang et al., 2011; Wen et al., 2008).

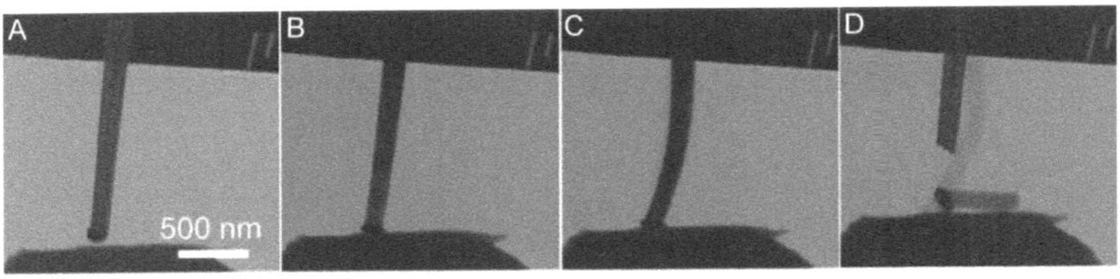

Fig. 4 (A)–(D) Different stages of GaAs NW deformation, from contact to failure. *Reproduced from Wang, Y.B., Wang, L.F., Joyce, H.J., Gao, Q., Liao, X.Z., Mai, Y.W., Tan, H.H., Zou, J., Ringer, S.P., Gao, H.J., et al., 2011. Super deformability and Youngs modulus of GaAs nanowires. Adv. Mater. 23 (11), 1356–1360 with permission from Wiley.*

For example Wang et al. (2011) found that GaAs of 50–150 nm in diameter could sustain 10%–11% strain before failure, a value significantly larger than measured in bulk material (Fig. 4). If so, NWs can withstand higher stresses, which in turn can result in increased piezoelectric response. Furthermore, there are reports of increased stiffness (Young's modulus) in NWs (Agrawal et al., 2008). Revisiting Eq. (10) for the piezotronic effect, an increased stiffness (reduced compliance) may result in an increased effect for a given strain.

Due to their aspect ratio, NWs are flexible, unconstrained by their environment, and inherently allow larger deformations. Furthermore, they allow the application of tensile and compressive stresses in the same orientation where the electrical properties are examined (d_{33} type effect) (Yang et al., 2009). This is unlike the bulk case, where generally the main deformation axis is orthogonal to the examined electrical properties (d_{31} type effect)—see Fig. 5. Generally $d_{33} > d_{31}$, and thus a larger response is to be expected from the former. The increased degree of mechanical freedom comes into play through elimination of substrate boundary condition for the inverse piezoelectric coefficient (Lefki and Dormans, 1994; Zhao et al., 2004)

$$d_{33,\,bulk}^{eff} = d_{33} - \frac{2S_{13}}{S_{11} + S_{12}} d_{31} \tag{11}$$

where $d_{33,\,bulk}^{eff}$ is the observed, clamped, coefficient, and S_{ij} are elastic compliance coefficients. For unclamped nanostructures the second-term vanishes

$$d_{33,\,NW}^{eff} = d_{33} \tag{12}$$

Zhao et al. (2004) have estimated the (negative) contribution of the substrate to ZnO piezoelectricity results in about 50% reduction in the effective piezoelectric coefficient.

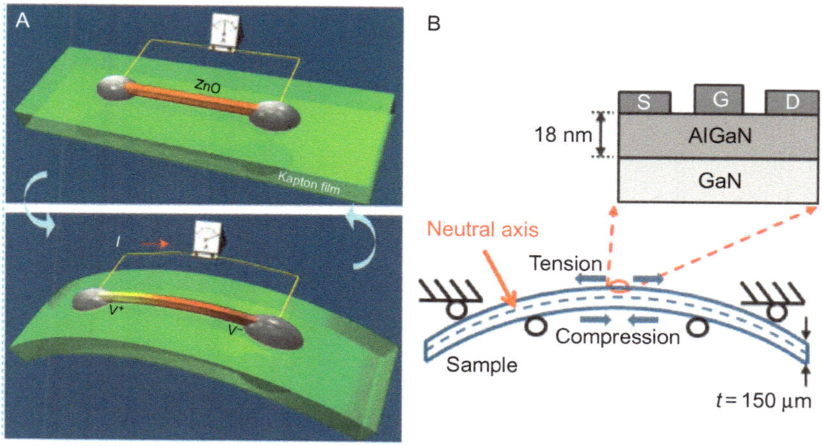

Fig. 5 (A) Schematic of an axial ZnO NW piezoelectric nanogenerator, experiencing axial deformation. (B) Schematic of a GaN/AlGaN HEMT experiencing bending, where the device contacts are transverse to the main strain orientation. *Panel (A) reproduced from Yang, R., Qin, Y., Dai, L., Wang, Z.L., 2009. Power generation with laterally packaged piezoelectric fine wires. Nat. Nanotechnol. 4 (1), 34–39 with permission from Nature Publishing Group. Panel (B) reproduced from Stutzmann, M., Steinhoff, G., Eickhoff, M., Ambacher, O., Nebel, C.E., Schalwig, J., Neuberger, R., Müller, G., 2002. GaN-based heterostructures for sensor applications. Diamond Relat. Mater. 11 (3), 886–891 with permission from Eslsevier.*

3.2 Electronic Properties of Piezoelectric Semiconducting Nanowires

Semiconductor NWs have an increased tendency for depletion as their diameter decreases, due to geometry and size related electrostatic effects. Briefly, the cylindrical geometry induces a shallow profile of the potential, compared to bulk (Calahorra and Ritter, 2013; Simpkins et al., 2008). When solving Poisson's equation in cylindrical coordinates, considering a depletion region extending from the NW surface toward its center, Eq. (7) takes the form Calahorra and Ritter (2013)

$$\frac{d^2\phi}{dr^2} + \frac{1}{r}\frac{d\phi}{dr} = -\frac{qN_D}{\epsilon_s}; \quad R_{dep} < r < R \tag{13}$$

where R is the NW radius and R_{dep} is the onset of the depletion region. Unlike the equation in Cartesian coordinates (used for bulk material), the additional $1/r$ term (second) requires the first derivative of the potential (i.e., the electric field) to decay slowly in order to avoid singularity—leading to increased depletion.

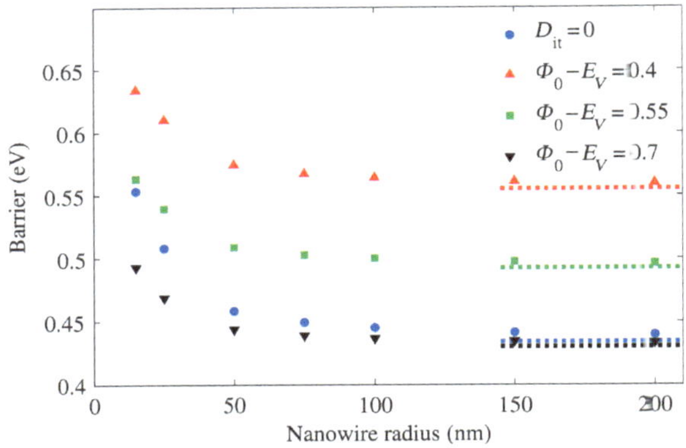

Fig. 6 Calculated Schottky barrier height for an N-type semiconducting NW as a function of diameter. Non-ideal effects lowering the barrier become less significant as size is reduced, resulting in a higher barrier for smaller NWs, and increased depletion. The barrier height approaches bulk value (*dashed*) with size. *Reproduced from Calahorra, Y., Yalon, E., Ritter, D., 2015. On the diameter dependence of metal-nanowire Schottky barrier height. J. Appl. Phys. 117 (3), 034308 with permission from AIP.*

NW electrostatics also have implications on Schottky barrier height, where non-ideal barrier-lowering effects are weakened in NWs (Fig. 6) (Calahorra et al., 2014, 2015). A possible implication of increased barrier height and depletion length is reduced screening of the piezoelectric potential formed within NWs in response to strain, leading to a more pronounced piezoelectric response.

3.3 Effect of Growth on the Piezoelectric Properties of Semiconducting Nanowires

Semiconductor NWs of the III–V family may crystallize in wurtzite as well as zinc-blende phase, unlike their bulk counterparts which crystallize exclusively as zinc-blende (non-nitride III–Vs) (Calahorra and Kar-Narayan, 2018; Lehmann et al., 2013). Fig. 7 shows the control achieved in switching growth from zinc-blende to wurtzite structure in InAs, GaAs, InP, and GaP NWs. Furthermore, it is possible to control growth orientation to grow NWs where the axial orientation is polar/non-polar. The piezoelectric coefficient matrix (Voigt notation) for wurtzite semiconductors contains three non-degenerate coefficients and is given by

$$d_{WZ} = \begin{pmatrix} 0 & 0 & 0 & 0 & d_{15} & 0 \\ 0 & 0 & 0 & d_{15} & 0 & 0 \\ d_{31} & d_{31} & d_{33} & 0 & 0 & 0 \end{pmatrix} \tag{14}$$

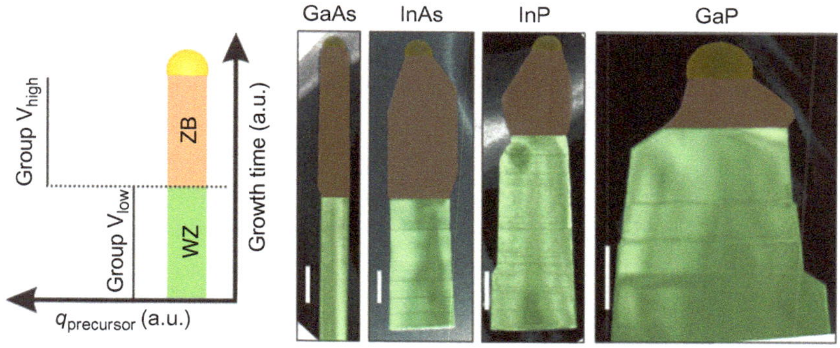

Fig. 7 Schematic of the group V modulation during NW growth (*left*) resulting in controlled crystal phase switching in several III–V NWs (*right*). *Reproduced from Lehmann, S., Wallentin, J., Jacobsson, D., Deppert, K., Dick, K.A., 2013. A general approach for sharp crystal phase switching in InAs, GaAs, InP, and GaP nanowires using only group V flow. Nano Lett. 13 (9), 4099–4105 with permission from ACS.*

This matrix is well known and used, considering the majority of work on piezoelectricity in semiconductors is related to ZnO and GaN—both wurtzite materials. It is easily and directly applied to NWs, since the growth orientation of the NW is usually [0001], corresponding to the polar axis (3-axis) of the matrix.

For zinc-blende materials (non-nitride III–Vs) a more complicated picture arises: the piezoelectric matrix is usually presented such that the 3-axis coincides with one of the main cubic axis

$$d_{ZB,[001]} = \begin{pmatrix} 0 & 0 & 0 & d_{14} & 0 & 0 \\ 0 & 0 & 0 & 0 & d_{14} & 0 \\ 0 & 0 & 0 & 0 & 0 & d_{14} \end{pmatrix} \tag{15}$$

This matrix has only one non-degenerate coefficient, the shear coefficient d_{14}. However, III–V NWs, either crystallizing in wurtzite or zinc-blende, have a tendency to grow in the [111]/[0001] orientation (for zinc-blende/wurtzite correspondingly). Considering the NW symmetry, it makes sense to rotate the piezoelectric matrix in Eq. (15) such that the 3-axis is [111]. When done in complete analogy to wurtzite, i.e., the 1-axis corresponding to $[11\overline{2}]/[10\overline{1}0]$ and the 2-axis corresponds to $[\overline{1}10]/[\overline{1}2\overline{1}0]$, the matrix takes the following form Calahorra et al. (2017a)

$$d_{ZB,[111]} = \begin{pmatrix} d_{11} & -d_{11} & 0 & 0 & d_{15} & 0 \\ 0 & 0 & 0 & d_{15} & 0 & d_{26} \\ d_{31} & d_{31} & d_{33} & 0 & 0 & 0 \end{pmatrix} \tag{16}$$

The distinctions between the piezoelectric matrices of zinc-blende and wurtzite, given by Eqs. (14) and (16) open up opportunities to engineer the piezoelectric properties of a single NW through formation of crystal phase heterostructures. For example, there is no radial symmetry in ZB since the response to a field applied along the 1-axis will result in a response different than a field applied along the 2-axis.

Demonstrating some of the principles mentioned in Sections 3.1–3.2 even in bulk material, is the work published by Gerngross et al. (2011), in which the piezoelectric response of chemically etched porous InP (zinc blende) was measured. The measured coefficients were found to be up to 30 times those of bulk InP, due to enhanced depletion in the material (increased surface area), and porosity induced softening (Cha et al., 2011). On the other extreme, quantum mechanical simulations have predicted two orders of magnitude increase in the piezoelectric coefficients of ultra-thin (1–2 nm) ZnO and GaN NWs (Agrawal and Espinosa, 2011).

The piezoelectric properties of ZnO, coupled with the relative ease of fabrication of nanostructures of this material, render it particularly useful and popular for applications in piezoelectric energy harvesting. ZnO is a direct wide band-gap (~3.3 eV) semiconductor with a wurtzite structure, inducing polarity along the *c*-axis of the structure (Vayssieres, 2003; Wang, 2011; Xu and Wang, 2011). Fig. 8A shows the wurtzite lattice structure of ZnO with $a = 3.25$ Å and $c = 5.20$ Å, and the origin of the dipole moment that gives rise to piezoelectric in this material (Onodera and Takes, 2012). Fig. 8B illustrates a typical growth morphology of ZnO nanowire or nanorod with a hexagonal crystal structure and the corresponding facets, where (001) is the polar surface and relatively more reactive while the side facets are non-polar and more stable (Vayssieres, 2003; Wang, 2011; Xu and Wang, 2011). Therefore, the ZnO

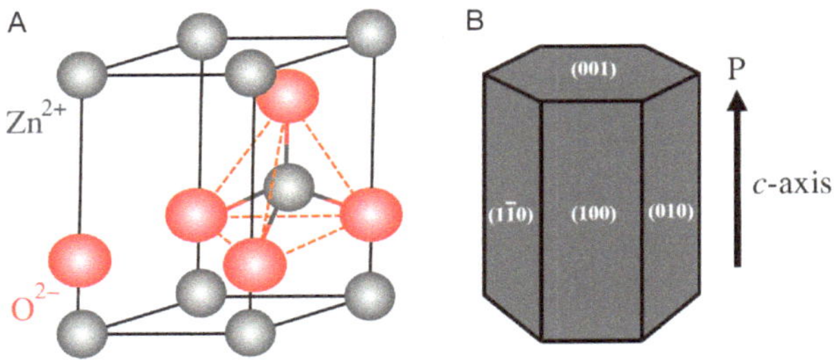

Fig. 8 (A) Wurtzite structure model of ZnO; (B) Typical growth morphology of ZnO with hexagonal crystal structure. The *arrow* indicates the polarization direction and the *c*-axis direction.

crystal tends to grow faster along *c*-axis, which is favorable particularly during nanowire growth as it results in a net dipole moment along the axis of the nanowire. ZnO has been shown to be easily grown into a variety of nanostructures with different morphologies and shapes, such as nanorods (Podrezova et al., 2014; Scrymgeour and Hsu, 2008; Vayssieres, 2003), nanotubes (Stassi et al., 2015; Sun et al., 2006), nanowires (Hu et al., 2012; Unalan et al., 2008; Vayssieres, 2003; Wang et al., 2006), nanobelts (Lucas et al., 2007), and nanorings, onto a variety of different substrates at relatively low temperatures (<100°C) compared to high-temperature-sintered piezoceramics.

Commonly used synthesis techniques include chemical vapor deposition (Schmidt-Mende and MacManus-Driscoll, 2007), physical vapor deposition (Wang, 2011; Xu et al., 2009), sol–gel synthesis (Ottone et al., 2014; Zhou and Wong, 2008), electrodeposition (Ottone et al., 2014; Prună et al., 2012; Sharma et al., 2010; Zhang et al., 2012), and hydrothermal synthesis (Cao et al., 2006; Chen et al., 2011; Cheng and Samulski, 2004; Liu and Zeng, 2003; Tian et al., 2011; Vayssieres, 2003; Xu and Wang, 2011). ZnO bulk and thin films are likely to suffer from poor piezoelectric performance due to defects both in the bulk and on the surface that may arise due to environmental degradation (Hu et al., 2012; Jalali et al., 2014, 2015), while the relatively high stiffness constant of ZnO renders it prone to mechanical failure (Espinosa et al., 2012). However, these limitations of ZnO could be overcome by incorporating nanostructures into nanocomposite systems, as nanostructured ZnO with higher aspect ratio and surface-to-volume ratio, possesses enhanced sensitivity to low-amplitude ambient vibrations as well as reduced fragility compared with their bulk or thin film counterparts (Boughey et al., 2016; Chen et al., 2010), making them particularly attractive for energy harvesting applications. In general, single crystal NWs are preferred over polycrystalline ones because there is a single domain of electric dipoles with one polarization direction, but even polycrystalline ZnO nanowires with preferred orientation along the nanowire axis have been shown to exhibit robust piezoelectric energy harvesting performance (Boughey et al., 2016; Ou et al., 2016).

4. CHARACTERIZATION OF PIEZOELECTRICITY IN SEMICONDUCTING NANOWIRES

Traditional macro-scale characterization of piezoelectricity is usually done by optical methods (see Figs. 2 and 9A) for displacement measurement. However, these methods are hardly applicable when considering

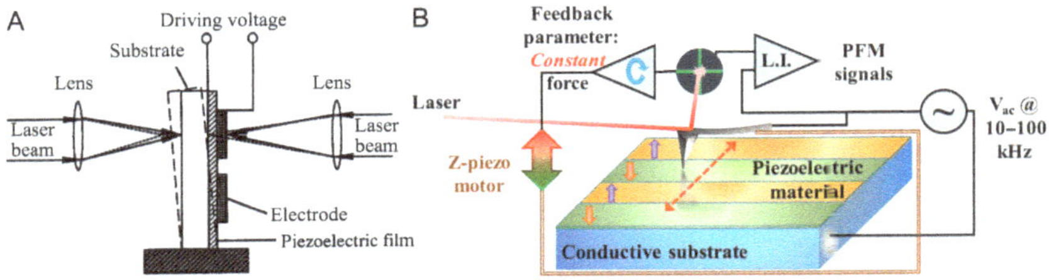

Fig. 9 (A) Schematic of a dual beam interferometer used to measure bulk material piezoelectric effects. (B) Schematic of a PFM setup: a conductive tip applies AC voltage to a piezoelectric sample, and simultaneously follows the resulting deformation. *Panel (A) reproduced from Kholkin, A.L., Wütchrich, Ch., Taylor, D.V., Setter, N., 1996. Interferometric measurements of electric field-induced displacements in piezoelectric thin films. Rev. Sci. Instrum. 67 (5), 1935–1941 with permission from AIP. Panel (B) reproduced from Calahorra, Y., Smith, M., Datta, A., Benisty, H., Kar-Narayan, S., 2017b. Mapping piezoelectric response in nanomaterials using a dedicated non-destructive scanning probe technique. Nanoscale 9 (48), 19290–19297 with permission from The Royal Society of Chemistry.*

nanostructures such as NWs, which are smaller than the characteristic wavelength. The prevalent tool for nanoscale piezoelectric characterization is piezoresponse force microscopy (PFM), which has been widely used for characterization of ferroelectric and piezoelectric materials (Gruverman et al., 2006). However, when it comes to NWs, either vertical or horizontal, the application of PFM is challenging, due to possible mechanical damage and/or dislodgment during contact mode operation, and reports of PFM-based NW characterization are not common. Other methods such as atomic force microscopy (AFM) based current generation or characterization of the direct piezoelectric effect are usually applied to NWs. These are discussed in the following sections.

4.1 Piezoresponse Force Microscopy

PFM works on the basis of the converse piezoelectric effect: a conductive AFM tip is scanned across a piezoelectric sample in contact mode, while an AC voltage is applied between the tip and the bottom electrode, through the sample. In response to the bias the sample deforms, and the deformation is monitored through a laser spot reflection from the tip, maintaining a constant contact with the sample through the feedback (Fig. 9B). Due to contact mode operation, application of PFM to nanostructures is challenging, and advanced PFM methods are be required for characterization (Calahorra et al., 2017a), as discussed later in detail.

The first report of PFM application to semiconductor nanostructures was published by Zhao et al. (2004), describing the measurement of piezoelectricity in ZnO nanobelts. It was found that the piezoresponse of the nanobelt is 2–3 times stronger than that of bulk material. This finding was explained by the unconstrained boundaries of the nanobelt (Eqs. 11 and 12).

Most PFM work revolved around characterization of relatively thick, vertical, NWs—predominantly ZnO and GaN (Fan et al., 2006; Scrymgeour and Hsu, 2008; Tamvakos et al., 2015; Tiwary et al., 2016). Scrymgeour and Hsu have correlated PFM and conductive AFM (c-AFM) measurements of ZnO nanorods (Fig. 10). Interestingly, they have found that thinner nanorods (~150 nm) yield higher piezoelectric responses, and these are coupled to lower conductivity, while the opposite is true for thicker nanorods (~500 nm). The size dependence could be explained by the increased

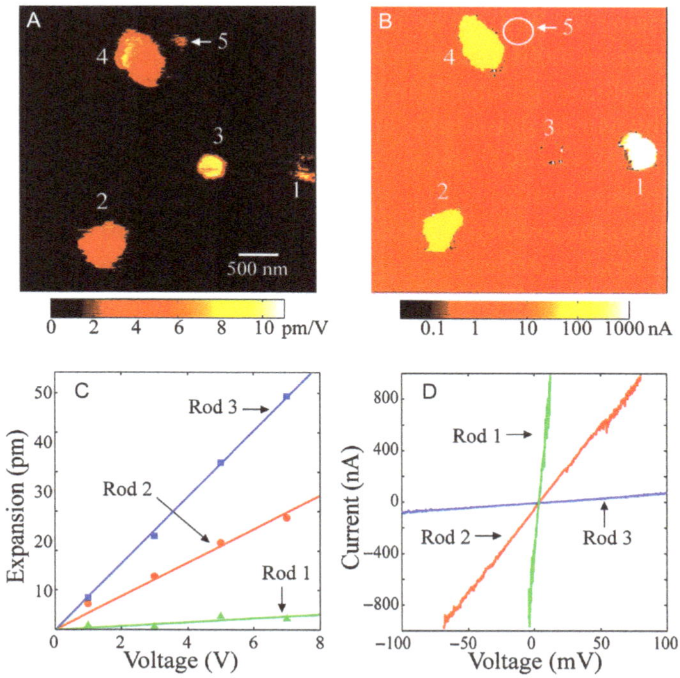

Fig. 10 (A) PFM and (B) c-AFM signals from the same area; (C) PFM expansion and (D) measured current from the NW shown in (A) and (B). *Reproduced from Scrymgeour, D.A., Hsu, J.W.P., 2008. Correlated piezoelectric and electrical properties in individual ZnO nanorods. Nano Lett. 8 (8), 2204–2209 with permission from ACS.*

depletion tendency of thin NWs, as discussed in Section 3.2. In is noteworthy that the NWs in this study underwent an annealing treatment to form ohmic contacts, and therefore the more common piezotronic effect is not observed in this case. Furthermore, the NWs were mechanically protected by embedment in a polymer layer, probably to overcome contact PFM induced mechanical damage for non-embedded NWs.

Reports of PFM measurements on horizontal NWs are scarce, mainly due to the challenge of using contact mode PFM on unconstrained objects. In a study by Minary-Jolandan et al. (2012) horizontal GaN NWs were studied by PFM. To avoid mechanical damage, the NWs were clamped to the surface at the edges. In this work two configurations were used to realize characterization of the three non-degenerate piezoelectric coefficients of the wurtzite structure—d_{33}, d_{31}, and d_{15}. For a horizontal NW in a standard PFM configuration, where the voltage is applied across the NW axis, the only measurable deformation is related to d_{15}, through a lateral signal (Fig. 11B). Characterizing d_{33} and d_{31} requires applying voltage along the NW axis (assuming it is in [0001] orientation), and therefore should be done by applying the signal through lateral electrodes to the NW (Fig. 11A). The results obtained by Minary-Jolandan et al. (2012) for the piezoelectric coefficients, $d_{33} = 12.8$, $d_{31} = -8.2$, and $d_{15} = 10.2$ pm/V, are 3–4 times higher than accepted bulk values.

Recently, the problems related to contact mode PFM characterization of nanomaterials have been recognized, and non-destructive PFM methods combining mechanical indentation modes with piezoresponse analysis have been developed (Fig. 12) (Calahorra et al., 2017b; Kalinin et al., 2018).

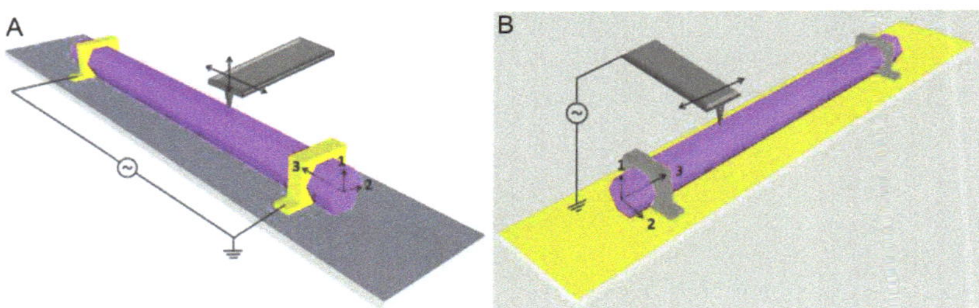

Fig. 11 Two configurations for complete characterization of piezoelectricity in horizontal GaN NWs: (A) voltage is applied along the NW [0001] axis for d_{33} and d_{31} measurement; (B) voltage is applied across the NW axis to measure d_{15}. *Reproduced from Minary-Jolandan, M., Bernal, R.A., Kuljanishvili, I., Parpoil, V., Espinosa, H.D., 2012. Individual GaN nanowires exhibit strong piezoelectricity in 3D. Nano Lett. 12 (2), 970–976 with permission from ACS.*

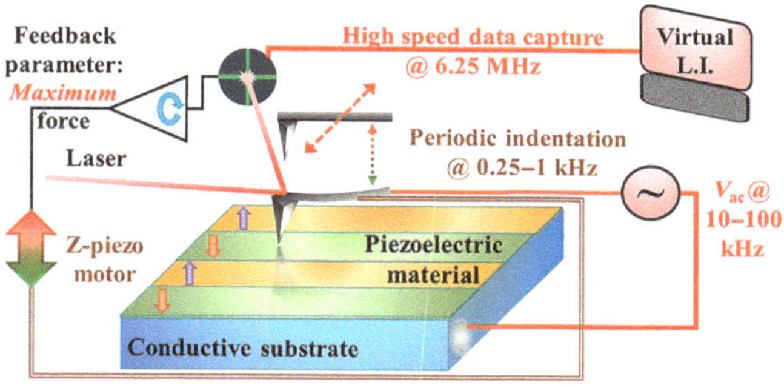

Fig. 12 Schematic of the ND-PFM setup: the tip periodically indents the sample, minimizing shear stress damage. An AC voltage is applied to induce the converse piezoelectric effect, and the piezoresponse is then extracted from the time periods the tip is in contact with the sample. *Reproduced from Calahorra, Y., Smith, M., Datta, A., Benisty, H., Kar-Narayan, S., 2017b. Mapping piezoelectric response in nanomaterials using a dedicated non-destructive scanning probe technique. Nanoscale 9 (48), 19290–19297 with permission from The Royal Society of Chemistry.*

At the heart of these methods lie two principles: (i) minimizing the contact mode mechanical damage to the sample using an intermittent contact mode; (ii) extraction of the PFM data from the period of time were significant tip/sample contact is achieved. Using ND-PFM the piezoelectric properties of non-clamped horizontal GaAs NWs were examined, as well as vertical InP NWs. This was the first direct PFM measurement of the d_{33} value of wurtzite InP. The value obtained, ~1 pm/V, was in the center of the theoretical predicted range of values (Calahorra et al., 2017a).

ND-PFM was further employed to characterize VLS grown horizontal GaAs NWs with a predominant zinc-blende phase (Calahorra et al., 2017a), and electrodeposited ZnO NWs (Boughey et al., 2016). Both materials allowed examination of the material in uncommon configurations: for GaAs—due to the horizontal configuration, and for ZnO—due to predominant polycrystalline nature, and a non-polar oriented NW axis. ND-PFM provided insight into intricacies of nanoscale piezoresponse, unavailable by contact PFM, which caused sample damage. In the case of GaAs NWs, the results were found to coincide with finite element simulations for the vertical and lateral deflections of the AFM due to the piezoelectric response of the NW, demonstrating the distinct contributions of different piezoelectric coefficients and side facets (Fig. 13C and D). For ZnO, a piezoresponse matching the polycrystalline nature of the sample was found (Fig. 13F).

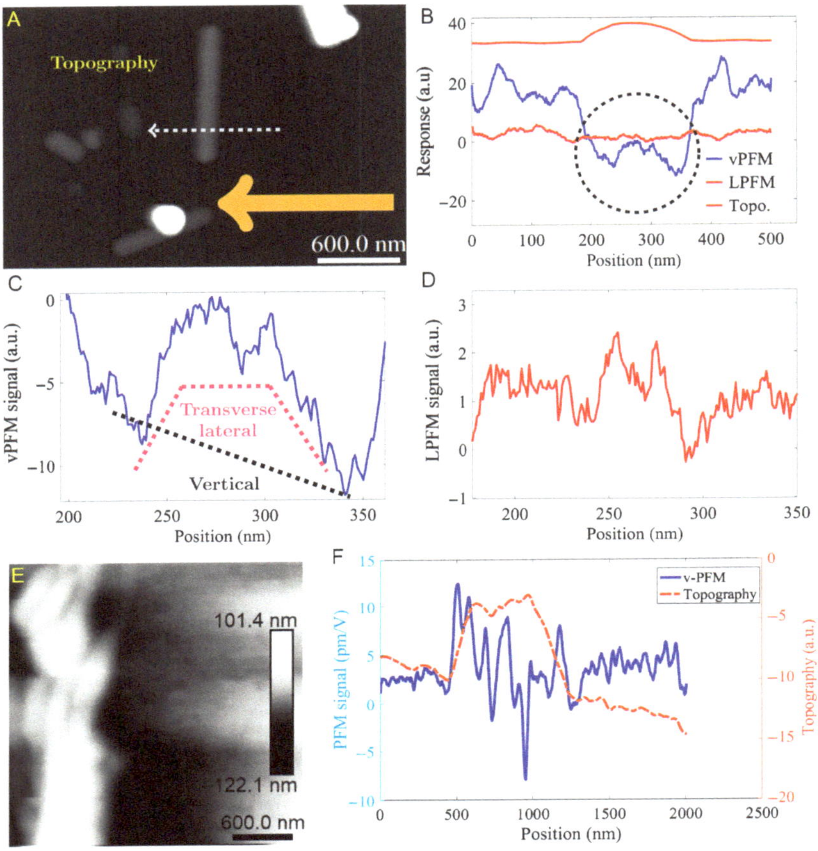

Fig. 13 ND-PFM measurements of a GaAs (A)–(D) and ZnO (E) and (F) NWs, taken across the NWs: (A) GaAs NW topography, with scan direction and cantilever orientation (*dashed* and *solid arrows* correspondingly); (B) PFM results, showing the vertical and lateral signals, alongside the scan topography. The *dashed circle* indicates the area of interest; (C) close-up to the vertical PFM signal, the *dashed lines* indicate the simulated signals which may explain the experimental results; (D) close-up to the lateral PFM signal, showing the signature of the side facets. (E) ZnO polycrystalline NW topography; (F) calibrated vertical PFM response from the NW, showing regions of ± 10 pm/V, in agreement with the poly-crystallinity and non-polar growth orientation. *Panel (D) reproduced from Calahorra, Y., Guan, X., Halder, N.N., Smith, M., Cohen, S., Ritter, D., Penuelas, J., Kar-Narayan, S., 2017a. Exploring piezoelectric properties of III-V nanowires using piezo-response force microscopy. Semicond. Sci. Technol. 32 (7), 074006 with permission from IOP.*

4.2 Conductive AFM-Based Measurements

A significant portion of piezoelectric NW characterization was performed using conductive AFM (c-AFM) techniques (Gao et al., 2007; Gogneau et al., 2014a,b; Jamond et al., 2016; Lu et al., 2009; Riaz et al., 2011;

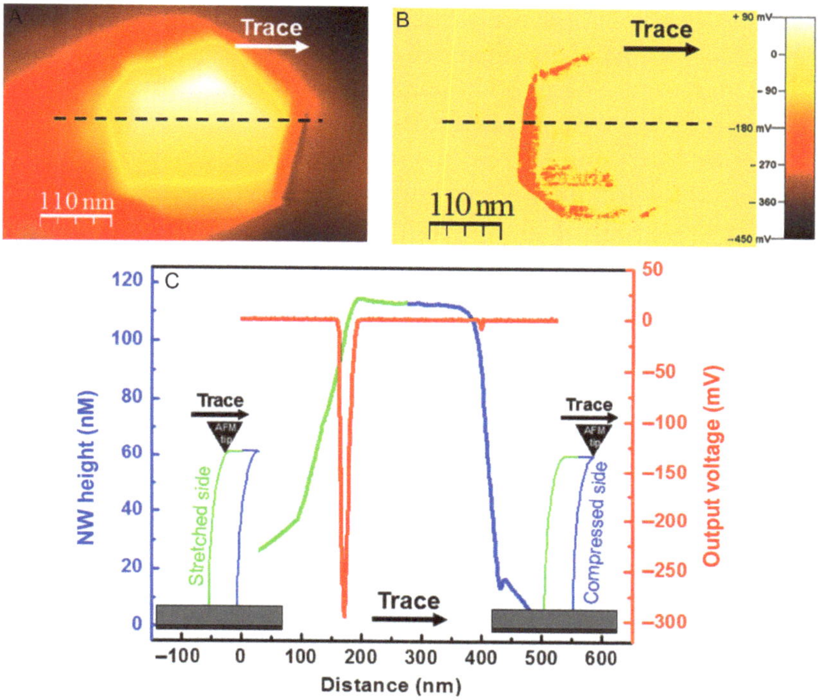

Fig. 14 Current asymmetry in the c-AFM measurements of a GaN NW (A) NW topography; (B) current map, showing the current arising in one side; (C) line scan of the current map and topography showing the current arising on the stretched side of the NW, and subsiding on the compressed side. *Reproduced from Gogneau, N., Chrétien, P., Galopin, E., Guilet, S., Travers, L., Harmand, J.-C., Houzé, F., 2014b. Impact of the GaN nanowire polarity on energy harvesting. Appl. Phys. Lett. 104 (21), 213105 with permission from AIP.*

Scrymgeour and Hsu, 2008). Generally, the tip is scanned or indented over vertically aligned NWs, grown on a conductive substrate (Fig. 14). Since current conduction takes place in this setup, electrical contacts to the NWs play a significant role in understanding the measurement results, and often there is a Schottky contact between the NW and the conductive AFM tip. As shown in Fig. 14 as the tip is scanned across the tip of the vertical NW, one side of the NW is strained while the other is compressed. The direction of the generated piezoelectric voltage is opposite in the two cases. Due to the piezotronic effect, in one instance the Schottky barrier between the NW and tip is lowered, while in the other it is increased; correspondingly, in the first case current flow (piezo-generated) is more efficient, while in the latter, significantly less current is measured. Lu et al. have measured current generation from both n- and p-type ZnO NWs. It was found that the measured voltage in the two cases is of opposite polarity, and that it originates from different sides of the NW (strained/compressed)—in accordance with the explanation that the piezotronic effect dominates the measurement.

Wang et al. (2014) have examined the influence of doping concentration in the range 10^{17}–10^{19} cm^{-3} on the performance of NW-based nanogenerators and piezotronic junctions. They have found that for nanogenerator operation, where there is a need to generate voltage and induce current flow, output current voltage was not a monotonic function of the doping, but that there is an optimal value of doping mitigating the decrease in resistivity on the one hand side, and screening of the piezoelectric potential on the other. For piezotronic applications, where the main effect is changing the Schottky barrier height, they have found that the lowest concentration yields the highest electromechanical gauge factor. This result stems from the increased depletion region associated with lower doping levels. A notable distinction between this method and PFM is the need to generate significant stresses (by bending), while due to the AFM deflection sensitivity (pm range) horizontal NWs can be examined as well as vertical.

4.3 Directly Measured Piezoelectric Properties of Semiconducting Nanowires

Both AFM-based methods open up an avenue for basic studies of NW piezoelectricity, however, the motivation for piezoelectric work on semiconductor NWs is mechanical energy harvesting and piezo-photo-tronic applications, and direct piezoelectric performance was frequently examined. Two configurations are usually considered: vertical (Fig. 15A), where the NWs are compressed and released to excite the piezoelectricity (Choi et al., 2009, 2010; Jamond et al., 2016; Xu et al., 2010), and horizontal (Fig. 15B), where the NWs are lying on a flexible substrate and are compressed/strained in accordance with the substrate (Xu et al., 2010; Yang et al., 2009; Zhu et al., 2010). Notably, in most of these cases, the NW experiences uniform strain/compression, unlike the conductive AFM case, where the NW experiences nearly pure bending. Furthermore, through rational design of growth and contacting, enhancement of the output voltage from mV to V range (serial connection of lateral devices), and current from nA to μA range (vertical array devices) was achieved.

The potential of piezoelectric semiconductor NWs for realizing further advanced applications such as piezoelectric modulated field effect transistors (FETs) (Fei et al., 2009; Kwon et al., 2008), or piezotronic logic (Wu et al., 2010), have also been demonstrated. Kwon et al. have shown that through piezotronic modulation of the gate-oxide-semiconductor junction, the

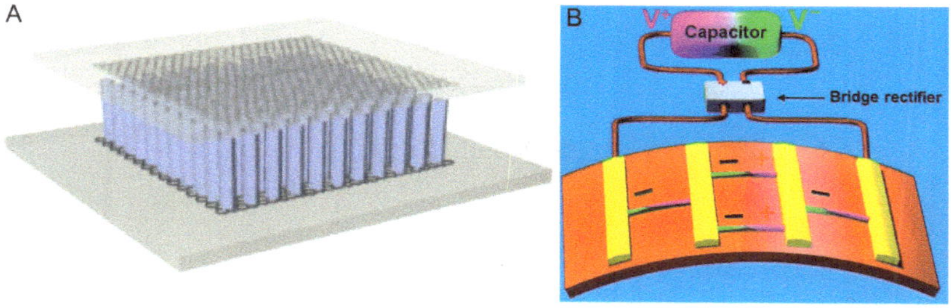

Fig. 15 Connecting NWs for energy applications. (A) parallel connections of a NW array, maximizing generated current. (B) serial connection of NWs, maximizing generated voltage. *Panel (A) adapted from Choi, D., Choi, M.-Y., Choi, W.M., Shin, H.-J., Park, H.-K., Seo, J.-S., Park, J., Yoon, S.M., Chae, S.J., Lee, Y.H., et al., 2010. Fully rollable transparent nanogenerators based on graphene electrodes. Adv. Mater. 22 (19), 2187–2192 with permission from Wiley. Panel (B) adapted from Zhu, G., Yang, R., Wang, S., Wang, Z.L., 2010. Flexible high-output nanogenerator based on lateral ZnO nanowire array. Nano Lett. 10 (8), 3151–3155 with permission from ACS.*

effective carrier mobility increases 4 times for 1.5% strain of the NW FET. They have ruled out piezoresistive effects by demonstrating an opposite effect on top- and bottom-gated devices, indicating that the crystal orientations determines the performance, and not the strain in-itself. Wu et al. (2010) used horizontal ZnO NWs located on both sides of a flexible substrate to realize a logic inverter. Bending the substrate results in compression of NWs on one side, and strain of the NW on the other. Due to the piezotronic effect, one Schottky barrier is reduce while the other increased leading to inverter operation (Fig. 16).

5. PIEZOELECTRICITY IN SEMICONDUCTOR NANOWIRES FOR ENERGY APPLICATIONS

5.1 II–VI Nanowires

Owing to a variety in synthesis methods (Leprince-Wang, 2015; Wang, 2004) and outputs (Gao et al., 2005; Wang, 2004) (NWs, nanoribbons) ZnO was the first semiconductor nanostructure to be vastly studied for nanoscale piezoelectric applications (Kong and Wang, 2003; Wang et al., 2004). Other members of the II–VI family such as CdSe, CdS, and ZnS were examined as well (Dong et al., 2012; Lin et al., 2008b; Rai et al., 2015). These materials generally have large, direct, band-gaps, and higher piezoelectric coefficients compared to III–Vs (also of opposite sign) (Arlt and Quadflieg, 1968; Miura et al., 1981), making them attractive candidates for piezotronic applications.

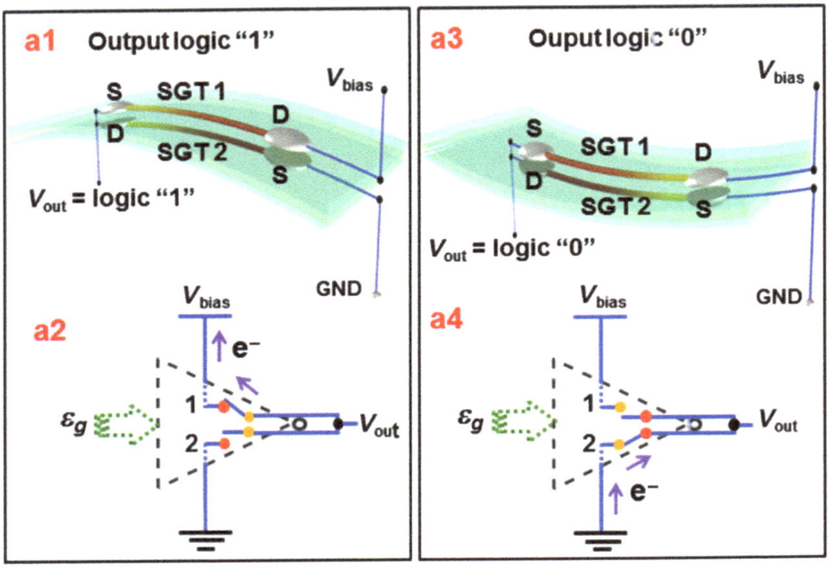

Fig. 16 Schematic of the two sided ZnO NW-based logic. The complementary response of the NWs on the opposite sides of the substrate, allows realizing an inverter operation. The stretched NW experiences reduced barrier and therefore when the device is bent upwards, the top NW is conductive (a1 and a2); when the device is bent downwards, the bottom NW is conductive (a3 and a4) *Reproduced from Wu, W., Wei, Y., Wang, Z.L., 2010. Strain-gated piezotronic logic nanodevices. Adv. Mater. 22 (42), 4711–4715 with permission from Wiley.*

5.1.1 ZnO NWs

ZnO NWs have been widely reported for their use in various energy applications, including as piezoelectric nanogenerators for mechanical energy harvesting. ZnO possesses good piezoelectric properties at nanoscale (Wang and Song, 2006; Zhang and al., 2003), which in some cases can be comparable to lead-free piezoceramic oxides. Other attractive properties include ease of nano-synthesis (Boughey et al., 2016), being environment friendly (Wang and Wu, 2012), biocompatible (Li et al., 2008), lightweight (Briscoe et al., 2012), scalable and low-cost (Greene et al., 2003), with good mechanical (Gao et al., 2009), and thermal stability (Zhou and Wong, 2008). In most cases, ZnO nanowires are grown on a conducting substrate which serves as the bottom electrode, and a top electrode is separately applied postgrowth. The device is then subjected to mechanical excitation and the resulting piezo-generated charge is collected to drive an external circuit. The presence of a substrate may introduce issues related to flexibility of the device and/or delamination of NWs during operation. To overcome this issue, ZnO nanowire-based nanogenerators have been developed where a polymer is introduced into the device design to improve mechanical stability as well as flexibility. Such device designs include dispersions of ZnO NWs within a soft polymeric matrix, however, this has the disadvantage of randomized polarization directions of the individual ZnO NWs. As a matter of fact, fatigue performance of ZnO-based nanogenerators is often not reported in the literature, as the material is additionally prone to environmental and mechanical degradation over time. However, studies where ZnO nanowires have been grown within nanoporous polycarbonate templates, and then incorporated into nanogenerators while still embedded and aligned, have resulted in long and reliable energy harvesting device lifetimes (Boughey et al., 2016; Ou et al., 2016). Table 1 summarizes key ZnO nanowire-based nanogenerators that have been reported over the years, highlighting inherent advantages and disadvantages of each.

5.1.2 Other II–VIs

An early demonstration of piezoelectric current generation from a 150–100 nm CdS NWs was published by Lin et al. in 2008 (Lin et al., 2008b). Using the conductive AFM method, different responses were achieved on the stretched and compressed sides of the NW, confirming piezoelectric dominated behavior. The output current was found to increase when pure WZ NWs were examined, compared to WZ/ZB mixed phase NWs. In a later work, the piezotronic effect was also demonstrated on CdS and CdSe NWs (Dai et al., 2017; Lin et al., 2008a), and energy harvesting with CdTe NWs (Hou et al., 2013).

5.2 III–V Nanowires

5.2.1 III-Nitride Nanowires

Alongside ZnO, III-Ns, and GaN in particular is the most widely studied piezoelectric NW material. The piezoelectric coefficients of GaN and AlN (about 3 and 5 pm/V (Guy et al., 1999)) are smaller than those of ZnO (10–12 pm/V), thus making them less attractive for energy harvesting applications (Fig. 17). However the large range of visible range band-gap tunability provided by III-N alloying, suggests that combining piezotronic and photonic applications (coined as piezo-phototronics) is worthwhile. The effect of strain on the performance of III-N-based optical devices has been examined (Du et al., 2015; Peng et al., 2015a,b; Song et al., 2017). As mentioned in Section 2.2, polarization-induced internal fields

Table 1 A Brief Summary of Different ZnO NW Structure Designs for Piezoelectric Nanogenerator Fabrication

NG Design	Type	Examples	Advantages	Disadvantages
Substrate based	ZnO NWs + rigid substrate	ZnO NWs on indium tin oxide (ITO)-coated glass substrate (Rivera et al., 2013; Scrymgeour and Hsu, 2008; Zhu et al., 2012) or Au-coated Si wafer (Han et al., 2013; Tian et al., 2011; Vayssieres, 2003; Xu et al., 2010)	Easy to obtain dense and well-aligned NWs	Not flexible and bendable; NWs are short (<4 μm), leading to low power output; reduced strain levels due to the brittleness of ZnO (Cauda et al., 2015)
	ZnO NWs + flexible substrate	ZnO NWs on ITO-coated polyester substrate (Briscoe and Dunn, 2015; Briscoe et al., 2012, 2013; Unalan et al., 2008), polyimide film (Hu et al., 2012), plastic film (Gao et al., 2007), or paper substrate (Nour et al., 2014)	Flexible and bendable; easy to obtain dense and well-aligned NWs	NWs are short, leading to low power output; prone to delamination when bent; reduced strain levels (Cauda et al., 2015)
Substrate less	ZnO NWs + dielectric polymer	ZnO NWs + poly(dimethylsiloxane) (PDMS) (Cauda et al., 2015; Shin et al., 2014), or poly(methyl methacrylate) (PMMA)(Hu et al., 2010)	Flexible, bendable, and robust	Low power output due to randomly oriented ZnO NWs and low proportion of NWs inside the composite; less control on the ratio of ceramic and polymer; not scalable, repeatable, and reproducible
	ZnO NWs + piezoelectric polymer	ZnO NWs + polyvinylidene fluoride (PVDF)(Lee et al., 2012)	Flexible, bendable, and robust	Randomly oriented ZnO NWs has lower piezoelectric coefficient than PVDF; this composite does not improve the overall power output
Template-assisted	ZnO NWs + polymeric template	ZnO NWs + polycarbonate (PC) template (Boughey et al., 2016; Ou et al., 2016; Prună et al., 2012; Sharma et al., 2010; Stassi et al., 2015; Zhou and Wong, 2008)	Easy to obtain long and well-aligned NWs; flexible, bendable, and robust; superior mechanical stability	Fatigue endurance limit relatively low output power due to the lower proportion of ZnO NWs

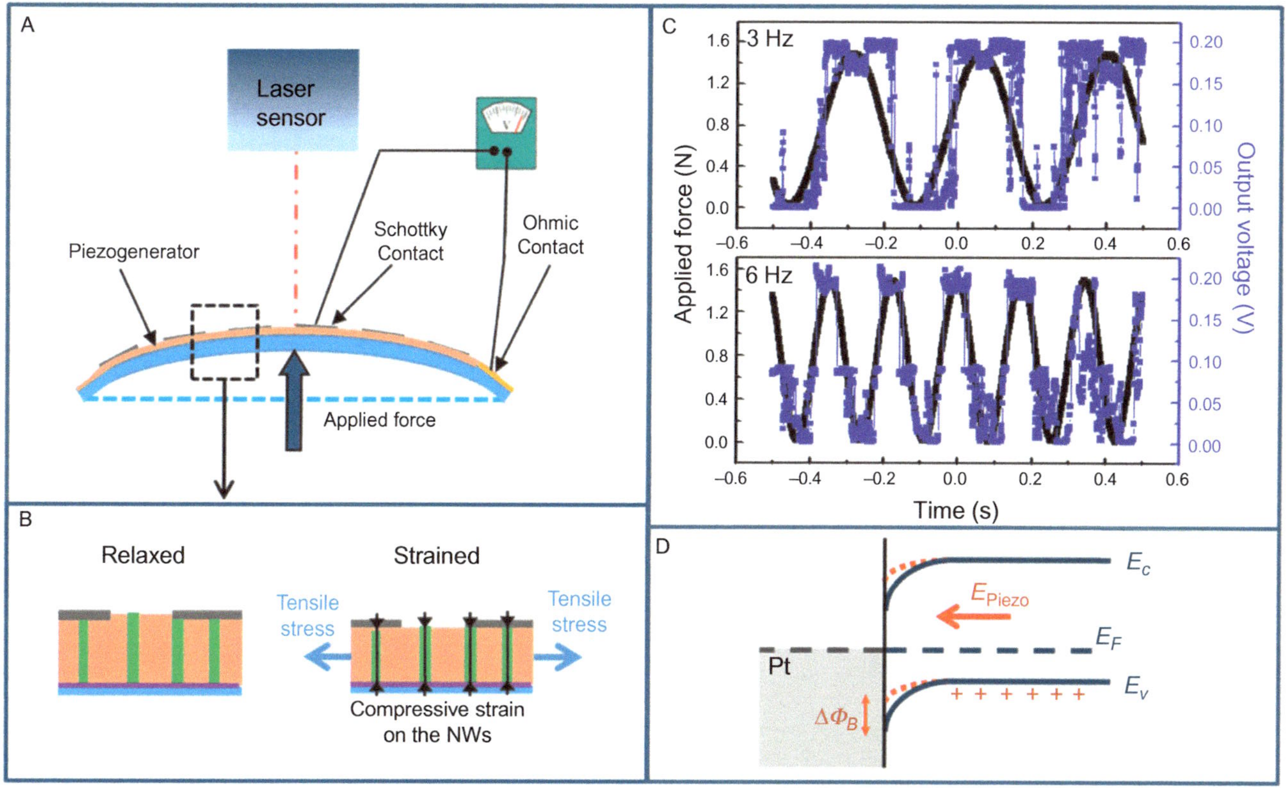

Fig. 17 Schematic and operation of a GaN array-based nanogenerator. (A) Schematic of the device and experimental setup; (B) schematic of the mode of operation, note that the NWs are compressed through transverse stretching; (C) nanogenerator output voltage; (D) schematic band diagram for the Schottky contact, which is responsible for the effect through modulation of the charges in the depletion region. *Reproduced from Jamond, N., Chrétien, P., Houzé, F., Lu, L., Largeau, L., Maugain, O., Travers, L., Harmand, J.-C., Glas, F., Lefeuvre, E., et al., 2016. Piezo-generator integrating a vertical array of GaN nanowires. Nanotechnology 27 (32), 325403 with permission from IOP.*

result in reduced emission efficiency. Peng et al. (2015a) have demonstrated the control of InGaN/GaN multiquantum well NW LED by application of stress. Subsequently, strain has been shown to improve the emission of bulk InGaN/GaN-based LEDs (Tsai et al., 2016).

Mechanical energy harvesting has also been successfully demonstrated using various types of III-N NWs, exploring issues such as doping and geometrical design and their influence on energy harvesting efficiency (Gogneau et al., 2014a; Huang et al., 2010; Jamond et al., 2016; Lin et al., 2011; Liu et al., 2016; Wang et al., 2014).

5.2.2 *Non-nitride III–V Nanowires*

The piezoelectric properties of III-As and III-P have not received significant attention, despite being known and utilized for electromechanical applications (Klier and Nedbal, 1994; Nedbal and Klier, 1995). The main reasons for that are the lower piezoelectric coefficients, and non-polar technology ({001} and {110} cubic orientations in zinc-blende), prevailing non-nitride III–V bulk electronic and optoelectronic technology (Singh, 2007; Wood and Jena, 2007). Nonetheless, the various nanostructures vs bulk distinctions described in Section 3, apply to non-nitride III–Vs as well (Calahorra and Kar-Narayan, 2018). Indeed high piezoelectric coefficients have been predicted for III–Vs in the wurtzite phase (Al-Zahrani et al., 2015), which is relevant for III–V NWs. Except from PFM work mentioned earlier (Calahorra et al., 2017a), piezoelectric generation from III–V were examined (Lysak et al., 2016; Soshnikov et al., 2011), pointing out the role of stress distribution from an upper contact to a single NW piezo-element, in increasing the piezoelectric efficiency. Recently, the piezoelectric properties of wurtzite GaAs were examined using the c-AFM apparatus (Alekseev et al., 2018), demonstrating the piezo-phototronic effect in increasing generated current under illumination.

6. SUMMARY

The piezoelectric properties of semiconducting nanowires have many applications in energy harvesting and sensing devices based on the interdependence of the mechanical, electrical, and optical properties of these materials. In most cases, nanowires have been found to exhibit superior electromechanical properties when compared to the bulk of the material, and these can be further tuned by controlling the growth method. The crystalline phases present in these semiconducting nanowires play a strong role in determining the nature of piezoelectricity in these systems, and hence advances in growth and characterization of these nanowires have paved the way for the development of novel devices based on this effect. Importantly, a greater level of understanding and control of piezoelectric properties of semiconducting nanowires will serve to underpin advances in the emerging fields of piezotronics and photo-piezotronics, as well as in the development of energy harvesters for small-power applications. Materials and device engineering at the nanoscale therefore offer new routes toward exploiting the piezoelectric properties of semiconducting nanowires for a wide range of energy applications, and there remains plenty of scope for innovation in this area.

REFERENCES

Agrawal, R., Espinosa, H.D., 2011. Giant piezoelectric size effects in zinc oxide and gallium nitride nanowires. A first principles investigation. Nano Lett. 11 (2), 786–790.

Agrawal, R., Peng, B., Gdoutos, E.E., Espinosa, H.D., 2008. Elasticity size effects in ZnO nanowires—a combined experimental-computational approach. Nano Lett. 8 (11), 3668–3674.

Al-Zahrani, H.Y.S., Pal, J., Migliorato, M.A., Tse, G., Yu, D., 2015. Piezoelectric field enhancement in III-V core-shell nanowires. Nano Energy 14, 382–391.

Alekseev, P.A., Sharov, V.A., Geydt, P., Dunaevskiy, M.S., Lysak, V.V., Cirlin, G.E., Reznik, R.R., Khrebtov, A.I., Soshnikov, I.P., Lhderanta, E., 2018. Piezoelectric current generation in wurtzite GaAs nanowires. Phys. Status Solidi Rapid Res. Lett. 12 (1), 1700358. ISSN 1862-6270. https://doi.org/10.1002/pssr.201700358.

Ambacher, O., Smart, J., Shealy, J.R., Weimann, N.G., Chu, K., Murphy, M., Schaff, W.J., Eastman, L.F., Dimitrov, R., Wittmer, L., et al., 1999. Two-dimensional electron gases induced by spontaneous and piezoelectric polarization charges in N-and Ga-face AlGaN/GaN heterostructures. J. Appl. Phys. 85 (6), 3222–3233.

Arlt, G., Quadflieg, P., 1968. Piezoelectricity in iii-v compounds with a phenomenological analysis of the piezoelectric effect. Phys. Status Solidi B 25 (1), 323–330.

Boughey, F.L., Davies, T., Datta, A., Whiter, R.A., Sahonta, S.-L., Kar-Narayan, S., 2016. Vertically aligned zinc oxide nanowires electrodeposited within porous polycarbonate templates for vibrational energy harvesting. Nanotechnology 27 (28), 28LT02. ISSN 0957-4484. https://doi.org/10.1088/0957-4484/27/28/28LT02. http://stacks.iop.org/0957-4484/27/i=28/a=28LT02?key=crossref.229fcb85100c2f8b963679491004fa4f.

Briscoe, J., Dunn, S., 2015. Piezoelectric nanogenerators—a review of nanostructured piezoelectric energy harvesters. Nano Energy 14, 15–29. ISSN 22112855. https://doi.org/10.1016/j.nanoen.2014.11.059.

Briscoe, J., Stewart, M., Vopson, M., Cain, M., Weaver, P.M., Dunn, S., 2012. Nanostructured p-n junctions for kinetic-to-electrical energy conversion. Adv. Energy Mater. 2 (10), 1261–1268. ISSN 16146832. https://doi.org/10.1002/aenm.201200205.

Briscoe, J., Jalali, N., Woolliams, P., Stewart, M., Weaver, P.M., Cain, M., Dunn, S., 2013. Measurement techniques for piezoelectric nanogenerators. Energy Environ. Sci. 6 (10), 3035. ISSN 1754-5692. https://doi.org/10.1039/c3ee41889h.

Calahorra, Y., Kar-Narayan, S., 2018. Piezoelectricity in non-nitride III-V nanowires: challenges and opportunities. J. Mater. Res. 33, 611–624. https://doi.org/10.1557/jmr.2018.29.

Calahorra, Y., Ritter, D., 2013. Surface depletion effects in semiconducting nanowires having a non-uniform radial doping profile. J. Appl. Phys. 114 (12), 124310.
Calahorra, Y., Mendels, D., Epstein, A., 2014. Rigorous analysis of image force barrier lowering in bounded geometries: application to semiconducting nanowires. Nanotechnology 25 (14), 145203.
Calahorra, Y., Yalon, E., Ritter, D., 2015. On the diameter dependence of metal-nanowire Schottky barrier height. J. Appl. Phys. 117 (3), 034308.
Calahorra, Y., Guan, X., Halder, N.N., Smith, M., Cohen, S., Ritter, D., Penuelas, J., Kar-Narayan, S., 2017a. Exploring piezoelectric properties of III-V nanowires using piezoresponse force microscopy. Semicond. Sci. Technol. 32 (7), 074006.
Calahorra, Y., Smith, M., Datta, A., Benisty, H., Kar-Narayan, S., 2017b. Mapping piezoelectric response in nanomaterials using a dedicated non-destructive scanning probe technique. Nanoscale 9 (48), 19290–19297.
Cao, H.L., Qian, X.F., Gong, Q., Du, W.M., Ma, X.D., Zhu, Z.K., 2006. Shape and size controlled synthesis of nanometre ZnO from a simple solution route at room temperature. Nanotechnology 17 (15), 3632–3636. ISSN 0957-4484. https://doi.org/10.1088/0957-4484/17/15/002.
Cauda, V., Stassi, S., Lamberti, A., Morello, M., Fabrizio Pirri, C., Canavese, G., 2015. Leveraging ZnO morphologies in piezoelectric composites for mechanical energy harvesting. Nano Energy 18, 212–221. ISSN 22112855. https://doi.org/10.1016/j.nanoen.2015.10.021.
Cha, S., Kim, S.M., Kim, H., Ku, J., Sohn, J.I., Park, Y.J., Song, B.G., Jung, M.H., Lee, E.K., Choi, B.L., et al., 2011. Porous PVDF as effective sonic wave driven nanogenerators. Nano Lett. 11 (12), 5142–5147.
Chen, X., Xu, S., Yao, N., Shi, Y., 2010. 1.6 V nanogenerator for mechanical energy harvesting using PZT nanofibers. Nano Lett. 10 (6), 2133–2137 ISSN 15306984. https://doi.org/10.1021/nl100812k.
Chen, L.-Y., Yin, Y.-T., Chen, C.H., Chiou, J.-W., 2011. Influence of polyethyleneimine and ammonium on the growth of ZnO nanowires by hydrothermal method. J. Phys. Chem. C 115 (43), 20913–20919. ISSN 1932-7447. https://doi.org/10.1021/jp2056199.
Cheng, B., Samulski, E.T., 2004. Hydrothermal synthesis of one-dimensional ZnO nanostructures with different aspect ratios. Chem. Commun. (8), 986. ISSN 1359-7345. https://doi.org/10.1039/b316435g.
Choi, M.-Y., Choi, D., Jin, M.-J., Kim, I., Kim, S.-H., Choi, J.-Y., Lee S.Y., Kim, J.M., Kim, S.-W., 2009. Mechanically powered transparent flexible charge-generating nanodevices with piezoelectric ZnO nanorods. Adv. Mater. 21 (21), 2185–2189.
Choi, D., Choi, M.-Y., Choi, W.M., Shin, H.-J., Park, H.-K., Seo, J.-S., Park, J., Yoon, S.M., Chae, S.J., Lee, Y.H., et al., 2010. Fully rollable transparent nanogenerators based on graphene electrodes. Adv. Mater. 22 (19), 2187–2192.
Curie, J., Curie, P., 1882. Phénomènes électriques des cristaux hémièdres à faces inclinées. J. Phys. Theor. Appl. 1 (1), 245–251.
Dai, G., Zou, H., Wang, X., Zhou, Y., Wang, P., Ding, Y., Zhang, Y., Yang, J., Wang, Z.L., 2017. Piezo-phototronic effect enhanced responsivity of photon sensor based on composition-tunable ternary $CdS_x Se_{1-x}$ nanowires. ACS Photon. 4 (10), 2495–2503.
Dong, L., Niu, S., Pan, C., Yu, R., Zhang, Y., Wang, Z.L., 2012. Piezo-phototronic effect of CdSe nanowires. Adv. Mater. 24 (40), 5470–5475.
Du, C., Jiang, C., Zuo, P., Huang, X., Pu, X., Zhao, Z., Zhou, Y., Li, L., Chen, H., Hu, W., et al., 2015. Piezo-phototronic effect controlled dual-channel visible light communication (PVLC) using InGaN/GaN multiquantum well nanopillars. Small 11 (45), 6071–6077.

Espinosa, H.D., Bernal, R.A., Minary-Jolandan, M., 2012. A review of mechanical and electromechanical properties of piezoelectric nanowires. Adv. Mater. 24 (34), 4656–4675. ISSN 09359648. https://doi.org/10.1002/adma.201104810.

Fan, H.J., Lee, W., Hauschild, R., Alexe, M., Le Rhun, G., Scholz, R., Dadgar, A., Nielsch, K., Kalt, H., Krost, A., et al., 2006. Template-assisted large-scale ordered arrays of Zno pillars for optical and piezoelectric applications. Small 2 (4), 561–568.

Fei, P., Yeh, P.-H., Zhou, J., Xu, S., Gao, Y., Song, J., Gu, Y., Huang, Y., Wang, Z.L., 2009. Piezoelectric potential gated field-effect transistor based on a free-standing ZnO wire. Nano Lett. 9 (10), 3435–3439.

Gao, P.X., Ding, Y., Mai, W., Hughes, W.L., Lao, C., Wang, Z.L., 2005. Conversion of zinc oxide nanobelts into superlattice-structured nanohelices. Science 309 (5741), 1700–1704.

Gao, P.X., Song, J., Liu, J., Wang, Z.L., 2007. Nanowire piezoelectric nanogenerators on plastic substrates as flexible power sources for nanodevices. Adv. Mater. 19 (1), 67–72.

Gao, Z., Ding, Y., Lin, S., Hao, Y., Wang, Z.L., 2009. Dynamic fatigue studies of ZnO nanowires by in-situ transmission electron microscopy. Phys. Status Solidi Rapid Res. Lett. 3 (7–8), 260–262. ISSN 18626254. https://doi.org/10.1002/pssr.200903235.

Gerngross, M.-D., Sprincean, V., Leisner, M., Carstensen, J., Föll, H., Tiginyanu, I., 2011. Porous InP as piezoelectric component in magneto-electric composite sensors. ECS Trans. 35 (8), 67–72.

Gogneau, N., Chrétien, P., Galopin, E., Guilet, S., Travers, L., Harmand, J.-C., Houzé, F., 2014a. GaN nanowires for piezoelectric generators. Phys. Status Solidi Rapid Res. Lett. 8 (5), 414–419.

Gogneau, N., Chrétien, P., Galopin, E., Guilet, S., Travers, L., Harmand, J.-C., Houzé, F., 2014b. Impact of the GaN nanowire polarity on energy harvesting. Appl. Phys. Lett. 104 (21), 213105.

Gordon, M.J., Baron, T., Dhalluin, F., Gentile, P., Ferret, P., 2009. Size effects in mechanical deformation and fracture of cantilevered silicon nanowires. Nano Lett. 9 (2), 525–529.

Greene, L.E., Law, M., Goldberger, J., Kim, F., Johnson, J.C., Zhang, Y., Saykally, R.J., Yang, P., 2003. Low-temperature wafer-scale production of ZnO nanowire arrays. Angew. Chem. Int. Ed. 42 (26), 3031–3034. ISSN 14337851. https://doi.org/10.1002/anie.200351461.

Gruverman, A., Kalinin, S.V., et al., 2006. Piezoresponse force microscopy and recent advances in nanoscale studies of ferroelectrics. J. Mater. Sci. 41 (1), 107–116.

Guy, I.L., Muensit, S., Goldys, E., 1999. Extensional piezoelectric coefficients of gallium nitride and aluminum nitride. Appl. Phys. Lett. 75 (26), 4133–4135.

Han, Z., Li, S., Chu, J., Chen, Y., 2013. Controlled growth of well-aligned ZnO nanowire arrays using the improved hydrothermal method. J. Semicond. 34 (6), 063002. ISSN 1674-4926. https://doi.org/10.1088/1674-4926/34/6/063002.

Hou, T.-C., Yang, Y., Lin, Z.H., Ding, Y., Park, C., Pradel, K.C., Chen, L.-J., Wang, Z.L., 2013. Nanogenerator based on zinc blende CdTe micro/nanowires. Nano Energy 2 (3), 387–393.

Hu, J., Beck, R.G., Deng, T., Westervelt, R.M., Maranowski, K.D., Gossard, A.C., Whitesides, G.M., 1997. Using soft lithography to fabricate GaAs/AlGaAs heterostructure field effect transistors. Appl. Phys. Lett. 71 (14), 2020–2022.

Hu, Y., Zhang, Y., Xu, C., Zhu, G., Wang, Z.L., 2010. High-output nanogenerator by rational unipolar assembly of conical nanowires and its application for driving a small liquid crystal display. Nano Lett. 10 (12), 5025–5031. ISSN 1530-6984. https://doi.org/10.1021/nl103203u.

Hu, Y., Lin, L., Zhang, Y., Wang, Z.L., 2012. Replacing a battery by a nanogenerator with 20 V output. Adv. Mater. 24 (1), 110–114. ISSN 09359648. https://doi.org/10.1002/adma.201103727.

Huang, C.-T., Song, J., Lee, W.-F., Ding, Y., Gao, Z., Hao, Y., Chen, L.-J., Wang, Z.L., 2010. Gan nanowire arrays for high-output nanogenerators. J. Am. Chem. Soc. 132 (13), 4766–4771.

Jalali, N., Woolliams, P., Stewart, M., Weaver, P.M., Cain, M.G., Dunn, S., Briscoe, J., 2014. Improved performance of p-n junction-based ZnO nanogenerators through CuSCN-passivation of ZnO nanorods. J. Mater. Chem. A 2 (28), 10945. ISSN 2050-7488. https://doi.org/10.1039/c4ta01714e.

Jalali, N., Briscoe, J., Tan, Y.Z., Woolliams, P., Stewart, M., Weaver, P.M., Cain, M.G., Dunn, S., 2015. ZnO nanorod surface modification with PDDA/PSS Bi-layer assembly for performance improvement of ZnO piezoelectric energy harvesting devices. J. Sol-Gel Sci. Technol. 73 (3), 544–549. ISSN 0928-0707. https://doi.org/10.1007/s10971-014-3512-4.

Jamond, N., Chrétien, P., Houzé, F., Lu, L., Largeau, L., Maugain, O., Travers, L., Harmand, J.-C., Glas, F., Lefeuvre, E., et al., 2016. Piezo-generator integrating a vertical array of GaN nanowires. Nanotechnology 27 (32), 325403.

Kalinin, A., Atepalikhin, V., Pakhomov, O., Kholkin, A.L., Tselev, A., 2018. An atomic force microscopy mode for nondestructive electromechanical studies and its application to diphenylalanine peptide nanotubes. Ultramicroscopy 185, 49–54

Kim, M.-H., Schubert, M.F., Dai, Q., Kim, J.K., Schubert, E.F., Piprek, J., Park, Y., 2007. Origin of efficiency droop in GaN-based light-emitting diodes. Appl. Phys. Lett. 91 (18), 183507.

Klier, E., Nedbal, J., 1994. Piezoelectric resonators of InP:Fe. Czechoslov. J. Phys. 44 (6), 575–584.

Kong, X.Y., Wang, Z.L., 2003. Spontaneous polarization-induced nanohelixes, nanosprings, and nanorings of piezoelectric nanobelts. Nano Lett. 3 (12), 1625–1631.

Kusaka, M., 1978. Electrical properties of metal-piezoelectric semiconductor interface under stress. Surf. Sci. 78 (1), 209–219.

Kusaka, M., Kanakura, M., Okazaki, S., 1974. Interface properties and Schottky barrier on the polar surfaces of CdS and GaP srystals. Jpn. J. Appl. Phys. 13 (S2), 437.

Kwon, S.-S., Hong, W.-K., Jo, G., Maeng, J., Kim, T.-W., Song, S., Lee, T., 2008. Piezoelectric effect on the electronic transport characteristics of zno nanowire field-effect transistors on bent flexible substrates. Adv. Mater. 20 (23), 4557–4562.

Łagowski, J., Gatos, H.C., 1972. Photomechanical effect in noncentrosymmetric semiconductors-CdS. Appl. Phys. Lett. 20 (1), 14–16.

Łagowski, J., Gatos, H.C., 1974. Photomechanical vibration of thin crystals of polar semiconductors. Surface Sci. 45 (2), 353–370.

Lee, M., Chen, C.-Y., Wang, S., Cha, S.N., Park, Y.J., Kim, J.M., Chou, L.-J., Wang, Z.L., 2012. A hybrid piezoelectric structure for wearable nanogenerators. Adv. Mater. 24 (13), 1759–1764. ISSN 09359648. https://doi.org/10.1002/adma.201200150.

Lefki, K., Dormans, G.J.M., 1994. Measurement of piezoelectric coefficients of ferroelectric thin films. J. Appl. Phys. 76 (3), 1764–1767.

Lehmann, S., Wallentin, J., Jacobsson, D., Deppert, K., Dick, K.A., 2013. A general approach for sharp crystal phase switching in InAs, GaAs, InP, and GaP nanowires using only group V flow. Nano Lett. 13 (9), 4099–4105.

Leprince-Wang, Y., 2015. ZnO nanostructure synthesis. In: Leprince-Wang, Y. (Ed.), Piezoelectric ZnO Nanostructure for Energy Harvesting Wiley Online Library, pp. 21–47.

Li, Z., Yang, R., Yu, M., Bai, F., Li, C., Wang, Z.L., 2008. Cellular level biocompatibility and biosafety of ZnO nanowires. J. Phys. Chem. C 112 (51), 20114–20117. ISSN 19327447. https://doi.org/10.1021/jp808878p.

Lin, Y.-F., Song, J., Ding, Y., Lu, S.-Y., Wang, Z.L., 2008a. Alternating the output of a CdS nanowire nanogenerator by a white-light-stimulated optoelectronic effect. Adv. Mater. 20 (16), 3127–3130.

Lin, Y.-F., Song, J., Ding, Y., Lu, S.-Y., Wang, Z.L., 2008b. Piezoelectric nanogenerator using CdS nanowires. Appl. Phys. Lett. 92 (2), 022105.
Lin, L., Lai, C.-H., Hu, Y., Zhang, Y., Wang, X., Xu, C., Snyder, R.L., Chen, L.-J., Wang, Z.L., 2011. High output nanogenerator based on assembly of GaN nanowires. Nanotechnology 22 (47), 475401.
Liu, B., Zeng, H.C., 2003. Hydrothermal synthesis of ZnO nanorods in the diameter regime of 50 nm. J. Am. Chem. Soc. 125 (15), 4430–4431. ISSN 0002-7863. https://doi.org/10.1021/ja0299452.
Liu, G., Zhao, S., Henderson, R.D.E., Leonenko, Z., Abdel-Rahman, E., Mi, Z., Ban, D., 2016. Nanogenerators based on vertically aligned InN nanowires. Nanoscale 8 (4), 2097–2106.
Lu, M.-P., Song, J., Lu, M.-Y., Chen, M.-T., Gao, Y., Chen, L.-J., Wang, Z.L., 2009. Piezoelectric nanogenerator using p-type ZnO nanowire arrays. Nano Lett. 9 (3), 1223–1227.
Lucas, M., Mai, W., Yang, R., Wang, Z.L., Riedo, E., 2007. Aspect ratio dependence of the elastic properties of ZnO nanobelts. Nano Lett. 7 (5), 1314–1317. ISSN 1530-6984. https://doi.org/10.1021/nl070310g.
Lysak, V.V., Soshnikov, I.P., Lahderanta, E., Cirlin, G.E., 2016. Piezoelectric effect in wurtzite GaAs nanowires: growth, characterization, and electromechanical 3D modeling. Phys. Status Solidi A 213 (11), 3014–3019.
Minary-Jolandan, M., Bernal, R.A., Kuljanishvili, I., Parpoil, V., Espinosa, H.D., 2012. Individual GaN nanowires exhibit strong piezoelectricity in 3D. Nano Lett. 12 (2), 970–976.
Miura, M., Murata, H., Shiro, Y., Iishi, K., 1981. Ionicity scale and piezoelectricity of crystals with zincblende-and wurtzite-type structure. J. Phys. Chem. Solids 42 (10), 931–936.
Nedbal, J., Klier, E., 1995. Piezoelectric resonators of semiinsulating GaAs. Phys. Status Solidi A 148 (1), 329–340.
Nour, E.S., Sandberg, M.O., Willander, M., Nur, O., 2014. Handwriting enabled harvested piezoelectric power using zno nanowires/polymer composite on paper substrate. Nano Energy 9, 221–228. ISSN 22112855. https://doi.org/10.1016/j.nanoen.2014.07.014.
Onodera, A., Takes, M., 2012. Electronic ferroelectricity in II-VI semiconductor ZnO. In: Advances in Ferroelectrics InTech, pp. 231–255 (chapter 11).
Ottone, C., Bejtka, K., Chiodoni, A., Farías, V., Roppolo, I., Canavese, G., Stassi, S., Cauda, V., 2014. Comprehensive study of the templating effect on the ZnO nanostructure formation within porous hard membranes. New J. Chem. 38 (5), 2058. ISSN 1144-0546. https://doi.org/10.1039/c3nj01135f.
Ou, C., Sanchez-Jimenez, P.E., Datta, A., Boughey, F.L., Whiter, R.A., Sahonta, S.-L., Kar-Narayan, S., 2016. Template-assisted hydrothermal growth of aligned zinc oxide nanowires for piezoelectric energy harvesting applications. ACS Appl. Mater. Interfaces 8 (22), 13678–13683. ISSN 1944-8244. https://doi.org/10.1021/acsami.6b04041.
Peng, M., Li, Z., Liu, C., Zheng, Q., Shi, X., Song, M., Zhang, Y., Du, S., Zhai, J., Wang, Z.L., 2015a. High-resolution dynamic pressure sensor array based on piezo-phototronic effect tuned photoluminescence imaging. ACS Nano 9 (3), 3143–3150.
Peng, M., Liu, Y., Yu, A., Zhang, Y., Liu, C., Liu, J., Wu, W., Zhang, K., Shi, X., Kou, J., et al., 2015b. Flexible self-powered GaN ultraviolet photoswitch with piezo-phototronic effect enhanced on/off ratio. ACS Nano 10 (1), 1572–1579.
Podrezova, L.V., Cauda, V., Stassi, S., Cicero, G., Abdullin, Kh.A., Alpysbaeva, B.E., 2014. Properties of ZnO nanorods grown by hydrothermal synthesis on conductive layers. Cryst. Res. Technol. 49 (8), 599–605. ISSN 02321300. https://doi.org/10.1002/crat.201300372.
Prună, A., Pullini, D., Mataix, D.B., 2012. Influence of deposition potential on structure of ZnO nanowires synthesized in track-etched membranes. J. Electrochem. Soc. 159 (4), E92. ISSN 00134651. https://doi.org/10.1149/2.003205jes.

Rai, S.C., Wang, K., Ding, Y., Marmon, J.K., Bhatt, M., Zhang, Y., Zhou, W., Wang, Z.L., 2015. Piezo-phototronic effect enhanced UV/visible photodetector based on fully wide band gap type-II ZnO/ZnS core/shell nanowire array. ACS Nano 9 (6), 6419–6427.
Riaz, M., Song, J., Nur, O., Wang, Z.L., Willander, M., 2011. Study of the piezoelectric power generation of ZnO nanowire arrays grown by different methods. Adv. Funct. Mater. 21 (4), 628–633.
Rivera, V.F., Auras, F., Motto, P., Stassi, S., Canavese, G., Celasco, E., Bein, T., Onida, B., Cauda, V., 2013. Length-dependent charge generation from vertical arrays of high-aspect-ratio ZnO nanowires. Chem. A Eur. J. 19 (43), 14665–14674. ISSN 09476539. https://doi.org/10.1002/chem.201204429.
Schmidt-Mende, L., MacManus-Driscoll, J.L., 2007. ZnO - nanostructures, defects, and devices. Mater. Today 10 (5), 40–48. ISSN 13697021. https://doi.org/10.1016/S1369-7021(07)70078-0.
Scrymgeour, D.A., Hsu, J.W.P., 2008. Correlated piezoelectric and electrical properties in individual ZnO nanorods. Nano Lett. 8 (8), 2204–2209.
Sharma, S.K., Rammohan, A., Sharma, A., 2010. Templated one step electrodeposition of high aspect ratio n-type ZnO nanowire arrays. J. Colloid Interface Sci. 344 (1), 1–9. ISSN 00219797. https://doi.org/10.1016/j.jcis.2009.12.026.
Shin, S.-H., Kim, Y.h., Lee, M.H., Jung, J.-Y., Seol, J.H., Nah, J., 2014. Lithium-doped zinc oxide nanowires–polymer composite for high performance flexible piezoelectric nanogenerator. ACS Nano 8 (10), 10844–10850. ISSN 1936-0851. https://doi.org/10.1021/nn5046568.
Simpkins, B.S., Mastro, M.A., Eddy Jr., C.R., Pehrsson, P.E., 2008. Surface depletion effects in semiconducting nanowires. J. Appl. Phys. 103 (10), 104313.
Singh, J., 2007. Electronic and Optoelectronic Properties of Semiconductor Structures. Cambridge University Press.
Song, W., Wang, X., Xia, C., Wang, R., Zhao, L., Guo, D., Chen, H., Xiao, J., Su, S., Li, S., 2017. Improved photoresponse of a-axis GaN microwire/p-polymer hybrid photosensor by the piezo-phototronic effect. Nano Energy 33, 272–279.
Soshnikov, I.P., Afanasev, D.E., Petrov, V.A., Cirlin, G.E., Bouravlev, A.D., Samsonenko, Y.B., Khrebtov, A., Tanklevskaya, E.M., Seleznev, I.A., 2011. Piezoelectric effect in GaAs nanowires. Semiconductors 45 (8), 1082.
Stassi, S., Cauda, V., Ottone, C., Chiodoni, A., Pirri, C.F., Canavese, G., 2015. Flexible piezoelectric energy nanogenerator based on ZnO nanotubes hosted in a polycarbonate membrane. Nano Energy 13, 474–481. ISSN 22112855. https://doi.org/10.1016/j.nanoen.2015.03.024.
Sun, Y., Riley, D.J., Ashfbld, M.N.R., 2006. Mechanism of ZnO nanotube growth by hydrothermal methods on ZnO film-coated Si substrates. J. Phys. Chem. B 110 (31), 15186–15192. ISSN 15206106. https://doi.org/10.1021/jp062299z.
Sze, S.M., Ng, K.K., 2006. Physics of Semiconductor Devices. John Wiley & Sons.
Tamvakos, D., Lepadatu, S., Antohe, V.A., Tamvakos, A., Weaver, P.M., Piraux, L., Cain, M.G., Pullini, D., 2015. Piezoelectric properties of template-free electrochemically grown ZnO nanorod arrays. Appl. Surf. Sci. 356, 1214–1220.
Tian, J.-H., Hu, J., Li, S.-S., Zhang, F., Liu, J., Shi, J., Li, X., Tian, Z.-Q., Chen, Y., 2011. Improved seedless hydrothermal synthesis of dense and ultralong ZnO nanowires. Nanotechnology 22 (24), 245601. ISSN 0957-4484. https://doi.org/10.1088/0957-4484/22/24/245601.
Tiwary, N., Sarkar, R., Rao, V.R., Laha, A., 2016. Piezoresponse force microscopy (PFM) characterization of GaN nanowires grown by plasma assisted Molecular beam epitaxy (PA-MBE). In: 2016 Joint IEEE International Symposium on the Applications of Ferroelectrics, European Conference on Application of Polar Dielectrics, and Piezoelectric Force Microscopy Workshop (ISAF/ECAPD/PFM), pp. 1–4.

Tsai, S.-C., Lu, C.-H., Liu, C.-P., 2016. Piezoelectric effect on compensation of the quantum-confined Stark effect in InGaN/GaN multiple quantum wells based green light-emitting diodes. Nano Energy 28, 373–379.

Unalan, H.E., Hiralal, P., Rupesinghe, N., Dalal, S., Milne, W.I., Amaratunga, G.A.J., 2008. Rapid synthesis of aligned zinc oxide nanowires. Nanotechnology 19 (25), 255608. ISSN 0957-4484. https://doi.org/10.1088/0957-4484/19/25/255608. http://stacks.iop.org/0957-4484/19/i=25/a=255608?key=crossref.4cbe421a88d4ebc2dc06b30270349592.

Vayssieres, L., 2003. Growth of arrayed nanorods and nanowires of ZnO from aqueous solutions. Adv. Mater. 15 (5), 464–466. ISSN 09359648. https://doi.org/10.1002/adma.200390108.

Wang, Z.L., 2004. Zinc oxide nanostructures: growth, properties and applications. J. Phys.: Condens. Matter 16 (25), R829.

Wang, Z.L., 2007. Nanopiezotronics. Adv. Mater. 19 (6), 889–892.

Wang, Z.L., 2011. Nanogenerators for self-powered devices and systems. In: Georgia Institute of Technology. SMARTech Digital Repository. ISBN: 9781450780162, pp. 1–131. http://hdl.handle.net/1853/39262.

Wang, Z.L., Song, J., 2006. Piezoelectric nanogenerators based on zinc oxide nanowire arrays. Science (New York, N.Y.) 312 (5771), 242–246. ISSN 1095-9203. https://doi.org/10.1126/science.1124005.

Wang, Z.L., Wu, W., 2012. Nanotechnology-enabled energy harvesting for self-powered micro-/nanosystems. Phys. Status Solidi Rapid Res. Lett. 51 (47), 11700–11721. ISSN 14337851. https://doi.org/10.1002/anie.201201656.

Wang, Z.L., Kong, X.Y., Ding, Y., Gao, P., Hughes, W.L., Yang, R., Zhang, Y., 2004. Semiconducting and piezoelectric oxide nanostructures induced by polar surfaces. Adv. Funct. Mater. 14 (10), 943–956.

Wang, X., Zhou, J., Song, J., Liu, J., Xu, N., Wang, Z.L., 2006. Piezoelectric field effect transistor and nanoforce sensor based on a single ZnO nanowire. Nano Lett. 6 (12), 2768–2772. ISSN 15306984. https://doi.org/10.1021/nl061802g.

Wang, Y.-B., Wang, L.-F., Joyce, H.J., Gao, Q., Liao, X.-Z., Mai, Y.-W., Tan, H.H., Zou, J., Ringer, S.P., Gao, H.-J., et al., 2011. Super deformability and Youngs modulus of GaAs nanowires. Adv. Mater. 23 (11), 1356–1360.

Wang, C.-H., Liao, W.-S., Ku, N.-J., Li, Y.-C., Chen, Y.-C., Tu, L.-W., Liu, C.-P., 2014. Effects of free carriers on piezoelectric nanogenerators and piezotronic devices made of GaN nanowire arrays. Small 10 (22), 4718–4725.

Wen, B., Sader, J.E., Boland, J.J., 2008. Mechanical properties of ZnO nanowires. Phys. Rev. Lett. 101 (17), 175502.

Wood, C., Jena, D., 2007. Polarization Effects in Semiconductors: From Ab Initio Theory to Device Applications. Springer Science & Business Media.

Wu, W., Wei, Y., Wang, Z.L., 2010. Strain-gated piezotronic logic nanodevices. Adv. Mater. 22 (42), 4711–4715.

Xu, S., Wang, Z.L., 2011. One-dimensional ZnO nanostructures: solution growth and functional properties. Nano Res. 4 (11), 1013–1098. ISSN 19980124. https://doi.org/10.1007/s12274-011-0160-7.

Xu, S., Adiga, N., Ba, S., Dasgupta, T., Wu, C.F.J., Wang, Z.L., 2009. Optimizing and improving the growth quality of ZnO nanowire arrays guided by statistical design of experiments. ACS Nano 3 (7), 1803–1812. ISSN 1936-0851. https://doi.org/10.1021/nn900523p.

Xu, S., Qin, Y., Xu, C., Wei, Y., Yang, R., Wang, Z.L., 2010. Self-powered nanowire devices. Nat. Nanotechnol. 5 (5), 366–373.

Yang, R., Qin, Y., Dai, L., Wang, Z.L., 2009. Power generation with laterally packaged piezoelectric fine wires. Nat. Nanotechnol. 4 (1), 34–39.

Zhang, X.Q., et al., 2003. Resonant exciton second-harmonic generation in self-assembled ZnO microcrystallite thin films. J. Phys.: Conden. Matter 15 (30), 5191.

Zhang, Y., Ram, M.K., Stefanakos, E.K., Goswami, D.Y., 2012. Synthesis, characterization, and applications of ZnO nanowires. J. Nanomater. 2012, 1–22. ISSN 1687-4110. https://doi.org/10.1155/2012/624520.

Zhao, M.-H., Wang, Z.-L., Mao, S.X., 2004. Piezoelectric characterization of individual zinc oxide nanobelt probed by piezoresponse force microscope. Nano Lett. 4 (4), 587–590.

Zhou, H., Wong, S.S., 2008. A facile and mild synthesis of 1-D ZnO, CuO, and α-Fe2O3 nanostructures and nanostructured arrays. ACS Nano 2 (5), 944–958. ISSN 19360851. https://doi.org/10.1021/nn700428x.

Zhou, J., Fei, P., Gu, Y., Mai, W., Gao, Y., Yang, R., Bao, G., Wang, Z.L., 2008. Piezoelectric-potential-controlled polarity-reversible Schottky diodes and switches of ZnO wires. Nano Lett. 8 (11), 3973–3977.

Zhu, G., Yang, R., Wang, S., Wang, Z.L., 2010. Flexible high-output nanogenerator based on lateral ZnO nanowire array. Nano Lett. 10 (8), 3151–3155.

Zhu, G., Wang, A.C., Liu, Y., Zhou, Y., Wang, Z.L., 2012. Functional electrical stimulation by nanogenerator with 58 V output voltage. Nano Lett. 12 (6), 3086–3090. ISSN 15306984. https://doi.org/10.1021/nl300972f.

CHAPTER EIGHT

Nanowire-Based Bulk Heterojunction Solar Cells

Olivia D. Hentz, Jayce J. Cheng, Paul H. Rekemeyer, Nina Andrejevic, Silvija Gradečak[1]
Department of Materials Science and Engineering, Massachusetts Institute of Technology, Cambridge, MA, United States
[1]Corresponding author: e-mail address: gradecak@mit.edu

Contents

1. INTRODUCTION

1.1 Thin Film and Nanowire-Based Solar Cells

Crystalline silicon solar cells based on planar heterojunction architecture (Fig. 1A) are currently the leading commercial photovoltaic (PV) technology, but there has been a significant effort to develop alternatives that overcome some of the limitations intrinsic to silicon photovoltaics. For example, there is a strong need to develop PV devices that are lightweight, flexible, and/or transparent and that can be produced on large scales using roll-to-roll processing. Of the alternative technologies, dye sensitized, organic, perovskite, and quantum dot (QD) solar

Semiconductors and Semimetals, Volume 98
ISSN 0080-8784
https://doi.org/10.1016/bs.semsem.2018.04.002

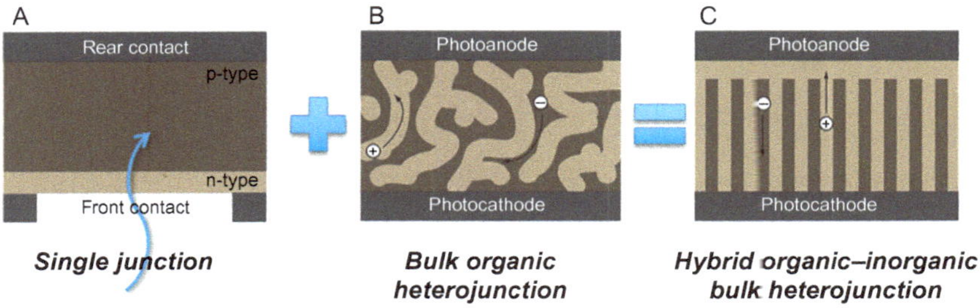

Fig. 1 Schematic of p–n junction solar cell architectures based on a (A) planar heterojunction, (B) bulk heterojunction, and (C) ordered bulk heterojunction.

cells have been extensively studied for their low-temperature and scalable synthesis. Several of these novel technologies, however, are limited in their efficiency by conflicting requirements posed by two important processes photogeneration and collection of charge. In this chapter, we cover recent advancements in the use of nanostructured electrodes to address these efficiency limitations, focusing on organic and QD solar cells as example systems.

Organic solar cells are based on solution processable conjugated polymers that can absorb incident light due to the energy difference between their highest occupied molecular orbital (HOMO) and their lowest unoccupied molecular orbital (LUMO). Generally, organic solar cells consist of two such polymers, one that acts as the electron acceptor and the other that serves as the electron donor, allowing for separation of the charged carriers generated by light absorption (Fig. 1B). Due to the high exciton binding energy in the conjugated polymers, carriers generally take the form of excitons, bound electron–hole pairs. In order for these carriers to be extracted, the exciton must be split into its constituent free electron and hole. This dissociation is achieved at the interface of the electron acceptor and the electron donor by electron transfer into the LUMO of the acceptor material, a process that is energetically favorable due to the offset between the acceptor and donor LUMO energies. However, the diffusion length of excitons in the conjugated polymers is very short (1–10 nm) compared to the device thicknesses required for full absorption of the incident light (hundreds of nm) (Marinova et al., 2017). As a result, many of the excitons generated by the incident light recombine before they diffuse to the electron-donor/electron-acceptor interface. This conflict between absorption and extraction is one of the leading challenges for organic solar cells.

QD solar cells face similar limitations, in this case caused by short minority carrier diffusion lengths. QD solar cells are based on a thin film of inorganic nanoparticles with dimensions smaller than their Bohr radius, resulting in quantum confinement of carriers. As a consequence of the quantum confinement, the band gap of QDs can be tuned and optimized for various device applications. In the case of QD solar cells, the size of QDs is selected to maximize the power output of the device based on the absorption of solar radiation and the amount of energy collected per photon absorbed. QD solar cells extract charge using either a Schottky contact or a p–n junction typically formed between a transparent n-type material, such as zinc oxide (ZnO), and the active QD layer. Due to the moderate exciton binding energy in these materials, carriers generally exist as free carriers and migrate due to both diffusion and drift. However, the sum of the minority carrier diffusion length and depletion width from the electronic junction is significantly smaller than the required QD layer thickness needed for optimal absorption of the incident solar light. Therefore, as with organic solar cells, requirements for maximum absorption conflict with requirements for efficient charge extraction.

For both organic and QD solar cells, it is apparent that improved device efficiency can be obtained by decoupling the absorption depth from the distance traveled by charge carriers for extraction. There have been a number of proposed methods to achieve this decoupling, primarily based on bulk heterojunction (BHJ) device architectures. Here we will discuss early work toward this goal and describe how ordered BHJs have emerged as a promising platform for efficient solar cells based on materials traditionally hindered by poor carrier diffusion.

1.2 Bulk Heterojunction Solar Cells

1.2.1 Traditional Bulk Heterojunction Solar Cells

The earliest efforts at overcoming the short carrier diffusion lengths in novel photovoltaic devices focused on organic solar cells. In 1995, Yu et al. introduced the idea of a "BHJ" solar cell in which the electron-acceptor and electron-donor materials were intermixed to form interpenetrating network with domain sizes comparable to the exciton diffusion length, as shown in Fig. 1 (Marinova et al., 2017).

In this device geometry, excitons need to travel relatively short distances to reach an interface between the electron acceptor and the electron donor, where they can then be separated and extracted to their respective contacts. However, these BHJ solar cells suffer from incomplete extraction pathways and relatively low mobility in the organic materials.

1.2.2 Ordered Bulk Heterojunction Solar Cells

To overcome the limitations of traditional BHJ organic solar cells, there have been efforts to integrate nanostructured inorganic materials into traditional conjugated polymers for a more structured heterojunction architecture, as shown in Fig. 1C. In particular, inorganic nanowires have been investigated for their superior electronic properties and the introduction of continuous pathways for charge transport. Additionally, internanowire spacing can be tuned to the exciton diffusion length of the relevant absorber. However, early work that used nanowire scaffolds of ZnO or TiO_2 resulted in poor filling of the polymer material into the inorganic network, which inhibited charge extraction (Greene et al., 2007).

Ren et al. took a novel approach to overcome this issue by directly blending inorganic nanowires (in this case, GaAs nanowires grown using metal–organic chemical vapor deposition) with the polymer active layer, poly(3-hexylthiophene-2,5-diyl) (P3HT), to form a hybrid organic–inorganic ordered BHJ (Fig. 2A) (Ren et al., 2011a). In addition to providing structured pathways for charge extraction, the presence of the GaAs nanowires induced order in the P3HT and thus improved transport through the polymer. A commonly observed adverse effect of BHJ architectures is the large interface area between materials, which allows for increased recombination from interfacial defects. In this study, Ren et al. overcame this limitation by passivating the surface of the GaAs nanowires with TiO_2, attaining a maximum efficiency of 2.36% (Fig. 2B). In this work, GaAs nanowires were selected to complement the absorption of P3HT, but it was shown that photocurrent generated by photons absorbed by nanowires is <4%; rather, nanowire contribution to the improved photocurrent mostly comes from the induced order in the polymer phase. This result had two major implications: (1) ordering or

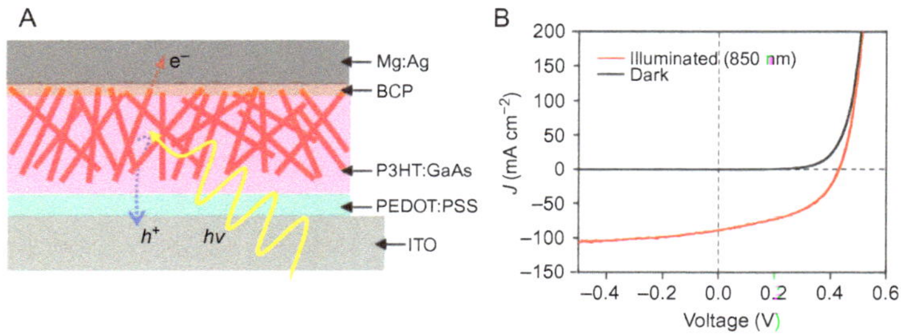

Fig. 2 (A) Schematic of an ordered bulk heterojunction solar cell based on GaAs nanowires in P3HT. (B) Current density–voltage measurements of a GaAs/P3HT solar cell after passivation of the GaAs nanowires with TiO_2. *Adapted with permission from Ren, S., et al., 2011. Heterojunction photovoltaics using GaAs nanowires and conjugated polymers. Nano Lett. 11, 408–413. Copyright 2011 American Chemical Society.*

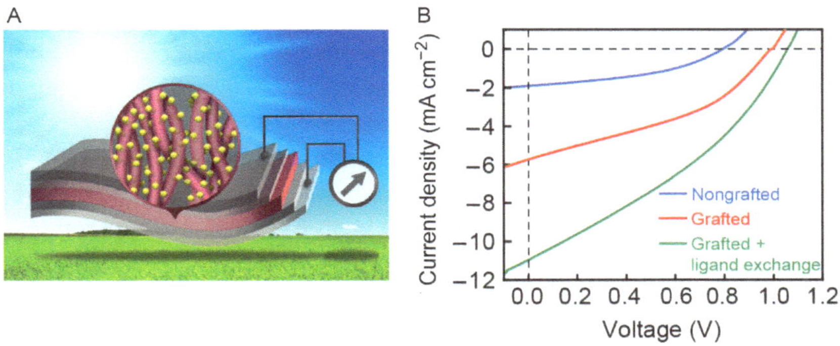

Fig. 3 (A) Illustration of an ordered bulk heterojunction solar cell based on P3HT nanowires coated in CdS quantum dots. Magnified region shows P3HT nanowires decorated with CdS quantum dots. (B) Current density–voltage measurements of solar cells based on P3HT nanowires with/without (grafted/nongrafted) CdS quantum dots and with CdS quantum dots with short EDT (ethanedithiol) ligands replacing the longer native ligands. *Adapted with permission from Ren, S., et al., 2011. Inorganic–organic hybrid solar cell: bridging quantum dots to conjugated polymer nanowires. Nano Lett 11, 3998–4002. Copyright 2011 American Chemical Society.*

crystallization of the polymer phase is critical in enhancing the device performance and (2) the choice of the nanowire materials can be relaxed to allow for easier, large-scale processing.

More recently, Ren et al. took this nanoscale ordering a step further by using nanostructured materials for both the electron-accepting and electron-donating materials in hybrid organic–inorganic ordered BHJ (Fig. 3A) (Ren et al., 2011b). Grafting CdS QDs onto P3HT nanowires allowed for good electronic coupling between the P3HT nanowires and the CdS QDs, leading to an overall power conversion efficiency (PCE) of 4.1% for devices with the optimum CdS loading (Fig. 3B).

1.3 Nanowire Scaffolds for Ordered Bulk Heterojunctions

The early studies of hybrid organic–inorganic ordered BHJ solar cells demonstrated that inorganic nanowire scaffolds show promise to expand the benefits of nanoscale ordering in solar cells due to their directional charge transport and the high conductivity. By improving the morphology of nanowire scaffolds and through careful materials selection for device active layers, limitations from short carrier diffusion lengths in organic materials and QDs can be dramatically improved. In particular, ZnO has been studied extensively for the synthesis of dense nanowire arrays due to a number of inherent advantages. ZnO nanomaterials are found in a variety of technology applications, from light-emitting diodes, photodetectors and photodiodes, and piezoelectric energy harvesters, to solar cells, owing to the compatibility of ZnO nanostructure synthesis with existing device fabrication techniques. An example of such an application is shown in Fig. 4. The wide band gap, chemical and thermal stability, and large exciton binding energy of ZnO have led to the incorporation of this II–VI semiconductor into several types of devices that leverage its optical, electronic, and mechanical properties (Vayssieres, 2003). ZnO nanoparticle films are used widely in photovoltaic devices as hole-blocking layers in hybrid polymer–ZnO nanoparticle devices (Beek et al., 2006; Lee et al., 2011), and as the n-type semiconductor in lead sulfide (PbS) QD devices (Chuang et al., 2014; Ip et al., 2012) and perovskite devices (Liu and Kelly, 2014). The nanoparticles are synthesized from precipitation in alcoholic solution and are spin coated onto the substrate to form a thin film, and the active layers may then be deposited on top of the ZnO nanoparticle film. Introducing nanowires into the device architecture can be accomplished by simply immersing the nanoparticle film into growth solution, which generates a dense array of nanowires. Nanowire scaffolds made of ZnO have been shown to enhance charge extraction as high n-type conductivity pathways in solar cell applications. However, further advancement is possible through optimization of the synthesis process to better control

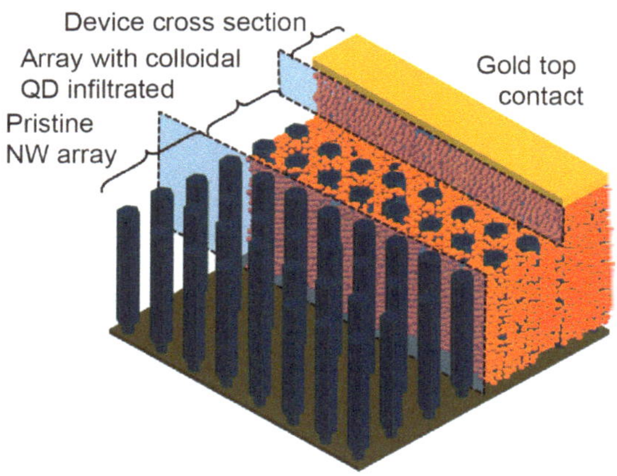

Fig. 4 Schematic of a quantum dot solar cell on a templated nanowire array, demonstrating an example application of well-controlled nanowire arrays. *Reprinted from Nicaise, S.M., Cheng, J.J., Kiani, A., 2015. Control of zinc oxide nanowire array properties with electron-beam lithography templating for photovoltaic applications. Nanotechnology 26, 075303–075310. doi:10.1088/0957-4484/26/7/075303. Copyright 2015 IOP Publishing. Reproduced with permission. All rights reserved.*

the nanowire array morphology. In the remainder of this chapter, we will first discuss the growth, processability, and scalability of ZnO nanowire scaffolds. We will then address efforts made to control the dimensionality of these nanowires to optimize them for solar cell applications. Finally, we will summarize how these wires have been used in state of the art solar cells from the standpoint of materials selection and characterization as well as the effect that these nanowires have on the optical and electronic properties of solar cell devices.

2. GROWTH, PROCESSABILITY, AND SCALABILITY

Engineering at the nanoscale has enabled advances in nanoscale technologies by providing access to new or enhanced device capabilities through tailored architectures that complement the length scales of relevant phenomena. In the context of nanostructured PV devices, some of the requirements include precise tailoring of the nanowire length and areal density to control light absorption and charge transport, respectively. However, fabrication of nanowire arrays for nanostructured PV poses several unique challenges as a result of their size and the complexity of the targeted morphology. First, the synthesis process should maintain material quality, particularly as the increased nanowire surface area to volume ratio leads to poorer surface quality compared to the bulk counterpart (Li et al., 2004; Lu and Lieber, 2006; Murray et al., 1993; Sun et al., 2016; Yin and Paul Alivisatos, 2005; Zhou et al., 2015). Second, parameters such as nanowire dimensions, spacing, orientation, and uniformity must be precisely controlled to ensure optimum device performance (Kovalenko et al., 2015; Lu and Lieber, 2006). Lastly, the synthesis methods should be scalable and compatible with other processing steps and materials. In this section, we discuss hydrothermal growth as a low temperature, scalable, and versatile synthesis method for the growth of high-quality ZnO nanowire arrays and address some advantages and challenges in the context of specific materials and device considerations.

2.1 Hydrothermal Synthesis of ZnO Nanowires

Hydrothermal synthesis is a low-temperature, solution-based method for the growth of ZnO nanostructures. First introduced by Vayssieres in 2003, the approach is suitable for a range of substrates, enables dimensional control through adjustment of relevant growth parameters, and can easily scale (Vayssieres, 2003). While alternative methods such as chemical vapor deposition (CVD) and molecular beam epitaxy (MBE) also yield high-quality nanowire arrays, their high-temperature requirements often exceed substrate tolerances, and the size and cost of these deposition systems may limit scale-up (Greene et al., 2003). Furthermore, these methods typically make use of metal seed particles that can act as recombination centers (Brewster et al., 2011). In contrast, hydrothermal growth of ZnO nanowire arrays is a low-temperature, scalable process compatible with a range of substrates, and since nanowire growth proceeds from ZnO nanoparticle seeds, no foreign material is introduced into the ZnO system using this approach.

The hydrothermal growth process begins with deposition of a zinc acetate (ZnAc) seed layer solution via sol–gel spin coating (Cheng et al., 2016; Greene et al., 2003) or sputtering (Cross et al., 2005; Wu et al., 2009), and subsequent annealing at 150–400°C converts the ZnAc into ZnO nanoparticles. Nanowires are then synthesized by immersion of the seeded substrate into a supersaturated aqueous solution of Zn^{2+} at 60–90°C (Vayssieres, 2003), as illustrated schematically in Fig. 5. As the relationship between the nanowires and underlying substrate is not epitaxial, seed layer deposition and wire growth have been demonstrated with comparable results on silicon (100) and (111) (Wei et al., 2010), indium tin oxide (ITO) (Greene et al., 2005), and Kapton polymer (Xu et al., 2008a) substrates, which are single crystal, polycrystalline, and amorphous, respectively. Furthermore, nanowire growth is the result of the vertical (*c*-facet) growth rate surpassing the lateral (*m*-facet) growth rate (Demianets et al., 2002; Vayssieres, 2003), so the final aspect ratio may be tuned by modification of the growth parameters, such as temperature, supersaturation, solvent, and presence of other species during synthesis, which directly influence the growth rates. Finally, since nanowire nucleation and growth proceed from the ZnO seed facets, the seed position and initial orientation determine the position and angle, respectively, of the nanowires relative to the substrate

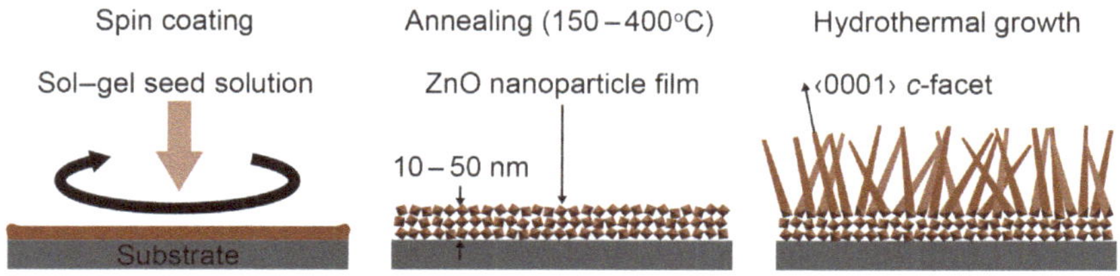

Fig. 5 Illustration of the process to produce hydrothermally grown ZnO nanowires from spin-coated ZnO seed layers. A sol–gel solution of zinc acetate is spin coated on the substrate and annealed for 10 min to form a film of ZnO nanoparticles. Substrates are then suspended face down in an aqueous growth solution to hydrothermally grow ZnO nanowire from the ZnO nanoparticles. *Reprinted with permission from Cheng, J.W.J., 2016. Hydrothermal synthesis of zinc oxide nanowire arrays for photovoltaic applications. Copyright 2016 Massachusetts Institute of Technology.*

(Huang and Lin, 2008; Wu et al., 2009); thus, the ZnO nanowire areal density and orientation depend strongly on the seed layer (Cheng, 2016). In the following subsections, we individually address four important aspects of the synthesis process—ZnO seed layer deposition, seed layer patterning for templated growth, substrate compatibility, and finally solution growth—alongside relevant challenges and advances made in these areas to control the final morphology of ZnO nanowire arrays.

2.1.1 The Influence of Seed Layer on ZnO Nanowire Morphology

As the first step of the ZnO nanowire growth process, seed layer deposition plays an important role in producing nanowires with controlled morphology. Seed nanocrystals are directly responsible for controlling nanowire placement and alignment, which indirectly affect other array properties like nanowire diameter and length. The two most common seeding techniques are sol–gel spin coating and RF sputtering, both of which yield good crystal quality under relatively benign processing conditions. Sol–gel spin coating can be further subdivided into continuous and discrete deposition methods, which produce seed layers in which adjacent seed crystals are in either contact (continuous) or isolated (discrete). Initial demonstrations of seeded ZnO nanowire growth relied on sol–gel spin coating of a colloidal suspension of zinc acetate in ethanol, generating a thin layer of discrete ZnO seeds (Greene et al., 2005). Although the phase transformation from ZnAc to ZnO occurs at 175°C, a high-temperature annealing step between 200°C and 350°C is necessary to effect vertical alignment of the seeds such that the *c*-axis, and thereby the growth direction, is oriented normal to the substrate. Annealing above 300°C drives out impurities, such as adsorbed solvent molecules, which would otherwise contribute to the poorer surface quality of hydrothermally synthesized nanostructures compared to their CVD-deposited counterparts (Joo et al., 2013). While the mechanism behind the nanocrystal realignment is not yet fully understood, a widely accepted explanation is that the *c*-facet is stabilized by impurities remaining from the sol–gel deposition process, which reduce its surface energy and thus render it energetically favorable (Greene et al., 2005). Since the seed layer consists of ZnO nanocrystals that are only tens of atomic layers thick, another proposed explanation is that the ZnO arranges itself into a layered, graphite-like crystal structure that is thermodynamically stable (Greene et al., 2005); however, stability of graphitic layers of ZnO nanocrystals has been shown theoretically and experimentally on Pt surfaces (Liu et al., 2015), but not on silicon, ITO, or glass. Finally, it has also been suggested that the *c*-facet is not atomically smooth, but has a characteristic roughness that compensates for the high *c*-facet energy by exposing other lower energy facets; however, this explanation has also not been experimentally validated (Greene et al., 2005). While the exact alignment mechanism is not yet elucidated, the effect of the high-temperature annealing step on alignment of ZnO nanocrystals remains clear. However, it is worth noting that lower temperature annealing is required for substrates like ITO and plastics, which cannot tolerate high temperatures, resulting in a less controlled nanowire alignment.

As mentioned previously, the seed layer also directly influences nanowire density and orientation, which are important parameters in optimizing device performance. These array properties are not independent; nanowires become more vertically aligned as nanowire density increases due to mutual impingement, whereby growth of off-axis nanowires is hindered by the presence of neighboring nanowires that interrupt their growth (Fig. 6). At high nanowire densities, enough off-axis nanowires are present that only those relatively perpendicular to the substrate continue growing, improving the vertical alignment of the array with growth time. The Herman's order parameter, *S*, is typically used to quantify the relative alignment of an ensemble of nanowires with respect to the substrate normal:

$$S = \frac{3}{2}\left\langle \cos^2 \varphi \right\rangle - \frac{1}{2} \tag{1}$$

where φ is the angle between an individual nanowire growth axis and the substrate normal (Nicaise et al., 2015a). *S* ranges from 1, which implies perfect ensemble alignment with the substrate normal, to -0.5, which implies perfect alignment perpendicular to the normal direction.

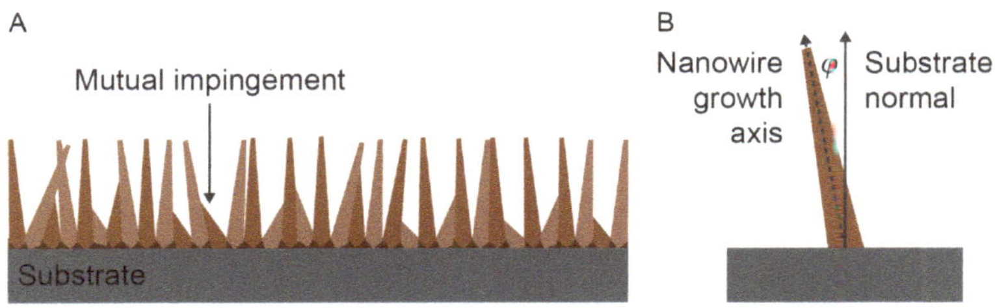

Fig. 6 (A) Illustration of the mutual impingement of nonvertically aligned ZnO nanowires, resulting in highly aligned nanowire arrays at high nanowire densities. (B) Illustration of how the angle (φ) of nanowires is measured from the substrate. *Reprinted with permission from Cheng, J.W.J., 2016. Hydrothermal synthesis of zinc oxide nanowire arrays for photovoltaic applications. Copyright 2016 Massachusetts Institute of Technology.*

Because the likelihood of mutual impingement increases with nanowire density, it follows that annealing, which reduces the seed density by causing coalescence of seeds through grain growth, is detrimental to nanowire alignment for randomly aligned seed layers (Ma et al., 2007). This behavior is characteristic of continuous spin-coated seed layers, and dense nanowire arrays grown from these seed layers show good alignment through the mutual impingement effect. The following section will address templated nanowire growth as it relates to the underlying seed layer. Top-down patterning of ZnO seed layers allows for fine control over spacing of nanowires and supplies a test system for device design through which optimal array properties may be determined.

2.1.2 Templated Growth of ZnO Nanowires

The capability to accurately control spacing, pitch, and alignment of ZnO nanowire arrays at length scales of 100–1000 nm critically influences the use of nanowires to mediate interaction between optical and electronic phenomena in nanostructured devices, including PV. Toward this end, seed layer patterning has had mixed effectiveness at controlling nanowire placement (Cheng, 2016). Relevant lithographic techniques for seed layer patterning include nanosphere lithography (Lee, 2013), electron-beam lithography (EBL) (Weintraub et al., 2007; Zhang et al., 2012), interference lithography (Kim et al., 2010), nanoimprint lithography (Jung and Lee, 2011), and traditional photolithography (Tak and Yong, 2005). The effectiveness of these techniques is limited both by pattern resolution and by deteriorating vertical alignment in increasingly sparse nanowire arrays. When template holes are too large, multiple seeds of the underlying seed layer are exposed to growth solution and thus produce multiple nanowires per hole (Xu et al., 2008b). At the same time, nanowires may suffer from misalignment caused by the seed layer deposition and annealing processes. To decouple the various parameters at play, Nicaise et al. examined how substrate choice, seed layer deposition and annealing, and template hole size affect the branching and alignment of ZnO nanowire arrays fabricated by EBL and hydrothermal synthesis (Nicaise et al., 2015b). Their findings suggest that (1) annealing of spin-coated seed layers improves nanowire array alignment for this deposition process; (2) arrays produced from sputtered seed layers have better alignment than those grown on spin-coated seed layers, while branching is comparable in both; and (3) the choice of substrate—silicon, ITO-coated glass, and poly(ethylene naphthalate) (PEN)—has minimal influence on nanowire array quality. Fig. 7 illustrates the fabrication process for this templated hydrothermal growth of ZnO nanowire

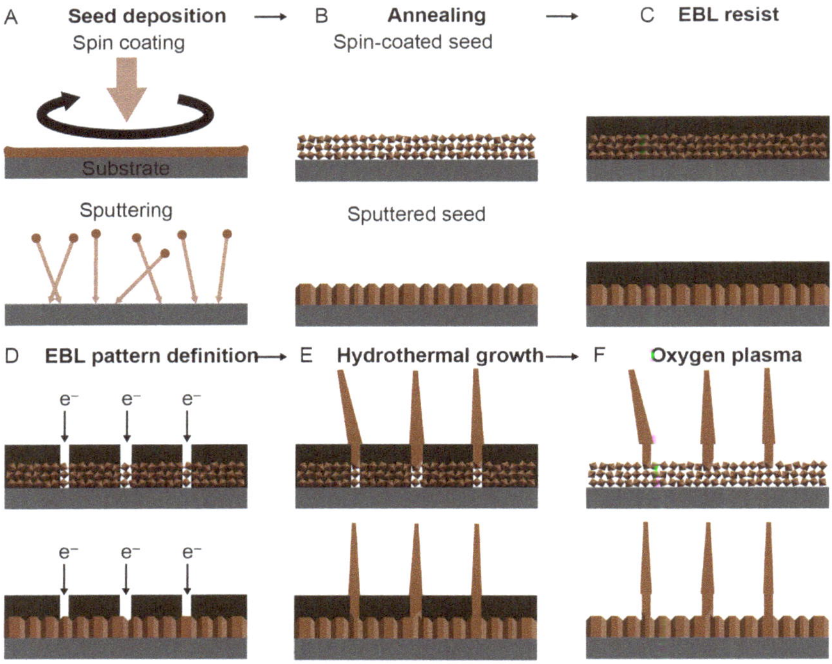

Fig. 7 Illustration of the process to create templated ZnO nanowire arrays from sputtered or spin-coated seed layers. Electron-beam sensitive resist is deposited on top of ZnO seed layers (A through C) and an electron beam is used to pattern holes of controlled size and spacing through the resist to expose the underlying ZnO (D). Nanowires are then hydrothermally grown through the template holes (E) and the resist is removed with oxygen plasma etching (F). *Adapted from Nicaise, S.M., Cheng, J.J., Kiani, A., 2015. Control of zinc oxide nanowire array properties with electron-beam lithography templating for photovoltaic applications. Nanotechnology 26, 075303–075310. doi:10.1088/0957-4484/26/7/075303. Copyright 2015 IOP Publishing. Reproduced with permission. All rights reserved.*

arrays. ZnO seed layer deposition was achieved via sputtering or spin coating from solution, and selected samples were annealed at 400°C for 2 h in ambient atmosphere, while others were left unannealed. Poly(methyl methacrylate) (PMMA) electron-beam resist was then spin coated and baked and templating holes were defined by EBL. Immersion of the patterned substrates in growth solution resulted in nanowire growth from the template holes, and oxygen plasma etching was used to remove the PMMA postgrowth. Fig. 8 shows examples of templated growth on different substrates with varied deposition and annealing conditions.

The authors found sufficiently low branching, acceptable for device fabrication, over a range of template hole sizes. However, annealing prompts a decrease in branching for sputtered seed layers and an increase for spin-coated seed layers. The increase in ZnO grain size for annealed, sputtered films relative to the template hole diameter exposes fewer grains to the growth solution and thus limits branching, whereas in spin-coated seed layers, the ZnO grains are significantly smaller than the hole diameter, even after annealing. In the case of highly misoriented spin-coated seed layers, annealing is found to improve grain alignment, so fewer nanowires impinge into the PMMA side walls which permit further branching; this effect is less prominent in sputtered samples where seed layers are highly aligned before annealing. Apart from highlighting the differences between seed films prepared using different deposition techniques, templated hydrothermal growth of ZnO nanowires offers a useful framework for both testing the effect of various deposition process parameters on branching and nanowire alignment and correlating specific nanowire morphologies to device performance. Since device requirements also often dictate substrate selection, an optimal seed layer is further compatible with a wide range of materials and permits growth of high-quality arrays independent of this growth substrate. The following section reviews important considerations for achieving uniform nanowire growth on various substrates and also details how simple substrate modification has enabled controlled nanowire growth on a more versatile range of materials.

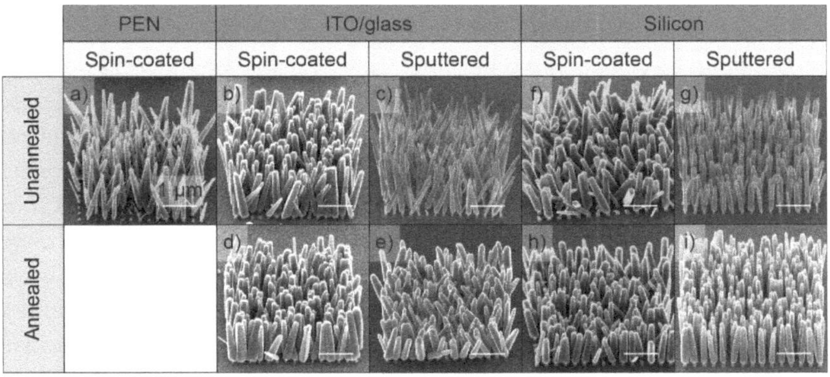

Fig. 8 SEM images of templated ZnO nanowires hydrothermally grown on varied substrates (PEN, ITO/glass, or silicon) from spin-coated (A, B, D, F, H) or sputtered (C, E, G, I) seed layers. Seed layers were either not annealed (A, B, C, F, G) or annealed at 400°C for 2h in ambient conditions (D, E, H, I) before nanowire growth. Images were obtained at a 45 degree tilt and were the most aligned and least branched regions observed for each experimental condition. *Reprinted from Nicaise, S.M., Cheng, J.J., Kiani, A., 2015. Control of zinc oxide nanowire array properties with electron-beam lithography templating for photovoltaic applications. Nanotechnology 26, 075303–075310. https://doi.org/10.1088/0957-4484/26/7/075303. Copyright 2015 IOP Publishing. Reproduced with permission. All rights reserved.*

2.1.3 Compatibility With Growth Substrate

Apart from meeting demands of large-scale application, a low-temperature and solution-based synthesis process such as hydrothermal growth is generally compatible with a broad range of substrates. This is in contrast to most CVD and MBE schemes, which are conducted at much higher temperatures. Greene et al. demonstrated hydrothermal growth of ZnO nanowire arrays on silicon, ITO glass, sapphire, titanium foil, and polymer surfaces using the same seed layer deposition and solution growth process (Greene et al., 2003). However, expanding the space of compatible growth substrates even further is of great interest in meeting evolving objectives for device capabilities. For instance, the push toward flexible and lightweight solar cells with nanostructured architectures has driven implementation of flexible electrode materials such as graphene. Progress in this direction will further depend on development of an effective interface between nanowires and the candidate electrode materials. A key advancement toward substrate-independent growth was demonstrated by Park et al. in the context of graphene cathode-based ZnO nanowire hybrid solar cells, using either QD or polymer (P3HT) active layers (Park et al., 2012). They initially observed poor wetting by the ZnO seed layer when deposited on a pristine graphene surface; however, modification of the graphene surface with conductive polymer interlayers that both wet the graphene surface and are chemically compatible with the ZnO layer enabled deposition of a uniform ZnO film while maintaining charge transfer between the ZnO and electrode. Two commercially available conducting polymers, poly(3,4-ethylenedioxythiophene)-block-poly(ethylene glycol) (PEDOT:PEG) doped with perchlorate (PC) in nitromethane and sulfonated poly(thiophene-3-[2-(2-methoxyethoxy)-ethoxy]-2,5-diyl) (Plexcore OC RG-1200) in ethylene glycol monobutyl ether/water (3:2), abbreviated PEDOT:PEG(PC) and RG-1200, respectively, were used in the study. As shown in Fig. 9, comparable uniformity and alignment were achieved for ZnO nanowire arrays grown under the same experimental conditions on ITO and on modified graphene. This method thus allows for substrate-independent growth of ZnO nanowires by simple substrate modification with a thin conductive polymer, a key step in the implementation of alternate electrodes in nanowire-based solar cells.

Alongside considerations of seed layer quality and substrate compatibility, the hydrothermal growth stage offers another parameter space which may be tuned to optimize nanowire growth. In the next section, we present an introduction to solution growth of nanowires from ZnO seed layers and how additives have been used to control facet growth rates and hence the final nanowire aspect ratio.

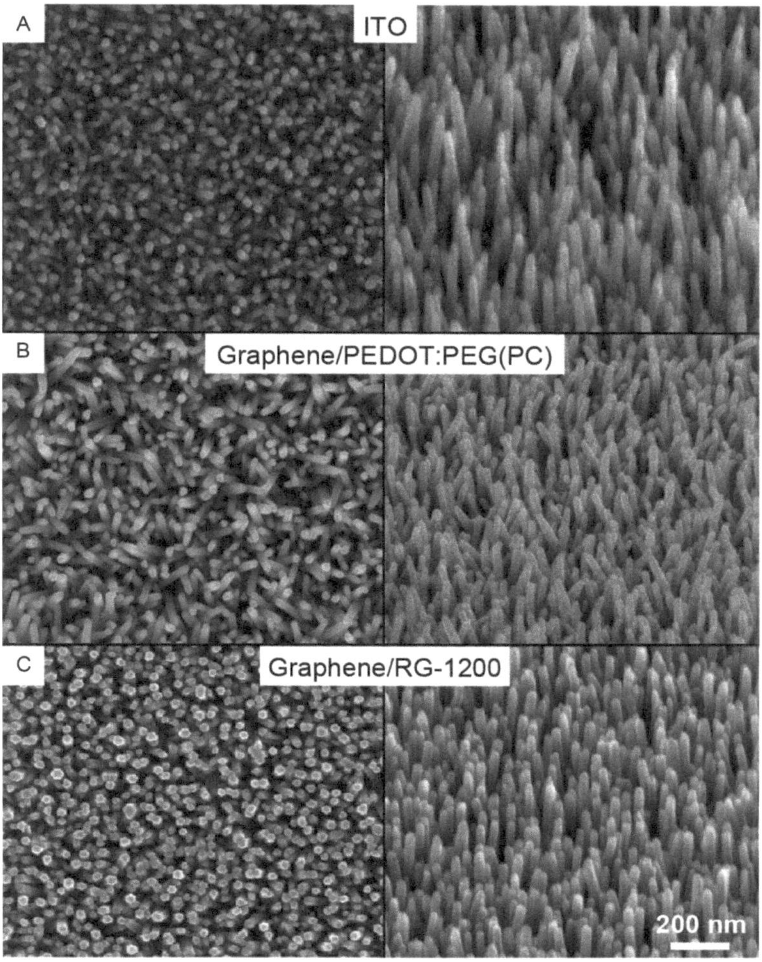

Fig. 9 SEM images of hydrothermally grown ZnO nanowires on (A) ITO, (B) graphene/PEDOT:PEG(PC), and (C) graphene/RG-1200 shown from the *top* (*left*) and from a 45 degree angle (*right*). Use of a polymer interlayer between graphene and the ZnO seed layer allows for the growth of uniform, well-aligned ZnO nanowires on graphene. *Reprinted with permission from Park, H., et al., 2012. Graphene cathode-based ZnO nanowire hybrid solar cells. Nano Lett. 13, 233–239. Copyright 2013 American Chemical Society.*

2.1.4 Solution Growth of ZnO Nanowires

The final stage of ZnO nanowire synthesis is the chemical bath deposition or hydrothermal growth step. In this step, the chemical precursors zinc nitrate ($Zn(NO_3)_2$) and hexamethylenetetramine (HMTA) are combined in deionized water and heated to a temperature in the range of 60–90°C (Parize et al., 2016).

The seeded substrate is then immersed in this supersaturated solution to facilitate nanowire growth from the seed particles. When crystallization of ZnO occurs via a reaction intermediate $Zn(OH)_2$, the following chemical reactions are expected to drive nanowire growth at the seed/solution interface (Nicholas et al., 2012):

$$\begin{aligned} &Zn^{2+}(aq.) + 2OH^{-}(aq.) \rightleftharpoons Zn(OH)_2(s) \\ &Zn(OH)_2(s) \rightleftharpoons ZnO(c) + H_2O(l) \end{aligned} \tag{2}$$

The ZnO nanowire growth has been shown to depend on a number of interrelated parameters, including temperature (Joo et al., 2011), zinc concentration (Govender et al., 2004; Xu et al., 2008a), pH (Baruah and Dutta, 2009; Joo et al., 2011; Tak and Yong, 2005; Tian et al., 2011), type of counterion (Govender et al., 2004; Wang et al., 2006), type of buffer solution (McPeak et al., 2011; Tak and Yong, 2005), and the presence of impurities (Joo et al., 2011). In the typical buffer chemistry used for nanowire growth, zinc ions participate in a variety of chemical reactions with associated reaction constants; thus, the relative concentration of each zinc complex is a function of temperature, pH, and zinc concentration (Cheng, 2016). Depending on the reaction conditions, one or more species will be dominant and primarily influence the observed crystal growth rate. When the individual reaction constants are known, changes in species concentration, or speciation, can be modeled as a function of pH and temperature. The speciation plots for zinc concentrations of 10 and 20 mM are depicted in Fig. 10 and show that Zn^{2+} is dominant in acidic solution from pH 0 to pH 7, while in basic solutions a number of complexes are present with various concentrations (Cheng, 2016).

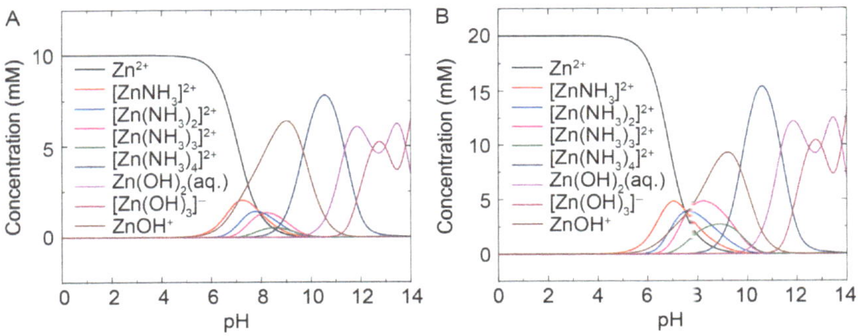

Fig. 10 Speciation plots for (A) 10 mM and (B) 20 mM solutions of zinc ions in an aqueous, equimolar, hexamethylenetetramine (HMTA) buffer solution. *Reprinted with permission from Cheng, J.J., Nicaise, S.M., Berggren, K.K., Gradečak, S., 2016. Dimensional tailoring of hydrothermally grown zinc oxide nanowire arrays. Nano Lett. 16, 753–759. Copyright 2016 American Chemical Society.*

Thus, nanowire growth is carried out in acidic solution to avoid the difficulty of separating effects of the different complexes.

HMTA is used as a buffer both to maintain the desired acidic conditions for the reaction (a pH of 5.5) and to serve as a source of HO^- ions, which are involved in the formation of the $Zn(OH)_2$ reaction intermediate. While the precise role of HMTA during the growth of ZnO nanowires is still under debate, it is well known that the gradual hydrolysis of HMTA with heat leads to the formation of NH_3 and formaldehyde (Parize et al., 2016). NH_3 then reacts with water to produce HO^- ions.

ZnO crystallization has further dependence on the purity of the growth solution, with profound effects on the final nanostructured morphology (Cheng, 2016). Additives and impurities affect the growth rate by adsorbing to the emergent nanowire surface, thereby inhibiting other adatoms from reaching adsorption sites (Kubota and Mullin, 1995). Since the nanowire facets differ in polarity and thus attract different ions in solution (Joo et al., 2011), different additives preferentially adsorb to the *c*- and *m*-facets of the crystal, providing opportunities for growth control. Citrate and polyethyleneimine were the first additives investigated for this purpose and, respectively, reduce growth rates on the *c*- and *m*-facets (Law et al., 2005; Tian et al., 2002). Furthermore, Joo et al. showed that metal ions with a +3 oxidation state in solution, such as Al^{3+} and Ga^{3+}, adhere to the negatively charged, lateral *m*-facet to produce long, tapered nanowires (Joo et al., 2011). On the other hand, ions with a +2 oxidation state, such as Mg^{2+}, adhere to the positively charged, top *c*-facet to form nanoplates (Joo et al., 2011). Careful selection of additives can be used to synthesize nanostructures with a wide range of aspect ratios, from less than 1 (nanoplates) to greater than 100 (nanowires) (Joo et al., 2011).

The sensitivity of nanowire growth to foreign species has both challenges and advantages. While solution purity is important for obtaining consistent and reproducible nanostructured morphologies, the ability to tune the facet growth rates with only small amounts of metal ion additives enables growth control without departing from the target zinc speciation regime. The following section expands this discussion with a focus on dimensional tailoring of ZnO nanowire arrays toward optimization of nanowire-based solar cell devices.

3. DIMENSIONAL TAILORING

Having described the hydrothermal growth process used for a scalable, low-temperature ZnO nanowire synthesis, in this section we will discuss the dimensional tailoring of these inorganic nanowires arrays for nanowire-based solar cells. While control over some nanowire array properties such as spacing (Cheng et al., 2016) and diameter (Yuan et al., 2011) has been demonstrated, progress in tuning nanowire arrays simultaneously over multiple parameters has been hindered by limited understanding of the ZnO nanowire growth mechanisms. We begin this section by identifying the major nanowire array parameters relevant for solar cell applications. We then discuss the areal density as an example of device-relevant properties of nanowire arrays that can be controlled. Next, we discuss work aimed at understanding ZnO nanowire growth mechanisms and how this information can be used to optimize nanowire arrays. Finally, we present an example of translating this fundamental understanding into the growth of nanowire arrays with uniform, controlled length independent of nanowire spacing.

3.1 Device-Relevant Parameters

As mentioned in Section 1, the ability to tune nanowire morphology on length scales relevant to internal optical and electronic processes has motivated the use of nanostructured materials in solar cell devices. There are several device-relevant parameters that can be tuned in the nanowire arrays, including length, alignment, and spacing. Additionally, nanowire array uniformity across these dimensions plays a critical role in device performance.

Nanowire length is tailored to suit the absorption and recombination properties of the active material (e.g., QDs or an organic semiconductor) and is directly related to the active material thickness. In general, the active material thickness is chosen to maximize absorption of the incident light while avoiding recombination of excited carriers. By integrating nanowires into the active material, bulk recombination is reduced—since carriers do not need to travel the thickness of the film to be collected—and the device thickness (and therefore the nanowire lengths) can be optimized for incident light absorption. Uniformity in nanowire length is also critical to device performance. Inconsistency in nanowire length may result either in low shunt resistance, whereby long nanowires create shunting pathways and increase the reverse leakage current, or in high series resistance, in cases where short nanowires fail to fully deplete the active layer film. Nanowire lengths can be tuned simply by changing the growth time; however, the nanowire growth rate is dependent on various nanowire array properties. As a result, achieving nanowire array uniformity over large areas has been challenging.

Nanowire spacing is another important parameter for nanowire arrays used in solar cell devices. In order to maximize carrier extraction, the spacing between nanowires should on average correspond to double the total distance that carriers can travel in the active material before recombining. In the case of organic solar cells, this distance corresponds to the exciton diffusion length, while in QD solar cells, it is the sum of the minority carrier diffusion length and depletion layer thickness in the QD film. Optimization of the nanowire spacing also depends on nanowire alignment relative to the substrate. In a device with poorly aligned nanowires, wire spacing in some device areas will be narrower than that required for efficient charge extraction, while, in other areas, wires will be too far apart, and carriers generated between them will recombine before they can be collected. Furthermore, in poorly aligned nanowire arrays, the gaps between the nanowires are blocked by the tops of the wires and therefore cannot be filled by the active material during deposition. Such inadequate infiltration of active material into the nanowire scaffold causes increased recombination at the resulting gaps in the device. This motivates inquiry into seed layer deposition techniques that give precise control of nanowire spacing and alignment, which was examined in the context of templated nanowire growth in the previous section. We extend this discussion by presenting a scalable alternative to this approach in the following section.

3.2 Controlling Nanowire Growth

3.2.1 Nanowire Spacing

As mentioned earlier, in an optimal solar cell architecture the ZnO nanowire spacing should be tuned to the carrier diffusion lengths and, in the case of free carrier devices, the depletion width in the active material. For example, depending on the choice of passivating ligand, the electron diffusion length in PbS QD films varies between 100 and 200 nm and the depletion region extends about 150 nm into the film (Liu et al., 2016). Theoretically, these devices could therefore be optimized with nanowire spacing between 200 and 400 nm; however, typical nanowire arrays grown by hydrothermal methods have a spacing closer to 70 nm.

In the previous section, we discussed templating as a mechanism for fine tuning of the nanowire spacing for sputtered and spin-coated films. This knowledge can be translated into more scalable methods, as demonstrated by Cheng et al. who introduced a novel spin-coating method termed *dynamic spin coating* as a scalable solution to control the ZnO nanowire spacing (Cheng et al., 2016). In this approach, ZnO seeds were dispersed uniformly and controllably over the substrate surface by dispensing the zinc acetate sol–gel seed solution onto a substrate that had already been ramped to the final spin speed; the seed solution was observed to immediately cover the full substrate and

Fig. 11 (A) Aerial density of ZnO seeds deposited using dynamic spin coating at varied spin speeds, *inset* shows an SEM image of one such sparse seed layer. (B) Scatter plot of ZnO nanowire density vs seed density used for nanowire growth, demonstrating that sparse seed layers can be used to hydrothermally grow sparse nanowire arrays. (C–F) SEM images of sparse ZnO nanowires grown from sparse seed layers of decreasing aerial density. *Reprinted with permission from Cheng, J.J., Nicaise, S.M., Berggren, K.K., Gradečak, S., 2016. Dimensional tailoring of hydrothermally grown zinc oxide nanowire arrays. Nano Lett. 16, 753–759. Copyright 2016 American Chemical Society.*

dewetting produced discrete ZnO seeds, ~5 nm in diameter (Fig. 11A, inset). The resulting seed density was found to be dependent on the spin speed (Fig. 11A) and could be controlled over an order of magnitude. To achieve a high degree of reproducibility, the spin coating was performed in a nitrogen glovebox with controlled humidity ($H_2O < 0.1$ ppm) in order to regulate the solvent evaporation rate. The seeded substrates were subsequently annealed in ambient atmosphere on a 400°C hot plate to achieve vertical *c*-plane texturing.

ZnO nanowires were grown from the sparse seed layers by hydrothermal synthesis, resulting in well-aligned, sparse nanowire arrays (Fig. 11B–F). The order parameter, which quantifies the relative alignment of the nanowire ensemble with respect to the substrate normal, was greater than 0.85 for all the nanowire arrays shown in this study. Recall that its value ranges from 1, which implies perfect ensemble alignment with the substrate normal, to −0.5, which implies alignment perpendicular to the normal direction. An order parameter greater than 0.85 suggests that the synthesized nanowires are well aligned vertically. This study thus revealed that high-quality ZnO nanowire arrays could be grown with controlled density using a highly scalable spin-coating method. However, while controlling the nanowire density, the authors observed changes in nanowire growth rate, requiring independent growth time optimization for each possible nanowire density. We will therefore next discuss work focused on understanding the growth mechanisms for ZnO nanowires to enable intentional design of ZnO nanowire arrays over multiple device-relevant parameters.

3.2.2 Mechanisms of Nanowire Growth

Historically, optimization of ZnO nanowire growth has been achieved through an iterative empirical process (Xu et al., 2009). To establish more deterministic approach, it is critical to understand the underlying growth mechanisms of the ZnO nanowire arrays and to determine the relative contribution of reaction-limited and diffusion-limited processes to growth rate. In particular, understanding the mechanism of *c*-plane growth would allow for control over dopant incorporation (Tay et al., 2012), growth rate (Joo et al., 2011), and impurity interactions (Kubota and Mullin, 1995; Pedro et al., 2006), which are points of emphasis in the following discussion.

Determination of solution-phase crystal growth mechanisms can be achieved either by measuring the relationship between growth rate and supersaturation (Chernov et al., 1986; Govender et al., 2004; Markov, 2004; Morin et al., 2010) or by characterizing surface features which indicate the mechanism at work (Baruah and Dutta, 2009; Li et al., 2005; Liu et al., 2006; McPeak and Baxter, 2009; Sun et al., 2006). However, in the first method, the supersaturation conditions must be constant throughout the growth process, which is not the case for traditional bulk synthesis methods in which reactants are consumed and therefore depleted during the course of the growth (Greene et al., 2003; Vayssieres, 2003; Xu et al., 2009). There have been a number of attempts at correlating reactant supersaturation to nanowire growth rate by accounting for the change in reactant concentrations throughout the growth (Govender et al., 2004; Mateo-alonso et al., 2012; Tak and Yong, 2005), but generally these attempts have been hampered by complicated Zn^{2+} speciation in the reaction conditions (Baruah and Dutta, 2009; Joo et al., 2011; Richardson and Lange, 2009). Continuous-flow (CF) growth, in which reactants are continuously supplied to maintain the reactant concentration throughout growth, has shown promise in overcoming these challenges by maintaining consistent growth conditions and it is therefore an ideal test bed for studying ZnO nanowire growth mechanisms (Chen and Yin, 2012; Morin et al., 2010).

However, in addition to the requirement that bulk reaction conditions remain constant throughout the nanowire growth process, the local environment of individual nanowires must also be considered (Lee et al., 2015). Since reactant transport and incorporation are serial events, either can be the rate-limiting step for crystal growth. The distinction is important, as we will find that operating in reaction rate-limiting regimes is critical to achieving uniform growth independent of pattern fill factor. Previous studies have demonstrated the importance of distinguishing between the growth of sparse vs dense nanowire arrays, indicating effects of nanowire density (Boercker et al., 2009) and diameter (Lee et al., 2015) on the rate of nanowire growth. Reactant transport to nanowire growth planes can vary based on these and other parameters and this variation can obscure the relationship between bulk reactant concentrations and growth rate.

To overcome these challenges in controlling the growth conditions for ZnO nanowire arrays, Cheng et al. used a CF setup in which precursors were continuously replenished to maintain homeostatic reactant supersaturation (Cheng et al., 2016). Based on the advancements made in growing nanowire arrays with varied areal densities, they were also able to compare the growth of wires across different densities to determine the effects of differences in local environment.

Nanowires were grown at identical temperature and Zn^{2+} concentration over a range of substrates, with seed density varying between $4 \times 10^{10}\,\mathrm{cm}^{-2}$ and $1.7 \times 10^{12}\,\mathrm{cm}^{-2}$ and dense seed layers deposited by spin coating or sputtering (to eliminate any possible effects of the seed deposition method (Nicaise et al., 2015a)). Growth rate was compared to both nanowire density and the total area ratio, S, of c-planes exposed to growth solution. The unitless ratio S was defined to be:

$$S = \rho A \tag{3}$$

where ρ is the nanowire areal density (cm^{-2}) and A is the mean nanowire c-facet area (cm^2). This value gives an instantaneous measurement of the substrate area fraction containing c-planes upon which Zn^{2+} growth units are impinging.

Unexpectedly, Cheng et al. found that nanowire density did not affect growth rate; rather growth rate was dependent solely on the c-plane area ratio S (Cheng et al., 2016). Therefore, a dense array of thin wires (Fig. 12A) was found to grow at the same rate as a sparse array of thick wires (Fig. 12B) as indicated in Fig. 12C as points A and B. This was further confirmed for

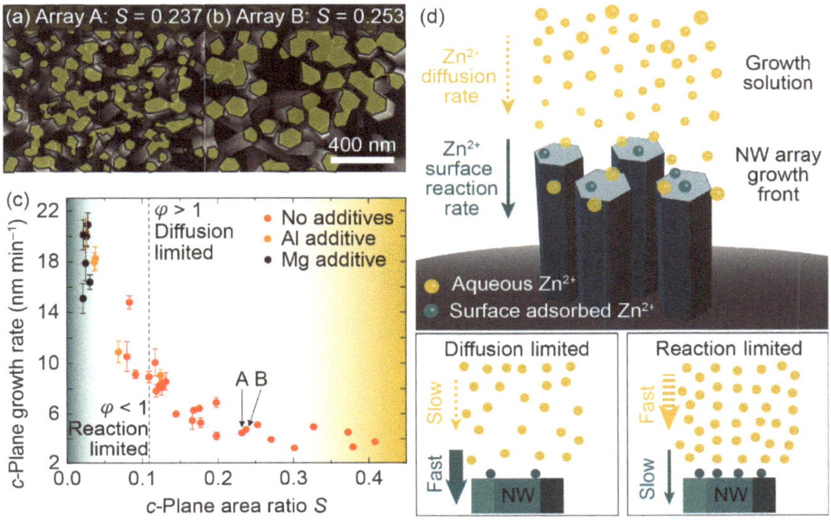

Fig. 12 Dense array of thin ZnO nanowires (A) and sparse array of thick ZnO nanowires (B) with the same *c*-plane area ratio and growth rate. (C) *c*-Plane growth rate vs *c*-plane area ratio. Nanowires from (A) and (B) are indicated as points A and B. The *dotted line* at *c*-plane area ratio of 0.11 indicates the transition from reaction-limited to diffusion-limited growth. (D) Illustration of diffusion- and reaction-limited growth processes for hydrothermally grown ZnO nanowires. *Reprinted with permission from Cheng, J.J., Nicaise, S.M., Berggren, K.K., Gradečak, S., 2016. Dimensional tailoring of hydrothermally grown zinc oxide nanowire arrays. Nano Lett. 16, 753–759. Copyright 2016 American Chemical Society.*

low S by growing samples with varied concentration of Al^{3+} or Mg^{2+} to inhibit m-plane growth (Joo et al., 2011) and reach S values as low as 0.02.

To explain this relationship, Cheng et al. adapted a diffusion/reaction-limited analysis (Boercker et al., 2009) that relates the consumption of Zn^{2+} at the c- and m-ZnO crystal planes to the diffusion of Zn^{2+} from the bulk solution (Fig. 12D) (Cheng et al., 2016). By assuming that (1) Zn^{2+} flux to the nanowire surface was equal to Zn^{2+} incorporation, (2) steady-state diffusion, and (3) first-order kinetics, the authors derived an equation for the c-plane growth rate, $R_{c\text{-plane}}$, in terms of external parameters:

$$R_{c-\text{plane}} = \frac{k_1 C_\infty D}{\rho_{\text{ZnO}} D + \rho_{\text{ZnO}} \delta k_1 S} \tag{4}$$

where k_1 is the surface reaction rate for dehydration of surface adsorbed $[Zn(H_2O)]^{2+}$ to crystal ZnO; C_∞ is the bulk solution concentration of Zn^2; ρ_{ZnO} is the ZnO molar density ($4.20 \times 10^{28}\,\text{m}^{-3}$); D is the aqueous Zn^{2+}

tracer diffusion coefficient (2.91×10^{-9} m^{2}s^{-1}) (Oelkers and Helgeson, 1988); δ is the stagnant layer thickness (the thickness of fluid adjacent to the substrate surface with only diffusive transport and no flow); and S is the *c*-plane area ratio. Since *c*-plane growth rate, $R_{c\text{-plane}}$, and *c*-plane area ratio, S, were readily available from the experimental measurements, the authors took the reciprocal of Eq. (4) to obtain a relationship that is linear with S to facilitate fitting:

$$\frac{1}{R_{c-\text{plane}}} = \frac{\rho_{\text{ZnO}}}{k_1 C_\infty} + \frac{\rho_{\text{ZnO}}\delta}{C_\infty D} S \tag{5}$$

The reciprocal growth rate vs *c*-plane area ratio for the nanowire arrays grown with CF synthesis at $[Zn^{2+}] = 10$ mM and 90°C, depicted in Fig. 12A–C, was fitted using Eq. (5) and yields a surface reaction rate $k_1 = 3.19 \times 10^{-6}$ ms^{-1} and a stagnant layer thickness $\delta = 1.6 \times 10^{-2}$ m.

From this analysis, the researchers were able to extract the Thiele modulus, ϕ, which measures the ratio of surface reaction rate to that of reactant diffusion to the crystal growth front (Eq. 6):

$$\phi = \frac{k_1 S \delta}{D} \tag{6}$$

Due to the serial nature of reactant diffusion and surface reaction (Fig. 12D), ϕ can be used to measure the degree to which each of these processes limit the nanowire growth. More specifically, when $\phi = 1$, the growth switches between reaction- and diffusion-limited. This crossover point is at $S = 0.11$ in the conditions used by Cheng et al. ($[Zn^{2+}] = 10$ mM at 90°C), as depicted in Fig. 12C. Therefore, when the *c*-plane area ratio is below 0.11, growth is limited by reaction at the nanowire surface due to the small area for reaction (Fig. 12D, right inset). When the *c*-plane area ratio is greater than 0.11, growth is instead limited by diffusion of reactant to the nanowire surface since the large *c*-plane surface area allows for rapid consumption of reactant (Fig. 12D, left inset).

As mentioned earlier, the growth mechanism can be identified by determining the relationship between *c*-facet growth rate and supersaturation. However, in the case of hydrothermal ZnO nanowire growth, changes in Zn^{2+} concentration result in changes to nanowire density and diameter and therefore affect the diffusion environment. As a result, the contribution of surface reaction and diffusion to growth rate under varied Zn^{2+} concentration cannot be differentiated. However, the diffusion/reaction analysis described above provided a tool for Cheng et al. to separate these two processes as Zn^{2+}

concentration is changed by accounting for the effect of variation in *S* on measured *c*-plane growth rates (Cheng et al., 2016). The diffusion-corrected *c*-plane growth rates ($R_{c\text{-plane}}\ S$) showed a linear relationship with Zn^{2+} concentration, in agreement with Eq. (7) and indicative of *c*-plane growth by direct addition of Zn^{2+}. Large crystals of ZnO grown using hydrothermal synthesis are known to grow via direct incorporation on the *c*-facet, so this finding is not unexpected (Janotti and Van de Walle, 2009). Growth via direct incorporation, also known as normal growth, has a linear relationship with concentration because it relies on adatoms falling directly into kinks on a rough crystal facet surface.

By identifying the *c*-facet growth mechanism, the authors were able to provide an understanding of how Zn^{2+} concentration controls the nanowire growth rate. Moreover, this result has important implications for the interaction of the *c*-facet with impurities or additives, which were mentioned in the previous section for their potential to enable further control of nanowire growth and final morphology. The effects of additives on nanowire growth rates are largely dependent on facet growth mechanism (Kubota and Mullin, 1995; Sangwal, 1996), and the *c*-facet is especially relevant because it is commonly the site of dopant incorporation due to its higher surface energy compared to the *m*-facet (Baruah and Dutta, 2009; Tay et al., 2012). Understanding these effects will allow further tuning of growth rates via impurity addition, which we discuss next.

3.2.3 Direct Control of Nanowire Growth Rates

Controlling the growth rate of ZnO nanowires independent of their spacing is important for achieving uniform nanowire length over large substrates with potential variation in spacing from edge effects and nonuniformities in the underlying seed layer. As discussed at the beginning of this section, uniformity in nanowire length is itself important for solar cell device performance. The diffusion/reaction analysis described above suggests that growth under reaction-limiting conditions yields growth rates independent of the nanowire diffusion environment, resulting in uniform nanowire growth across varied nanowire densities. To test this hypothesis, Cheng et al. used electron-beam lithography to define a set of nanowire rows with varied distances (Fig. 13)—thereby establishing a locally tuned nanowire environment—and used citrate ion adsorbents to achieve reaction-limited growth by restricting incorporation of Zn^{2+} into the *c*-facet surface (Cheng et al., 2016; Tian et al., 2003). As can be seen in Fig. 13, each individual row of nanowires had a different local environment

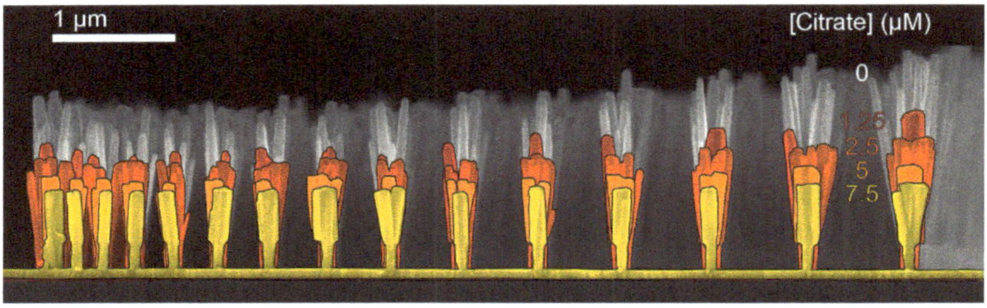

Fig. 13 ZnO nanowires templated by electron-beam lithography to create rows of nanowires with varied spacing between the rows. Cross-sectional SEM images of templated arrays grown with varied citrate concentrations from 0 to 7.5 μM (wires are artificially colored to indicate citrate concentration). SEM images of nanowires grown with increasing citrate concentrations are superimposed on an SEM image of nanowires grown with no citrate ions, demonstrating decreased nanowire length and increased uniformity with increased citrate concentration. *Reprinted with permission from Cheng, J.J., Nicaise, S.M., Berggren, K.K., Gradečak, S., 2016. Dimensional tailoring of hydrothermally grown zinc oxide nanowire arrays. Nano Lett. 16, 753–759. Copyright 2016 American Chemical Society.*

due to the graded row spacing—*S* peaks toward the left of the array, where rows are most closely packed, and decreases as the pitch is increased.

Cheng et al. grew four samples under varying citrate concentrations from 0 to 7.5 μM, with increased citrate concentration resulting in shorter nanowires, i.e., reduced growth rate (Fig. 13) (Cheng et al., 2016). Specifically, as citrate concentration was increased, the surface reaction rate constant k_1 (calculated using Eq. 5) was found to decrease, suggesting that *c*-plane growth was inhibited by citrate adsorbents. As further evidence that this change in growth rate was based on a reaction-limiting mechanism, the Thiele modulus and stagnant layer thickness also decreased with increased citrate concentration. As a consequence of this reaction-limited growth, nanowire uniformity across the array of differently spaced wires was improved (length variation dropped from 41% to 16%) (Fig. 13), suggesting a method for controlling nanowire uniformity in device-relevant applications. Therefore, the work by Cheng et al. demonstrates that rate-limiting conditions, dictated by growth conditions, are important to precisely controlling the growth rate of ZnO nanowires. More specifically for solar cell applications, these results imply that nanowires of controlled and uniform length with spacing tailored to the active material can be grown precisely without use of experimental trial and error by operating in a reaction-limited regime.

In this section, we have addressed the important nanowire array parameters for optimizing the use of ZnO nanowires in photovoltaic devices and discussed recent research which has facilitated control over nanowire spacing, length, uniformity, and growth rate. In the next section, we will discuss how these nanowires can be used to increase the efficiency of solar cells which suffer from short minority carrier diffusion lengths.

4. DEVICE CONSIDERATIONS

In this section, we discuss the application of ZnO nanowires to PV devices to enhance device performance. We focus on QD solar cells due to the promising efficiency gains and relatively high stability observed for these systems; however, the concepts discussed here can be directly applied to organic and other nanostructured solar cells. We first summarize early work in applying ZnO nanowires to QD solar cells and how materials selection has been used to improve these devices. We then cover recent attempts at understanding the complex optical and electronic effects of ZnO nanowire scaffolds. Particularly, we discuss the mechanisms by which ZnO nanowires improve device performance as well as future avenues for improvement.

4.1 Enhanced Efficiency With ZnO Nanowires

One of the most common architectures for efficient QD devices is a p–n junction made up of PbS nanoparticles and ZnO. PbS QDs are attractive as a PV material due to their compatibility with low-cost solution deposition techniques, which can also be achieved on a small thermal budget. Furthermore, because PbS QDs have a direct band gap, the strong absorption coefficient of $\alpha > 10^4\,\mathrm{cm}^{-1}$ for most photon energies above the band gap implies that only ~1 μm film thickness is needed for complete absorption. Due to their relative doping and band alignment, PbS and ZnO form a suitable p–n junction, and, since ZnO is largely transparent to visible and infrared light, devices based on ZnO and PbS can be illuminated through the semitransparent ZnO layer with minimal losses from parasitic absorption. However, as mentioned previously in this chapter, QD-based solar cells suffer from poor charge extraction due to the short carrier diffusion lengths. While the majority of PbS QD solar cells made on ZnO use a compact film of ZnO or ZnO nanoparticles, Jean et al. demonstrated that ZnO nanowires enhance charge extraction (Fig. 14) (Jean et al., 2013). By replacing the compact ZnO film with a dense array of ZnO nanowires, the authors were able to improve device efficiencies from 3.2% to 4.3% (Jean et al., 2013). More recently, Wang et al. used ZnO nanowires over 1 μm in length with larger QDs (smaller band gap) than used by Jean et al. and demonstrated efficiencies

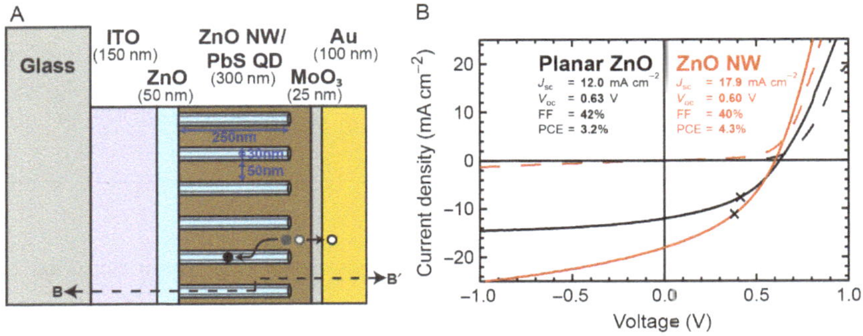

Fig. 14 (A) Schematic of an ordered bulk heterojunction solar cell made up of ZnO nanowires and PbS quantum dots. (B) Current density–voltage measurements of planar and nanowire-based ZnO/PbS quantum dot solar cells indicating enhanced performance with ZnO nanowires. *Adapted with permission from Jean, J., et al., 2013. ZnO nanowire arrays for enhanced photocurrent in PbS quantum dot solar cells. Adv. Mater. 25, 2790–2796. Copyright 2013 John Wiley and Sons.*

of up to 6.1% (Wang et al., 2013). These devices showed large short-circuit current due to their thick active layer and the small band gap of the absorbing material; however, they had a poor fill factor and open-circuit voltage. These deficits were likely related to the long nanowires, causing increased recombination due to the large number of interfacial defects and increased depletion volume, as well as the small band gap QDs, which limit the energy extracted per photon absorbed (Wang et al., 2013).

To overcome these challenges, Rekemeyer et al. further demonstrated a combination of ligand selection and nanostructured electrodes to push the efficiency of nanowire-QD PV devices up to 9.6% while using relatively short nanowires and larger band gap QDs (Fig. 15) (Rekemeyer et al., 2016). In a typical PbS QD solar cell device, a single ligand exchange treatment is used for all QD layers, forming a film with uniform conduction and valence band energy levels. However, since ligand selection can be used to modify band edge energies of the QDs, QDs with varied ligands can be layered to modify the electronic nature of the QD film (Chuang et al., 2014). Rekemeyer et al. applied this principle to ZnO nanowire-based PbS QD solar cells to assist in charge extraction and to reduce carrier recombination at the anode. Specifically, the authors selected the architecture shown in Fig. 15 that consists of 12–15 layers of PbS QDs with tetrabutylammonium iodide (TBAI) used to replace the native QD ligands and 2 layers with 1,2-ethanedithiol (EDT) used for this replacement (Rekemeyer et al., 2016). As a result, a 0.2-eV

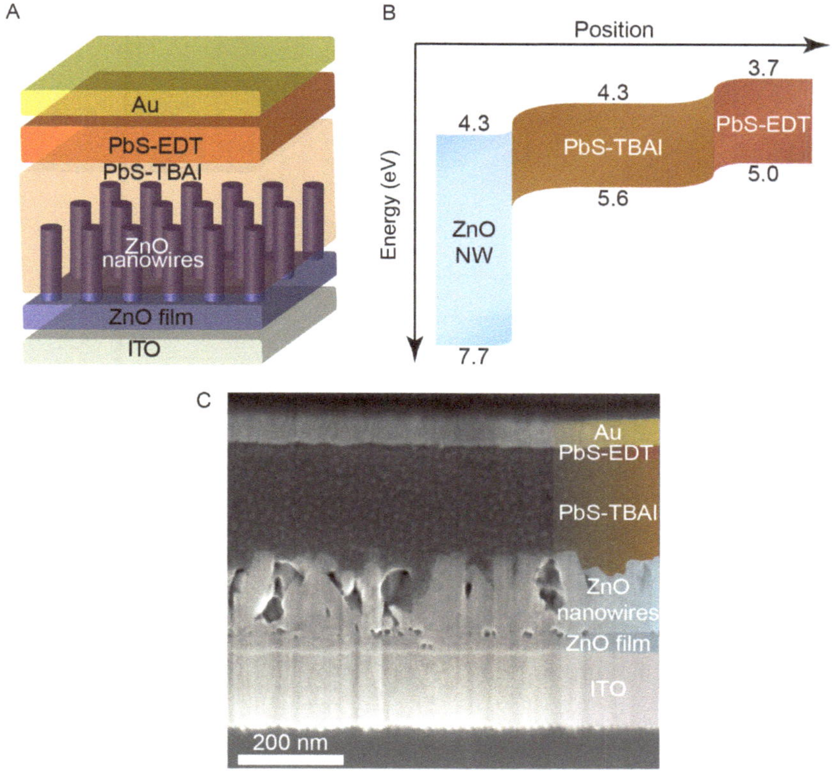

Fig. 15 (A) Schematic of ZnO nanowire-based PbS quantum dot solar cells with 12 layers of PbS quantum dots coated in TBAI ligands and 2 layers of PbS quantum dots coated in EDT ligands. (B) Energy-level diagram qualitatively indicating the energetic band alignment of ZnO and PbS quantum dots with TBAI and EDT ligands. (C) Cross-sectional SEM image of ZnO nanowire-based PbS quantum dot solar cell with layers indicated. *Adapted with permission from Rekemeyer, P.H., et al., 2016. Enhanced photocurrent in PbS quantum dot photovoltaics via ZnO nanowires and band alignment engineering. Adv. Energy Mater. 6, 1600848. Copyright 2016 John Wiley and Sons.*

conduction band offset was formed at the intersection of these sets of layers, promoting charge extraction from the QD film (Chuang et al., 2014).

To quantify the benefit of incorporating ZnO nanowire arrays in these devices, the authors compared devices on planar ZnO to the ones containing ZnO nanowires that are otherwise identical (Rekemeyer et al., 2016) (Fig. 16). On average, ZnO nanowire devices showed an 11% improvement in PCE over planar devices with a particularly large increase in short-circuit current. The highest performing nanowire devices reached 9.6% PCE and set the

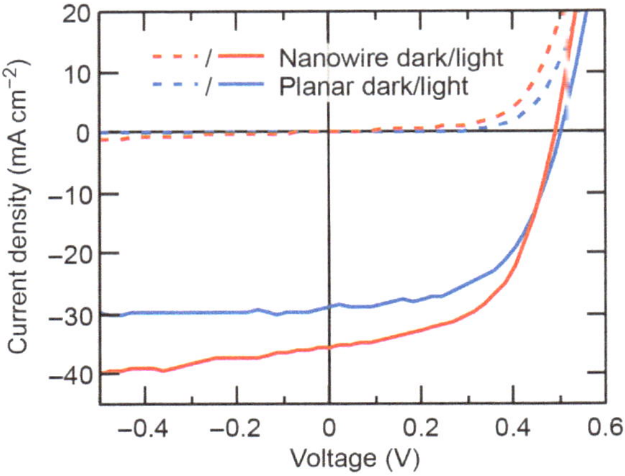

Fig. 16 Current density–voltage measurements of planar and nanowire-based ZnO/PbS-TBAI/PbS-EDT solar cells prepared under identical conditions demonstrating enhanced current extraction in nanowire-based devices. Measurements were performed after 20 days of storage in ambient conditions. *Adapted with permission from Rekemeyer, P.H., et al., 2016. Enhanced photocurrent in PbS quantum dot photovoltaics via ZnO nanowires and band alignment engineering. Adv. Energy Mater. 6, 1600848. Copyright 2016 John Wiley and Sons.*

record short-circuit current (>30 mA cm^{-2}) for PbS QD solar cells with the ideal 1.3 eV band gap (Carey et al., 2015; Wang et al., 2013).

Additionally, long-term stability testing was performed to ensure that the nanowire device architecture maintains the same performance stability in air as planar band alignment engineered QD solar cells (Chuang et al., 2014; Rekemeyer et al., 2016). The PCE was found to decrease only ~20% from the peak values after 3 months of storage in air without encapsulation. The increasing device performance over the first 20 days of aging may be attributed to oxidation of the PbS-EDT layer, which increases the hole doping (Klem et al., 2008), and therefore enhances the strength of the electric field at the junction with the PbS-TBAI film.

4.2 The Role of Nanowires

Having demonstrated the benefits of ZnO nanowires in improving the efficiency of PbS QD-based photovoltaic devices, we now discuss progress that has been made in understanding the causes of this enhancement. As mentioned earlier in this chapter, the motivation for applying ZnO nanowires to existing QD or organic solar cells comes from the short minority carrier diffusion lengths observed in these materials. We will therefore first discuss work focused on quantifying QD carrier diffusion lengths in relationship with the common nanowire dimensions.

While ZnO nanowire arrays are primarily used to decouple the directions of light absorption and carrier extraction, the ZnO nanowires have additional unintentional effects on these devices. For example, the presence of the nanowire array in the device has the potential to modify the absorption of light in the active layer due to light scattering, and the existence of a large interfacial area is likely to affect recombination. We have already alluded to this second effect as a possible cause of the decreased fill factor and open-circuit voltage for nanowire-based devices. Therefore, in this section we will also discuss work that has been done to identify the primary causes of benefits to the PbS QD layer by incorporating ZnO nanowires, specifically separating effects of absorption and extraction. We will then cover recent work that aims to identify possible detriments from these ZnO nanowires, pointing to possible areas of improvement in these devices.

4.2.1 Quantification of Carrier Diffusion Lengths

As mentioned previously, nanowire arrays are primarily used in solar cell devices to overcome limitations of short minority carrier diffusion lengths, and the dimensions of these arrays can be tailored based on the electronic properties of the active material to maximize charge extraction. It is therefore important to quantify the minority carrier diffusion lengths in device-relevant materials. Rekemeyer et al. developed a framework for measurement of short carrier diffusion lengths using electron beam-induced current (EBIC) measurements and PbS QDs as a model system (Rekemeyer et al., 2017). EBIC measurements measure minority carrier diffusion in full solar cell devices and therefore reflect the electronic properties of the active film after all fabrication steps. This is particularly important in the case of QD PV devices, where fabrication can include multiple chemical treatments that affect charge transport.

Cross-sectional EBIC measurements involve scanning an electron beam across the layers of an electronic device (Fig. 17). When the electron beam is scanned over regions with an internal electric field, carriers generated by the electron beam are swept across the device and measured as current. As a result, cross-sectional EBIC measurements can identify active junctions as well as provide information about the electronic nature of these junctions and the charge carriers in the material.

More specifically, cross-sectional EBIC measurements reveal the carrier collection probability, ϕ, through the thickness of a device. In the one-dimensional case (which can be obtained by treating carrier generation in

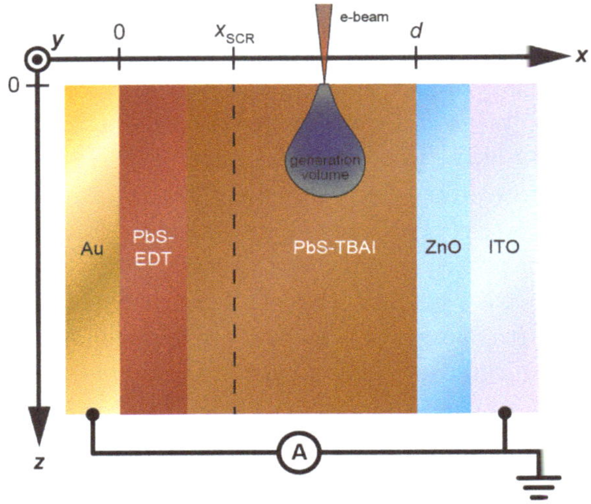

Fig. 17 Schematic of cross-sectional EBIC measurements on PbS quantum dot solar cell devices. *Reprinted with permission from Rekemeyer, P., Chuang, C.-H.M., Bawendi, M.G., Gradecak, S., 2017. Minority carrier transport in lead sulfide quantum dot photovoltaics. Nano Lett. 17, 6221–6227. Copyright 2017 American Chemical Society.*

each dimension as independent), this carrier collection probability is related to the minority carrier diffusivity, D, and lifetime, τ, by the relationship (Donolato, 2012):

$$D\frac{d^2\phi(x)}{dx^2} - \frac{\phi(x)}{\tau} = 0 \tag{7}$$

Therefore, measuring the 1D carrier collection probability profile $\varphi(x)$ is equivalent to measuring the diffusion of carriers.

The EBIC current at a given point in a sample is dependent on both the carrier collection probability at that point and the nature of the electron beam used to excite carriers in the sample. More specifically, the electron beam at any given point will excite carriers within a certain volume, termed the generation volume, and EBIC current will reflect the carrier collection probability within this generation volume. Rekemeyer et al. developed a quantitative model of the generation volume in the materials of interest by performing Monte Carlo simulations and modifying parameters of an axisymmetric generation model to best fit the simulated data (Rekemeyer et al., 2017). This axisymmetric model, based on the more commonly used Donolato model, was created by the authors to separate the axial and radial components of the generation profile. Having determined the form of the generation function, Rekemeyer et al. deconvoluted the experimental collection probability function from the EBIC carrier collection profile using the equation:

$$\vec{I} = G\vec{\phi} \tag{8}$$

where G is a matrix representing the discretized generation function, and $\vec{I}$ and $\vec{\phi}$ are vectors representing the one-dimensional discrete EBIC and carrier collection probability profiles, respectively (Rekemeyer et al., 2017). The form of $\phi(x)$ could then be solved for in terms of the effective diffusion length of minority carriers by applying appropriate boundary conditions to obtain Eq. (7) (Scheer and Schock, 2011):

$$\phi(x) = \frac{\cosh\left(\frac{x-d}{L_{\text{eff}}}\right) - s_i \sinh\left(\frac{x-d}{L_{\text{eff}}}\right)}{\cosh\left(\frac{d-x_{\text{SCR}}}{L_{\text{eff}}}\right) - s_i \sinh\left(\frac{d-x_{\text{SCR}}}{L_{\text{eff}}}\right)} \tag{9}$$

where τ_{eff} is the effective carrier lifetime, d is the device thickness and x_{SCR} is the position of the edge of the space charge region in the device. This carrier collection probability (Eq. 9) was then fit to the deconvoluted carrier collection profile to extract the value of the effective diffusion length.

The full process of extracting an effective diffusion length from raw EBIC data on a cross-sectional device sample is shown in Fig. 18.

As measured diffusion lengths are highly affected by surface recombination due to the shallow injection depth of carriers into the sample, and therefore were highly dependent on the accelerating voltage (higher accelerating voltage results in deeper carrier injection and higher diffusion lengths). The bulk minority carrier diffusion length relevant for PV devices was therefore extrapolated according to the equation:

$$L_{\text{eff}} = L_{\text{bulk}} \sqrt{1 - \exp\left(-\frac{\overline{z}}{L_{\text{bulk}}}\right)} \tag{10}$$

where $\overline{z}$ is the mean carrier injection depth which is determined by the accelerating voltage of the electron beam. Using this method, a lower bound on the bulk carrier diffusion length of the PbS QDs was found to be 110 nm (Rekemeyer et al., 2017) (Fig. 19).

The EBIC measurements and analysis described in this section can be applied to other materials with short carrier diffusion lengths that are difficult

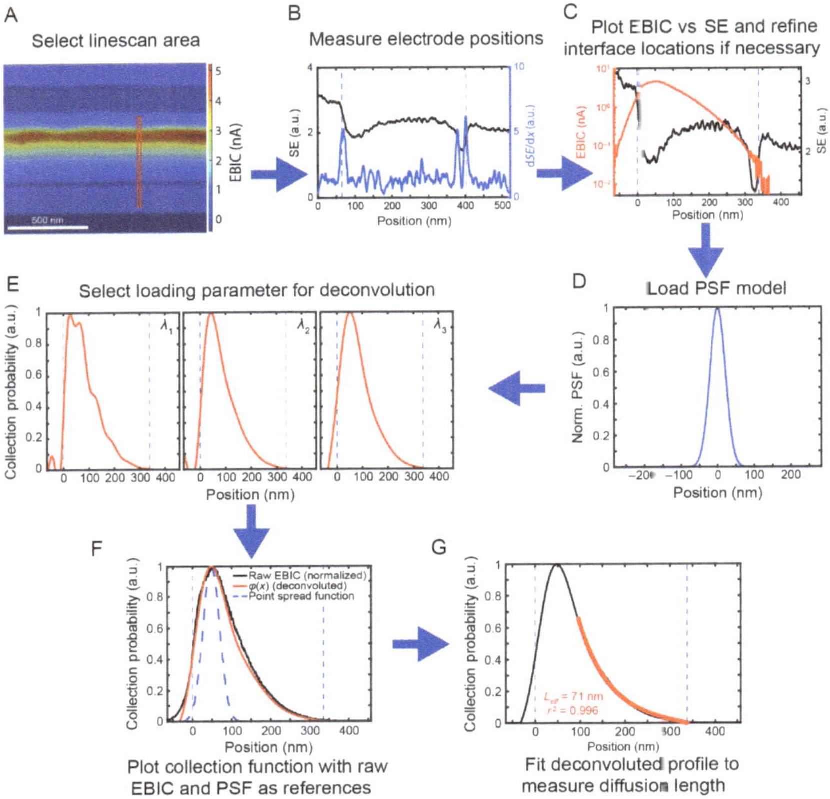

Fig. 18 Process flow for quantitative EBIC analysis to extract minority carrier diffusion lengths from cross-sectional EBIC line scans. A thin slice selected from the EBIC map (A) and SE linescan (B) are overlayed (C) to establish the position of the electrodes. Next, the point-spread function (PSF) of the electron-hole pair generation volume along the x-axis is loaded (D) and a suitable loading parameter must be selected (E). The deconvoluted carrier collection profile, raw EBIC linescan, and PSF are then compared (F) and the effective diffusion length can be estimated by fitting the quasi-neutral region of the deconvoluted carrier collection profile (G). *Adapted from Rekemeyer, P.H., 2017. Nanostructured photovoltaics: improving device efficiency and measuring carrier transport. Copyright 2017 Massachusetts Institute of Technology.*

to measure by other means. By applying such methods to the case of lead sulfide QDs, it is clear that the minority carrier diffusion lengths are in fact a limiting factor for these devices. Additionally, the authors of this study found that the active junction in these devices was not the ZnO/PbS-TBAI interface as expected, but rather the PbS-TBAI/PbS-EDT interface, suggesting that the PbS-TBAI layer was in fact n-type rather than p-type

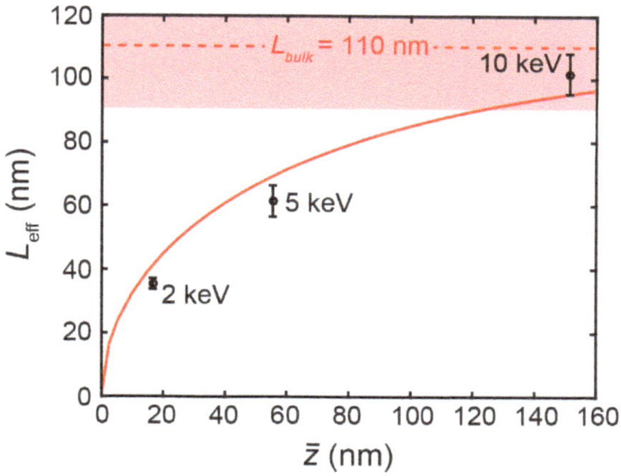

Fig. 19 Effective minority carrier diffusion length in PbS quantum dot solar cells vs mean carrier injection depth ($\bar{z}$), measured through cross-sectional EBIC measurements at varied accelerating voltages. Extrapolation to infinite $\bar{z}$ (where the role of surface recombination is minimized) gives an estimated bulk diffusion length of 110 nm. *Reprinted with permission from Rekemeyer, P., Chuang, C.-H.M., Bawendi, M.G., Gradecak, S., 2017. Minority carrier transport in lead sulfide quantum dot photovoltaics. Nano Lett. 17, 6221–6227. Copyright 2017 American Chemical Society.*

as was previously assumed (Rekemeyer et al., 2017). This realization brings into question the role of the n-type ZnO nanowires in enhancing the performance of these QD devices, which will be discussed in the next section.

4.2.2 Benefits: Absorption vs Extraction

While ZnO nanowires are generally applied to short minority carrier diffusion length devices to assist in carrier extraction, as mentioned previously in this section, there remain questions regarding the relative contributions of optical and electronic effects of these nanowires on improvements to device efficiency. To separate the effects of these potential electronic and optical processes, Rekemeyer et al. used a combination of quantum efficiency and light-harvesting efficiency (LHE) measurements (Rekemeyer et al., 2016).

Quantum efficiency of optoelectronic devices refers to the ratio of charge to photon conversion (and vice versa). In PV devices, external quantum efficiency is the product of efficiencies corresponding to various processes that occur between light incidence and charge extraction. Apart from being an important metric for comparing various device architectures and active layers, performing different types of quantum efficiency measurements isolates specific processes, which are depicted in Fig. 20. In the inverted device architecture typical to ZnO nanowire-based PbS QD solar cells, incident AM1.5G

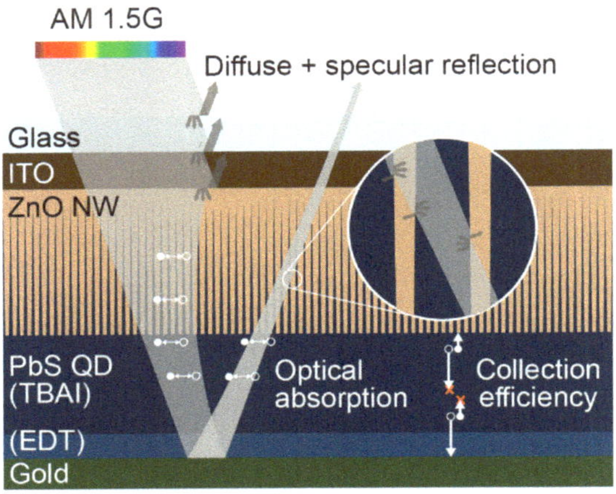

Fig. 20 Illustration of optical and electronic processes which occur under the illumination of ZnO nanowire-based PbS solar cells. *Reprinted with permission from Cheng, J.W.J., 2016. Hydrothermal synthesis of zinc oxide nanowire arrays for photovoltaic applications. Copyright 2016 Massachusetts Institute of Technology.*

illumination enters the device through the glass superstrate. At each material interface (namely air–glass, glass–ITO, ITO–ZnO, ZnO–QD, and QD–Au) light is transmitted, reflected specularly, or reflected diffusely. Light that is transmitted into the QD active layer is absorbed; light intensity roughly decreases as it proceeds on its particular path. Upon reaching the back electrode, the light is almost fully reflected at the QD–Au interface, thus completing a double pass before exiting the active layer. Absorption of light in the active layer causes generation of electrons and holes via the photovoltaic effect. The charge generation rate is highly spatially nonuniform—shorter wavelengths have a lower penetration depth and are absorbed near the ZnO–QD interface, while longer wavelengths are absorbed toward the QD–Au interface. The ratio of charge generation rate to incident photon flux, irrespective of spatial distribution, is the LHE. Generated carriers are then separated by the junction and collected by their respective electrodes, and the ratio of this collection rate to the total charge generation is the internal quantum efficiency (IQE). Finally, the external quantum efficiency of the device is given by the product of the LHE and IQE, quantifying the ratio between the total current collection rate and the flux of incident photons. By consequence, IQE reflects the efficiency of extracting photo-generated carriers, while EQE reflects the overall efficiency of absorbing incident photons and converting that excitation into collected charge.

Comparison of EQE spectra (Fig. 21A) for planar and ZnO nanowire-based PbS solar cells has shown an improved external quantum efficiency

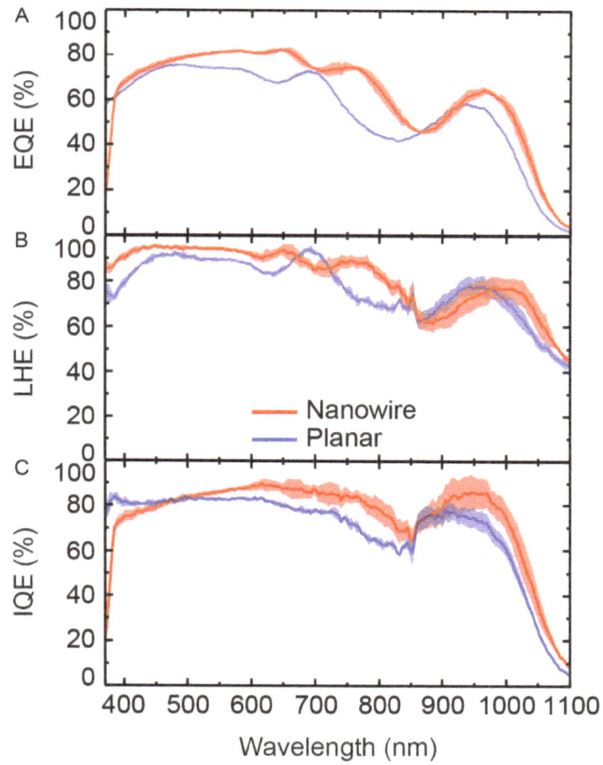

Fig. 21 (A) External quantum efficiency (EQE) and (B) light-harvesting efficiency (LHE) measurements of planar and nanowire-based ZnO/PbS quantum dot solar cells. (C) Internal quantum efficiency (IQE) was calculated by dividing the EQE spectra by the LHE spectra. *Shaded regions* indicated 95% confidence interval of the measurements. *Adapted with permission from Rekemeyer, P.H., et al., 2016. Enhanced photocurrent in PbS quantum dot photovoltaics via ZnO nanowires and band alignment engineering. Adv. Energy Mater. 6, 1600848. Copyright 2016 John Wiley and Sons.*

for nanowire-based devices over most wavelengths; however, effects of absorption and charge extraction cannot be differentiated based solely on EQE (Rekemeyer et al., 2016). Therefore, Rekemeyer et al. measured the IQE and LHE of both device architectures to separately observe differences in optical absorption and carrier extraction (Rekemeyer et al., 2016).

LHE measurements revealed light-trapping effects caused by the ZnO nanowire array. In the visible portion of the spectrum, modest wavelength-dependent differences in the LHE of these two architectures (Fig. 21B) were attributed to thin-film interference effects. However, in the near-IR (at photon wavelengths greater than 700 nm), there was a clear improvement in LHE for nanowire-based devices. This enhancement was particularly clear over

1000 nm, near the QD absorption edge. The two architectures were fabricated with equal amounts of active material, so differences in absorption cannot be due to different device thicknesses. Additionally, ZnO is transparent in this wavelength range and therefore parasitic absorption from the nanowires could not have caused the observed increase in absorption. The authors therefore concluded that the enhanced LHE at long wavelengths was caused by improved light trapping. Due to the low absorption coefficient of the QDs in this spectral region, this long wavelength light is not fully absorbed by the film even after reflection off the gold contact (and therefore two optical passes through the film) is accounted for. In the case of weakly absorbing near-IR photons, wave guiding from individual nanowires or scattering by the nanowire array could increase the optical path length of these photons in the active material (Mokkapati and Catchpole, 2012).

In addition to the increase in LHE in the near-IR for nanowire devices, Rekemeyer et al. observed a red shift in the excitonic peak position for nanowire devices, which they attributed to the thin-film interference effects like those observed in the visible wavelength range. Transfer matrix simulations (Burkhard et al., 2010) for a planar device stack consisting of 145 nm ITO/variable ZnO (50–250 nm)/400 nm PbS-TBAI/45 nm PbS-EDT/150 nm Au are shown in Fig. 22. These simulations show a qualitative agreement with the empirical quantum efficiency data in that thin-film interference

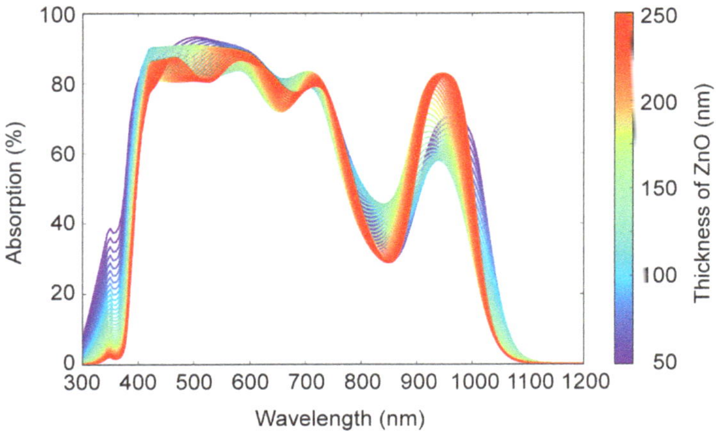

Fig. 22 Simulated absorption profiles for planar ZnO/PbS quantum dot solar cells with varied ZnO thickness showing a shift in the excitonic peak due to thin-film interference effects. *Simulations courtesy of Chia-Hao Chuang and reprinted with permission from Rekemeyer, P.H., 2017. Nanostructured photovoltaics: improving device efficiency and measuring carrier transport. Copyright 2017 Massachusetts Institute of Technology.*

effects shift the absorption profile, even with no change in the active layer thickness, and that the apparent peak position of the excitonic peak shifts by tens of nanometers.

IQE spectra (Fig. 21C) revealed an increase in carrier collection for nanowire devices over planar devices for wavelengths longer than 500 nm, which can be attributed to the increase in volume within the active QD layer which is within a diffusion length of a junction. The interpenetration of the ZnO nanowires in the PbS QD layer produces a 3D homo-type junction that helps funnel holes from the front of the device to the PbS-TBAI/PbS-EDT rectifying junction at the rear of the device. This effect is more pronounced for shorter wavelengths, for which the absorption coefficient is higher, and therefore the generation rate is higher, near the front of the device. However, the enhancement in charge collection could be compromised by increased interfacial recombination in the nanowire architecture. Moreover, the dominant recombination pathway in the QD film is bimolecular. In this case recombination in the depletion regions is significant, and the additional electric field from the homo-type junction could improve the efficiency of electron extraction (Rekemeyer, 2017).

Taken together, the results presented by Rekemeyer et al. indicate that the improvement in PbS QD solar cells observed when switching from a planar ZnO to a ZnO nanowire array comes from a combination of optical (light-harvesting) and electric (carrier collection) effects and therefore further clarify the role of ZnO nanowires in efficient solar cell devices.

4.2.3 Challenges: Surface Defects and Recombination Pathways

After evaluating the contribution of optical and electronic effects to the increase in efficiency in nanowire-based QD solar cells, we now turn to the limitations imposed by such nanowire scaffolds. As mentioned in Section 3 of ZnO nanowires for photovoltaic devices, the optimal nanowire length for PbS QD solar cells is expected to be around 400–600 nm. However, in work aimed at optimizing nanowire length and device thickness in PbS QD solar cells, Cheng et al. demonstrated that device performance begins to decrease for nanowires longer than 200 nm (Fig. 23) (Cheng, 2016). In contrast, the integrated external quantum efficiency, which indirectly measures the short-circuit current density of a device at lower light intensities, does not show this drop-off with increased nanowire length and shows the expected increase due to increased absorption of the incident light. In this section, we discuss the causes for this discrepancy as they relate to the challenges posed by ZnO nanowires in photovoltaic devices.

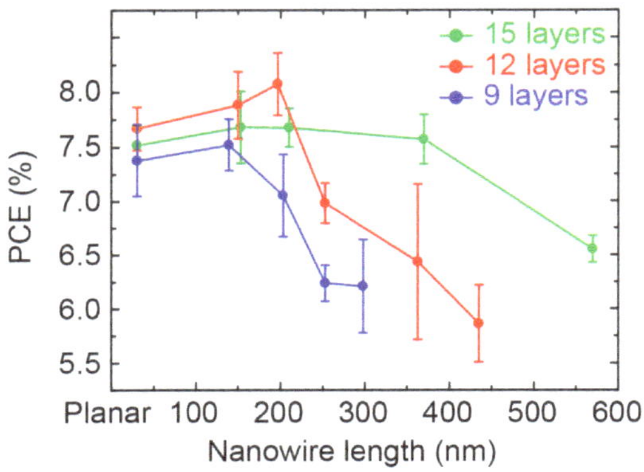

Fig. 23 Power conversion efficiency for ZnO/PbS quantum dot solar cells with varied ZnO nanowire length and thickness of the PbS quantum dot layer (i.e., number of deposited layers of PbS quantum dots). The maximum performance is obtained with nanowire lengths around 200 nm. *Reprinted with permission from Cheng, J.W.J., 2016. Hydrothermal synthesis of zinc oxide nanowire arrays for photovoltaic applications. Copyright 2016 Massachusetts Institute of Technology.*

In order to determine the cause of the decreasing short-circuit current density in nanowire-based devices with increasing nanowire length, Cheng et al. performed a number of measurements to separately measure the influence of optical absorption and collection efficiency (Cheng, 2016). By applying EQE and LHE methods (similar to those used by Rekemeyer, 2017) to devices with varied nanowire length, the authors were able to deduce the source of efficiency drop-off for devices with long nanowires.

Comparing the general shape of the LHE spectra and the EQE spectra for 15-layer devices (Fig. 24), the authors found clear qualitative similarities: both LHE and EQE spectra for the planar device exhibit interference peaks. However, the LHE and EQE spectra for nanowire devices are much smoother, which suggests that scattering is preventing optical modes from being set up. Additionally, nanowire devices had generally higher LHE over the visible and NIR regions. This supports claims by Rekemeyer et al. that part of the enhancement seen in EQE comes from nanowire optical scattering in the PV device (Rekemeyer et al., 2016).

However, since measured LHE was almost identical for all nanowire devices regardless of the nanowire length, it is unlikely that the discrepancy between $J_{SC,EQE}$ and J_{SC} can be explained by increased light harvesting from nanowire scattering. Additionally, the effects of enhanced light harvesting due to scattering are not expected to be intensity dependent in the range of

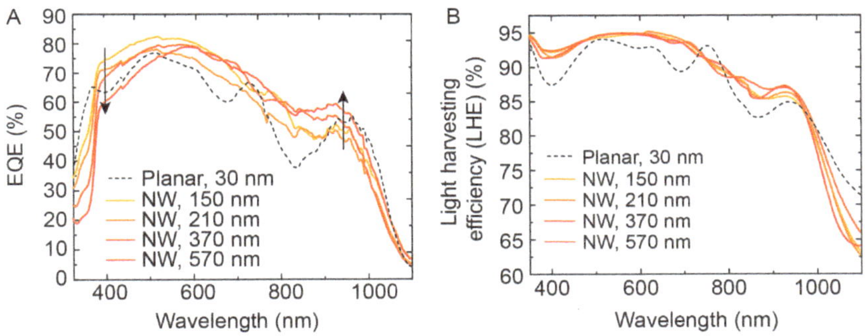

Fig. 24 (A) EQE and (B) LHE measurements for ZnO/PbS quantum dot solar cells with varied ZnO nanowire length. LHE is not dependent on nanowire length, however EQE shows wavelength-specific dependence indicative of effects related to current extraction. *Adapted with permission from Cheng, J.W.J., 2016. Hydrothermal synthesis of zinc oxide nanowire arrays for photovoltaic applications. Copyright 2016 Massachusetts Institute of Technology.*

intensities used for EQE and J–V testing. Therefore, the authors concluded that current extraction rather than optical absorption causes the intensity-dependent effects observed as the differing $J_{SC,EQE}$ and J_{SC} with increased nanowire length (Cheng, 2016).

Cheng et al. proposed two possible explanations for the dependence of current extraction on light intensity and the increase of this dependence with increasing nanowire length (Cheng, 2016). The first possible explanation considered the optical interaction of the incident light with the nanowire array. As nanowire length increases, scattering within the active layer increases, thereby extending the optical path length. The spatial optical generation profile also changes, with generation taking place deeper into the device. The altered generation profile could cause increased recombination rates in certain parts of the device where light is unintentionally concentrated, as the carrier concentration is higher than for the equivalent planar device. These rates could increase with increased light intensity and therefore reduce collection efficiency. The second possible explanation was related to the increased interface area between ZnO and the active material with increasing nanowire length. Increases in nanowire length are associated with greater interfacial recombination at the ZnO nanowire–QD interface since the total interfacial area increases. While subband gap state recombination dominates in planar devices (Chuang et al., 2015), interfacial recombination may dominate in nanostructured devices due to the large interface area between the ZnO and the active material

(Ehrler et al., 2013). In support of this explanation, the relationship between photocurrent and illumination intensity has been observed to be sublinear for devices with high recombination at the ZnO–QD interface due to space-charge-limiting conditions (Ehrler et al., 2013).

To test the first of these explanations, the authors performed frequency-domain time difference (FDTD) optical simulations to predict the effects of ZnO nanowires on light concentration in these devices (Cheng, 2016). The $J_{SC,FDTD}$ predicted by FDTD increased with nanowire length (indicated at the top of each modeled device architecture in Fig. 25), which was in agreement with the EQE results. Furthermore, some notable differences between the planar and nanowire device cross sections can be observed in the corresponding generation profiles. First, in the case of the nanowire devices, the spatial generation rate profile is not uniform along the *x*-axis; rather, carrier generation was concentrated at nanowire interstices. This contrasts the case of planar devices in which the generation profile remained independent of the *x*-coordinate. Second, optical scattering from the nanowires caused optical generation deeper (along the *z*-axis) into nanowire devices as compared to planar counterparts. Lastly, the total carrier generation rate increased by about 20% with the addition of nanowires, suggesting that carrier concentration could be increased by the light concentration observed at the nanowire interstices. This could result in increased recombination and lower collection

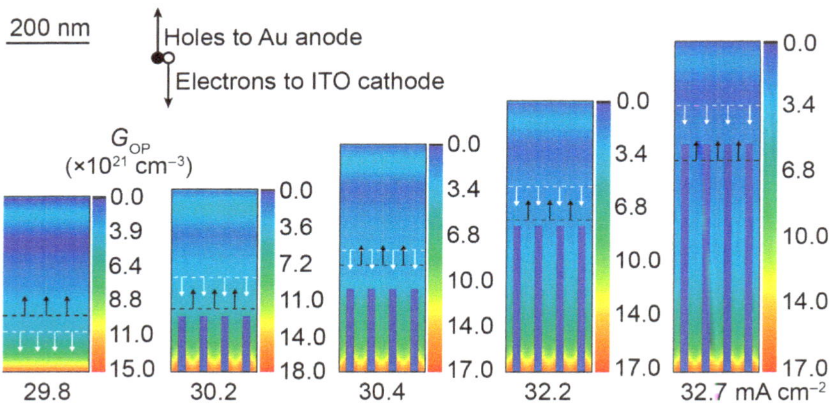

Fig. 25 Frequency-domain time difference optical simulations showing the spatial distribution of the optical generation rate through the cross-section of ZnO/PbS quantum dot solar cells with varied ZnO nanowire length and constant volume of PbS QDs. *Adapted with permission from Cheng, J.W.J., 2016. Hydrothermal synthesis of zinc oxide nanowire arrays for photovoltaic applications. Copyright 2016 Massachusetts Institute of Technology.*

efficiency at these locations. However, the J_{SC} calculated from these models tracked the $J_{SC,EQE}$ closely, suggesting that these potential losses are not large enough to explain the discrepancy between J_{SC} and $J_{SC,EQE}$. The authors therefore proceeded to test the possibility that interfacial recombination was responsible for the observed difference in J_{SC} and $J_{SC,EQE}$ (Cheng, 2016).

The electronic properties of ZnO nanowires have been shown to be particularly sensitive to their surface structure due to the propensity of the surface to interact with adsorbed molecules. In addition, the exposed crystal facets have been found to be atomically rough following the hydrothermal synthesis process (Cheng et al., 2016). It is therefore likely that a significant concentration of surface states exists within the band gap that could contribute to interfacial recombination—indeed, surface states in QD PV devices have been correlated with changes in band offset and Fermi-level pinning (Chuang et al., 2015). These surface states reduce carrier collection efficiency and therefore need to be passivated to improve J_{SC}. In order to determine if surface states contributed to the current deficit observed for devices with increasing nanowire length, Cheng et al. passivated the surface of ZnO nanowire arrays using titania (TiO_2) and measured the change in short-circuit current and integrated EQE with increasing nanowire length in PbS QD solar cells (Cheng, 2016). As shown in Fig. 26, both passivated and unpassivated devices exhibited a peak in J_{SC} at a nanowire length of 250 nm, but the

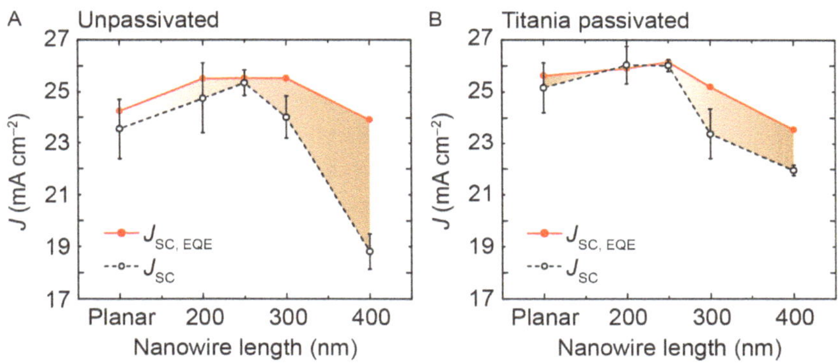

Fig. 26 Short-circuit current density measured from current density–voltage measurements (J_{SC}–V) and from integrated EQE measurements ($J_{SC,EQE}$) for ZnO/PbS quantum dot solar cells with varied ZnO nanowire length using unpassivated (A) and TiO_2-passivated (B) ZnO. The *shaded region* highlights the difference in short-circuit current density measured from these two methods. *Reprinted with permission from Cheng, J.W.J., 2016. Hydrothermal synthesis of zinc oxide nanowire arrays for photovoltaic applications. Copyright 2016 Massachusetts Institute of Technology.*

passivated devices had consistently superior J_{SC} and showed less of a decline in J_{SC} for devices with longer nanowires, indicating that titania successfully improves current collection. Surface passivation was specifically found to halt the decline in V_{OC} with increasing nanowire length, which was observed for unpassivated devices, and these results suggested that interfacial recombination is a significant source of carrier loss that may be successfully overcome by passivating the ZnO nanowire surface. This work by Cheng et al. revealed the importance of surface states as a primary limiting factor in nanowire-based lead sulfide solar cells (Cheng, 2016). Overcoming the effects of surface recombination is an ongoing area of interest in the application of nanowires to electronic devices due to the increased surface area formed in these devices.

5. CONCLUSIONS AND OUTLOOK

The ordered bulk heterojunction (OBHJ) has emerged as a promising architecture to address conflicting requirements for maximum absorption and efficient charge extraction in alternative solar cell technologies, such as organic and QD solar cell devices. Inorganic nanowire scaffolds have the potential to fulfill the principal goals of the OBHJ architecture due to their structural uniformity, high conductivity, and control over the directionality of charge transport. In this chapter, we focused on ZnO nanowire scaffolds due to the favorable materials properties of ZnO for solar cell applications and the versatility and scalability of the hydrothermal synthesis process.

The ability to tune major parameters of the nanowire array, including length, spacing, and alignment, on length scales relevant to the internal optical and electronic processes in solar cell devices is a critical aspect of the optimization process. To identify the factors that influence the final nanowire morphology and render the nanowire synthesis process more deterministic, dynamic spin coating and CF synthesis were discussed as instructive approaches to study nanowire growth from the standpoint of seed layer deposition and solution-phase growth processes, respectively. Identification of the ZnO nanowire growth mechanism through diffusion/reaction analysis revealed the importance of tuning the growth conditions to operate under reaction-limited conditions for precise growth rate control.

Having discussed avenues for dimensional control of nanowire arrays, we then examined how ZnO nanowires have been applied to existing photovoltaic devices to enhance performance, concentrating on PbS QD solar cells due to their promising efficiency gains and high stability. EBIC measurements identified the minority carrier diffusion length as a limiting factor for these devices, which suggests that improvements to current extraction are critical for enhancing device performance; however, the study also revealed that the dominant junction was not at the interface with ZnO as expected, motivating further investigation into the precise role of ZnO nanowires through quantum efficiency measurements. Taken together, these quantum efficiency measurements indicated that the improved photocurrent density in nanowire devices is due to improvement in both light-harvesting and carrier collection efficiency.

Continued research into device limitations and routes to overcome these challenges will be essential to pushing the efficiencies of these devices to be competitive with existing technology. We discussed in this chapter how surface recombination has to be addressed for ZnO nanowire devices, due to the large surface area between the ZnO nanowires and active material, and there has been early work demonstrating the promise of various passivation methods to overcome these limitations (Chang et al., 2015). Additionally, developing methods for finely tuning the dimensions of these wires over multiple device-relevant parameters will allow for deliberate device architecture design. While some nanowire array parameters have been effectively tuned, continued research into the growth mechanisms of ZnO nanowire will facilitate wider control over nanowire array growth. Finally, ZnO doping presents opportunities for enhanced electronic properties in devices based on these nanowires. For PbS QD solar cells like those discussed in this chapter, n-type doping of the ZnO could enhance the homojunction with the active quantum dot layer, while p-type doping could facilitate a p–n junction between the ZnO and quantum dot layer to increase carrier extraction at the front of these devices. Overall, there remains an untapped potential in the application of ZnO nanowires to solar cell devices, driving continued research and interest in these nanostructured devices.

REFERENCES

Baruah, S., Dutta, J., 2009. pH-dependent growth of zinc oxide nanorods. J. Cryst. Growth 311, 2549–2554.

Beek, W.J.E., Wienk, M.M., Janssen, R.A.J., 2006. Hybrid solar cells from regioregular polythiophene and ZnO nanoparticles. Adv. Funct. Mater. 16, 1112–1116.

Boercker, J.E., Schmidt, J.B., Aydil, E.S., 2009. Transport limited growth of zinc oxide nanowires. Cryst. Growth Des. 9, 2783–2789.

Brewster, M.M., Zhou, X., Lim, S.K., Gradečak, S., 2011. Role of Au in the growth and nanoscale optical properties of ZnO nanowires. J. Phys. Chem. Lett. 2, 586–591.

Burkhard, B.G.F., Hoke, E.T., Mcgehee, M.D., 2010. Accounting for interference, scattering, and electrode absorption to make accurate internal quantum efficiency measurements in organic and other thin solar cells. Adv. Mater. 22, 3293–3297.

Carey, G.H., Levina, L., Comin, R., Voznyy, O., Sargent, E.H., 2015. Record charge carrier diffusion length in colloidal quantum dot solids via mutual dot-to-dot surface passivation. Adv. Mater. 27, 3325–3330.
Chang, J., et al., 2015. High reduction of interfacial charge recombination in colloidal quantum dot solar cells by metal oxide surface passivation. Nanoscale 7, 5446–5456.
Chen, L., Yin, Y., 2012. Facile continuous flow injection process for high quality long ZnO nanowire arrays synthesis. Cryst. Growth Des. 12, 1055–1059.
Cheng, J. W. J. Hydrothermal synthesis of zinc oxide nanowire arrays for photovoltaic applications, MIT PhD thesis. (2016).
Cheng, J.J., Nicaise, S.M., Berggren, K.K., Gradečak, S., 2016. Dimensional tailoring of hydrothermally grown zinc oxide nanowire arrays. Nano Lett. 16, 753–759.
Chernov, A.A., Rashkovich, L.N., Mkrtchan, A.A., 1986. Solution growth kinetics and mechanism: prismatic face of ADP. J. Cryst. Growth 74, 101–112.
Chuang, C.-H.M., Brown, P.R., Bulović, V., Bawendi, M.G., 2014. Improved performance and stability in quantum dot solar cells through band alignment engineering. Nat. Mater. 13, 796–801.
Chuang, C.H.M., et al., 2015. Open-circuit voltage deficit, radiative sub-bandgap states, and prospects in quantum dot solar cells. Nano Lett. 15, 3286–3294.
Cross, R.B.M., De Souza, M.M., Narayanan, E.M.S., 2005. A low temperature combination method for the production of ZnO nanowires. Nanotechnology 16, 2188–2192.
Demianets, L.N., Kostomarov, D.V., Kuz'mina, I.P., Pushko, S.V., 2002. Mechanism of growth of ZnO single crystals from hydrothermal alkali solutions. Crystallogr. Rep. 47, S86–S98.
Donolato, C., 2012. A reciprocity theorem for charge collection. Appl. Phys. Lett. 270, 2–5.
Ehrler, B., et al., 2013. Preventing interfacial recombination in colloidal quantum dot solar cells by doping the metal oxide. ACS Nano 7, 4210–4220.
Govender, K., Boyle, D.S., Kenway, P.B., O'Brien, P., 2004. Understanding the factors that govern the deposition and morphology of thin films of ZnO from aqueous solution. J. Mater. Chem. 14, 2575–2591.
Greene, L.E., et al., 2003. Low-temperature wafer-scale production of ZnO nanowire arrays. Angew. Chem. Int. Ed. 42, 3031–3034.
Greene, L.E., et al., 2005. General route to vertical ZnO nanowire arrays using textured ZnO seeds. Nano Lett. 5, 1231–1236.
Greene, L.E., Law, M., Yuhas, B.D., Yang, P., 2007. ZnO–TiO_2 core–shell nanorod/P3HT solar cells. J. Phys. Chem. C 111, 18451–18456.
Huang, J.-S., Lin, C.-F., 2008. Influences of ZnO sol-gel thin film characteristics on ZnO nanowire arrays prepared at low temperature using all solution-based processing. J. Appl. Phys. 103, 14304.
Ip, A.H., et al., 2012. Hybrid passivated colloidal quantum dot solids. Nat. Nanotechnol. 7, 577–582.
Janotti, A., Van de Walle, C.G., 2009. Fundamentals of zinc oxide as a semiconductor. Rep. Prog. Phys. 72, 126501–126529.
Jean, J., et al., 2013. ZnO nanowire arrays for enhanced photocurrent in PbS quantum dot solar cells. Adv. Mater. 25, 2790–2796.
Joo, J., Chow, B.Y., Prakash, M., Boyden, E.S., Jacobson, J.M., 2011. Face-selective electrostatic control of hydrothermal zinc oxide nanowire synthesis. Nat. Mater. 10, 596–601.
Joo, J.H., Greenberg, K.J., Baram, M., Clarke, D.R., Hu, E.L., 2013. Aqueous epitaxial growth of ZnO on single crystalline Au microplates. Cryst. Growth Des. 13, 986–991.
Jung, M.-H., Lee, H., 2011. Selective patterning of ZnO nanorods on silicon substrates using nanoimprint lithography. Nanoscale Res. Lett. 6, 159.

Kim, T.-U., Kim, J.-A., Pawar, S.M., Moon, J.-H., Kim, J.H., 2010. Creation of nanoscale two-dimensional patterns of ZnO nanorods using laser interference lithography followed by hydrothermal synthesis at 90 °C. Cryst. Growth Des. 10, 4256–4261.

Klem, E.J.D., et al., 2008. Impact of dithiol treatment and air annealing on the conductivity, mobility, and hole density in PbS colloidal quantum dot solids. Appl. Phys. Lett. 92, 212105–212107.

Kovalenko, M.V., et al., 2015. Prospects of nanoscience with nanocrystals. ACS Nano 9, 1012–1057.

Kubota, N., Mullin, J.W., 1995. A kinetic model for crystal growth from aqueous solution in the presence of impurity. J. Cryst. Growth 152, 203–208.

Law, M., Greene, L.E., Johnson, J.C., Saykally, R., Yang, A.P., 2005. Nanowire dye-sensitized solar cells. Nat. Mater. 4, 455–459.

Lee, Y. S. Defect engineering of cuprous oxide thin-films for photovoltaic applications, MIT PhD thesis. (2013).

Lee, T.-H., Sue, H.-J., Cheng, X., 2011. Solid-state dye-sensitized solar cells based on ZnO nanoparticle and nanorod array hybrid photoanodes. Nanoscale Res. Lett. 6, 517.

Lee, J.M., No, Y., Kim, S., Park, H., Park, W.I., 2015. Strong interactive growth behaviours in solution-phase synthesis of three-dimensional metal oxide nanostructures. Nat. Commun. 6, 6325–6331.

Li, Q.H., Wan, Q., Liang, Y.X., Wang, T.H., 2004. Electronic transport through individual ZnO nanowires. Appl. Phys. Lett. 84, 4556–4558.

Li, Q., et al., 2005. Fabrication of ZnO nanorods and nanotubes in aqueous solutions. Chem. Mater. 17, 1001–1006.

Liu, D., Kelly, T.L., 2014. Perovskite solar cells with a planar heterojunction structure prepared using room-temperature solution processing techniques. Nat. Photonics 8, 133–138.

Liu, X., Jin, Z., Bu, S., Zhao, J., Liu, Z., 2006. Growth of ZnO films with controlled morphology by aqueous solution method. J. Am. Ceram. Soc. 89, 1226–1231.

Liu, B.-H., Boscoboinik, J.A., Cui, Y., Shaikhutdinov, S., Freund, H.-J., 2015. Stabilization of ultrathin zinc oxide films on metals: reconstruction versus hydroxylation. J. Phys. Chem. C 119, 7842–7847.

Liu, M., et al., 2016. Double-sided junctions enable high-performance colloidal-quantum-dot photovoltaics. Adv. Mater. 28, 4142–4148.

Lu, W., Lieber, C.M., 2006. Semiconductor nanowires. J. Phys. D Appl. Phys. 39, 387–406.

Ma, T., Guo, M., Zhang, M., Zhang, Y., Wang, X., 2007. Density-controlled hydrothermal growth of well-aligned ZnO nanorod arrays. Nanotechnology 18, 35605.

Marinova, N., Valero, S., Delgado, J.L., 2017. Organic and perovskite solar cells: working principles, materials and interfaces. J. Colloid Interface Sci. 488, 373–389.

Markov, I.V., 2004. Crystal Growth for Beginners. World Scientific Publishing Co. Pte Ltd, Singapore.

Mateo-alonso, A., Elwenspoek, M., Zacharias, M., 2012. Homoepitaxial branching: an unusual polymorph of zinc oxide derived from seeded solution growth. ACS Nano 6, 7133–7141.

McPeak, K.M., Baxter, J.B., 2009. ZnO nanowires grown by chemical bath deposition in a continuous flow microreactor. Cryst. Growth Des. 9, 4538–4545.

McPeak, K.M., et al., 2011. Chemical bath deposition of ZnO nanowires at near-neutral pH conditions without hexamethylenetetramine (HMTA): understanding the role of HMTA in ZnO nanowire growth. Langmuir 27, 3672–3677.

Mokkapati, S., Catchpole, K.R., 2012. Nanophotonic light trapping in solar cells. J. Appl. Phys. 112, 101101–101119.

Morin, S.A., Bierman, M.J., Tong, J., Jin, S., 2010. Mechanism and kinetics of spontaneous nanotube growth driven by screw dislocations. Science 328, 476–480.

Murray, C.B., Noms, D.J., Bawendi, M.G., 1993. Synthesis and characterization of nearly monodisperse CdE (E = S, Se, Te) semiconductor nanocrystallites. J. Am. Chem. Soc. 115, 8706–8715.

Nicaise, S.M., Cheng, J.J., Kiani, A., 2015a. Control of zinc oxide nanowire array properties with electron-beam lithography templating for photovoltaic applications. Nanotechnology 26, 075303–075310.

Nicaise, S.M., Cheng, J.J., Kiani, A., Gradečak, S., Berggren, K.K., 2015b. Control of zinc oxide nanowire array properties with electron-beam lithography templating for photovoltaic applications. Nanotechnology 26, 75303.

Nicholas, N.J., et al., 2012. The mechanism for hydrothermal growth of zinc oxide. CrystEngComm 14, 1232–1240.

Oelkers, E., Helgeson, H.C., 1988. Calculation of the thermodynamic and transport properties of aqueous species at high pressures and temperatures: aqueous tracer diffusion coefficients of ions to 1000 °C and 5 kb. Geochim. Cosmochim. Acta 52, 63–85.

Parize, R., et al., 2016. Effects of hexamethylenetetramine on the nucleation and radial growth of ZnO nanowires by chemical bath deposition. J. Phys. Chem. C 120, 5242–5250.

Park, H., et al., 2012. Graphene cathode-based ZnO nanowire hybrid solar cells. Nano Lett. 13, 233–239.

Pedro, M., Fernando, A., Rein, P., 2006. The influence of impurities on the crystal growth kinetics according to a competitive adsorption model. Cryst. Growth Des. 6, 2814–2821.

Rekemeyer, P. H. Nanostructured photovoltaics: improving device efficiency and measuring carrier transport, MIT PhD thesis. (2017).

Rekemeyer, P.H., et al., 2016. Enhanced photocurrent in PbS quantum dot photovoltaics via ZnO nanowires and band alignment engineering. Adv. Energy Mater. 6, 1600848.

Rekemeyer, P., Chuang, C.-H.M., Bawendi, M.G., Gradecak, S., 2017. Minority carrier transport in lead sulfide quantum dot photovoltaics. Nano Lett. 17, 6221–6227.

Ren, S., et al., 2011a. Heterojunction photovoltaics using GaAs nanowires and conjugated polymers. Nano Lett. 11, 408–413.

Ren, S., et al., 2011b. Inorganic–organic hybrid solar cell: bridging quantum dots to conjugated polymer nanowires. Nano Lett. 11, 3998–4002.

Richardson, J.J., Lange, F.F., 2009. Controlling low temperature aqueous synthesis of ZnO. 1. Thermodynamic analysis. Cryst. Growth Des. 9, 2570–2575.

Sangwal, K., 1996. Effects of impurities on crystal growth processes. Prog. Cryst. Growth Charact. Mater. 32, 3–43.

Scheer, R., Schock, H.W., 2011. Chalcogenide Photovoltaics. Wiley-VCH, pp. 9–128.

Sun, Y., Riley, D.J., Ashfold, M.N.R., 2006. Mechanism of ZnO nanotube growth by hydrothermal methods on ZnO film-coated Si substrates. J. Phys. Chem. B 110, 15186–15192.

Sun, Y., Liu, N., Cui, Y., 2016. Promises and challenges of nanomaterials for lithium-based rechargeable batteries. Nat. Energy 1, 16071.

Tak, Y., Yong, K., 2005. Controlled growth of well-aligned ZnO nanorod array using a novel solution method. J. Phys. Chem. B 109, 19263–19269.

Tay, C., et al., 2012. Low temperature aqueous solution route to reliable p-type doping in ZnO with K: growth chemistry, doping mechanism, and thermal stability. J. Phys. Chem. C 116, 24239–24247.

Tian, Z.R., Voigt, J.A., Liu, J., Mckenzie, B., Mcdermott, M.J., 2002. Biomimetic arrays of oriented helical ZnO nanorods and columns. J. Am. Chem. Soc. 124, 12954–12955.

Tian, Z.R., et al., 2003. Complex and oriented ZnO nanostructures. Nat. Mater. 2, 821–826.

Tian, J.-H., et al., 2011. Improved seedless hydrothermal synthesis of dense and ultralong ZnO nanowires. Nanotechnology 22, 245601.

Vayssieres, L., 2003. Growth of arrayed nanorods and nanowires of ZnO from aqueous solutions. Adv. Mater. 15, 464–466.

Wang, J.X., et al., 2006. Hydrothermally grown oriented ZnO nanorod arrays for gas sensing applications. Nanotechnology 17, 4995–4998.

Wang, H., Kubo, T., Nakazaki, J., Kinoshita, T., Segawa, H., 2013. PbS-quantum-dot-based heterojunction solar cells utilizing ZnO nanowires for high external quantum efficiency in the near-infrared region. J. Phys. Chem. Lett. 4, 2455–2460.

Wei, Y., et al., 2010. Wafer-scale high-throughput ordered growth of vertically aligned ZnO nanowire arrays. Nano Lett. 10, 3414–3419.

Weintraub, B., Deng, Y., Wang, Z.L., 2007. Position-controlled seedless growth of ZnO nanorod arrays on a polymer substrate via wet chemical synthesis. J. Phys. Chem. C 111, 10162–10165.

Wu, W.-Y., Yeh, C.-C., Ting, J.-M., 2009. Effects of seed layer characteristics on the synthesis of ZnO nanowires. J. Am. Ceram. Soc. 92, 2718–2723.

Xu, S., Lao, C., Weintraub, B., Wang, Z.L., 2008a. Density-controlled growth of aligned ZnO nanowire arrays by seedless chemical approach on smooth surfaces. J. Mater. Res. 23, 2072–2077.

Xu, S., et al., 2008b. Patterned growth of vertically aligned ZnO nanowire arrays on inorganic substrates at low temperature without catalyst. J. Am. Chem. Soc. 130, 14958–14959.

Xu, S., et al., 2009. Optimizing and improving the growth quality of ZnO nanowire arrays guided by statistical design of experiments. ACS Nano 3, 1803–1812.

Yin, Y., Alivisatos Paul, A., 2005. Colloidal nanocrystal synthesis and the organic-inorganic interface. Nature 437, 664–670.

Yuan, Z., Yu, J., Jiang, Y., 2011. Growth of diameter-controlled ZnO nanorod arrays by hydrothermal technique for polymer solar cell application. Energy Proced. 12, 502–507.

Zhang, D., et al., 2012. Controllable fabrication of patterned ZnO nanorod arrays: investigations into the impacts on their morphology. ACS Appl. Mater. Interfaces 4, 2969–2977.

Zhou, B., Shi, B., Jin, D., Liu, X., 2015. Controlling upconversion nanocrystals for emerging applications. Nat. Nanotechnol. 10, 924–936.

INDEX

Note: Page numbers followed by "*f*" indicate figures, and "*t*" indicate tables.

N

Q

R

S

U

V

W

X

Y

Z

FSC
www.fsc.org
MIX
Papier | Fördert gute Waldnutzung
FSC® C083411

Kolibri 360 GmbH
Ferdinand-Jühlke-Straße 7
99095 Erfurt, Deutschland
produktsicherheit@kolibri360.de